Saline Lakes V

Developments in Hydrobiology 87

Series editor

H. J. Dumont

Saline Lakes V

Proceedings of the Vth International Symposium on Inland Saline Lakes,
held in Bolivia, 22–29 March 1991

Edited by

Stuart H. Hurlbert

Reprinted from Hydrobiologia, vol. 267 (1993)

Springer Science+Business Media, B.V.

Library of Congress Cataloging-in-Publication Data

Saline lakes V / edited by Stuart H. Hurlbert.
    p.   cm. -- (Developments in hydrobiology ; 87)
    Papers from the 5th International Symposium on Inland Saline
Lakes, held 22-29 March 1991, at Hotel Titikaka, Lake Titicaca,
Bolivia.
    ISBN 978-0-7923-2416-4   ISBN 978-94-011-2076-0 (eBook)
    DOI 10.1007/978-94-011-2076-0
    1. Salt lakes--Congresses. 2. Salt lake ecology--Congresses.
3. Aquatic animals--Effect of salt on--Congresses. 4. Plants,
Effect of salt on--Congresses.   I. Hurlbert, Stuart H.
II. International Symposium on Inland Saline Lakes (5th : 1991 :
Hotel Titikaka, Bolivia) III. Title: Saline lakes 5. IV. Series.
QH95.9.S24 1993
574.5'2636--dc20   93-24192

ISBN 978-0-7923-2416-4

v

Contents

vi

Hydrobiologia **267**, 1993.
S.H. Hurlbert (ed.), Saline Lakes V.

Editor's preface

The Fifth International Symposium on Inland Saline Lakes was held 22–29 March 1991 at Hotel Titikaka on the shores of that lake, a short distance from the town of Huarina, Bolivia. There were 52 participants representing 16 countries, including six countries (Argentina, Bolivia, Chile, China, Peru and Russia) unrepresented at previous symposia in this series. Following the symposium, 43 of the participants participated in a 6-day excursion to visit the salars and salt lakes of central and southern Bolivia, travelling in the backs of trucks, on the tops of buses, and in our own private one-car train. Highlights included a visit to an enchanted isle in a 9000 km^2 'sea' of halite, Salar de Uyuni; inspection of the pollution-threatened Lago Uru-Uru; the hot springs and megapisoliths of Salar de Pastos Grandes; and the ice islands and flamingos of Laguna Colorada. In the May 1991 issue of the newsletter *Salinet*, W.D. Williams recounts some of the less formal moments of the symposium and excursion.

Twenty-three papers presented by the participants, plus an additional one reporting a microcosm study on salinity effects, constitute the present volume. Seven manuscripts had to be rejected or were withdrawn by their authors and numerous others, including my own on flamingos, failed to materialize in time. Nevertheless, the 24 papers cover the wide array of subject matters and scales characteristic of our 'inter-discipline' and represent the symposium well.

With sadness we report that David Frey died while revising his paper for this volume. We have completed that task for him. David had to cancel his plans to attend the symposium for health reasons, so Brian Timms had read his paper there. As would be expected, it is a valuable one giving us much new information on the distribution of southern hemisphere chydorids relative to salinity.

Numerous individuals must be thanked for contributions to the organization of the symposium and excursion and the production of this volume. First would be the Organizing Committee, consisting of Carlos E. Arze (Chairman), Alexandra Sanchez de Lozada, Simone Servant-Vildary, Maximo Libermann-Cruz and Jorge Quintanilla Aguirre. They did a flawless job in handling the considerable logistics of this meeting and in creating for all participants so many pleasant memories of Bolivia and her people. Rosamaria Ruiz and Justo Zapata provided efficient on-site coordination of symposium and excursion activities, respectively. Even to an old antiplano traveller, it remains a mystery how Justo was able to whisk four dozen of us so efficiently from one magic spot to another on our 1800 km trajectory through the altiplano. Guido Solis is thanked for an illuminating talk in the city of Oruro on pollution threats to Lago Uru-Uru. Special thanks are also extended to Gonzalo Rico Calderon, General Manager of the Empresa Nacional de Electrificacion (ENDE), for allowing the group to lodge for three days at the ENDE's very comfortable workers camp at Laguna Colorada. Finally, our appreciation to the many other Bolivians who served us with such efficiency and friendliness in so many ways.

Major financial support of the symposium was provided by grants from the National Science Foundation (U.S.A.) and the Office de Recherche Scientifique et Technique Outre-Mer (France). Additional support was provided by the Liga de Defensa de Medio Ambiente (Bolivia), Universidad Mayor de San Andres (Bolivia), San Diego State University (U.S.A.), The SOROS Foundation (U.S.A.), and the International Society for Limnology.

All manuscripts submitted for these proceedings were critically reviewed by at least three referees. With only a few exceptions, all manuscripts underwent major revision. This often involved extensive reanalysis of data and redrafting of tables and figures. I thank all authors for the amazing grace with which they acceded to my often overbearing 'suggestions' and my requests for second and third revisions. W.D. Williams gave invaluable help in redrafting text and figures for three manuscripts, and reviewing a couple of others as well. Other scientists providing critical reviews included I.A.E. Bayly, K. Bertine, S. Bowen, G. Brunskill, G. Chong, B. Collier, F. Comín, P. De Deckker, D. Dexter, H. Dumont, A. Echelle, D. Farris, D. Frey, D. Galat, E. Goldberg, J. Green, R. Green, M. Guerlesquin, U.T. Hammer, B. Hann, R. Hecky, D. Herbst, R. Jellison, B. Javor, B. Jones, S. Kilham, W. Last, J. Melack, P. Micklin, W. Minckley, A. Nissenbaum, T. Northcote, J. O'Brien, U. Pick, R. Renaut, J. Reuter, M. Rosen, A. Sheldon, I. Soulie-Märsche, G. Sprules, D. Stephens, S. Twombly, R. Wetzel, and W. Wurtsbaugh.

Finally, special thanks to Dawn Hamilton for her efficient assistance over the last three years with the extensive correspondence this undertaking has involved.

San Diego, California
April 1993

S.H. HURLBERT

Large saline lakes of former USSR: a summary review

N. V. Aladin & I. S. Plotnikov
Zoological Institute, Russian Academy of Science, St. Petersburg, Russia

Abstract

Seven of the largest lakes in central Asia (former USSR) are saline: Caspian Sea, Aral Sea, lakes Balkhash, Issyk-Kul, the Chany complex, Alakul and Tengiz. They range in salinity from sometimes < 3 g l^{-1} to 19 g l^{-1}. The paper provides a summary review of their major physico-chemical and biological features. Several are threatened by activities in their drainage basins, particularly diversion of inflowing waters.

Introduction

There are more than 2 800 000 lakes in the CIS (former USSR), with a total surface area of nearly 980 000 km^2. Most (99.2 per cent) are small and shallow, with an area of less than 1 km^2, but sixteen are amongst the world's largest lakes, each with an area in excess of 1000 km^2. Of these very large lakes, seven are saline: the Caspian and Aral seas, and lakes Balkhash, Issyk-Kul, Chany, Alakul and Tengiz. The total surface area of these saline lakes is nearly 523 000 km^2, *i.e.* more than half (53.4 per cent) of the total lacustrine area in the CIS. Table 1 gives their main characteristics at the highest recorded water-levels.

Rather little information on these important lakes is freely available outside the CIS. The present paper aims to introduce these lakes by means of a summary review of their major limnological features.

Caspian Sea

The Caspian Sea is the largest body of inland water in the world. Situated in the CIS and Iran, this endorheic lake stretches north-south some 1200 km, and has a mean width of 320 km. The length of its shoreline is nearly 7000 km. Its present (1991) area is 371 000 km^2, and present elevation ± 28.5 m below sea-level. Maximum

Table 1. Major physico-chemical features of the largest salt lakes in the CIS. Values for each lake are for the time during the 20th century when it had its highest water level.

Lake	Area (km^2)	Volume (km^3)	Elevation above sea level (m)	Max. depth (m)	Salinity (g l^{-1})
Caspian	422 000	79 000	− 26	1072	10–12
Aral	66 100	1064	53	69	8–10
Balkhash	22 000	122	343	27	0.5–7
Issyk-Kul	6 280	1730	1608	702	5–6
Chany	3 245	7.1	106	10	1.5–4
Alakul	2 650	N.A.	343	45	5–7
Tengiz	1 590	N.A.	304	8	3–19

N.A. not available.

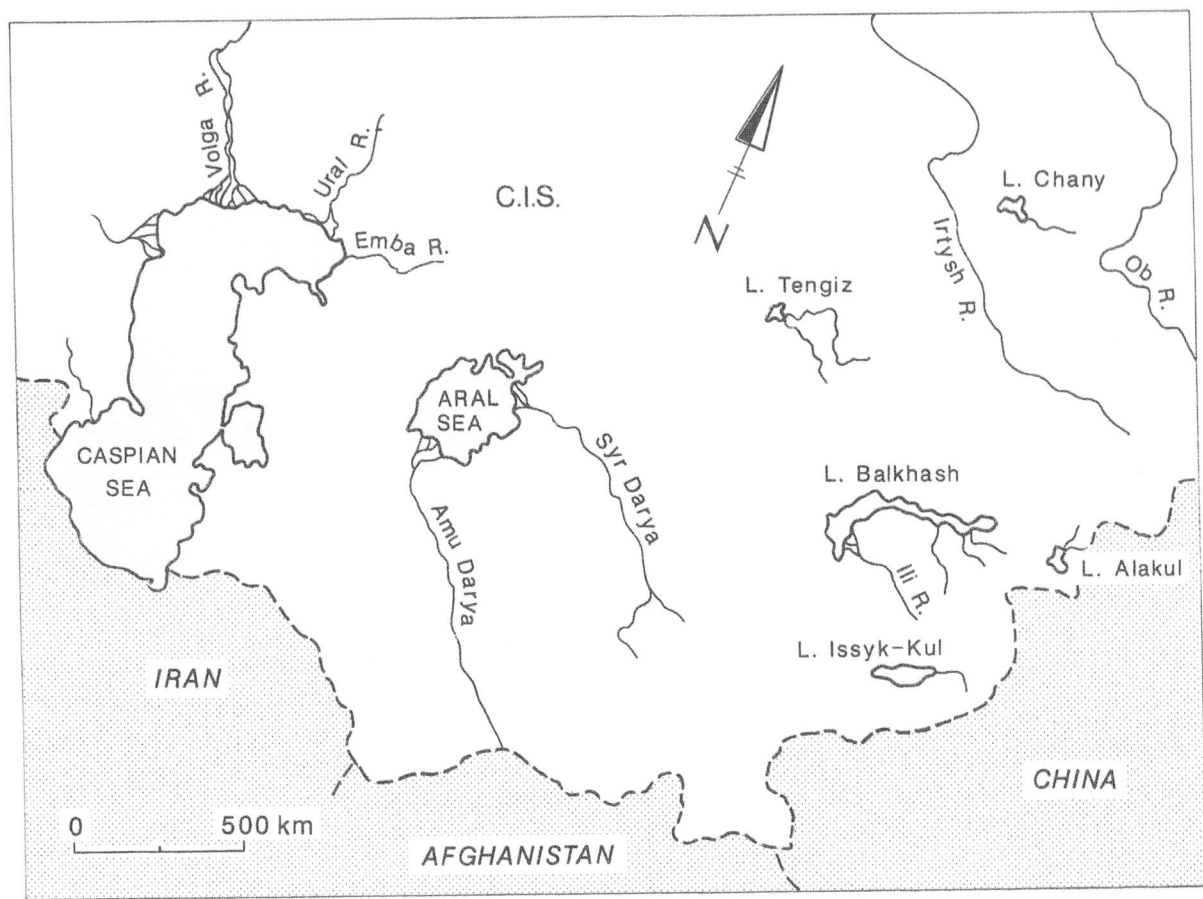

Fig. 1. Location of lakes discussed in the text.

depth is 1025 m. Since 1929, its depth has decreased because of a decrease in the supply of inflow waters resulting from the building of hydroelectric power stations on the Volga. By 1969, its level had fallen some 2.5 m, and its area had diminished by 51 000 km^2.

The largest embayments of the Caspian Sea are: Kizlyarskiy and Komsomolets in the north, Mangyshlakskiy, Kenderly, Kazakhski, Kara-Bogaz-Gol and Krasnovodskiy in the east, and Agrakhanskiy and Bakinskiy in the west. There are fifty islands with a total surface area of 350 km^2, the largest being Kulaly, Tyuleniy, Chechen', Artem, Zhiloy and Ogurchinskiy islands. Morphometrically, the lake can be divided into a northern, middle and southern part. Into the northern part flow the Volga, Embra, Ural and Terk rivers; these contribute 88 per cent of

the total annual river inflow. On the west coast the rivers Sulak, Samur and Kura give 7 per cent of the total inflows. The remaining inflows (5 per cent) are provided by rivers from Iran. There are no regular streams on the east coast.

Moderate and high winds are frequent so that a large number of days has significantly rough water. In summer, the average temperature of surface waters is 24–26 °C, but in the south it reaches 29 °C, and in Krasnovodskiy Gulf, 32 °C. On the east shore, the upwelling of cold water from the deeper parts of the lake often results in temperatures as low as 10–12 °C in June and August. In winter, the temperatures are quite different: in the north, water temperature is –0.5 °C, but in the middle it is 8–10 °C. The northern Caspian Sea is covered with ice for 2–3 months.

Average salinity is 12.7 to 12.8 g l^{-1}. Maximum values are recorded in the east – up to 13.2 g l^{-1}g. Minimum values occur in the north – to 1–2 g l^{-1} – and reflect the freshening effect of the Volga inflow.

The flora and fauna of the Caspian is rather small. However, significant biomass is present. Even so, there are ± 500 plant species, and at least 629 invertebrate and 78 fish species. Species have been introduced either intentionally or accidentally (Karpevich, 1975; Zenkevich, 1963).

The Caspian has long been an important fishery where many valuable species are caught. Thus, it provides 82 per cent of the world catch of sturgeons. However, because the water-level has fallen and because of river flow regulation, many spawning grounds have disappeared, and habitats for reproduction of migrating fish decreased. This, in turn, has caused a decrease in the fish catch (Zenkevich, 1963).

At present, the water-level has stopped falling and has even increased slightly. However, water pollution is becoming more intense especially because of increasing pollution from the Volga River. As a result, the existence of sturgeons is endangered; for example, the toxic effect of pollution gives rise to different pathologies, including lamination of muscles.

Aral Sea

The Aral Sea, formerly the second largest lake in the CIS in area, is now rapidly decreasing in size. In the past thirty years, its water-level has dropped 15 m, its salinity has increased from 8–10 g l^{-1} to 30 g l^{-1}, and its volume has decreased by more than 60 per cent. The desiccation is the result of reduced inflows caused primarily by water diversions for irrigation from two large inflowing rivers, the Syrdarya and Amudarya. From the middle of the nineteenth century until 1960–1962, the average salinity was 8–10 g l^{-1}. During that time, most of the biota was of freshwater origin and was only able to regulate hyperosmotically. In 1974–1976, however, the average salinity of the Aral increased by more than 14 g l^{-1} and

most species of freshwater origin became extinct. From 1974–1976 to 1985–1987, *i.e.* when the average salinity was < 24 g l^{-1}, most species were of brackish water origin and were osmoconformers or hypoosmotic regulators. In 1985–1987, when the average salinity increased to > 24 g l^{-1}, nearly all of these populations also became extinct. At present, the average salinity is 28–30 g l^{-1}, and most species are of marine origin and are either osmoconformers or hypoosmotic regulators. In the future, when average salinity will exceed 42–44 g l^{-1}, nearly all of these marine species also will become extinct, and the only inhabitants will be hypoosmotically regulating species typical of hypersaline water-bodies.

Principal sources of information on this lake are: Williams & Aladin, 1991; Aladin & Potts, 1992.

Lake Balkhash

Lake Balkhash (*Ak-Dengiz* in Kazakh) is a terminal lake in eastern Kazakhstan, located in the Balkhash-Alakol desert at an altitude of 343 m above sea level. Its area varies with its water level and is 17 000–22 000 km^2. The lake extends east-west by *ca* 588–614 km, and is from 9–19 km wide in its eastern section and 74 km wide in its western section. In the 1960s, maximum depth was 26.5–27.0 m, and the volume was 122 km^3. Volume was 111.5 km^3 in the 1920s. Northern shores are high and rocky; southern ones are low-lying, sandy, and fringed by dense reeds and swamp. Climate in the region is arid, sharply con-tintental, and annual evaporation over the lake is 950–1200 mm and annual precipitation is 150 mm. The Balkhash catchment has an area of 500 000 km^2 and contains > 52 600 km of rivers and streams, the largest of which are the rivers entering the western part and the rivers Karatal, Lepsy, Aksu and Ayaguz entering the eastern part. The River Ili contributes some 78 per cent of total annual inflow to the lake, which averages 15.6 km^3 (Anon, 1984; Domrachev, 1933).

According to A. A. Tursunov (personal communication, 1991), the water balance of Balkhash under natural conditions (*i.e.* before 1970) was maintained by mountain rivers with a discharge of 23.8 km^3 per year. Most of this, 17.4 km^3 per year, was provided by the River Ili, and the rest, 6.4 km^3 per year, by the eastern rivers (Karatal, Aksu, Lepsy, Ayaguz), all except the last flowing from Dzungaria Ala-Tau. However, only 14.9 km^3 of the annual discharge actually reached Lake Balkhash where it completely evaporated; the rest, 8.9 km^3, was expended in the basin to maintain marshes of the unique delta of the River Ili (Domrachev, 1940; Tarasov, 1965), tugays along the river, and small irrigated areas in the basins of the River Ili (Domrachev, 1940; Kipshakbaev *et al.*, 1985) and eastern rivers (3.2 km^3 per year).

In 1967, an extension of irrigated areas began, and in 1970 the Kapchagay Reservoir, the largest reservoir in the region, began to fill. As a result, water withdrawal from the river basin increased sharply, and from about 1974 a new regressive period began which caused a sharp decrease in the level of Lake Balkhash and intensified earlier anthropogenic pressures on the lake ecosystem and the River Ili (Tarasov, 1965; Abrosov, 1963).

Since 1970, the basin's water resources have also decreased to 22.4 km^3 per year due to natural climate aridity. This resulted in the flow of the River Ili decreasing to 16 km^3. Water losses have also increased, including losses for filling and maintaining Kapchagay Reservoir, for irrigation, and for other reasons. As a result, the level of Balkhash fell sharply. Relative increases in the amounts of water during the period 1979–1981 did not reverse the trend and by the beginning of 1983 the level of the lake was at *ca* 341.0 m, its historical minimum. Continued falls in its level will lead to the complete degradation of the basin ecosystem, as has happened in the Aral Sea basin (Kipshakbaev *et al.*, 1985; Tursunov *et al.*, 1986).

Further events have completely confirmed predictions. Moreover, aridity in the basin has intensified. Thus, by the end of 1986, the level of Balkhash had decreased to 340.5 m. Acceptance of scientific recommendations, however, has led to some stability in the level of the lake at its 100 year minimum, and preparation for the times when full-flowing rivers would ensure increased inputs. This occurred in 1988 when inflows into Kapchagay Reservoir were > 20 km^3 (the 100 year maximum [1960] was 22.7 km^3, and the annual mean is 14.7 km^3). In summary, conservation measures precluded any further filling of Kapchagay Reservoir. In 1988, inputs into Balkhash from the eastern rivers were nearly 50 km^3, and from all sources were > 194 km^3. Of these inputs, *ca* 12 km^3 evaporated (mean evaporation, 15 km^3), but the volume in the lake increased by 7.0 km^3 and the level by nearly 100 cm.

Thus it appears that Lake Balkhash and the Ili-Balkhash basin are not now threatened: the lake level is significantly above the minimum, the terrestrial environment and the River Ili have improved ecologically, and the most serious forms of human pressure (Kapchagay Reservoir, irrigation) have eased. This should enable the maintenance of the lake's water level at 341 m. Even so, it would be unwise to assume that the threat has gone. The recommendations accepted enabled some improvement in the position of the lake so far as its biota was concerned, but river pollution continues, salinization proceeds, and the concentration of certain toxic substances such as nitrates, pesticides, heavy metals (copper, zinc and cadmium) and carcinogenic substances increases.

Lake Balkhash is divided into two relatively independent sections: a wide and shallow western section, stretching from south-west to northeast, and a deep (to 27 m) and narrow eastern section, stretching from west to east. These sections are connected by the Strait of Uzun-Aral, a narrow (3.8–4.2 km), shallow (maximum depth 2.8–3.3 m) channel. Because of the lake's division into two sections of unequal size, with most inflow into the western section, salinity in west Balkhash is very low (1.1 g l^{-1}), whereas in east Balkhash, salinity is much higher (4.3 g l^{-1}). Water transparency is 5.5 m. Surface temperatures fall to 0 °C in December, but reach 28 °C in June. Differences in summer between surface and hypolimnetic temperatures are < 3.3 °C. Generally, aerobic conditions prevail, but an oxygen deficiency sometimes develops in vegetated shallow areas. Bottom sediments are mainly silty, but peat and balkhashite also occur.

An increase in lake salinity has also been observed, resulting not only from changes in the hydrological balance, but also from a rise in the salinity of the River Ili (from 0.25–0.37 to 0.42 g l^{-1}) after the regulation of its flow. By 1978, the salinity of Balkhash had increased from 1.12–4.31 to 1.42–5.14 g l^{-1}.

Ionic composition is distinctive. The proportion of chloride (9–21 equiv. percent) is 2–3 times lower than the proportion of chloride in the sea. However, the proportions of potassium, calcium, magnesium, sulphate and carbonate/bicarbonate ions are significantly higher. In eastern Balkhash, the proportion of potassium ions (2.9 equiv. percent) is very high in comparison to other waters (e.g. 0.6 equiv. percent in the ocean and the Aral Sea). The lower proportion of calcium ions, especially in comparison with the Aral and Caspian seas, also is notable (Anon., 1984; Panov, 1933).

Two hundred and six species of algae have been recorded. Half are diatoms, and 67 per cent are planktonic. The benthic algae are predominantly diatoms. Most algae in Balkhash are freshwater forms (oligohalobionts) or euryhaline forms (77 per cent), but there are a few halophiles (9 per cent), halobionts (4 per cent) and mesohalobionts (10 per cent) (Karpevich, 1975; Anon., 1984). Cryophilous diatoms predominate in spring and autumn, but with increasing temperatures green algae develop. Cyanobacteria predominate in summer. In spring, phytoplankton biomass is 1 g m^{-3} and in summer up to 47 g m^{-3}. In autumn, it decreases to 0.6 g m^{-3}. The least saline western part of the lake has the most dense zooplankton (annual average value, 3.5 g m^{-3}), whereas in the more saline eastern part the average is significantly lower (0.6 g m^{-3}). A regular change in zooplankton species composition occurs from west to east. In the western part, freshwater and euryhaline forms predominate, but freshwater forms disappear eastwards and are replaced by halophiles and mesohalobionts. Because of the increased salinity, some halophiles have now spread from the east to the west. By the end of the 1970s, the average phytoplankton biomass in

western Balkhash had decreased by more than 50 percent, possibly because of the increased salinity. Freshwater species are rarer and their place has been taken by saline forms (Anon., 1984).

Zooplankton has consisted of 54 species: 5 protozoans; 28 rotifers (the most common being *Filinia* (= *Triarthra*) *longiseta* (Ehrenb.), *Polyarthra platyptera* Ehrb, *Pompholyx sulcata* Hud., *Keratella* (= *Anuraea*) *cochlearis* (Gosse), *Pedalion oxyuris* (Zernov); 11–18 cladocerans (the most common being *Diaphanosoma brachyurum* Lievin and *Daphnia cristata* (Sars) (= *Cephaloxus cristatus*); 8–11 copepods (the most common being *Thermocyclops crassus* (Fischer), *Mesocyclops leuckarti* (Claus), and *Arctodiaptomus* (= *Diaptomus*) *salinus* (Daday)) (Rylov, 1933; Karpevich, 1975). In the zooplankton, *A. salinus* and *D. brachyurum* comprised 60–70 per cent of the biomass. In the more saline eastern Balkhash, the predominant forms were *A. salinus*, *Polyphemus pediculus* (L.) and *Sida crystallina* (Muller) (Saduakasova, 1972). In the 1970s, zooplankton biomass attained 1 g m^{-3} (Anon., 1984).

The recent zoobenthic fauna of Balkhash basically consists of introduced species. In the 1950s and 1960s, some species were introduced successfully from the Caspian Sea and the Sea of Azov. They included the Caspian polychaete worms *Hypania invalida* (Grube) and *Hypaniolla kowalevskyii* (Grimm), the mysids *Paramysis lacustris* (Czern), *P. intermedia* (Czern), *P. ullskyi* (Czern), and *P. baeri* (Czern), the amphipod *Corophium curvispinum* Sars, the mollusc *Hypanis colorata* (Eichw.) (= *Monodacna colorata*), as well as the accidently introduced molluscs *Dreissena polymorpha* (Pall.), *Anodonta cygnea* (L.), and *A. cellensis* (Schroter) from the River Ural. These introduced forms are now predominant in the zoobenthos (Karpevich, 1975).

The composition of the zoobenthos is different in the eastern and western parts of the lake. In the less saline western part, the biomass is highest (mean of 4.2 g m^{-2}), and the main species are the oligochaetes *Tubifex tubifex* (Muller), *Nais pardalis* Piquet and *Stylaria lacustris* (L.), the mollusc *H. colorata*, the mysids *P. intermedia*, *P. lacustris* and *P. baeri*, and the amphipod *C. curvispinum*.

In eastern Balkhash, on the other hand, the zoobenthos biomass is lower 1.4–2.1 g m^{-2}), and there are no polychaetes or amphipods. Also absent is *H. colorata*, but the endemic mollusc *Bithynia caerulans* West. occurs. As one moves eastward, mysids disappear in the order *P. baeri*, *P. lacustris*, *P. ullskyi*, and, last, *P. intermedia*; and not even the last species is found in the eastern end of the lake. The dominant components of the zoobenthos in the east are chironomids, oligochaetes, and molluscs. The main chironomid taxa under conditions of increased salinity are *Chironomus salinarius* Kieff. and *Procladius* sp. (Anon., 1984; Mikulin, 1933).

Of special note, so far as the biota is concerned, is the unfavourable ionic composition of Balkhash water (high concentrations of potassium and magnesium; Karpevich, 1975) compared to other large saline contintental water bodies. Thus, any increase in salinity quickly affects its biota and causes a major reduction in biomass.

There is a valuable fishery in Lake Balkhash, with 21 species of fish present. The original fish fauna, however, consisted of only 4 species in the lake itself: the Balkhash marinka (*Schizothorax argentatus* Kessler), the Ili marinka (*S. pseudaksaiensis* Hen.), the Balkhash perch (*Perca schrenki* Kessler), and the spotted stone loach (*Nemachilus strauchi* (Kessler)). Other fish inhabited the deltas of the rivers. In the lake, representatives of the high-Asian fauna predominate. From 1930 to the 1960s, a series of introductions was made. Of these, several have become the principal species of the fishery and provide up to 99–100 per cent of the total catch: bream (*Abramis brama orientalis* Berg), roach (*Rutilus rutilus caspicus* Berg), carp (*Cyprinus carpio* L.), sander (*Lucioperca lucioperca* L.), wels (*Silurus glanis* L.), and asp (*Aspius aspius* (L.)).

Fish introductions have increased fish productivity by 20–30 per cent: In the 1960s and 1970s, the catches were 6700–16500 tons per year (Karpevich, 1975; Anon., 1984). However, the fall in water level and the progressive increase in salinity have diminished the suitability of conditions for the reproduction of most fish present (Anon, 1984).

Presently, there is a project to maintain the hydrological regime in western Balkhash by separating it from eastern Balkhash by a dam and sluice in the Uzun-Aral Strait. In this way, the water supply to eastern Balkhash will be limited. In other words, it is proposed to sacrifice the eastern part of this unique water body which will become a shallow hypersaline lake (Anon., 1984).

Issyk-Kul

Issyk-Kul is a terminal, endorheic, mountain lake. Situated in north-eastern Kirghizia, it is located in the mountains of the northern Tien Shan between the ranges Kungey-Alatau to the north and Terskey-Alatau to the south. Its elevation is 1608 m above sea level. Major morphometric parameters are: area, ±6236 km^2; length 178 km; width, 60 km; maximum depth, 668 m; average depth, 278 m; and volume, 1738 km^3. More than 50 rivers discharge into the lake, with a total annual discharge of >3 km^3. The largest are the Dzhergalan and Tyup rivers, which flow into the eastern part of the lake. Rivers are snow-fed. Water from underground sources is of great significance (40 per cent of inflow, according to some) in the hydrological budget of Issyk-Kul.

Like all endorheic lakes, the water-level fluctuates. In the 17 and 18th centuries, the water level was 10–12 m higher than it is at present, and an outflow from the western part of the lake into the River Chu existed. Conversely, a thousand years ago the waterlevel was lower than the present one. Evidence for the considerable amplutude of Issyk-Kul transgressions and regressions during climate changes is given by the presence of ancient lake terraces of 8–10 m height, and by ruins of submerged settlements at a depth of 8 m. During the past two centuries, the level of Issyk-Kul has decreased, and since 1886 it has fallen by 4 m (7 m according to some sources) (Anon., 1984).

The regional climate is warm, temperate and dry. Average annual precipitation is 250 mm, but annual evaporation from the lake surface is nearly 700 mm. Surface water temperatures in January are not less than 2–3 °C, and in July extend to 19–20 °C. At depths of more than 100 m, the water temperature remains stable all year at 3.5–4.0 °C. The lake is not covered with ice in winter, but in cold winters ice can appear in some bays.

Water colour is blue, and transparency, > 12 m. The salinity at present is 6 g l^{-1}, but in 1930 it was 5.8 g l^{-1}. It cannot be used for drinking or irrigation. Compared to sea-water, sulphate concentrations are higher (44 equiv. per cent), chlorides are lower (45 equiv. per cent), and there is more magnesium, calcium and (bi)carbonate but less sodium and potassium (68.5 equiv. per cent). Issyk-Kul can be regarded as a mesohaline sodium-sulphate-chloride water-body.

Because of a general regional tendency towards decreased humidity and increased water withdrawals, the fall in the water-level of the lake level will accelerate. However, the large lake volume will prevent any rapid increase in salinity.

Bottom sediments are argillaceous silts, with narrow marginal strips of sand.

The lake is oligotrophic. The concentration of dissolved organic matter is low: permanganate oxidation in central regions may reach 2.7 mg O_2 l^{-1}. Oxygen concentrations are always high (7.1 mg l^{-1} or more), and a phytoplankton photosynthetic rate of 171 mg C m^{-2} d^{-1} has been recorded (Anon., 1984).

The phytoplankton consists of 299 species, of which 149 are diatom species in the nannoplankton, cyanobacteria and green algae predominate. There are many periphytic diatoms. Two seasonal peaks of phytoplankton abundance occur: May and October–November, minimum abundance is in January–February. Phytoplankton biomass in open parts of the lake is up to 211 mg m^{-3}, but in the bays it is higher (up to 1.7–5.3 g m^{-3}). Maximum numerical and biomass densities occur at 15 to 60 m.

Angiosperms occur down to 2 m, and charophytes are also present. The latter form an unbroken carpet down to depths of 30–40 m, and dominate the aquatic plant communities of the lake. The highest production of charophytes occurs at depths from 15 to 20 m. Charophyte biomass can reach 60 g m^{-2}. Of the angiosperms, *Phragmites communis* Trin. is most important but

various species of *Potamogeton* are widespread. Annual macrophyte production is estimated at 1.72 × 10^6 tons (Anon., 1984).

There are 154 species recorded from the zooplankton: 76 protozoans, 78 rotifers, 11 cladocerans, and 8 copepods (Anon. 1984; Foliyan, 1973). Only 15 species occur in large numbers, namely, 9 species of rotifer (*Eosphora ehrenbergi* Weber, *Synchaeta cecilia* Rouss., *S. gyrina* Hood, *Euchlanis oropha* Gosse, *Brachionus quadridentatus* Herm., *B. urceus* (L.), *Keratella quadrata* (Muller), *Hexarthra oxyuris* (Zernov), and *H. fennica* (Lev.); two species of cladocerans (*Alona rectangula* Sars, *Alonella nana* (Baird)), and one species of copepod (*Arctodiaptomus salinus* (Daday)). *A. salinus* is the dominant member of the zooplankton but has a low productivity (production: biomass = 1.7). On the whole, the zooplankton biomass is low, with the annual average value ± 20 mg m^{-3}, and the maximum value up to 163 mg m^{-3} (Anon., 1984).

The zoobenthos has 176 species: 35 protozoans, 33 annelids, 20 crustaceans, 84 non-crustacean arthropods, and 4 molluscs. It is reasonably uniform down to 30–35 m, but at deeper levels it sharply declines in abundance. The dominant forms are chironomid larvae, molluscs, amphipods and mysids. These comprise up to 80–90 per cent of the biomass. The most numerous are 9 chironomid species: *Chironomus plumosus* L., *C. anthracinus* Zett., *C. cineulatus* Mg., *C. tentans* F., *Glyptotendipes gripecoveni* Kieff., *G. barbipes* Ean., *Stictochironomus pictulus* M., *Tanytarsus* sp., and *Cricotopus bicinctus* Meig. The mollusc *Radix auriculata obliquata* (M.) is widespread down to 60 m. Also widespread are the amphipods *Gammarus negri* M. and *G. ocellatus* M.. In summer the benthic biomass in the bays is 5–14 g m^{-2} (in some localities it reaches 20 g m^{-2}). Below the charophyte zone, zoobenthic biomass falls from 8.0–10.0 g m^{-2} to 2.5–3.5 g m^{-2} at depths of 60–70 m. In the profundal region, *i.e.* most of the lake bottom, zoobenthic biomass is < 0.2–0.3 g m^{-2}. Here, only the endemic oligochaetes *Enchytraeus przewalskii* Hrabe and *E. issykulensis* Hrabe and the amphipod *Issykogammarus hamatus* Chev. occur. In the 1960s, the mysids *Paramy-*

sis kowalevskyi (= *P. lacustris*), *P. baeri* and *P. intermedia* were introduced from Lake Balkhash. These spread during the 1970s over shallow areas, particularly in bays and fresher localities near rivers inflows. Their biomass reaches 1.5–2.5 g m^{-2}.

The fish fauna is represented by 27 taxa in five families. The long isolation of the lake has resulted in the development of several endemic species: Schmidt's dace (*Leuciscus schmidti* (Herz.), Issyk-Kul dace (*L. bergi*), gudgeon (*Gobio gobio latus*), Issyk-Kul marinka (*Schizothorax issykkuli* Berg), spotted stone loachs (*Nemachilus strauchi ulacholicus* Anikin, and *N. strauchi dorsaloides*), and the scaleless osman (*Dyptychus dybovskii* Kessler). In all, there were twelve original species. Since the 1930s, additional species have been introduced. The first to be introduced was the Sevan trout (*Salmo ischan gegarkuni* Kessler); this was followed by the introduction of the Aral bream (*Abramis brama aralensis* Berg), carp (*Cyprinus carpio*), and the sander (*Lucioperca lucioperca*). The commercial fish catch is from 250 to 400 tons per year. The main fish in this catch are daces, sander and trout (Anon., 1984).

Finally, in the bays of Issyk-Kul nearly 20 000–50 000 individual water birds over winter. To protect them, a reserve was established in 1958. This also protects pheasants and the mountain fauna of Kirghizia. The muskrat *Ondatra zibethicus* was introduced in 1944 and has spread in some localities along the shores in marsh vegetation.

Lake Chany complex

The Lake Chany complex, located in the Novosibirsk region between the rivers Ob and Irtish, lies in the Barabinsk steppe at an elevation of 106 m above sea-level. Its area fluctuates according to water-level changes but is between 1990 and 3245 km^2. The average depth is 2.2 m and the maximum, 10 m (presently 8.5 m). Depths of less than 3 m comprise 30 per cent of the total area, and depths of less than 2 m, 80 per cent. The lake complex is 82 km long, 36 km wide, and is characterised by a very indented shore line with many

consists of a large Lake Chany (consisting of four sectors and Lake Yarkul) and a small lake Chany. One section of the larger lake is now isolated by a dam. The endorheic catchment area is 29 935 km^2. Inflow from rivers is small, and the main ones are the rivers Chulym and Kargat. These are mainly (91 per cent) supplied by snowmelt.

Water-level depends closely on hydrological fluctuations in southwestern Siberia. Since the end of the nineteenth century, when regular hydrological observations began, a constant fall in the water-level has occurred. This correlates with a decrease in regional humidity. Even so, periodic falls and rises in water-level occur. Thus, by 1914 the level had risen by 1.6 m. It then fell 2.8 m by 1937. It rose again by 2 m until 1950, but by 1971 had again fallen some 1.8 m. The average annual amplitude in water-level is 0.8 m. There are also short term local fluctuations caused by winds; these may reach 0.7–0.8 m (Anon., 1984).

The climate in the Barabinsk steppe is continental, with warm summers and cold winters. The average winter water temperature is 2.0 °C. In summer (July), the water temperature reaches 20.6 °C. The lake is ice-covered by the end of October or beginning of November, but ice disappears by the end of April or the beginning of May. The shallow regions of the lake can be frozen to the bottom. Because of high winds and the shallow depth of the lake, the water-column is always isothermal.

The bottom sediments comprise various silts. Sands cover not more than 12 per cent of the bottom.

The lake is moderately saline, but values differ in different parts of the lake. In 1976, values were from 1.1 to 14.7 g l^{-1}. The ionic composition is dominated by sodium and chloride ions, and, compared to sea-water, proportions of magnesium and sulphate are higher, and (bi)carbonate much higher, but sodium, potassium and chloride proportions are lower. Salinity is rising, and from 1948–1978 increased from 1.40–4.05 g l^{-1} to 8.2–11.8 g l^{-1}. As ice develops in winter, the salinity increases (Anon., 1984).

the small Lake Chany is eutrophic. Macrophytes include thirteen species of submerged and emergent plants, but only *Phragmites australis* (Cav.) Trin. ex. Steud. and *Potamogeton pectinatus* L. are common. The reed communities are situated at the shore margins and may be to 1 km wide. Due to recent falls in water-level, a significant fraction of the reed swamp is now dry land.

There are 135 species of phytoplanktonic algae, including 69 species of green algae and 46 species of cyanobacteria that together constitute most of the phytoplankton biomass. Phytoplankton is most diverse in the least saline parts of the lake. The phytoplankton biomass (1978) reaches 2.42–62.27 mg l^{-1} in the large Lake Chany and 168 mg l^{-1} in the smaller one (Ermolaev, 1986). As the salinity increases, the proportion of euryhaline species rises, and freshwater forms disappear progressively. Green algae persist to a salinity of 7 g l^{-1}, and cyanobacteria to 10 g l^{-1} (Anon., 1984).

According to Vizer (1986), 57 zooplankton species occur (most in the smaller lake): 27 rotifer species, 21 cladoceran species, and 9 copepod species. The lake is characterized by the development of a freshwater Copepoda-Cladocera complex, species of which develop high biomass and numerical densities. They include *Daphnia longispina* (O.F.M.), *Chydorus sphaericus* (O.F.M.) and *Mesocyclops leuckarti*. The most frequently found rotifers are *Keratella quadrata* and *Brachionus angularis* Gosse. *Arctodiaptomus salinus* occurs in the most saline parts of the lake. The dominant forms in saline parts are the rotifers *Hexarthra mira* (Hudson) and *Filinia terminalis* (Plate), and the cladocerans *Ceriodaphnia reticulata* (Jurine), *Diaphanosoma brachyurum* and *Moina microphthalma* Sars.

Average zooplankton biomass in summer reaches 15 g m^{-3}. Most comprises crustaceans with rotifers constituting <2 percent of the biomass. High zooplankton producton is the basis for a productive commercial fishery (Anon., 1984), but water-level fluctuations have had a negative influence on zooplankton biomass. The sharp fall in levels to 1971 led to a threefold decrease.

The zoobenthos of the Lake Chany complex contains 114 species (Miseyko *et al.*, 1986). Of these, 68 species are insects (45 species of chironomids), 13 are molluscs, 2 are crustaceans, and 3 are annelids. On silts, the larvae of *Chironomus plumosus* predominate, whereas on sand, *C. defectus* Kieff., *Tanytarsus* sp. and *Procladius* sp. predominate. Ceratopogonids and oligochaetes are also numerous. Mollusc numerical and biomass densities are low. In the more saline parts of the lake, the zoobenthos biomass is not high (2.8 g m^{-2} against 23.7 g m^{-2} in less saline parts). The most salt-sensitive forms are oligochaetes. *Chironomus anthracinus* becomes the dominant form in the saline parts (Anon., 1984).

The Lake Chany complex has an important commercial fishery, with annual average catches of 4400 tons. The original fish fauna is represented by the roach (*Rutilus rutilus* Pall.), perch (*Perca fluviatilis* L.), pike (*Esox lucius* L.), ide (*Leuciscus idus* (L.)), and carp (*Carassius carassius* L., *C. auratus*) – all of commercial significance – and by the dace (*Leuciscus leuciscus* L.), gudgeon (*Gobio gobio*), and lake minnow (*Paraphoxinus percnurus* (Pall.)). Roach formerly made up as much as 90 per cent of the catch. In the 1950s and 1960s, several other fish were introduced: carp (*Cyprinus carpio*), tench (*Tinca tinca*), bream (*Abramis brama* (L.)). sander (*Lucioperca lucioperca*), European cisco (*Coregonus albula* L.), peled (*C. peled* (Gmelin)), muksun (*C. muksun* (Pallas)), arctic cisco (*C. autumnalis* (Pallas)) and sheefish (*Stenodus leucichthys*), but of these introductions only bream, sander and carp were successful. The size of the fish catch depends on lake volume. When water-levels are low, catches fall to 300 tons, but at high levels they increase to 9800 tons. Moreover, when water-levels are low, fish reproduction is inhibited: the increased salinity reduces roe survival and there is a reduced area of spawning grounds (most of which are in the smaller lake and the river valleys). Additionally, at low levels, the rate of biomass increase diminishes, and natural mortality increases, especially in winter. For most of the fish the highest salinity at which ova develop and larvae survive is ±4 g l^{-1}, but for perch this value is 7 g l^{-1} (Anon., 1984).

To retain the commercial fishery it is necessary to maintain the present water-level, or regain former levels. With this in mind, the most saline part of the lake, which had little fishery significance, was isolated from the main lake by a dam. The aim was to maintain lake levels in the reduced area despite a decrease in regional water abundance. There are also plans to withdraw water from the River Ob and divert it into Lake Chany and to establish an outflow from the lake into the River Irtish.

Lake Alakul

Lake Alakul (or *Alakol*) is a terminal lake in the semi-desert zone of Kakakhstan within the Balkhash-Akakul basin at an elevation of 343 m above sea-level. The River Emel discharges into it. Extending northwest to southeast, it is 104 km long, 52 km wide and has an area of 2650 km². Maximum depth is 54 m and average depth, 22.1 m. From 1952–1962, water-level rose by 4.25 m. Average annual fluctuations in water-level are 1.2 m. The thermal regime is characteristic of that for deep lakes: in summer, the lake stratifies, but is mixed in autumn following the occurrence of high winds. In summer, surface temperatures reach 24–26 °C, but ice cover is present from January to April (Filonets, 1965; Kurdin, 1965).

Water transparency varies from 0.6–0.8 m in shallow areas to >6 m in central parts. Salinity is highest in the central and eastern parts of the lake (up to 8–10 g l^{-1}). The least saline water is in the northwest (nearly 5 g l^{-1}). Salinity of surface waters is lower (1.1–2.0 g l^{-1}) in winter when the ice cover on the lake suppresses the wind-induced mixing of river inflows with the saline lake water. In other seasons, currents prevent such stratification. In the early 1960s, there was a decrease in salinity following an inflow of fresh water. Ionic composition varies in different parts of the lake (Kurdin & Shilnikovskaya, 1965).

Macrophytes occur throughout the lake, but are most developed in the northwest. The phytoplankton has 156 species, of which diatoms are the most numerous (Logvinovskikh, 1965).

The zooplankton is unevenly distributed. It is most diverse in the least saline parts of the lake in the northwest where shallow depth, indented shorelines and abundant macrophyte vegetation favour zooplankton development. In this region, the dominant forms are *Arctodiaptomus salinus*, *Cyclops* sp., *Ceriodaphnia reticulata*, *Diaphanosoma brachyurum*, *Brachionus angularis*, *B. calyciflorus* Pallas, *Keratella cochlearis* and *Asplanchna herricki* Guerne. Zooplankton biomass here is 1.8–2.9 g m^{-3}. In the southeast, where depths are greater and vegetation less well-developed, zooplankton biomass falls to 0.4–1.0 g m^{-3}. In the central and northern regions, *Arctodiaptomus salinus* and *Eudiaptomus graciloides* Lill. become dominant, the cladocerans decrease in abundance and the rotifers increase in abundance. In the southeast, where rotifers are dominant, the zooplankton biomass decreases to 0.5 g m^{-3}. *Brachionus plicatilis* Muller and *Hexarthra oxyurus* are the main species involved (Loginovskikh & Dyusengaliev, 1972).

The zoobenthos comprises oligochaetes, leeches, insect larvae, and (rarely) molluscs. The principal group is larval chironomids (45 species). Zoobenthos development depends on substrate type and salinity. It is highest on grey silts (3.88 g m^{-2}), and least on silty sands (0.78 g m^{-2}). The highest zoobenthic biomass is found in the northwest (5.73 g m^{-2}), but it is also high in central areas at depths of 30–40 m (5.04 g m^{-2}). The lowest values occur in eastern bays (1.16 g m^{-2}). Overall, the biomass density is 2.61 g m^{-2}, and chironomids dominate everywhere, comprising 96 percent of the total biomass (Loginovskikh, 1965).

Fish are represented by marinka (*Schizothorax argentatus*), carp, (*Cyprinus carpio*) loach (*Nemachilus strauchi*) and perch (*Perca schrenki*). Sander (*Lucioperca lucioperca*) was introduced in the 1960s. Marinka, carp and perch are the commercial species (Nekrashevich, 1965).

Lake Tengiz

Lake Tengiz (*Dengiz*) is an alkaline lake situated in a tectonic basin in the north of Kazakhskiy Melkosopochnik, in the Kurgaldzhi reserve in the

Tselinograd region of Kazakhstan. Its area is 1590 km², it is 75 km long, 40 km wide and 8 m deep, and it is surrounded mainly by desert. Some islands occur just off the eastern shoreline. There is a shallow gulf in the northeast. The lake derives most inflow from snow-melt. The bottom is even and partly covered by black silts. In very dry years the lake partly dries out. The dominant ions are sodium and sulphate, and the salinity is 12.7 g l^{-1}, reaching 18.2 g l^{-1} in the gulf. The lake is ice-covered in December, but the ice breaks up in April. Inflowing rivers are the Nura and Kulanutpes. Little is known of the biological features of this lake.

Acknowledgements

We thank Professor W. D. Williams, University of Adelaide, for editorial assistance, and at the same institution, Miss Kelly Maurice-Jones and Mrs J. Read for artwork and typing, respectively.

References

Abrosov, V. N., 1963. Ozero Balkhash. The Lake Balkhash. Leningrad, 'Nauka', 127 pp. (Russian)

Aladin, N. V. & W. T. W. Potts, 1992. Changes in the Aral Sea ecosystems during the period 1960–1990. Hydrobiologia 237: 67–79.

Anon., 1984. Prirodnie Resursy Bolshikh Ozer SSSR i Veroyatnie ikh Izmeneniya. Natural Resources of Large Lakes in the USSR and their Probable Changes. Leningrad, 'Nauka'. 228 pp. (Russian)

Bolshaya Sovetskaya Encyclopedia. 3-e izd. The Large Soviet Encyclopedia. 3rd edition. vols. 1, 2, 25. Moscow, 'Sovetskaya Entsiklopedia'. (Russian)

Domrachev, P. F., 1933. Materialy k fiziko-geograficheskoy kharakteristike ozera Balkhash. Materials on the physico-geographical characteristic of the Lake Balkhash. In Issledovaniya ozer SSSR, v. 4. Leningrad. Studies of the USSR lakes, v.4: 31–56. (Russian)

Domrachev, P. F., 1940. Ozero Balkhash kak obyekt geograficheskogo izucheniya i issledovatelskie raboty, provodivshiesya na nem za polednee desyatiletie. Lake Balkhash as object for geographical studies and researches carried out on it in the last decade (1928...1938). In Izvestiya Vsesoyuznogo Geograficheskogo Obschestva, v. 72, No. 6. Proceedings of the All-Union Geographical Society: 651 pp. (Russian)

Ermolaev, V. I., 1986. Planktonnie fitotsenozy ozera Chany. Plankton phytocenoses of Lake Chany. In Ecologiya ozera Chany. Moscow 'Nauka'. Ecology of Lake Chany: 76–87. (Russian).

Filonets, P. P., 1965. Morfometria Alakolskikh ozer. Morphometry of the lakes of the Alakol system. In Alakolskaya vpadina i ee ozera. Alma-Ata, 'Nauka'. The Alakol hollow and its lakes: 79–87. (Russian)

Foliyan, L. A., 1973. Zooplankton ozera Issyk-Kul (kachestvenny sostav). Zooplankton of Lake Issyk-Kul (qualititative composition). In Ikhtiologischeskie i gidrobiologicheskie issledovahiya v Kirgizii. Frunze, 'Ilim'. (Ichthyological and hydrobiologycal studies in Kirgizia: 3–11. (Russian)

Karpevich, A. F., 1975. Teoria i praktika akklimitaztsii vodnykh organizmov. Theory and practice of acclimatization of aquatic organisms. Moscow, 'Pischevaya Promyshlennost' 432 pp. (Russian).

Kipshakbaev, N. K. & Zh. E. Baygisiev et al., 1985. Sistemniy analiz Ili-Balkhashskoy problemy i kontseptsiya ravnovesnogo priridopolzovaniya. System analysis of the Ili-Balkhash problem and conception of balanced nature managment. In Problemy komplexnogo ispolzovaniya vodnykh resursov Ili-Balkhashskogo basseyna. Izd. Kazakhskogo Gos. Universiteta, Alma-Ata. Problems of complex nature resources management in the Ili-Balkhash basin: 80 pp. (Russian)

Kurdin, R. D., 1965. Termicheskiy rezhim Alakolskikh ozer. Thermic regime of the Alakol system lakes. In Alakolskaya vpadina i ee ozera. Alma-Ata, 'Nauka'. The Alakol hollow and its lakes: 182–195. (Russian)

Kurdin, R. D. & L. S. Shilnikovskaya, 1965. Gidrokhimicheskiy rezhim Alakolskikh ozer. Hydrochemical regime of the Alakol system lakes. In Alakolskaya vpadina i ee ozera. Alma-Ata, 'Nauka'. The Alakol hollow and its lakes: 209–222. (Russian)

Loginovskikh, E. V., 1965. Kormovaya baza Alakolskikh ozer i ee ispolzovanie rybami. Food basis of the Alakol lakes and its use by fishes. In Alakolskaya vpadina i ee ozera. Alma-Ata, 'Nauka'. The Alakol hollow and its lakes: 223–235. (Russian)

Loginovskikh, E. V. & T. Dyusengaliev, 1972. Kolichestvennaya kharakteristika zooplanktona Akakolskikh ozer. Quantitative characteristics of the zooplankton of the Alakol lakes. In Rybnie resursy vodoemov Kazakhstana i ikh ispolzovanie, vyp. 7. Alma-Ata, 'Kaynar'. Fish resources of Kazakhstan waterbodies, v. 7: 89–94. (Russian)

Miseyko, G. N., L. L. Sipko & V. V. Kryzhanovskiy, 1986. Zoobenthos ozera Chany. Zoobenthos of Lake Chany. In Ecologiya ozera Chany. Novoskibirsk, 'Nauka'. Ecology of Lake Chany: 128–146. (Russian)

Nekrashevich, N. G., 1965. Materialy po ikhtiofaune Alakolskikh ozer. Materials on the ichthyofauna of the Alakol lakes. In Alakolskaya vpadina i ee ozera. Alma-Ata, 'Nauka'. The Alakol hollow and its lakes: 236–268. (Russian)

Panov, A. P., 1933. Khimicheskaya otsenka vody ozera

12

Balkhash. Chemical appreciation on the Lake Balkhash water. In Issledovaniya ozer SSSR, v. 4. Leningrad. Studies of the USSR lakes, v. 4: 105–112. (Russian)

Rylov, V. M., 1933. K svedeniyam o planktone ozera Balkhash. Information on the lake Balkhash zooplankton. In Issledovaniya ozer SSSR, v. 4. Leningrad. Studies of the USSR lakes, v. 4: 57–70. (Russian)

Saduakosova, R. E., 1972. Zooplankton ozera Balkhash. Zooplankton of Lake Balkhash. In Rybnie resursy vodoemov Kazakhstana i ikh ispolzovanie, vyp. 7 Alma-Ata, 'Kaynar'. Fish resources of Kazakhstan water-bodies, v. 7: 97–100 (Russian)

Tarasov, M. N., 1965. Gidrokhimiya ozera Balkhash. Hydrochemistry of the Lake Balkhash. Leningrad, Gidrometeoizdat', 371 pp. (Russian).

Tursunov, A. A. *et al.*, 1986. O sostoyanii problem Ili-Balkhashskogo basseyna. On the Ili-Balkhash basin state problem. In: Voprosy gidrologii oroshaemykh zemel Kazakhstana. Alma-Ata, 'Nauka'. Problems of hydrology of irrigated lands in Kazakhstan, 161 pp. (Russian).

Vizer, L. S., 1986. Zooplankton ozera Chany. Zooplankton of Lake Chany. In: Ecologiya ozera Chany. Novosibirsk, 'Nauka'. Ecology of Lake Chany: 105–114. (Russian).

Voskoboynikov V. A., A. N. Gundrizer, B. G. Ioganzen, S. F. Kononov, V. M. Kraynov, G. M. Krivoschekov, N. A. Nesterenko, Yu. F. Malyshev, M. I. Feoktistov, V. A. Schenev, 1986. Obschiy ocherk ikhtiofauny ozera Chany. General feature of the Lake Chany ichthyofauna. In: Ecologiya ozera Chany. Novosibirsk, 'Nauka'. Ecology of Lake Chany: 158–196. (Russian).

Williams, W. D. & N. V. Aladin, 1991. The Aral Sea: Recent limnological changes and their conservation significance. Aquatic Conservation 1: 3–17.

Zenkevich, L. A., 1963. Biologiya morey SSSR. Biology of the seas of the USSR. M., Izd. AN SSSR 739 pp. (Russian).

Zhandaev M. Zh., 1972. Geomorfologiya Zailiyskogo Alatau i problemy formirovaniya rechnykh dolin. Geomorphology of Zailiyskiy Alatau and problems of river valleys forming. Alma-Ata, 'Nauka', 159 pp. (Russian).

Mongolian salt lakes: some features of their geography, thermal patterns, chemistry and biology

A. N. Egorov

Institute of Limnology, Academy of Sciences, Sevastyanov Street 9, 196199 St. Petersburg, Russia

Key words: saline, lakes, heliothermal, limnology, hydrobiology

Abstract

There are > 3500 lakes of area > 0.1 km^2 in Mongolia. Most have salinities > 1 g l^{-1}. Of these saline lakes, 12 have areas > 50 km^2 each, 10, areas > 100 km^2, and 2, areas > 1000 km^2. Limnological investigations of salt lakes include morphometric studies, and investigations of bottom sediments, thermal regimes, chemical features, and biological (including fish) characteristics. Data from these investigations now provide a much clearer picture of the limnology of saline lakes in central Asia.

Introduction

The limnology of saline lakes in Mongolia, despite their widespread occurrence and abundance, has not been well studied. Even so, several authors have contributed to our knowledge of them. Potanin (1883) was amongst the first to investigate Mongolian salt lakes. He investigated the chemistry of lakes Ureg (Mongolian Altai), Durgon, Hyargas (the Valley of the Great Lakes in the Gobi region), Sangiyn Dalay (Khangai), Uvs (west of the Gobi region), and lakes of eastern Mongolia (Angart, Davsan, Djirlin-Tzagan and others). Smirnov (1932) investigated the chemistry of saline lakes and springs (lakes Uvs, Hyargas, Durgon, Sangiyn-Dalay and Tzokhor). He concluded that formerly there were large bodies of fresh water with a combined area of ± 50000 km^2 whose remnants are now the fresh and saline lakes of the Valley of the Great Lakes in the Gobi region. Between 1924 and 1926, Polynin (1930) investigated salt accumulation in Lake Tukhum in central Mongolia. Murzaev (1948, 1952) and, later, Kuznetzov & Murzaev (1963)

completed a detailed geographical study of Mongolia and this included a study of saline lakes.

A conclusion from these previous studies is that lakes were more abundant and larger in former periods (Tertiary and Quarternary) in central Asia than they presently are. Thus, most lakes of the Gobi region of Mongolia are relictual. Modern data on long-term climatic fluctuations confirm this conclusion.

Studies of salt lakes by Mongolian scientists began with those of Tzegmid (1955) and Davasuren (1961). They studied the paleolimnology and evolution of lakes in west Mongolia (Valley of the Great Lakes). Avirmed (1959) described the chemical composition and location of Lake Devter (western Mongolia, Gobi region). Luvsandorzh (1959) also contributed to our knowledge of the chemical composition and economic value of Mongolian salt lakes: He studied the salt lakes and their deposits in the Valley of Great Lakes and lakes Suzh and Gantzun (Gobi)j. His monograph (Luvsandorzh, 1967) was the result of many years study. Of other works on lake mineral resources, the investigations of Tzend (1966) on lakes of the east-Mongolian plain and Khangai region should be mentioned. Namnandorzh, Tzeren & Njmdorzh (1966) described salt lakes with mud of medical use, such as lakes Toson, Ikh-Tukhum, Dolon and Davst. Tzerensodnom (1971) described many physical, chemical and morphometric features of the largest Mongolian lakes.

Many studies of Mongolian salt lakes were related to the evolution of mineral resources and their possible use to the chemical, food, tanning and other industries. As a result, much of the limnological information is dispersed, employed different techniques of investigation, and is not very useful in a scientific sense. More rigorous knowledge is provided by Luvsandorzh (1967), Tzerensodnom (1971) and Dulma (1974). Investigations by the present author, in collaboration with others during a study of Mongolian lakes by Soviet-Mongolian geological and biological expeditions (1986–90), now permit much firmer knowledge of the principal features of Mongolian salt lakes. The recent limnological review by Williams (1991) of Chinese and Mongolian salt lakes has also contributed. Bulyon (1985) has outlined the principal limnological issues to be addressed.

Geographical features

More than 80 per cent on Mongolian lakes (± 3500) of area $> 0.1 \, km^2$ have salinities $> 1 \, g \, l^{-1}$. The total volume of saline water approaches 30 km^3, a volume some three times that of Mongolian freshwater lakes, if Hubsugul Lake, which contains most Mongolian fresh water (380 km^3), is left out of account. In Dornod Ajmak alone (eastern Mongolia), there are ± 1500 permanent and ephemeral lakes with salinities up to 320 $g \, l^{-1}$. The maximum salinity recorded is $> 400 \, g \, l^{-1}$.

The largest salt lakes are situated mainly in the north-west: lakes Uvs (13 350 km^2), Hyargas (1407 km^2), Dorgon (305 km^2), Telmen (194 km^2), Sangiyn Dalay (165 km^2), Oygon (61.3 km^2), Dzugny (27.0 km^2), Bust (21.9 km^2),

and Ureg (238 km^2). Their salinity ranges from 3.3 $g \, l^{-1}$ (Dzugny) to 27.3 $g \, l^{-1}$ (Oygon).

In the total Mongolian land area of approximately $1565 \times 10^6 \, km^2$, 12 salt lakes have areas $> 50 \, km^2$, 10 have areas $> 100 \, km^2$, and 2 have areas $> 1000 \, km^2$. The total lacustrine area (including freshwater lakes) is 15 600 km^2, i.e. 1 per cent of the country (Tzerensodnom, 1971).

Natural habitats and environments are diverse: They include coniferous regions in the north to sandy and stony-desert environments in the south. However, highlands cover most of the country. In all, only 15.2 per cent of Mongolia lies at an elevation of < 1000 m.a.s.l., whereas 40 per cent lies between 1000 and 1500 m.a.s.l., 20 per cent between 1500 and 2000 m.a.s.l., 22.4 per cent between 2000 and 3000 m.a.s.l., 2.4 per cent between 3000 and 4000 m.a.s.l., and 0.02 per cent > 4000 m.a.s.l. The average elevation of Mongolia is 1580 m.a.s.l., but 44.7 per cent of the land area is situated at higher elevations. Mongolia can truly be considered as mountainous.

The lakes are distributed irregularly but certain limnological features can be distinguished in particular geographical regions and elevation zones. Thus, four principal geographical regions can be distinguished (Tzegmid, 1975): The Altai mountain region, Khangai-Khentei mountain region, the Gobi region, and the eastern-steppe region (Fig. 1). Each region contains a different number of lakes. As Table 1 shows, the greatest number occurs in the Gobi region; this is an endorheic region with many terminal lakes.

An analysis of how lakes are distributed according to geographical regions shows that 34 per cent of them are located in the Altai and Khangai-Khentei mountainous regions, and 66 per cent in plain-steppe desert regions. Indeed, 36.5 per cent of all lakes are located in the Gobi region despite its low precipitation ($\pm 150 \, mm \, y^{-1}$), great evaporation (900–1000 mm y^{-1}), scarcity of rivers, and low relief. The largest lakes also occur here (e.g. Lake Hyargas and many others), as well as many smaller lakes, mostly with high salinities. The large number of lakes here has been explained as the result of two phenomena. First, the discharge into the Gobi of many rivers from the Altai

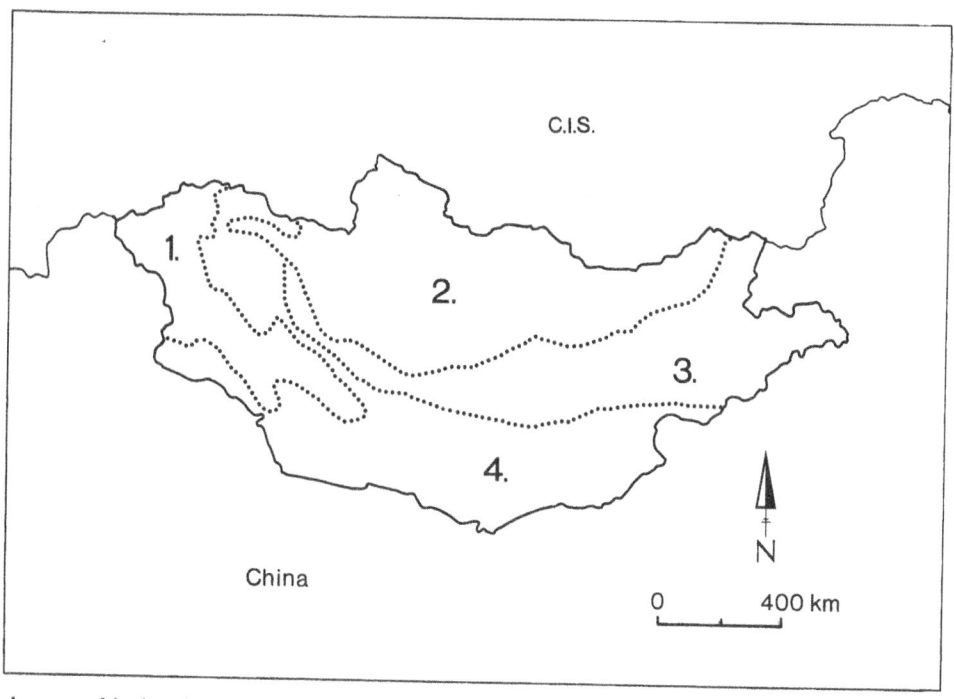

Fig. 1. Principal geographical regions of Mongolia. 1, Altai Mountain region; 2, Khangai-Khentei Mountain region; 3, Gobi region; 4, eastern-steppe region.

Table 1. The distribution of lakes according to geographical region (from Tzerensodnom, 1971). Data are only for lakes of area > 0.1 km².

Geographical region	Distribution	
	As per cent of total number (N = 3500)	As per cent of total area (A = 15 600 km²)
Altai mountains	13.3	10
Khangai-Khentei	20.8	30
Eastern-steppe	29.4	11
Gobi	36.5	54

and Khangai Mountains (the Gobi lies in a tectonic trough between these ranges). Second, the presence of tectonic basins in which lie large lakes representing the relics of ancient and much larger lakes. All lakes in this region are part of the central Asian endorheic basin and fed by rivers flowing from the southern part of the Khangai and the eastern part of the Mongolian Altai.

So far as size is concerned, Mongolian lakes vary from very small to extremely large. However, most have areas from 0.1 to 10 km². Such lakes comprise 98.1 per cent of all lakes numerically but only 14 per cent of the total area of Mongolian lakes.

The elevational distribution of lakes has a number of distinctive characteristics. First, the range is very wide, some 3050 m. Second, most lakes are located between 500 and 2000 m.a.s.l.: About 74 per cent of all lakes are in this semi-arid region, many being closed basins with unstable water regimes and most highly saline. Third, at about 2000 m.a.s.l., there are a few lakes (about 2 per cent) of stable hydrology but without outflow. Finally, from 2000 to 4374 m.a.s.l., there are many lakes (more than 24 per cent) which have outflows and are fresh.

Thermal patterns

Thermal patterns in Mongolian salt lakes have been poorly studied thus far. Most information relates to observations obtained during expeditions, and continuous observations are practically absent. Analysis of this information, however, indicates that thermal patterns are determined mainly by climatic factors, geographical features, lake morphometry, local and geomorphological characteristics, and salinity. Such features show some regional distribution, of course.

The seasonal pattern is driven by the continental climate. This leads to a rapid and intense warming of lakes in spring and summer, and in winter to the formation of an ice cover up to $2 \pm$ m thick (with some shallow lakes frozen to the bottom. The period of ice cover is from 7 to 8 months in lowland lakes. In the north, ice formation begins at the end of October, and in the south about mid-November.

The annual range of temperature in salt lakes is characteristically wide: in highland lakes it is 20–25 °C, and in lowland ones, 30 °C or more.

Warming begins in March when, after snow has disappeared from lake ice, the temperature of surface water layers begins slowly to rice. After the ice melts in May, the water temperature quickly rises above 4 °C, whereupon isothermal conditions develop in shallow lakes and thermal stratification develops in deep lakes (e.g. Hyargas). Maximum rates of heating take place in July and August, and after August, water temperatures begin to fall.

Deviations from the typical thermal pattern for temperate lakes may develop according to specific morphometric features, basin morphology, and salinity. Thus, it appears that the thermal pattern in Lake Hyargas, with its trough-like form and considerable depth (92–95 m), involves the development of a so-called 'thermal bar' in spring and autumn, i.e. the development of a ring-shaped and relatively narrow transitional zone of water with a temperature *ca* 4 °C dividing the lake into two zones, one below and one above 4 °C. This phenomenon has been well-studied in large lakes (Tikhomirov, 1982).

A second factor promoting the formation of 'anomalous' thermal patterns is lake salinity. In several lakes of the Erkhel-Trans-Hubsugul region, the Dund-Shavar-east steppe region, the Oygon-Khangai mountainous region, and the

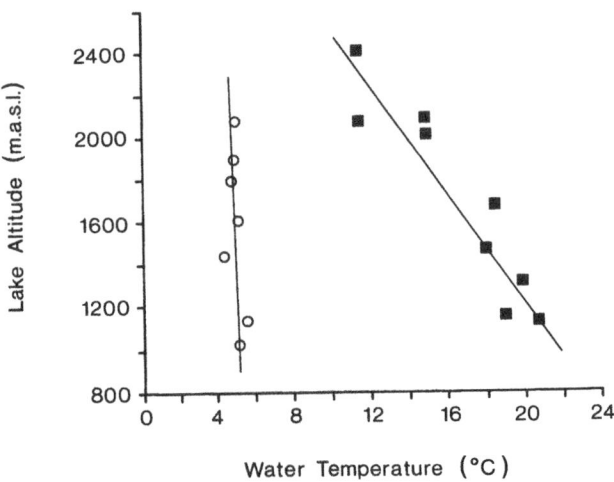

Fig. 2. The relationship between lake bottom water temperature and elevation for 9 shallow (■; mean depths of 2–9 m) and deep lakes (○; mean depths of 12–50 m) of the Altai Mountain, Khangai-Khentei, and Gobi regions.

Tzavdan-west Gobi region, the so-called 'greenhouse effect' has been discovered. Here, a thin layer of fresh water overlies highly saline water and produces a temperature inversion so that the difference in water temperature at the surface and near bottom may be 20 °C or more. The difference between surface and bottom water salinities may exceed $20 \, g \, l^{-1}$. The phenomenon may be transitory, seasonal, or persist all year. Such lakes, termed heliothermal, are known elsewhere. They may be utilized to obtain energy.

A third factor influencing thermal properties is elevation. Since our expeditions were at the period of maximal heat content for the lakes, a correlation between water temperature and lake elevation was observed (Fig. 2). The bottom water temperature of lakes decreases with altitude, while the temperatures near the bottom in deep lakes remain approximately constant.

Chemical features

All Mongolian lakes are of continental origin, and have ionic compositions dominated by Na^+, K^+, Cl^-, and $SO_4 =$. The lakes of the arid Gobi region and elsewhere are also relics of previously more extensive internally drained water-bodies (Murzaev, 1952; Devjatkin, 1981). These features are the main determinants of the present chemical composition.

Thus, the chemical composition of lake waters largely reflects the nature of watershed geology. However, the importance of climate is indicated by regional differences in lake composition. In general, there is a gradual increase in salinity and a change in chemical composition from north to south that parallels the increasing aridity along this gradient. Thus, the wetter regions of northern Mongolia (Transkhubsugul, alpine Khangai, Khentei, and two northwestern regions of the Mongolian Altai) contain mainly fresh or only slightly saline water, while in the more arid regions of the Gobi and east-steppe, lakes are more saline.

There is also a correlation with elevation such that with a decrease in elevation there is usually an increase in salinity. The least saline lakes $(0.08–0.2 \text{ g l}^{-1})$ lie high in the mountains of the Mongolian Altai and Khangai, and the most saline ones lie in arid regions of the Gobi and east-steppe regions where salinity can reach 308 g l^{-1} (Lake Davsan), 321 g l^{-1} (Lake Suuzh), and even higher. High salinities in these regions result largely from high evaporation (ca 1000 mm p.a.) and widely distributed soil salinization (Table 2). Even so, the relation between elevation and salinity is not simple. Regression analysis of data relating the salinity and elevation of 42 lakes provided a value for r of only 0.003! This is partly explained by the fact that lakes of atypical salinity are frequently present in the various lake regions: for example in the mountain-forest region of Darkhat (north Mongolia), Lake Shishkid has a salinity of 5.5 g l^{-1}, and Lake Khar in the Gobi region has a salinity of 0.2 g l^{-1}.

The general increase in salinity from north to south and from west to east is associated with changes in chemical composition (Table 2). Data on ionic composition in the largest saline lakes (Table 3) and a factor analysis of such data indicate a wide variety of chemical types and points to the difficulty of designating most lakes to be of the sulphate, chloride or carbonate type (Rasskazov & Abramov, 1987). As Table 3 indicates, Cl^-, $SO_4^=$ and HCO_3^- are equally predominant in many of the large lakes, while the cations are strongly dominated by $Na^+ + K^+$.

Horizontal and vertical variations in the salinity of salt lakes reflect freshwater inflow and wind-induced mixing and generally do not exceed $0.1–0.2 \text{ g l}^{-1}$ (Tzend, 1966; Tzerensodnom, 1971; Egorov, unpubl.; Williams, 1991). The principal exception would be heliothermal lakes in which vertical salinity gradients may exceed 20 g l^{-1}.

So far as seasonal changes are concerned, salinity decreases in July–August when precipitation is maximal and sometimes also in May when snow melt is large. Salinity increases when water levels are low, as at the end of autumn. Seasonal and long-term salinity fluctuations reflect the continental climate of Mongolia. Since 1850 the climate of central Asia has shown a tendency toward increasing aridity (Shnitnikov, 1975). As a result, rivers have decreased flows and lakes have decreased water levels and increased salinities – as manifest, for example, in high elevation terminal lakes of the Tianshan and Pamir Mountains in the CIS. Superimposed on the long-term trend are shorter alternating wet and dry climatic cycles with corresponding changes in hydrologic regimes. These phenomena have also been observed for the largest lakes in the western Gobi Region (Fig. 3).

Table 2. Distribution of lake salinity (g l^{-1}) and soil salinization (in per cent of area) in geographical regions of Mongolia.

Region	Lake salinity (g l^{-1})	Soil salinization (as per cent of region)	Chemical class of lake water
Altai mountains	0.08–4.0	2.5	Bicarbonate
Khangai-Khentei	0.19–22.0	4.0	Bicarbonate-sulphate
Gobi and eastern-steppe region	0.30–320.9	30.5	Bicarbonate-sulphatechloride

Table 3. Major ionic composition of some large Mongolian salt lakes.

Lake	Value[a]	Salinity ($g\,l^{-1}$)	Cations			Anions			Source of data[b]
			Na + K	Ca	Mg	Cl	SO$_4$	HCO$_3$	
Gobi region									
Boon Tsagaan	A	5.3	1.58	0.04	0.17	1.17	1.68	0.67	Egorov & Sevastyanov
	B		80	2	18	42	44	14	(1986)
Orog	A	50.5	18.0	0.02	0.41	16.9	6.25	8.91	Egorov & Sevastyanov
	B		95.5	0.1	4.4	63.3	17.3	19.4	(1987)
Hyargas	A	8.5	2.38	0.01	0.30	1.35	2.22	2.20	Egorov & Sevastyanov
	B		79.2	0.4	20.4	32	38	30	(1986)
Uvs	A	18.78	5.47	0.01	0.84	6.15	4.93	1.21	Davaasuren
	B		76.8	0.17	23.0	57.6	34.1	6.5	(1961)
Dorgon	A	5.14	1.41	0.00	0.16	0.56	1.11	1.90	Egorov & Sevastyanov
	B		81	0.4	18.6	22	33	45	(1986)
Khangai-Khentei region									
Sangiwyn Dalay	A	3.76	1.18	0.01	0.09	1.14	0.29	1.05	Egorov & Sevastyanov
	B		86	1	13	58	11	31	(1986)
Oygon	A	27.3	9.20	0.03	0.55	7.69	8.30	1.56	Egorov & Sevastyanov
	B		88.7	0.4	10.9	52.2	41.7	6.1	(1987)
Arhel	A	22.7	6.82	0.04	0.68	2.77	11.0	1.39	Egorov & Sevastyanov
	B		82.6	0.6	16.8	23.7	69.4	6.9	(1986)
Tsagan	A	17.4	5.66	0.01	0.17	1.62	7.26	2.67	Egorov & Sevastyanov
	B		94.0	0.2	5.8	19.0	62.8	18.2	(1987)
Darday	A	21.4	7.17	0.01	0.05	3.07	4.99	6.15	Egorov & Sevastyanov
	B		98.3	0.2	1.5	29.8	35.6	34.6	(1987)
Dzugny	A	3.3	0.98	0.02	0.10	0.89	0.46	0.82	Egorov & Sevastyanov
	B		81.6	2.2	16.2	52.2	20.0	27.8	(1987)
Altai mountain region									
Talmen	A	6.8	1.8	0.01	0.37	1.43	1.78	1.57	Egorov & Sevastyanov
	B		70	0	30	39	36	25	(1986)
Ureg	A	5.6	1.01	0.01	0.52	0.74	2.04	1.23	Egorov & Sevastyanov
	B		48.3	0.6	51.1	25.0	50.8	24.2	(1987)
Duro	A	5.1	1.41	0.00	0.16	0.56	1.11	1.90	Egorov & Sevastyanov
	B		81	0.4	19	22	33	45	(1986)
Steppe region									
Dund-Shavar	A	310.0	–	–	–	–	–	–	Egorov & Sevastyanov (1989)
Bulan-Shavar	A	330.0	–	–	–	–	–	–	Egorov & Sevastyanov

[a] A = $g\,l^{-1}$; B = equivalent percentage of cations or anions.
[b] Cited works by Egorov & Sevastyanov are unpublished sets of data.

In closed lakes without permanent inflow, such as lakes Dund-Shavar and Bulan-Shavar (eastern-steppe region), and Tatzyn and Ulan (Gobi region), aridity results in the formation of brine

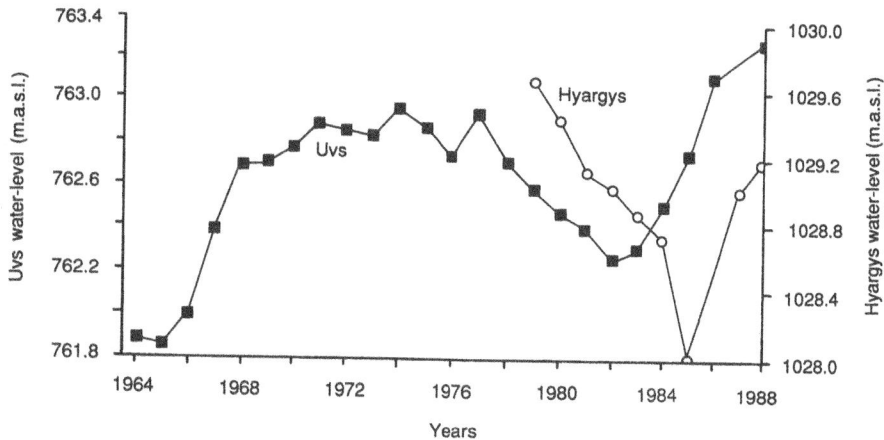

Fig. 3. Fluctuations in the water-level of two salt lakes in the Gobi region during recent decades.

with a salinity of $> 100 \, g \, l^{-1}$ or the total disappearance of the lakes. In lakes with a permanent inflow, the general tendency to desiccation may be interspersed with periods when levels increase and salinity decreases. Such fluctuations are typical of lakes Orog, Bon-Tzagan, Hyargas, Uvs and others in the Gobi basin. For example, the salinity of Lake Bon-Tzagan was $5.74 \, g \, l^{-1}$ in 1941, $4.80 \, g \, l^{-1}$ in 1962, $3.86 \, g \, l^{-1}$ in 1968, and $5.31 \, g \, l^{-1}$ in 1986. One cause of such change may be increased precipitation in mountain regions, followed by increased water-levels in mountain lakes and increased groundwater discharge into the lakes of the Gobi region (lakes Bon-Tzagan and Hyargas). Because it obstructs atmospheric water movement in a southerly direction, the Khangai-Khentei mountain massif causes intensive precipitation over these regions.

The other factor determining river discharge into lakes of the Gobi region is human use of water, e.g. for irrigation. If deprived of significant inflows, lakes can change from permanent to ephemeral bodies of water. Lake Orog in the Gobi region provides an example. In this way, the influence of natural climatic aridity can be reinforced by human use.

Some biological features of the largest salt lakes

The special geomorphological features of Mongolia, combined with its distinctly continental and arid climate and geographical position are the essential features which explain the nature of the lacustrine biota. So far as its geographical position is concerned, a few elements of the Chinese fauna penetrate in the west, whilst some Mediterranean species penetrate in the east. The total result is that the fauna comprises a mixture of forms of Arctic, European, Siberian, Tibetan and local origin.

The fauna, terrestrial and aquatic, of Mongolia is especially rich in bird (376) and mammal (133) species. This reflects the diverse geographical conditions and habitats. Smaller numbers of amphibians (8) and reptiles (11) reflect the arid and distinctly continental climate. Waterfowl populations inhabit practically all lakes and exemplify adaptation to climatic conditions (Dulma, 1974).

The biology of the aquatic biota is poorly investigated. Most information concerns large freshwater lakes with useful fisheries. The biota of salt lakes has been studied only incidentally. The biota of the numerous small lakes is practically unstudied.

On the basis of what is known, the general features of the biota are as follows:

(1) The plankton and fish faunas fall into three zoogeographical regions. First, a Central Asian endorheic region (the Valley of Great Lakes, Altai mountain lakes, lakes of the

Khangai plateau, Gobi lake basin). Second, a region which discharges into the Arctic Ocean (the Bulgan, Shishkid, Selenga rivers). And third, a region which discharges into the Pacific Ocean (the Onon, Kerulen and Khalkhingol Rivers).

(2) The zoogeographical boundaries roughly coincide with physical-geographical features.

(3) On the whole, the freshwater lakes with the greatest fish diversity are in the Arctic Ocean drainage system. The salt lakes here are represented only by small water bodies located in intermontane closed basins.

Particular features of the largest salt lakes are as follows. All biomass data presented are for wet weights in the summer.

Lake Uvs. This is the largest lake in Mongolia (3350 km^2). It lacks an outlet and has a maximum depth of 20 m. Secchi disk transparency may reach 6 m. Salinity is *ca* 19 g l^{-1}. Zooplankton biomass is *ca* 0.11 g m^{-3}. Fish include *Oreoleuciscus pewzowi* (Herzenstein) and *O. potanini* (Kessler). The zooplankton and benthos are unstudied.

Lake Hyargas. The second largest salt lake in area (1407 km^2), it has the deepest basin with a maximum depth of >90 m. Salinity is 8.5 g l^{-1}. In summer, copepods dominate the plankton, but details of species composition of the zooplankton and benthos remain unknown. Zooplankton biomass is *ca* 0.31 g m^{-3}. Rotifers dominate in winter under ice. *Ceratium hirundinella* O. F. Muller and *Pediastrum* sp. are abundant in the phytoplankton. *Oreoleuciscus pewzowi* and *O. potanini* are important fish species present.

Lake Bon Tzagan. This is a closed oligo-mesotrophic water-body in the Valley of Great Lakes. Its area is 252 km^2 and its maximum depth <18 m. Its salinity is 5.3 g l^{-1}. The zooplankton biomass reaches 2.1 g m^{-3}. Rotifers are common in the plankton. Phasganophora brevipennis *NAV.,* Diura nanseni *Kempny,* Aeschna affinis *Lind.,* Sympetrum flaveolum *L.,* Limnophilus ab-

strusus *McL. and* Oecetis ochracea *Curtis dominate the benthos.*

Lake Ureg. Located in the Mongolian Altai at an altitude of 1425 m.a.s.l. (Table 1), this lake has an area 237.6 km^2 and a maximum depth of 42 m (average depth is 15 m). Secchi disk transparency is to 8 m, and salinity reaches 5.1 g l^{-1}. Macrophyte beds cover up to 20 per cent of the lake area, with the common cane sedges and horsetails dominant. The benthic fauna is poor, and only single specimens of molluscs and amphipods are met. The zooplankton consists mainly of rotifers and copepods with a biomass of *ca* 0.34 g m^{-3}. The ichtyofauna is represented by *Oreoleuciscus pewzowi.*

Lake Sangiyn Dalai. Located in Khangai, this closed oligotrophic lake has an area of 165.3 km^2 and its depth reaches 30 m. The salinity is *ca* 4 g l^{-1}. Zooplankton biomass does not exceed 0.7 m m^{-3}. Hydrogen sulphide occurs in the profundal layers. Fish present include *Oreoleuciscus pewzowi.*

Lake Telmen. Another large oligotrophic lake of the Khangai plateau, this lake is 194 km^2 in area, and has a depth up to 27 m. Salinity is 7 g l^{-1}. Water transparency is <5 m. The only phytoplankter so far recorded is *Asterionalla formosa* Hass. Copepods and rotifers dominate the zooplankton, with a biomass of ca 0.2 g m^{-3}.

Acknowledgements

I thank Professor W. D. Williams, University of Adelaide, for editorial assistance, and at the same institution, Miss Kelly Maurice-Jones and Mrs J. Read for artwork and typing, respectively.

References

Avirmed, A., 1959. (The analysis of the salt composition of Lake Devter. Transactions of the Institute of Chemistry, Academy of Sciences, Mongolian Peoples' Republic, Ulan-Bator, 3–4: 42–52. (In Mongolian)

Bulyon, B. B., 1985. Limnological issues of Mongolia. Ulan-Bator-Moscow, 104 pp. (In Russian)

Davasuren, D., 1961. Physical-chemical structure of Lake Uvs. Scientific Transactions of State University, Ulan-Bator: 27–36. (In Mongolian)

Devjatkin, E. V., 1981. Cainozoic of Central Asia. Nauka, Moscow, 196 pp. (In Russian)

Dulma, A., 1974. Biology of Lakes in the Mongolian Peoples' Republic. Autoreferat of dissertation. University of Irkutsk, Irkutsk, 52 pp. (In Russian)

Kuznetzov, N. G. & E. M. Murzaev, 1963. Lakes developmental stages of Central Asia in Quaternary Time. In: S. V. Kalesnik (ed.), Lakes of the Semiarid Zone. Academy of Sciences, USSR, Leningrad-Moscow: 157–173. (In Russian)

Luvsandorzh, S., 1959. The chemical composition of Lake Hyargas. Transactions of the Institute of Chemistry, Academy of Sciences, Mongolian Peoples' Republic, Ulan-Bator, 3–4: 53–61. (In Mongolian)

Luvsandorzh, S., 1967. Mineral lakes of Mongolia. Academy of Sciences, Mongolian Peoples' Republic, Ulan-Bator, 112 pp. (In Mongolian).

Murzaev, E. M., 1948. The Valley of the Great Lakes of western Mongolia and the origin of her landscapes. Transactions of the Second All-Union Geographical Congress 1: 367–378. Geographical Publishing House, Moscow.

Murzaev, E. M., 1952. Mongolian Peoples' Republic. Geographical Publishing House, Moscow, 472 pp. (In Russian)

Namnandorzh, O., S. Tzeren & O. Njmdorzh, 1966. The medical use of mud of some salt lakes. Scientific Transactions of State University, Ulan-Bator: 211–216. (In Mongolian)

Polynov, B. B., 1930. Preliminary report of pedogeographical expedition to North Mongolia in 1926. Data of committee on the investigations of the Mongolian and Tannu-Tuvinsky Peoples' Republics and Buryt-Mongolian ASSR. Academy of Sciences, USSR, Leningrad, 9: 149 pp. (In Russian)

Potanin, G. N., 1883. Essays on North-West Mongolia. Results of Travel in 1879–1883. Journal and Proceedings of Russian Geographical Society 3: 1–425. (In Russian)

Rasskazov, A. A. & A. V. Abramov, 1987. On sodium lakes of the M.P.R. Journal of Lithology and Mineral Resources 6: 88–99. (In Russian)

Shnitnikov, A. V., 1975. Fluctuations in the climate and general humidity of the 18th–20th centuries and the future. Proceedings of the All-Union Geographical Society, 107: 473–484. (In Russian)

Smirnov, V. A., 1932. Progress reports on hydrochemical investigations of the Mongolian expedition in 1926. Transactions of the Mongolian Committee 1: 1–23. (In Russian)

Tickhomirov, A. I., 1982. Thermal characteristics of large lakes. Nauka, Leningrad, 232 pp. (In Russian)

Tzegmid, S., 1975. The natural regions of the Mongolian Peoples' Republic and problems concerning their rational use and the protection of their resources. Transactions of the Institute of Pathobiology, Ulan-Bator, 2: 50–69. (In Mongolian)

Tzend, N., 1966. The chemical composition of some lakes in east Mongolia. Transactions of the Institute of Stockbreeding, Ulan-Bator, 15: 167–176. (In Mongolian)

Tzerensodnom, Z., 1971. Lakes of Mongolia. Academy of Sciences, Mongolian Peoples' Republic, Ulan-Bator, 202 pp. (In Mongolian)

Tzerensodnom, Z., Z. Sanzhmtav, Tz. Sugar & O. Tzerev, 1986. The hydrology, chemical composition and physics of Lake Uvs. Academy of Sciences, Mongolian Peoples' Republic, Ulan-Bator, 3: 1–19. (In Mongolian)

Williams, W. D., 1991. Chinese and Mongolian saline lakes: A limnological overview. Hydrobiologia 210: 39–66.

Chinese saline lakes

Zheng Mianping, Tang Jiayou, Liu Junying & Zhang Fasheng
Chinese Academy of Geological Sciences, Research and Development Center for Epithermal and Saline Lakes, Baiwanzhuang Road, Postcode 100037, Beijing, People's Republic of China

Key words: China, saline lakes, palaeolimnology, *Artemia*, *Dunaliella*, geoecology

Abstract

China has many saline lakes. Most occur in the west and north-east. Four main regions can be distinguished: Qinghai-Tibet Plateau, North western, North-central and Eastern. All types of chemical composition occur, but some regionalization of types is found. The Palaeolimnology of many saline lakes in China has been investigated, and a variety of dating techniques indicate ages between the Quaternary and the Recent. Organisms studied include *Artemia*, *Dunaliella salina* and some halophilic bacteria. The important role of organisms in many processes of geochemical and geological interest is stressed. Geoecology, as a combination of geology, mineral deposition and ecology, is a subject worth greater attention.

Introduction

China has many saline environments: Lakes and underground interstitial brine are discontinuously distributed in western and north eastern China, regions representing nearly half of China's total area. Saline lakes occur in Xinjiang, Qinghai, Tibet, Gansu, Inner Mongolia, Shan'xi, Shanxi, Ningxia, Hebei, Heilongjiang, Jilin and Liaoning. Underground brines are present at the northern and southern ends of the Bohai Sea between Liaoning and Shandong. There are > 1000 saline lakes (salinity > 3.5 g l^{-1}) of which 534 have been investigated (Table 1). They have a total area of $> 50\,000$ km^2. Of these, *ca* 50 have been explored in detail, including, among others: Qarhan Salt Lake, Da Qaidam Lake and Dalangtan Lake in the Qaidam Basin of Qinghai; Bongkog Lake, Zabuye (Chabyer) Lake and Zhacang Caka Lake in Tibet; Lop Nur in Xinjiang; Ebinur Lake and Barkol Lake in the Junggar Basin; Jartai Salt

Pond, Yabrai Salt Pond, Chaganmenlinuo'er Lake and Dalad Mirabilite Lake in Inner Mongolia (recently discovered under sand); and the Xiechi Lake in Shan'xi.

There is a long history of exploitation of saline lakes in China. Rock salt was first exploited for human consumption, and Xiechi Lake in Yuncheng of Shanxi, on the margin of the Central Plains of China, is one lake from which salt was mined for this purpose from early times. According to archaeological and radiocarbon studies, the utilization of the lake began 4500 y ago. A rhyme entitled the 'South Wind', from before the Xia Dynasty (*ca* 21st–16th B.C.), says that when a south wind blows, the evaporation of brine is strong, favourable for the making of salt in the sun. And *The Book of Water* of the Northern Wei Dynasty (386–534) has a detailed description of the area of the Yuncheng (Xiechi) Lake, the behaviour of the brine, the structure of salt pans and methods of salt exploitation.

Table 1. Chinese saline lakes and playas ($>3.5\,g\,l^{-1}$ salinity) classified according to tyoes of mineral deposits of potential commercial value.

Region	Abundant deposits of potassium or magnesium salts	High concentrations of B, Li, Mg or B, Li, K, Cs, Br, as brines or evaporites [a]	Abundant deposits of halite, mirabilite, trona, etc.
Qinghai-Tibet Region (I)	15	48	63
Northwestern Region (II)	1	1	288
North-central Region (III)	–	–	116
Eastern Region (IV)	–	–	7
Totals	16	49	494

[a] Lakes with high boron and lithium concentrations fall into 2 categories: those belonging to the sulphate type with high Mg^{2+} concentrations, and those belonging to the carbonate type with high concentrations of CO_3^{2-} or HCO_3^- and of Br^-, Cs^+ and K^+.

Because of space limitations, only a few results of present research on Chinese saline lakes are described in this article. Omitted from our discussions is any reference to research on structural geochemistry, mineral deposits, mineralogy, and hydrogeology.

Salt lake regions

The main region of saline lakes in China extends roughly between 28° and 53° N (Zheng, 1989). Its origins reflect regional geological structures, geomorphology, climate and other major factors. Four salt lake regions can be distinguished (Fig. 1): a Qinghai-Tibet Plateau region (I), a Northwestern region (II), a North-central region (III), and an Eastern region (IV).

Qinghai-Tibet plateau region (I)

This plateau, at an average elevation >4000 m, is bounded in the south by the Himalaya, in the north by the Kunlun and Qilian Mountains, and in the east roughly by the Yanbajain-Qinghai Lake area. Lakes are widespread and cover a total area of about $50\,900$ km² or more (including playas) (Zheng *et al.*, 1989). Among them are approximately 352 saline lakes, totalling $21\,465$ km², and accounting for slightly more than one third of total lake area in the region (Table 2).

The region is characterized by saline lakes with almost all types of chemical sediments and composition: halite, mirabilite, K, Mg-salt (e.g. Qarhan Salt Lake), trona, as well as lakes of unusual composition, e.g. those rich in B, Li, Cs, Br (e.g. Zabuye Lake) (Zheng *et al.*, 1985a)

The unique variety in chemical composition is caused by specific geographical, geological and chemical factors. Especially important is the cold arid climate at high elevation. Since the beginning of the Quaternary, the dry cold climate has evidently tended to expand from north to south (Zheng, 1989). Four main modern climate zones may be recognized from south to north: a sub-frigid plateau zone, a semi-arid to arid zone, a temperate plateau zone, and an extremely arid zone (Lin & Wu, 1981).

Based on geomorphological and evolutionary features, salt lake basins of the plateau can roughly be grouped into the following types:

A. Those developed in the north. These are large intermontane tectonic basins controlled by large faults, including the Qaidam and Kumkol Basins, which are large, strongly subsided areas formed by the uplifting of the plateau, and characterized by a strong Himalayan (Cenozoic) tectonism. They have over 2000 m of Quaternary lacustrine deposits in some lakes, and are at elevations of about 2520 m (Qarhan Lake) to 3876 m (Ayakum Lake).

B. Those developed on the plateau surface.

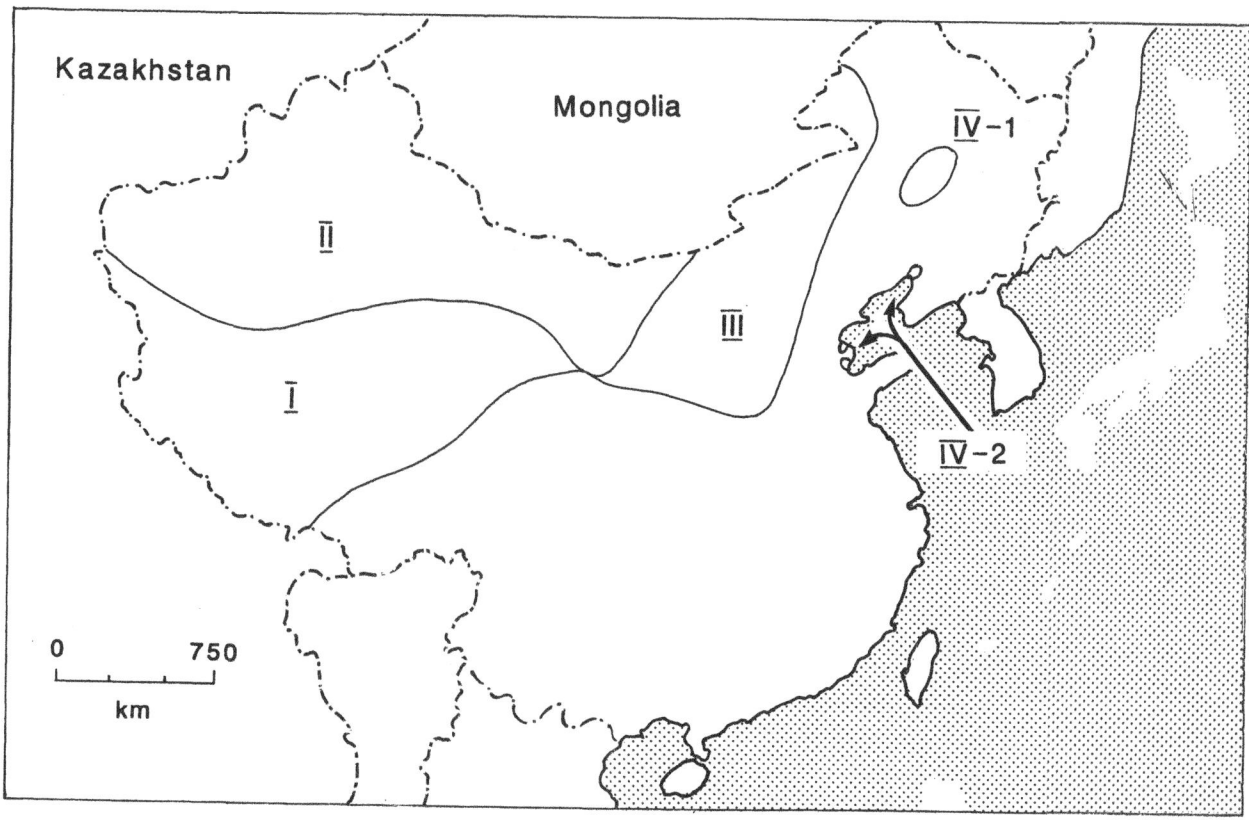

Fig. 1. Saline lake regions of China.

These are basins controlled by faults of different sizes, or are small lakes formed by glaciers and rivers. Most large and medium-sized lake basins of this type are downfaulted or downwarped ones. Lakes are at high elevations, extending from 4300 m (Cocholong Lake) in the south to 5057 m (Ninghu Lake). During the Early-Middle Quaternary, all the lakes, except those in the central part of the plateau, had outflows. This period is called the pan-fluviolacustrine period and lake deposits are all freshwater ones (Zheng *et al.*, 1986, 1989). During the Late Pleistocene, the climate became drier and colder and most of the lakes became closed basins, losing their outflows.

The third factor is the source of the salts. Since the beginning of the Quaternary, the region has been under a continental environment, and the source of salts is not associated with any seawater recharge. Common salt constituents, such at Ca, Mg, Na, Cl and SO_4, mainly derive from weathering and leaching of the regional lithology. But rare elements, such as B, Li, K, Cs, Th, Nd and As, are highly concentrated in some saline lakes of the plateau, and these have specific sources. Recent studies show that the source of these rare salts is related to external magmatic and geothermal belts and in these abundant constituents such as B, Li, K, Cs, As and Cl are transported in solution from deep strata. Some flow directly into saline lakes (Zheng *et al.*, 1989). The northern hot spring of the Da Qaidam Lake, which formed $24\,720 \pm 1750$ y ago (according to the ^{14}C dating of the tufa from the hot spring), is now discharging about 78 t yr^{-1} Li into the Lake.

North western region (II)

This region is located north of the Qinghai-Tibet Plateau region. Here, salt lake basins are gener-

Table 2. Lakes of the Qinghai-Tibet Plateau.

Province	Area of lake province (km²) A₁	Total area of lakes and associated playas (km²) A₂	Number of lakes and associated playas, by size category (km²)					Saline lakes and associated playas[d]		Weakly saline lakes[d]		Fresh lakes[d]		$\frac{A_2}{A_1}$	$\frac{A_3}{A_2}$	$\frac{A_4}{A_2}$	$\frac{A_5}{A_2}$
			≥1,000	<1,000–500	500–100	100–10	10–0.5	No.	Area (km²) A₃	No.	Area (km²) A₄	No.	Area (km²) A₅			(×100)	
Qilian[a]	62631	631	0	1	0	0	9	0	0	0	0	10	631	1.01	0	0	100
Qaidam[a]	208161	11813	3	1	6	15	10	26	11422	6	137	3	254	5.67	96.69	1.16	2.15
Kumkol[a]	74619	973.5	0	1	1	5	10	3	864.5	11	97	3	12	1.30	88.80	9.97	1.23
Hohxil[a]	151956	3087.7	0	0	7	42	208	65	992.9	126	1762.3	66	332.5	2.03	32.16	57.07	10.77
Qiangtang[a]	432263.75	9196.5	0	0	17	134	308	177	5146.3	137	3277.6	145	772.6	2.13	55.96	35.64	8.40
Northern Tibet[a]	36715	15524.3	3	2	23	99	191	72	2989.5	96	8886	150	3739.8	4.24	18.67	57.24	24.09
Southern Tibet[b]	141588	1502.1	0	0	4	11	43	2	14.0	9	880	47	608.1	1.06	0.93	58.58	40.48
Qinghai Hu-Gonghe[b]	120078	4796	1	0	1	1	3	3	113	1	463.5	2	48	3.99	2.36	96.64	1.00
Bayan Har[c]	419517	1885.2	0	2	1	12	303	4	14	2	46	312	1825.2	0.45	0.74	2.44	96.82
Southeastern Tibet[b]	469197	1502.5	0	1	1	11	121	0	0	3	129	131	1373.5	0.32	0	8.59	91.41
Total	2445725.75	50911.8	7	4	61	330	1206	352	21465.2	391	19849.9	869	9596.7	2.08	42.16	38.99	18.85

[a] Entire province is endorheic.

[b] Province includes both endorheic and exorheic regions.

[c] Entire province is exorheic.

[d] Salinity categories are defined as follows: fresh, <0.1 g l⁻¹; weakly saline, 0.1–3.5 g l⁻¹; and saline, >3.5 g l⁻¹.

ally at lower elevations, 2000–5000 m, with one or two (e.g. Aydingkol Lake in Turpan) at 154 m below sea-level. Huge basins alternate with mountains, vast deserts and plateaus. These include the Tarim Basin (generally at 800–1000 m above sea-level), Hexi Corridor (1000–1500 m) and the desert belt (1000–2000 m) of the Alxa Plateau (Anon., 1979). The region adjoins the USSR to the north and west and the Helan Mountains to the east. Some 290 saline lakes occur of which the largest is Lop Nur (20 000 km^2 if the salt crust is included). Some saline lakes lie beneath the surface and have been investigated only slightly.

The saline lakes contain predominantly halite, mirabilite and trona. Potassium concentration in Lop Nur and Ebinur Lake (western Alxa) is relatively high. Abundant kuluzite or trace amounts of carnallite and halo-sylvite have also been found in these lakes. A few saline lakes have relatively high concentrations of U. The Turpan Hami Basin (Tianshan Basin) contains considerable nitrate.

Tectonically, the region belongs to the relatively stable fault block province (Tarim and Alxa fault blocks) and fold belt (Tianshan Mountains, Qilian Mountain and Longgu Mountain). The Tarim Basin climate in the Quaternary was similar to that in the Qaidam Basin but, the former has a thinner Quaternary lacustrine sedimentary sequence (500–1000 m in drill hole) than the latter, suggesting that the lake basins in the region were less active than in region . Northward and north-eastward, modern climate zones are: an extremely arid area of the south temperate zone, a semi-arid area of the mid-temperate zone, and another semi-arid area in the southern temperate zone. Since the Quaternary began, the region has been endorheic. Some Pleistocene downfaulted basins are the result of large faults and lower-middle Pleistocene salt layers (e.g. at Wanyao in the Hexi Corridor) have been found in several downfaulted basins, indicating that the climate was dry and arid in Early-Middle Quaternary. Some recent saline or weakly saline lakes are also the result of faults. For instance, saline lakes easily form in the cross-points of two perpendicular series of faults (network fault structures). Depressions occur in such positions. Examples are given by lakes on the margins of the Alxa Desert which are arranged in a paternoster, chessboard pattern (Fig. 2).

North-central region (III)

The north-central region lies north-east of regions I and II, adjoining Mongolia to the north, the Taihang Mountains-Greater Khingan Mountains to the east, and the Baiyu Mountain to the south [including the Ordos Plateau (generally at 1000–1500 m a.s.l.), the Inner Mongolia Plateau (at 1000–2000 m a.s.l), and Hulum Buir Basin (at 200–500 m a.s.l)]. The region has 116 saline lakes. Modern saline lakes are smaller in size and simpler in composition than those in regions I and II, and predominantly contain trona, mirabilite, halite, and other common salts. Trona saline lakes (e.g. Chaganlimenuo'er Lake) are larger, and some lakes have a high concentration of Br (e.g. Erliannuoer Lake has 300 mg l^{-1} Br).

Tectonically, the region is relatively stable. In the central part of the fault block area, Quaternary deposition was rather weak, and the saline lakes are mostly of eolian or sedimentary origin, having very thin deposits with salt layers only a few cm to 1.2 m thick. The Quaternary sedimentary sequence, however, may be from 500 m to more than 1000 m thick and contain sulphate sediments. For example, the lower part (Q$_2$; see legend to Fig. 4) of Yuncheng (Xiechi) Lake (belonging to the Fen-Wei Graben) has glauberite and loeweite-bearing mirabilite layers (Tang, 1988), and the Middle Pleistocene in Dalad, in Hetao Downfaulted Basin, has 2–3 thick layers of pure mirabilite. In the lower part of the mirabilite layers are beds containing foraminiferan fossils. The above facts show that tectonic movements which occurred during the collision of India and the Eurasia Plates were of great significance to the formation of salt lake basins in China. The longitudinal enlargement of the Qinghai-Tibet Plateau caused by compression between two plates resulted in the genesis and development of the Fen-Wei Graben and Hetao Downfaulted

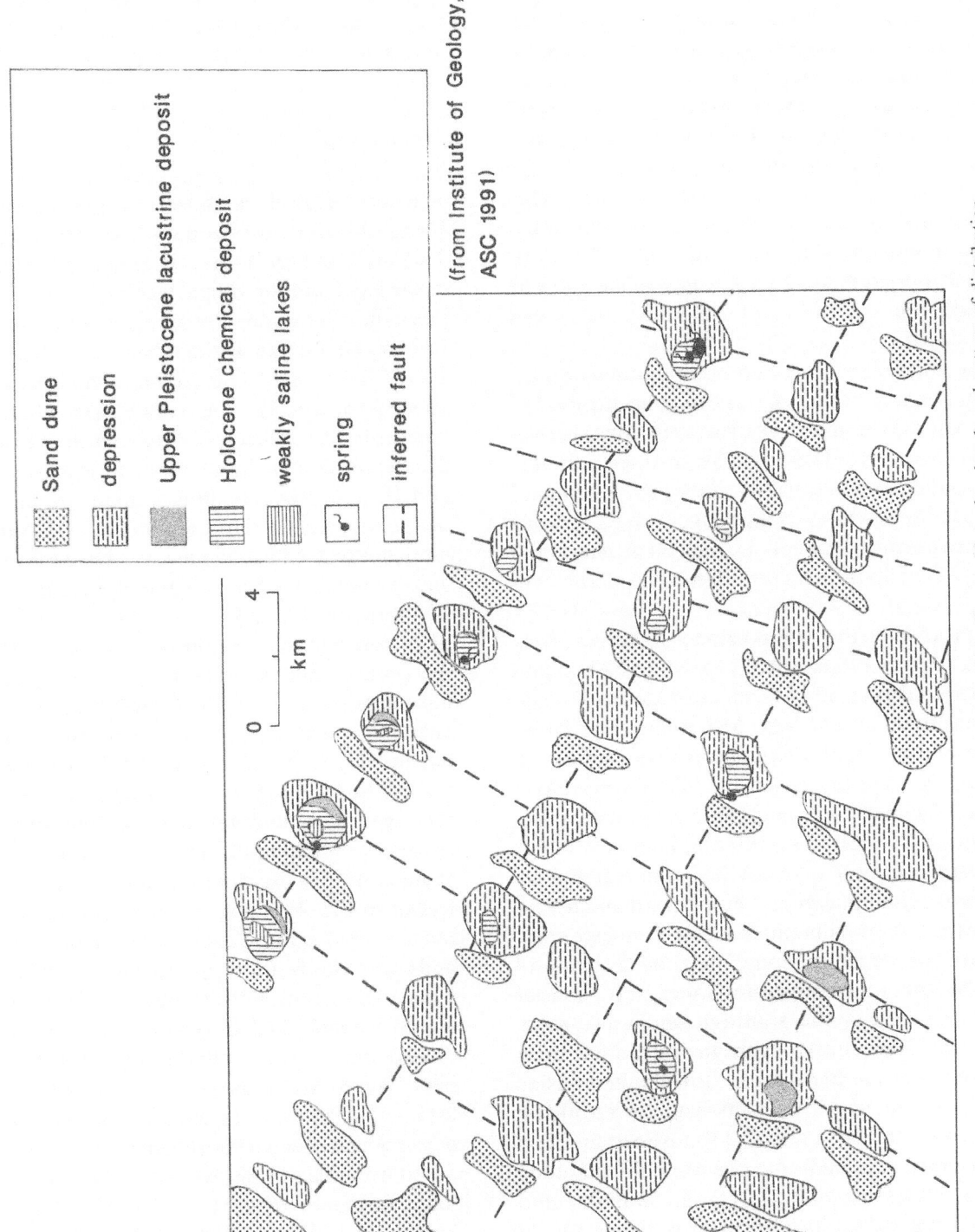

Sand dune

depression

Upper Pleistocene lacustrine deposit

Holocene chemical deposit

weakly saline lakes

spring

inferred fault

(from Institute of Geology, ASC 1991)

0 — km — 4

Fig. 2. Saline lakes near the margin of the Alxa Desert. Note the chessboard pattern of distribution.

Basin (Tapponier, 1982), and intermittent marine invasion. This is probably an important cause of the enrichment in $MgSO_4$ (bloedite) of the Q_2 salt layers of the Yuncheng (Xiechi) Salt Lake (Anon., 1979). During the Early-Middle Quaternary, the arid belt of China expanded several times into eastern China from the dry, salt-forming centre in the Golmud-western Qaidam area. On a climatic basis (Anon., 1979), the region belongs to the semi-arid to arid province of the temperate zone.

Eastern Region (IV)

In small closed basins of the semi-arid to semi-humid area of the temperate zones of eastern China, there are some scattered, small lakes and areas of saline groundwater. They are found mainly in two areas: the Nenjiang River basin of western Jilin where they occur in silted depressions and are mostly recent small lakes of low salinity (Fig. 1, IV-I); and the Xiayun River basin at the northern shore of the Bohai Gulf and the coastal belt of the Penglai Bay in Shandong Province where they occur in littoral clastic strata and comprise saline groundwater at depths of $0-100$ m, with a salinity of $50-218$ g l^{-1}, formed in a marine environment, and are dated as Late Pleistocene-Holocene (Fig. 1, IV-2 Zheng et al., unpublished, 1990).

Chemical typology

The chemical types of brines in saline lakes are variously classified, but no one classification is entirely satisfactory. In China and the USSR the method mostly used was developed by Valyashk (1955). Some authors (Zheng et al., 1986, 1989; Chen et al., 1985) subdivide the carbonate type into strong, moderate and weak carbonate subtypes, on the basis of the K_c value ($K_c = (Na_2CO_3 + NaHCO_3/total$ salt$) \times 100\%$, where concentrations are expressed as weight percent). The subtype is regarded as strong when $K_c > 29$; as moderate when $K_c = 8-29$; and as weak when $K_c = 0.1-8$. Another coefficient,

$R = \gamma SO_4 2 - /\gamma Mg 2^+$, has been proposed that is based on sulphate concentrations where γ are expressed as normality: sulphate brines are sub-divided into a Na sulphate subtype ($R > 1.5$), a Na-Mg sulphate subtype ($R = 1.5-0.5$), and a chloride transitional subtype ($R = 0.01-0.075$). There is another main type, the chloride type ($R = 0.01-0$).

Based on chemical data for 350 lakes, the distribution of Chinese salt lakes is indicated in Fig. 3. Some general conclusions follow.

A. All types of chemical composition are present in Chinese salt lakes (Zheng et al., 1989).

B. From the Quaternary arid center of the Qaidam Basin-Tarim Basin outwards the following zones occur: a chloride-sulphate subtype zone, a chloride-bearing Mg-sulphate subtype zone or Na-sulphate zone, and a carbonate type + Na-sulphate subtype or carbonate type zone. This agrees well with tectonic, geochemical and climatic differences between the different subregions.

C. Local exceptions simply reflect the influence of complex local geographical and geological conditions. For instance, lakes with interstitial brines east of the Bohai Gulf are of the sodium sulphate type and are closely related in origin to seawater recharge. Northwest of zone III on the Qinghai-Tibet Plateau (Fig. 3), the carbonate type occurs. In zone I_2, the presence of the carbonate type reflects recent volcanism and hydrothermal activity. In the Alxa fault block (the Great Bend of the Yellow River), small shallow lakes are mainly of the carbonate type, and were formed by recent eolian erosion, sedimentation under a steppe climate, and recharge of waters dominated by $-$ $Na-HCO_3$ or $Ca-Na-HCO_3$.

Palaeolimnology

About 30 saline lake sequences have been dated by radiocarbon techniques and 10 by palaeomagnetic techniques. A few (e.g. Qarhan) have been investigated using thermoluminescence dating and lanthanide series dating techniques (Chen et al., 1985). Micropalaeontological, sedimento-

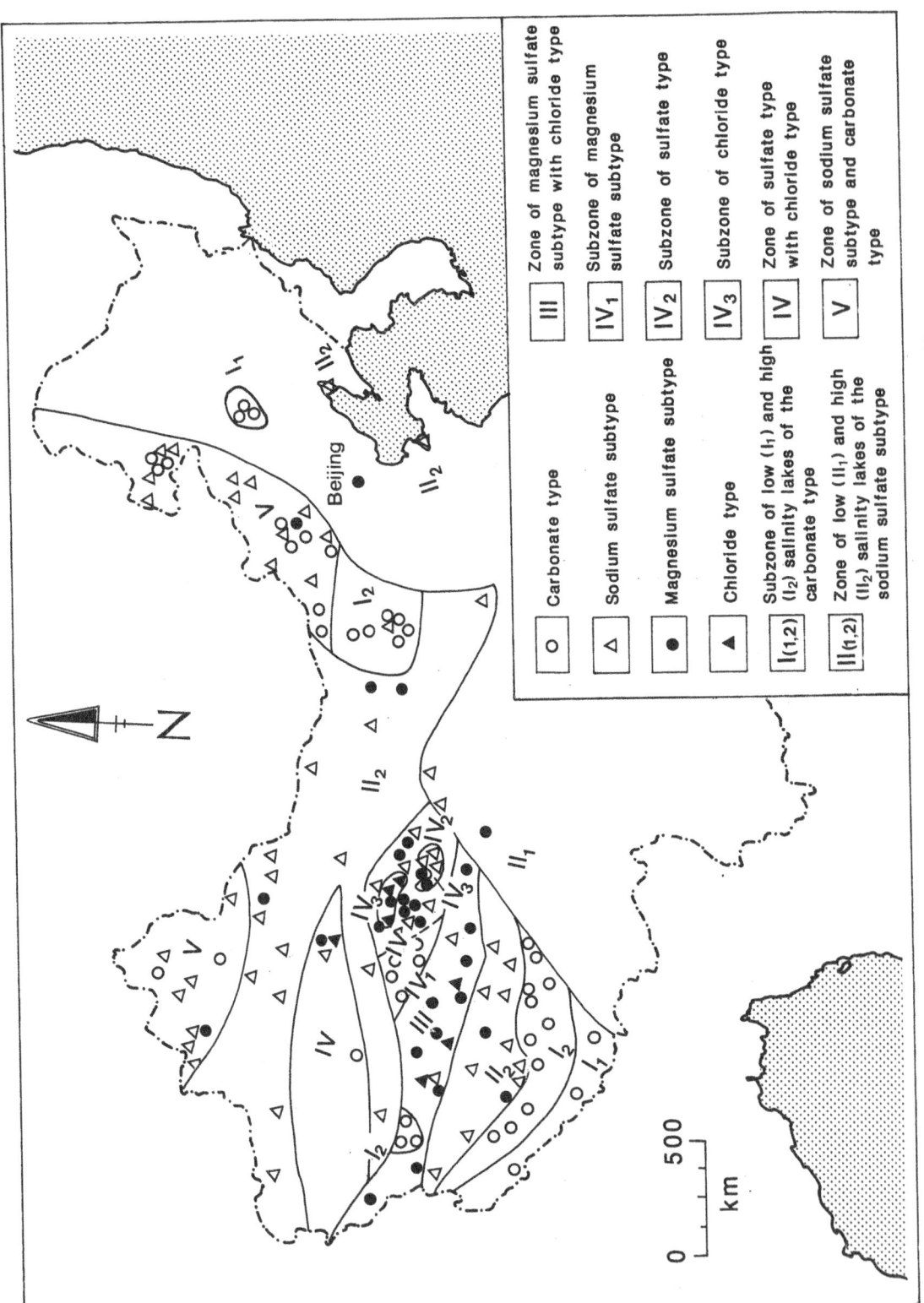

Fig. 3. Distribution of Chinese salt lakes according to type of chemical composition.

logical, stratigraphic and geochemical studies have also been conducted. According to many data, Dalangtan Lake of western Qaidam (*i.e.* in the Quaternary acid center) has experienced 3 stages of development (Q_1, Q_2 and Q_3–Q_4; see legend to Fig. 2 for explanation) since the very Early Pleistocene (2.48 my B.P.), with the deposition of salt at each stage. The general trend is for the average thickness of salt layers, the ratio of salt to rock, and the volume of salt gradually to increase from stage Q_1 to Q_3–Q_4. Southeastwards and southwards from Dalangtan Lake, the Quaternary salt-forming period tends to shorten and occur later, and the salt layer to be thinner. In Qardam Lake, the earliest salt was formed about $22\,400 \pm 510$ y ago. From the Qaidam Basin southwards to the vast lake area on the plateau surface, the earliest salt (e.g. in Baingoin Lake) was formed in the Middle-late Pleistocene (Q_3^2), with a ^{14}C age of *ca* $18\,900$ y B.P. In Zhacang Caka Lake, the salt is *ca* $10\,000$ y old, as evidenced by ^{14}C dating. Further south, in the southern Tibetan Lake area, borax of the Late Holocene appears (Fig. 4).

Eastwards from Dalangtan, the Quaternary salt-forming period tends to be shortened, but trends are different from those southwards. Here, Early-Middle Pleistocene salt deposits of smaller thickness still occur. For instance, in Dalad Lake and Yuncheng Lake (about 1200–1600 km east of Dalangtan), deposits of mirabilite, glauberite and even astrakhanite are found of approximately Middle Pleistocene age and several to 10 m thick (Zheng *et al.*, unpublished, 1990). However, no Early-Middle Late Pleistocene and Early Pleistocene deposits are present (Fig. 5).

Eastwards from the Lop Nur-Wuzongbu Lake area (*i.e.* west of the arid salt-forming center), the Late Pleistocene salt-forming period is known also to be shorter and later (Fig. 6). In Wuzongbu Lake, drilling has revealed thick (several m to 20 m) halite deposits as recent as Middle Late Pleistocene age. In Barkol Lake, mirabilite layers 20 000 y old have been found; eastward in Jartai and Chaganlimenno'er Lake, the salt deposits are mainly Holocene rather than Late Pleistocene; further eastwards in Bayannuo'er Lake, the salt-

forming period has extended from 8000–9000 y B.P. to the present; in Tarigen Lake of Hulum Buir, the salt-forming period was shorter, extending from 4000 to ± 2900 y B.P. Since ± 2900 y B.P., the lakes have become fresh.

Because of periodic changes in climatic and tectonic conditions, cyclothemic or repetitive depositional sequences involving the alternations of freshened and salinized phases, as well as more complex cyclotherms, commonly occur in Quaternary saline lakes. These cyclotherms are of three: seasonal-perennial, multiyear (10–20 y or centennial), and long-term (hundreds to tens of thousands of years). These can be referred to as ranks III, II and I, respectively (Zheng *et al.*, 1985a, 1986, 1989). Here, only variations in rank I cyclotherms are briefly described where there is cyclic alternation of a thicker argillaceous bed and a layers of salt deposits. In Dalangtan, as the preliminary palaeomagnetic analysis and ^{14}C age data from drill holes show, there are about 8–10 cyclotherms of rank I in the Early Pleistocene, which are generally tens of thousands of years old, with the maximum $> 270\,000$ years of age; Middle Pleistocene cyclotherms number 7–15 and are generally tens of thousands of years old, with the maximum $> 130\,000$ years old; Late Pleistocene cyclotherms number 2–7, with ages ranging from tens of thousands to hundreds of years; and Holocene cyclotherms which are 2–3 in number and are dated at thousands of years (Zheng *et al.*, 1986, 1989). Eastwards of Dalad Lake and Yuncheng Lake, the whole Quaternary has only 2–3 salt cyclotherms of rank I (Fig. 5).

Halophilic organisms and biological mineralogenesis

Since the 1980s, halophilic organisms (algae, bacteria, *Artemia* sp.) have been studied in more than 10 saline lakes (Zheng *et al.*, 1985b, 1989, Holland *et al.*, 1991). These include Da Qaidam Lake in Qaidam, Zabuye Lake in Tibet, Ebinur Lake in Xinjiang, Ejinuo'er Lake in Inner Mongolia, and Yuncheng Lake in Shanxi. Biological mineralogenesis in hypersaline environments, pointing to

32

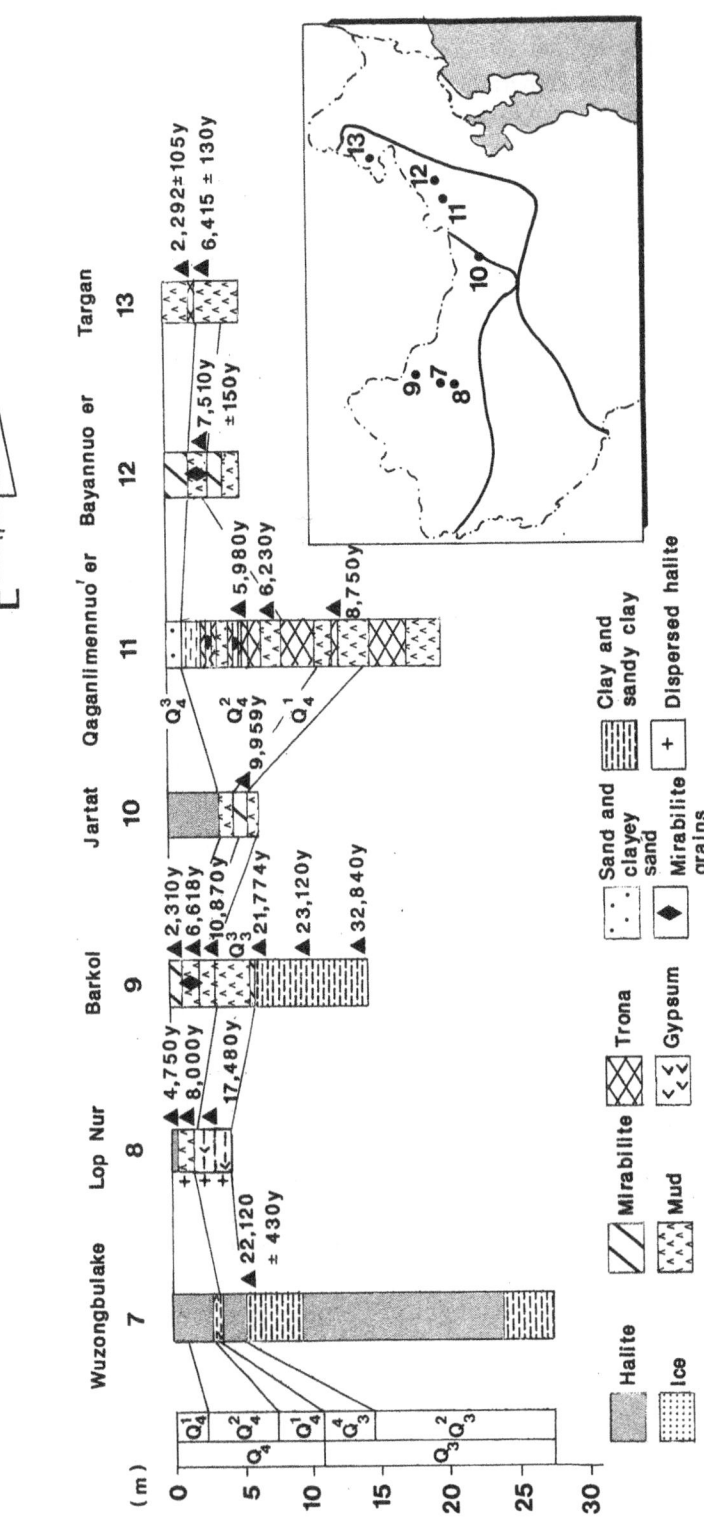

Fig. 4. Correlation of $Q_3^3 - Q_4$ salt-bearing sections of saline lakes on the Qinghai-Tibet Plateau. N_2-Pliocene; Q_1-Early Pleistocene; Q_2-Middle Pleistocene; Q_3 -ate Pleistocene; Q_4-Holocene.

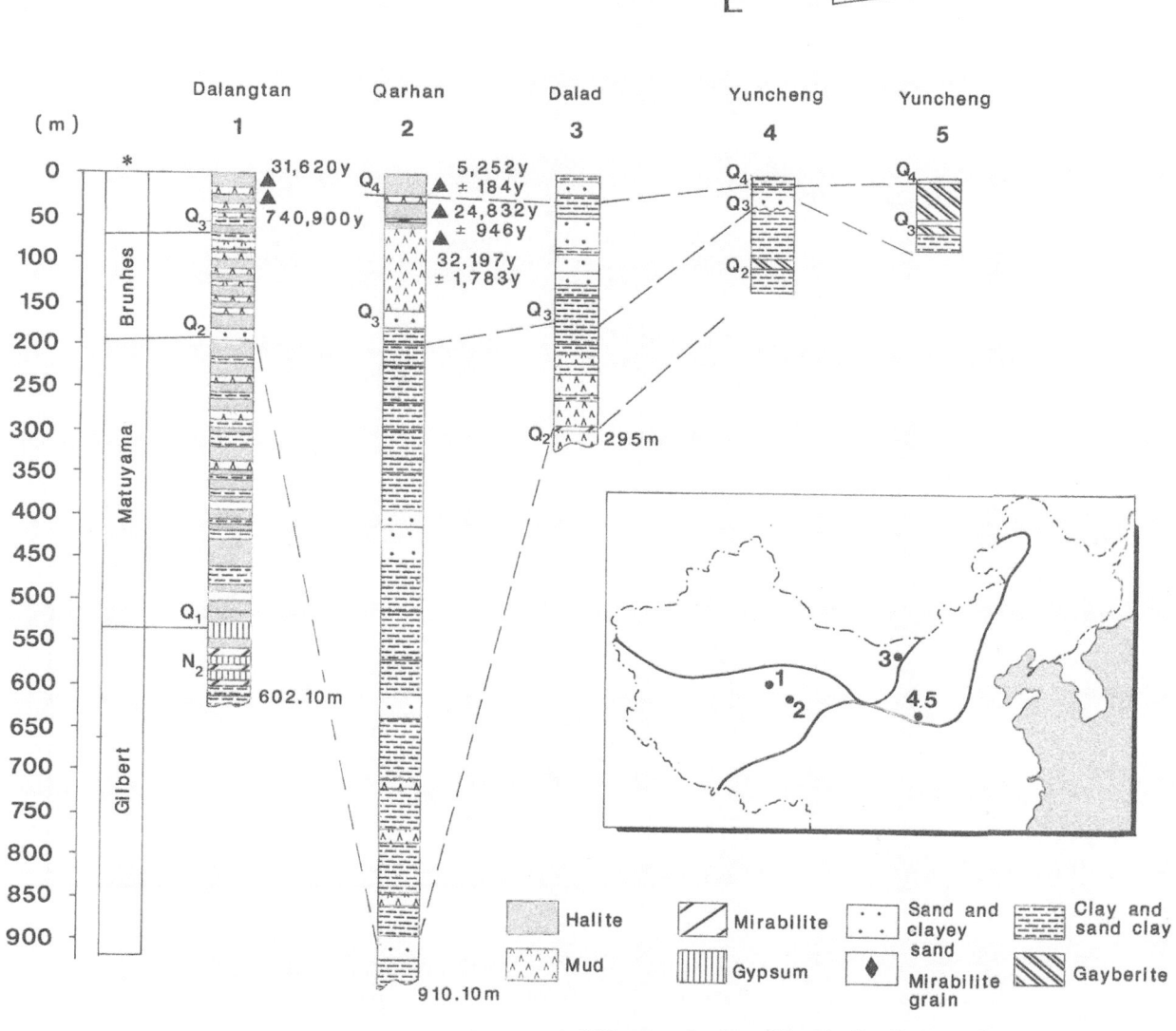

Fig. 5. East-west comparative profiles of Quaternary saline lakes in China.

directions for geoecological research, has also been studied (Zheng *et al.*, 1989).

Dunaliella salina Teod. has been discovered in the northern Zabuye Lake (Zheng *et al.*, 1975b). The north lake has hundreds of scattered travertine islands. Data from ^{14}C dating show that since 22 000 y B.P. the north lake has been recharged by fresh spring water from nearby glaciers, and an appropriate salinity near the islands (150–200 g kg^{-1}) has been maintained enabling the long-term multiplication of halophilic algae. According to ^{14}C dating of algae, such algae have been present in the travertine from at least 8500 y B.P.

Investigations suggest that the high light intensity on the plateau (generally 10 000–130 000 lux), together with an appropriately high salinity and inorganic matter concentration, favours the development of β-carotene in *Dunaliella salina*. From 1986 to the present, however, freshwater

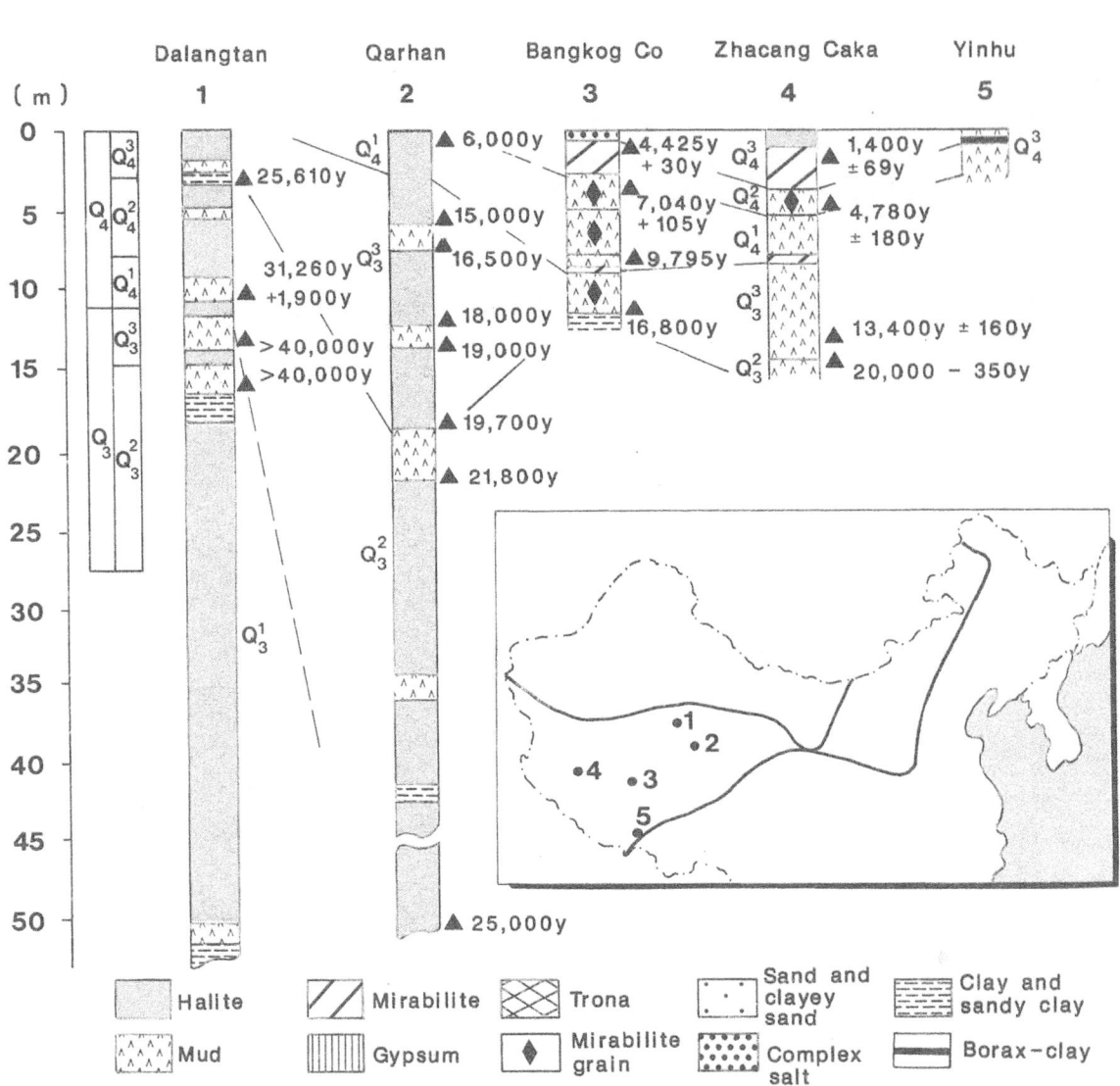

S ⟶

	Dalangtan	Qarhan	Bangkog Co	Zhacang Caka	Yinhu
(m)	1	2	3	4	5

Fig. 6. East-west comparative profiles of Q_3^2 – Q_4 saline lakes in China.

recharge to the northern part of Zabuye Lake has sharply decreased and this has caused the water-level to fall 50 cm, the salinity to increase to 310–360 g kg^{-1}, and the algae to withdraw to the periphery of the travertine islands. In 1989, *Dunaliella salina* multiplied and was shown to be fairly cryophilic (growing even at *ca* 0 °C).

Two species of halophilic bacteria have been identified: *Halobacterium dachaidanensis* and al-

kalotrophic *Halococcus dachaidanensis* from the Da Qaidam Lake in the Qaidam Basin (Zheng, 1989). In addition, two halophilic and pigmented bacteria, *Natronobacterium* and *Ectothiorhodospira*, have been successfully separated from Chabyer Lake sediment.

Basis research on *Artemia* has shown that a great number of bisexual strains as well as a few parthenogenetic strains occur in the interior saline lakes. Lakes where *Artemia* occurs have salinities of 26–341 g l^{-1} (Table 3). Some are considerably more saline than the usual upper limit for *Artemia*. Nau Co has a salinity of 215.5 g l^{-1} and the *Artemia* present is red. The chemical compositions of lakes where *Artemia* occurs are of the sodium sulphate subtype, and the magnesium sulphate and carbonate type. Trace elements concentrations are very different: for example, B, Li, Rb, Cs, As, Br, etc. have higher concentrations in lakes of the Qinghai-Tibet plateau than in salt lakes elsewhere in China.

Saline lakes with *Artemia* are located in various climatic zones and at a range of elevations, from the semi-arid region of the subfrigide plateau zone to the subhumid region of the warm temperate zone, and from 190 to 4436 m a.s.l. The mean annual air temperature is <0 °C at the five salt lakes of the plateau subfrigid zone where *Artemia* has been found. The absolute maximum air temperature is 42.7 °C at Yuncheng Lake, and the absolute minimum air temperature is −41.5 °C at Barkol Lake. Further studies of *Artemia* in China are needed.

The many specialized organisms of hypersaline environments play an important role in several biogeochemical cycles such as those of carbon and oxygen. Additionally, they play an important role in the formation of some mineral deposits such as oil, trona, and even boron. Their significance in salt deposits, however, has not received much attention from salt geologists (Zheng *et al.*, 1985b; Zheng *et al.*, 1989). The present authors once suggested that research be conducted on geoecology in hypersaline environments pointing out that, in evaporative environments, geoecological research is important in understanding biological mineralogenesis. In other words, the relationship between geological conditions and the biological multiplication and activity of organisms in saline lakes should be studied within a geological setting as well as in relation to sedimentation and mineralogenesis. The ecology of halophilic organisms, the origin and significance of ecological conditions, the influence of biological activity and biochemical evolution on the origin, composition, and deposition of minerals in saline lakes, and the palaeoecological environment of halophilic organisms and its effect should all be explored. Geoecology represents the combined study of geology, mineral deposition, and ecology.

Table 3. Distribution of *Artemia* in inland saline lakes in China.

Province	Lake	Position		Salinity (g l^{-1})	pH	Major ions (g l^{-1})							
		Longitude (E)	Latitude (N)			Na	K	Ca	Mg	Cl	SO$_4$	SO$_3$	HCO$_3$
Xizang	Zhangchaka	82° 23′	32° 47′	350.0	8.0	77.6	11.4	0.4	8.5	123.9	43.1		2.1
(Tibet)	Nau co	82° 01′	32° 08′	215.5	9.5	71.0	5.0	0.004	0.26	48.7	71.8	13.1	3.1
	Jibuchaka	84° 10′	32° 02′	82.7		23.4	4.8	0.03	1.4	33.8	13.2	0.8	2.0
	Dong Co	84° 07′	32° 01′	114.3	8.7	22.3	9.0	0.056	4.2	31.1	38.5	0.8	6.0
	Bangkog Co	89° 05″	31° 07′	258.2	8.7	80.2	21.3	0.0	0.11	114.4	35.3	1.4	1.3
Qinghai	Suban	94° 00′	38° 09′	32.3	8.9	8.1	0.4	0.07	1.9	9.8	10.7	0.2	0.94
	Da Qaidam	95° 03′	37′ 08′	340.6	8.0	102.3	5.1	0.5	15.7	186.2	27.7		
	Xiao Qaidam	95° 06′	37° 04′	239.5	7.8	81.2	1.1	0.4	4.4	108.5	41.0	0.28	1.1
	Ga Hal	97° 47′	37° 02′	90.6	8.3	27.3	0.3	0.3	4.2	45.3	13.1	0.14	0.034
Xinjian	Ebinur	82° 08′	45° 00′	78.8		22.7	0.9	0.16	3.0	22.8	28.2		0.39
	Barkol	92° 08′	43° 07′	276.7		72.2	1.0	163.2	16.0	97.2	6.5	7.6	0.193
Nei Mongol	Jartal	105° 42′	39° 07′	310.0		111.0		0.65	11.1	189.8	17.4		
	Ejinuoer	115° 55′	45° 10′	341.0		138.9	7.8		219.7	18.3			
Shaanxl	Dingbian	107° 06′	37° 05′										
Shanxi	Yuncheng	111° 00′	35° 00′	26.3		7.7		0.37	1.2	4.2	16.0		0.055
Hebal	Zhangbei	114° 12′	41° 10′										

Acknowledgements

Professor W. D. Williams, University of Adelaide, is thanked for editorial assistance. The paper was typed by Miss J. Read and the figures were redrawn by Miss K. Maurice-Jones, also of the University of Adelaide. Our special gratitude goes to Professor S. H. Hurlbert, San Diego University.

References

Anon., 1979. Atlas of the People's Republic of China. The Publications of the Maps, Beijing. (in Chinese)

Chen Kezao, 1955. The sedimentary characteristics and palaeoclimate evolution of Qarhan Salt Lake. Scientia Sinica B. 5: 463–473. (in Chinese)

Holland, H. D., G. I. Smith, H. W. Jannasch, A. G. Dickson, Zheng Mianping & Ding Tiping, 1991. Lake Zabuye and climatic history of the Tibetan Plateau. Die Gewissenschaften 9: 37–44.

Lin Zhenyao & Wu Xianding, 1981. The Climatic Divisions of the Qinghai-Tibet Plateau. Acta Geogr. Sinica 36: 22–33.

Tang Chunzhen, 1988. Brief account of historical development of Yuncheng Salt Lake. Sanxi Geology 3: 10–12 (in chinese)

Tapponnier, P., 1982. Propagating extrusion tectonics in Asia; New insights from simple experiments with plasticine. Geology 10: 611–616.

Valyashk, M. G., 1955. Basic chemical types of natural waters and the conditions producing them. Records of the Academy, USSR, 102: 315–318. [in Russian]

Zheng Mianping, 1989. Geological research on saline lakes and their general developments prospects. Mineral Deposits and Geology (Special publication on saline lake geology), No. 3–4: 1–34. Institute of Mineral Deposits, Chinese Acad Geol. Sci. [in Chinese]

Zheng Mianping, Liu Wengao & Xiang Jun, 1985a. Study of Zabuye saline lake, Tibet. Scientific papers on geology for international exchange. Prepared for the 27th International Geological Congress. Geological Publishing House, Beijing: 1731–1784. (in Chinese)

Zheng Mianping, Liu Wengao & Xiang Jun, 1985b. The discovery of halophilic algae and halobacteria at Zabuye Salt Lake, Tibet, and preliminary study on the geoecology. Acta Geolog. Sinica 2: 162–171. (in Chinese)

Zheng Mianping, Liu Wengao & Xiang Jun, 1990. Preliminary study on the geoecology of halophilic algae and halobacteria found in Chabyer (Zabuye) Caka salt lake, Xizang (Tibet). In Geology of the Himalayas, vol. 2: 499–509. Geological Publishing House, Beijing. (in Chinese)

Zheng Mianping, Liu Wengao, Xiang Jun & Jiang Zhongti, 1986. On Saline Lakes in Tibet, China. Proc. Sympos. on Mesozoic and Cenozoic Geology in connection with the 60th anniversary of the Geological Society of China. Geological Publishing House, Beijing: 693–706. (in Chinese)

Zheng Mianping, Xian Jun et al., 1989. Saline Lakes on the Qinghai-Xizang (Tibet) Plateau. Beijing Scientific and Technical Publishing House, Beijing. (in Chinese)

The major ion chemistry of some southern African saline systems

J. A. Day

Freshwater Research Unit, Zoology Dept, University of Cape Town, Rondebosch 7700, South Africa

Key words: Southern Africa, Namib Desert, hydrochemistry, saline waters, major ions

Abstract

Africa south of about 23°S has few natural athalassic lakes, saline or freshwater. South Africa, however, is rich in temporary pans, many of which are saline, while permanent saline springs occur along the coastal strip of the Namib Desert in Namibia. This paper examines the chemistry of the major ions in 67 Namibian waters, 47 of which have not previously been reported in the literature, and compares them with 66 South African waters, five of which have not previously been reported, and with saline lakes in East Africa.

The highest value for total dissolved solids in South African waters was 276 g l^{-1} (Koekiespan, south-western Cape) and the highest for Namibian waters were 160 g l^{-1} (Hosabes, a small spring on a gypsous crust) and 302 g l^{-1} (a salt pan at Oranjemund at the mouth of the Orange River). The dominant ions in fresh waters in the region are Ca^{2+} and HCO_3^-/CO_3^{2-} in the interior and in Namibia, and Na^{2+} and Cl^+ on the south and east coasts. Regardless of the geochemistry of their substrata, the dominant ions in the saline waters throughout the region are Na^+ and Cl^-. Thus differential precipitation of $CaCO_3$ and $CaSO_4$, as a result of evaporative concentration at high salinities, appears to be the determinant of the proportions of the major ions in these systems.

The permanent springs on gypsous crusts along the coast of Namibia, although dominated by Na^+ and Cl^- ions, contain considerable quantities of both Ca^{2+} and SO_4^{2-} ions.

Introduction

One of the remarkable limnological features of southern Africa is the virtual absence of natural athalassic lakes south of about 23°S. Indeed, Lake Fundudzi in the far northern Transvaal is the only permanent natural lake in South Africa (none occurs in Namibia), although coastal lagoons (relictual estuaries) and man-made lakes on regulated rivers are common. Southern Africa is very rich in small (often < 1 ha), shallow depressions that fill with water after rain and are known as pans or, in the south-western Cape and Namibia, as vleis. Some of the larger of these, such as Lake Chrissie in the eastern Transvaal and Barber's Pan in the western Transvaal, dry up only occasionally. Others in the more mesic eastern part of the subcontinent tend to be seasonal, while those in the west may fill episodically at intervals of several years. Although some of these pans contain fresh or only slightly brackish water, many are saline (taken here as TDS > 3 g l^{-1}: Williams, 1964).

Literature on these systems, particularly on details of water chemistry, is scanty. The most complete analyses of southern African inland salt waters are those of Hutchinson *et al.* (1932) (which also deals with non-saline systems) and a recent synthesis by Seaman *et al.* (1991). Various aspects of the physical, chemical or biological limnology of a few individual systems have been described for Pretoria Salt Pan (Prior, 1978; Schoeman & Ashton, 1982; Ashton & Schoeman, 1983, 1988); for the invertebrate fauna of a salt works at Wintersdam in the Orange Free State (Mitchell & Seaman, 1988); for floodplain pans of the Pongola River (Musil *et al.*, 1973); and for the pans of the southern Kalahari (Lancaster, 1978) and the western Orange Free State (de Bruiyn, 1972). The origins and geomorphology of southern African pans have been described by Shaw (1988), amongst others. Various aspects of the chemistry and biology of Zimbabwean pans have been described by Weir (1966, 1969). Namibia, particularly the coastal Namib Desert, is much more arid than South Africa or Zimbabwe. Despite this, a number of small (usually 1–100 m²) springs and waterholes persist even during long periods of drought. Only one or two of them can be dignified by being called lakelets but they are of interest in that they are often saline and, except after heavy rains, are the only surface waters available for humans or wildlife to drink. With the exception of a paper by Berry (1971) on Etosha Pan, very little is known about the few isolated standing waters in the interior of Namibia. Some aspects of the water chemistry of several of those in the central Namib Desert itself have been described by Gevers *et al.* (1963), Grobbelaar (1976), Grobbelaar & Seely (1980), Kok & Grobbelaar (1980, 1985) and Day & Seely (1988). What little is known of the biota of such systems is reviewed by Day (1990). (A note on terminology: the whole country that used to be known as South-West Africa is now known as Namibia; the Namib is the coastal desert that stretches along the entire coast of Namibia and into southern Angola.)

Although chemical data are now available for a number of Namibian systems, no attempt has yet been made to draw general conclusions about the chemistry of Namibian waters. This paper analyses the proportions of major ions in these systems and in some previously unexamined athalassic waters of the Skeleton Coast and Namib-Naukluft National Parks. In particular, the proportions of the major ions are examined in relation to geology and groundwater chemistry, and an attempt is made to characterise the waters of springs on gypsous crusts in the coastal Namib Desert.

The dominant ions in most salt lakes worldwide are probably Na^+ and Cl^- (see e.g. Kilham, 1990). In Africa, however, the best-known salt lakes are the soda lakes of East Africa, which are almost exclusively dominated by Na^+ and HCO_3^-/CO_3^{2-} ions (Talling & Talling, 1965; Wood & Talling, 1988). The saline pans of South Africa, on the other hand, are dominated by Na^+ and Cl^- ions (Seaman *et al.*, 1990; Silberbauer & King, 1991). The results of the present analyses are compared with those for both East Africa and South Africa, including a few previously unpublished data sets for lentic systems in South Africa, made available by the South African Department of Water Affairs (DWA).

In order to see the extent to which ionic proportions in surface waters are determined by geological processes, as opposed to evaporative concentration, it is necessary to examine waters unaffected by evaporation. Thus a number of analyses of water from boreholes and groundwater excavations is included in the Namibian data set. The only extensive study of the chemistry of ground water in the region (Bond, 1946: South African borehole water) does not provide complete analyses of the major ions. Previously published data for hot springs in South Africa (Hoffman, 1979) and Namibia (Ashton & Schoeman, 1984) are therefore used as a further indication of the chemical composition of groundwater in the region. It is acknowledged that these waters are not ideal for comparative purposes, but they are the best available.

The study areas

Namibia

The region for which most of the data were obtained is the northern coastal strip of the Namib

Desert from Sossusvlei in the south to the Cunene River on the Angolan border in the north (Fig. 1). Namibia lies in the rainshadow of the entire subcontinent, since prevailing winds and rains come from the east. Localised onshore rain is virtually non-existent. The western part can be classified as extreme desert, precipitation being <10 mm y^{-1} at the coast at Walvis Bay. The rainfall gradient is steep, mean annual precipitation reaching about 20 mm at Gobabeb, 85 mm at Ganab and increasing inland to about 200 mm y^{-1} at Windhoek. Rainfall is extremely erratic and unpredictable: in this century, 'good' rains have fallen in 1930/31, 1976 (nearly 300 mm at Ganab) and 1978. The Namib is not entirely arid

owing to the regular fogs that originate from the cold seas of the Benguela current and roll inland as far as Ganab for up to 100 days y^{-1}. It has been estimated that precipitation from this source may be as great as 100 mm y^{-1}. The climate of the central Namib Desert is detailed in Lancaster, Lancaster & Seely (1984).

Topographically the coastal strip can be divided into a southern dune field and a northern gravel plain with scattered inselbergs and dunes along the coast. Although a number of river beds dissect the coastal plain, all are at most seasonal, running for no more than a couple of weeks a year. Subsurface water in some river beds allows the development of vegetation; indeed, some support narrow belts of trees that act as longitudinal oases, allowing large mammals that are normally adapted to mesic conditions to penetrate through the desert to the coast. Some of these mammals excavate waterholes in the dry river beds.

Geologically (Fig. 2) the northern part of the region consists of patches of largely igneous basement complex (>2000 my), a variety of schists, dolomites and gneisses of the Damara system (1000–750 my), sedimentary rocks of the Nama system (650–550 my) and much younger sedimentary strata of the Karoo system (<210 my) (Martin, 1965; Tankard et al., 1982). Gypsous crusts are common in patches within about 50 km of the coast (Watson, 1978). It appears that a series of springs arises from a fault lying more or less parallel to the coast (Day & Seely, 1988). As a result of the intense solar radiation, and consequent evaporation, the water from these springs becomes very saline. Where these springs occur on gypsous crusts, the beds of the springs may consist of thick deposits of gypsum-impregnated sands.

The waters analysed in the present study can be divided into (a) rainwater held in temporary pools and lakes on the plains and in the southern dune field (Sossusvlei), or left as temporary pools in river beds as flood waters recede; (b) ground water collected from excavations made during sampling or from boreholes and (c) permanent springs and waterholes (excavations, usually in

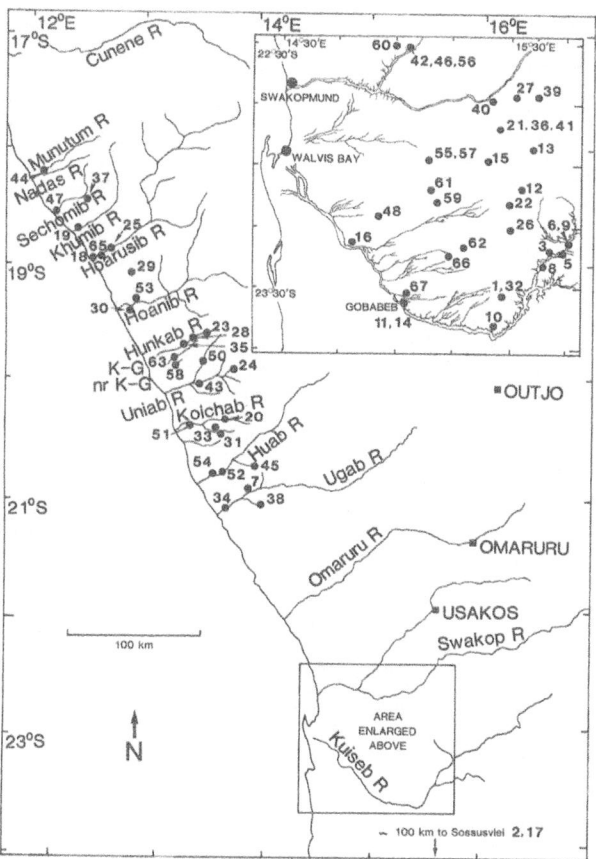

Fig. 1. Central and northern Namibia showing the localities of all but three of the sites. Consult Table 1 for names of localities. See Fig. 3 for the localities of Sossusvlei, Oranjemund Salt Pan and Etosha Pan, and for the location of this map in relation to the rest of southern Africa.

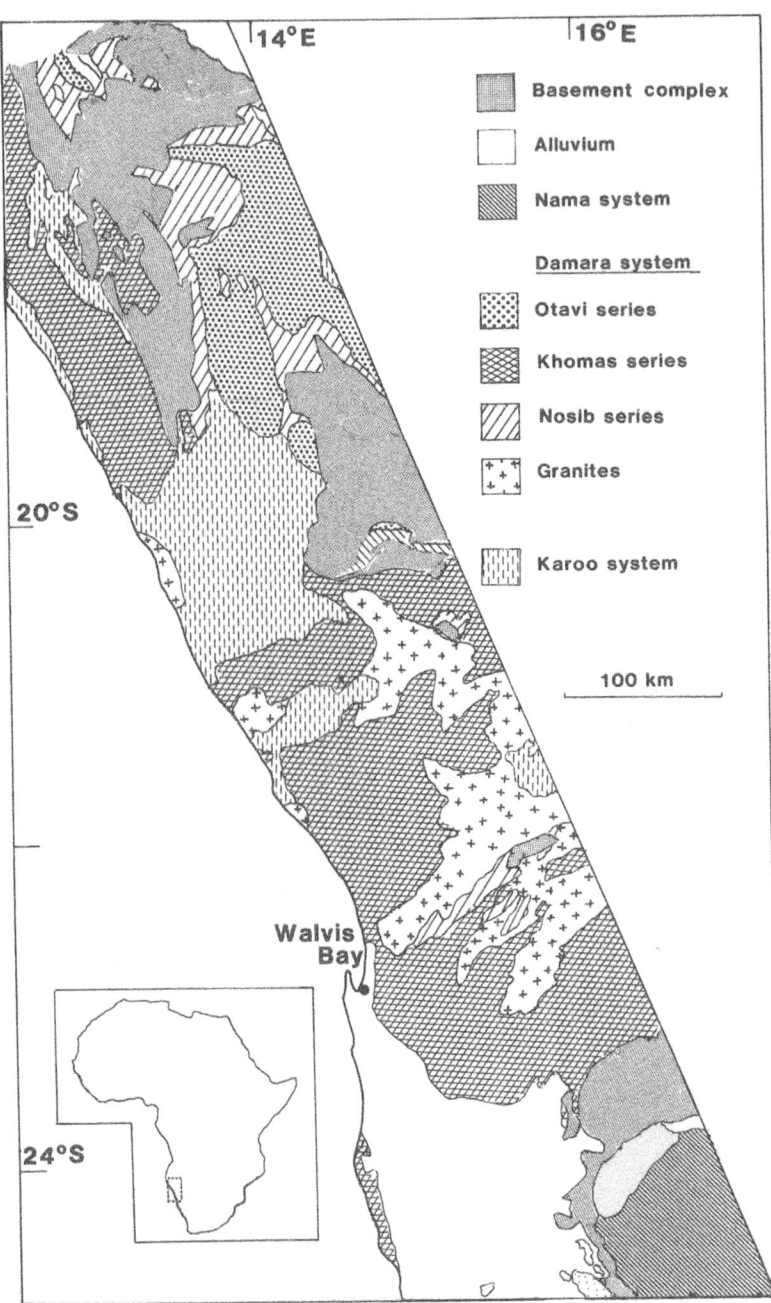

Fig. 2. Simplified geological map of the coastal strip of northern Namibia (modified from Martin, 1965).

river beds, made by large mammals, including man). Waterbodies in this last category are open to the surface and are thus subject to evaporation. They include two moderately fresh running streams (Gemsbokbron and Uniab 1) and two lakelets (Ausis & Oasis).

South Africa

The climate of South Africa is classified as subtropical, although the western half is arid (mean annual precipitation <400 mm y^{-1}), and rainfall exceeds 2500 mm y^{-1} only in the mountains of the south-western Cape and Natal. Rainfall exceeds evaporation only in the far north (which may explain why Lake Fundudzi is the only permanent lake in the country). Pans in the east (summer rainfall, precipitation > 1000 mm y^{-1}) fill seasonally in summer; pans in the south-west (winter rainfall, precipitation ca 500 mm y^{-1} for the areas relevant to this study) fill seasonally in winter. Rainfall is not only much lower (< 400 mm y^{-1}) but is also much less predictable in the central and western parts. Thus pans fill intermittently, usually in summer. Details of the climate of South Africa can be found in Tyson (1986).

Topographically, South Africa consists of an elevated plain, tilted downwards to the west and surrounded by mountains to the south and east, from which a steep escarpment drops to the coast. Geologically, the plain consists largely of an ancient (ca 210–190 my) sedimentary basin (the rocks of the Karoo system) surrounded by a variety of older rocks, both igneous and sedimentary. Further details of the geology and topography of South Africa can be found in Truswell (1970) and Tankard et $al.$ (1982).

The greatest concentration of pans in South Africa occurs in the north-western Cape and the western Transvaal and Orange Free State. These areas, all of which fall within the more arid part of the drainage basin of the large, westward-flowing Orange River, are essentially endorheic, although some lie on old drainage lines, becoming exorheic after rare heavy rains. Most of these pans lie on Ecca and Dwyka Karoo sediments, which are fairly rich in salts (Bond, 1946; Tankard et $al.$, 1982), or on the sands and sediments of the Kalahari basin. The surface sands in the Kalahari often cover Ca^{2+}-rich caliche, dolomite and marl and, where sand has been removed by deflation, these materials may form hardpan beds that become pans: Etosha Pan in Namibia is an example. Other concentrations of pans occur on the alluvial floodplain of the Pongolo River in coastal northern Natal, while a number are also found on the sandy coastal forelands of the south-western Cape; these sands are of marine origin.

Methods

To avoid including samples with gross analytical error, each data set has been included only if the sum of the major anions and the sum of the major cations agree within 10% (silicates and nitrates often account for at least 5% of the anions in Namibian waters). TDS values are analytical, as opposed to being calculated, unless otherwise stated. Conductivity values are generally given in mS m^{-1} at 25 °C; asterisks in the Tables indicate values for which the standard temperature was not stated in the literature. Since pH values were usually recorded in the laboratory, they may bear little relationship to ambient values, and have not been included.

Namibian data

The Namibian data consist of analyses of 116 water samples taken from 67 different localities (Fig. 1). The origins of the data taken from the literature are given in Table 1. Of the remainder, 44, including several time-series, were collected by officials of the (then) South-West African Department of Nature Conservation during a year-long survey of the surface waters of the Namib-Naukluft National Park in 1977/8 and 30, each from a different locality, by the author in the Skeleton Coast and Namib-Naukluft National Parks in 1981 and 1982. Ten samples were collected during an environmental survey of the area under the control of the Rössing Uranium Mine in 1982. Of the last, the sums of the anions and the cations agreed within 10% for only four (Piet's Spring, South-western Panner Pool, Boulder Pool and Arandis); the other data have not been included. Analytical details are not available for these samples. Although only the data set with the highest TDS value per site is listed in Table 1, monthly data are available for Ubib, Soutwater, Springbokwater, and Poacher's Fountains 1 and 2 (Table 2) and these have been used in various of the analyses.

A word of caution is needed regarding samples taken from the small, more saline springs. The water in these systems is seldom more than 100–

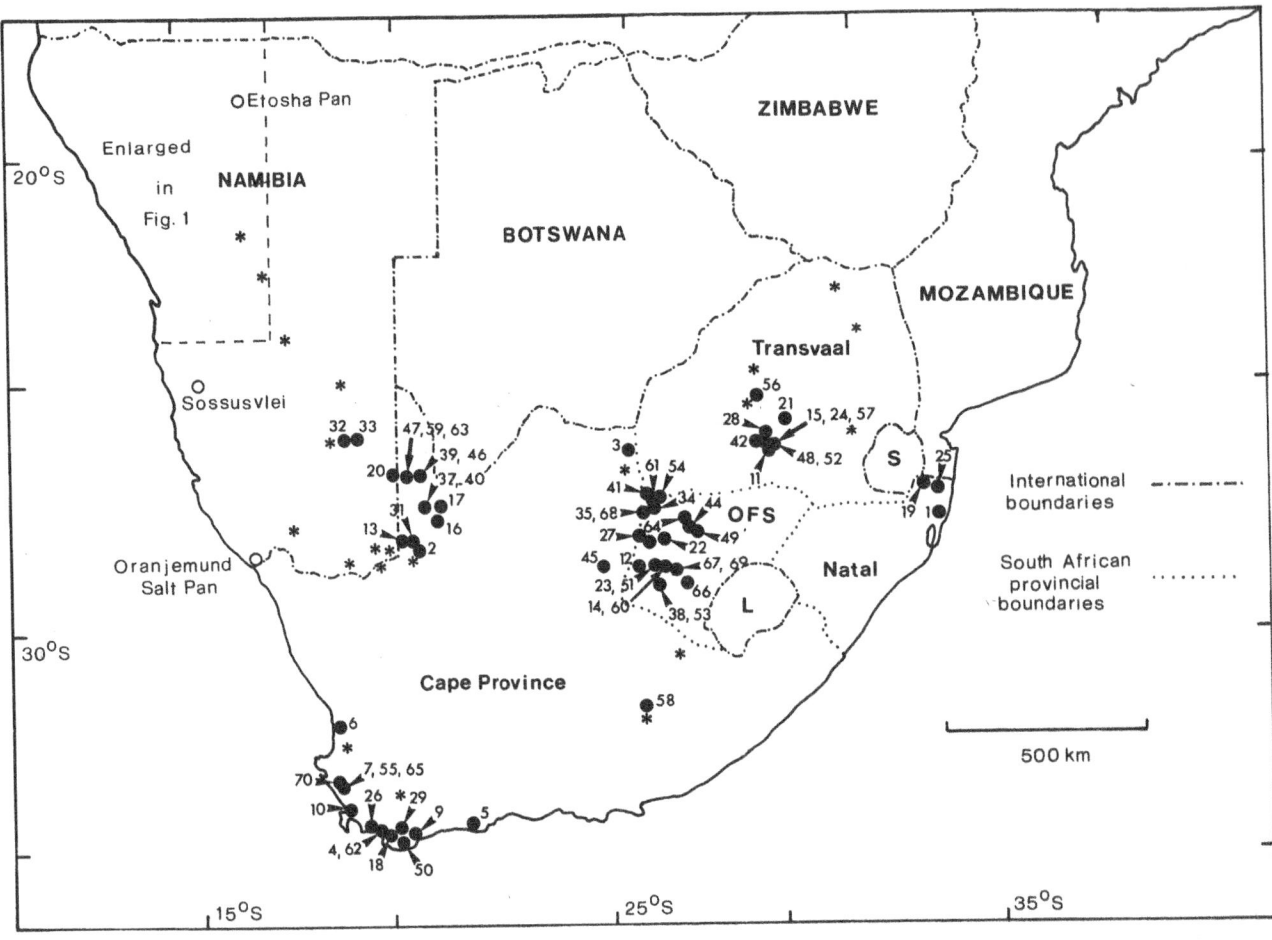

Fig. 3. Southern Africa showing the localities of the South African sampling sites (numbered dots: for key, see Table 3) and all of the hot springs (stars). L = Lesotho; OFS = Orange Free State; S = Swaziland.

200 mm deep and is often strongly stratified chemically (see Day & Seely, 1988 for an example). Thus in removing a one-litre sample it is inevitable that the stratification will be disturbed and, depending on the degree of disturbance, a variable mix of layers will be included in the sample. This may explain some of the apparent anomalies in the results.

All of the original samples (and most of the published ones) were analysed by the (then) South-West African DWA in Windhoek. Details of the analytical methods can be found in Day & Seely (1988). Samples were not preserved and were often kept unchilled for several days or weeks before analysis. It is assumed that the val-

ues for the major ions were not seriously affected by this procedure.

South African data

The South African data were taken mainly from Seaman *et al.* (1991) and Silberbauer & King (1991) (Table 3). Previously unpublished data consist of analyses of the waters of five lentic systems performed by the South African DWA. Verlorevlei (5 samples, May 1970 – September 1972) is a coastal lagoon occasionally in contact with the sea; De Hoop Vlei (44 samples, July 1965 – September 1977) and Lake Sibayi (44

Table 1. The data for the Namibian systems. Conductivity is in mS cm^{-1} at 25 °C, except where asterisks indicate that no temperature was indicated in the original data. Total dissolved solids are in g l^{-1}: those with asterisks have been calculated from the sum of the major ions; the others were determined gravimetrically. Ions are in meq l^{-1}. ALK = HCO$_3^-$ + CO$_3^{2-}$. B/H = borehole water; E = endorrheic; G/W = ground water; SP = spring; T = temporary. Lower-case letters refer to different samples from the same site. Localities are indicated in Fig. 1 according to the numbers in the left-hand column of this table.

Locality	Cond.	TDS	Na$^+$	K$^+$	Ca^{2+}	Mg^{2+}	Cl$^-$	SO$_4^{2-}$	ALK	Ref[a]	Date
1 Zebra Pan (T/E)	0.19	0.1	0.1	0.2	1.3	0.2	0.2	0.1	1.3	1	07.06.78
2 Sossusvlei a (T/E)	0.39*	0.3*	2.0	0.4	0.9	0.8	0.4	0.2	2.8	2	26.07.76
Sossusvlei b (T/E)	2.32*	1.7*	17.4	1.8	1.3	2.5	5.4	4.6	12.5	2	26.07.76
3 Kuiseb R. Pool C (T)	0.71	0.4*	1.7	0.4	4.0	1.8	2.3	1.0	2.9	3	11.77
4 Sossusvlei G/W	0.65*	0.5*	1.9	0.0	3.5	2.7	0.6	1.6	4.3	2	06.08.75
5 Kuiseb R. Pool B (T)	0.65	0.5*	2.4	0.0	2.0	2.6	1.6	0.5	5.6	3	01.78
6 Kuiseb R. Pool A (T)	0.73	0.5*	4.0	0.0	1.8	1.2	3.2	0.3	4.4	3	11.77
7 Ugab R. Pool 2 (T)	0.95	0.6	4.7	0.4	2.9	1.5	4.8	1.5	2.7	1	29.01.83
8 Kuiseb R. Pool D (T)	0.84	0.7*	4.7	0.5	3.7	1.4	3.7	1.1	5.1	3	01.78
9 Kuiseb R. Pool (T)	1.02	0.7	6.1	1.0	1.8	2.8	6.5	1.6	3.4	1	05.12.78
10 Kuiseb R. Pool E (T)	1.10	0.8*	6.6	0.6	4.0	1.0	5.9	1.5	4.9	3	01.78
11 Gobabeb old B/H	1.14	0.8	7.6	0.7	4.2	1.9	5.2	1.8	7.1	1	07.06.78
12 Ganab B/H	1.23	0.9	4.4	1.0	5.9	2.9	8.5	2.6	2.4	1	11.77
13 Gemsbokwater B/H	1.48	1.1	10.6	0.9	2.9	3.0	7.9	2.6	4.5	1	11.77
14 Gobabeb new B/H	1.68	1.2	11.8	0.8	5.5	2.2	8.7	2.7	8.5	1	15.03.78
15 Hotsas B/H	1.98	1.5*	13.2	1.4	5.6	2.7	15.4	4.0	2.2	4	11.77
16 Swartbank (SP)	2.25	1.6	21.0	1.0	3.1	4.2	17.2	4.5	6.6	1	27.06.82
18 Hoarusib R. Spring	2.50	1.7	25.2	0.8	1.4	3.3	23.1	0.5	7.6	1	21.01.83
19 Khumib R. (SP)	3.40	2.1	26.5	0.2	3.3	3.9	18.7	8.7	4.7	1	21.01.83
20 Springbokwasser (SP)	4.00	2.6	28.3	0.2	12.5	0.9	25.8	8.6	6.4	1	27.01.83
21 Groot Tinkas B/H	4.15	3.0*	24.5	1.6	14.0	7.6	37.7	6.7	2.6	4	22.12.77
22 Tumasberg W/H	4.00	3.1*	29.3	1.9	11.7	7.8	37.0	6.4	5.8	4	1977–8
23 Hunkab R. Pool 1 (T)	5.50	3.4	45.2	0.4	4.4	4.9	41.7	7.3	3.0	1	25.01.83
24 Graafwater (SP)	5.10	3.4	43.0	1.3	4.3	6.0	33.8	5.3	12.8	1	27.01.83
25 Leylandsdrift (T/R)	5.50	3.4	50.4	0.7	2.9	8.3	51.3	2.2	6.5	1	23.01.83
26 Amichab W/H	4.40	3.5*	26.4	2.2	17.5	10.6	41.4	7.8	4.1	4	1977–8
27 Hebron (SP)	5.70	3.7	27.4	0.4	20.5	9.6	41.7	6.2	6.3	1	07.06.78
28 Hunkab R. Spring	5.70	3.7	52.6	0.4	4.5	4.3	46.2	8.5	4.7	1	25.01.83
29 Ganias (SP)	6.50	3.9	52.6	0.8	10.0	2.8	52.1	7.3	4.6	1	24.01.83
30 Oasis (SP/Lakelet)	5.50	4.1	53.9	1.1	0.7	14.2	50.7	13.3	2.2	1	21.01.83
31 Gemsbokbron (SP)	6.60	4.2	60.9	0.1	1.6	3.6	47.9	9.8	6.0	1	27.01.83
32 Zebra Pan B/H	6.45	4.2*	55.0	1.3	7.2	8.1	44.5	12.4	5.8	4	25.11.7
33 Wolfwasser (SP)	7.50	4.5	71.7	0.1	0.3	0.4	53.5	10.7	5.5	1	27.01.83
34 Ugab R. Pool 1 (T)	7.60	4.6	56.1	0.8	15.0	7.7	67.6	7.6	2.6	1	20.01.83
35 Hunkab R. Pool 2 (T)	8.00	5.1	73.5	0.8	4.8	3.3	65.4	12.5	3.7	1	25.01.83
36 Groot Tinkas 1 (SP)	8.50	5.6	63.9	4.0	16.3	7.1	74.9	11.3	2.4	1	22.12.77
37 Ogams (SP)	10.0	7.1	108	1.4	3.3	4.9	99.2	18.8	0.7	1	23.01.83
38 Dead end (SP)	11.5	7.2	90.4	3.2	22.5	8.8	96.9	21.9	2.0	1	29.01.83
39 N–E Corner B/H	16.2	11.5*	97.4	6.7	54.6	27.4	159	23.2	2.0	4	20.12.77
40 Rietfontein (SP)	16.2	11.6*	92.8	6.5	57.8	39.2	161	21.6	8.2	4	07.06.78
41 Groot Tinkas 2 (SP)	17.1	11.9*	127	7.7	37.1	25.9	185	13.4	1.1	4	22.12.77
42 Piet's Spring	17.2	12.8	117	2.1	45.0	37.9	141	52.3	16.6	1	16.01.85
43 Uniab R. Pool 2 (T)	17.5	13.7	152	0.3	67.5	1.6	182	32.1	0.9	1	26.01.83
44 Okau (SP)	20.5	15.1	208	2.3	31.0	21.7	225	39.8	1.6	1	22.01.83
45 Gai-Ais (SP)	30.0	17.9	276	0.7	0.5	27.5	251	40.8	6.0	1	28.01.83
46 Panner pool (SP)	25.9	18.5	184	3.3	80.0	45.3	237	55.2	17.6	1	16.01.85
47 Sechomib R. (SP)	30.0	18.5	221	3.1	50.0	33.3	265	34.0	1.6	1	22.01.83
48 Qunchochoab (SP)	25.4	19.5*	236	7.6	54.6	31.4	270	48.9	2.1	4	17.01.78
49 Etosha Pan (T/E) a	26.0*	20.2	340	2.8	0.1	0.1	293	18.5	26.7	5	14.07.71
Etosha Pan (T/E) b	67.0*	61.4	985	7.7	0.1	0.2	841	77.1	89.0	5	14.07.71
50 Obob R. (SP)	35.0	20.3	235	0.5	55.0	42.5	276	31.5	1.7	1	26.01.83
51 Uxieb (SP)	33.0	20.3	313	0.3	0.5	7.7	259	45.8	8.8	1	27.01.83
52 Huab R. Pool 3 (T)	35.0	20.7	309	1.3	0.3	23.3	276	31.3	17.6	1	28.01.83
53 Ausis (SP/Lakelet)	32.0	21.2	283	4.4	2.2	73.3	265	100	3.4	1	24.01.83
54 Huab R. Pool 2 (T)	37.0	21.8	328	0.8	0.2	23.2	287	29.6	32.2	1	28.01.83
55 Poacher's Spring 1	32.2	22.3	220	14.8	111	32.8	321	33.4	1.5	1	20.04.78
56 Boulder Pool (SP)	34.6	27.8	295	6.7	133	53.5	383	75.2	18.0	1	16.01.85
57 Poacher's Spring 2	45.0	28.9	319	20.5	134	59.4	407	40.6	1.4	1	26.01.79
58 Nr Kharu-Gaiseb (SP)	52.0	29.7	396	3.1	60.0	26.7	423	50.0	1.3	1	26.01.83
58 Foram Spring	48.0	35.3	477	43.6	38.8	45.2	513	77.3	2.9	1	09.11.78
60 Arandis (SP)	48.4	36.0	407	12.7	45.0	45.3	504	90.7	16.2	1	16.01.85
61 Ubib (SP)	66.0	42.1	565	29.5	77.5	74.2	634	81.3	2.1	1	26.06.82
62 Springbokwater (SP)	55.6	45.0	591	36.4	37.1	61.9	545	188	5.7	1	27.04.78
63 Kharu-Gaiseb (SP)	85.0	50.3	700	6.2	90.0	56.7	789	40.0	2.3	1	26.01.83
65 Klein Oasis (SP)	104.0	64.7	952	23.3	0.2	127	1000	54.2	29.4	1	21.01.83
66 Ordino (SP)	176.7	112	1680	57.4	49.4	161	1507	324	4.1	1	22.03.78
67 Hosabes (SP)	117.1	160	2580	56.4	2.9	142	2113	479	4.6	6	06.03.84
68 Oranjemund salt pan		302	4370	53.2	198	703	5076	143	5.0	7	?

[a] Ref 1 = Original data; 2 = Grobbelaar (1976); 3 = Kok & Grobelaar (1980); 4 = Kok & Grobbelaar (1985); 5 = Berry (1972); 6 = Day & Seely (1988); 7 = Seaman *et al.* (1991).

Table 2. The time-series for five Namibian gypsous springs. Conductivity is in mS cm^{-1} at 25 °C and total dissolved solids (determined gravimetrically) in g l^{-1}. Ions are in meq l^{-1}. ALK = HCO$_3^-$ + CO$_3^{2-}$.

Locality & Date*	Cond.	TDS	Na$^+$	K$^+$	Ca^{2+}	Mg^{2+}	Cl$^-$	SO$_4^{2-}$	ALK
Foram Spring									
24.02.78	36.6	29.0	407	26.7	36.1	30.1	423	54.4	1.8
25.03.78	36.3	26.7	359	25.6	35.0	49.0	386	51.3	4.5
24.04.78	36.6	26.8	336	25.6	33.0	34.0	352	48.1	2.2
07.06.78	34.6	24.8	324	23.1	42.0	17.6	349	44.4	1.8
06/7.78	31.5	24.1	319	32.3	33.3	26.7	342	51.8	3.1
31.07.78	33.7	23.3	324	19.3	31.2	24.8	330	50.1	1.7
07.08.78	35.9	23.0	296	21.0	30.9	27.1	328	54.8	2.0
29.09.78	30.9	20.7	243	30.8	29.6	23.4	292	35.0	3.0
09.11.78	48.0	35.3	477	43.6	38.8	45.2	513	77.3	2.9
30.11.78	35.0	22.4	307	17.9	26.9	27.1	311	44.4	2.4
12.01.79	35.0	21.3	285	23.1	25.5	26.5	301	38.9	2.2
25.01.79	35.1	23.5	317	24.9	34.0	27.0	338	44.2	2.6
Poacher's Spring 1									
09.01.78	40.0	29.6	264	13.3	154	55.5	425	42.3	1.3
26.01.78	28.6	22.5	230	13.1	107	35.8	327	38.6	2.2
24.02.78	33.6	27.8	274	15.4	138	42.0	401	34.0	1.1
22.03.78	29.2	22.1	212	14.2	109	36.8	313	35.5	1.7
20.04.78	32.1	22.3	220	14.8	111	32.8	321	33.4	1.5
07.06.78	27.7	20.2	195	13.7	96.6	32.4	282	31.5	2.9
06/7.78	25.6	19.1	185	17.7	94.6	26.4	276	34.1	2.3
31.07.78	24.3	17.5	174	15.9	84.2	24.8	254	29.0	2.2
08/ç.78	24.7	18.5	174	22.1	86.7	25.3	261	35.0	1.5
08/9.78	24.7	18.3	152	22.1	90.8	28.2	265	30.7	0.9
Poacher's Spring 2									
20.12.77	35.0	28.2	274	17.4	117	45.6	414	36.7	2.7
09.01.78	100.0	60.4	550	30.8	317	121.0	944	40.6	1.7
24.02.78	35.9	30.4	284	16.0	32	29.6	445	37.0	1.9
22.03.78	34.9	26.8	245	14.4	146	57.7	409	38.5	1.6
20.04.78	42.4	31.7	296	17.4	165	60.2	468	31.8	1.8
07.06.78	36.1	30.3	255	16.0	160	52.4	420	31.8	1.8
06/7.78	36.2	28.3	245	18.9	140	49.6	397	37.1	1.1
31.07.78	35.0	27.3	237	18.7	134	47.8	378	37.1	2.3
04.08.78	33.4	23.2	180	15.4	117	47.6	337	30.4	1.8
19.09.78	33.4	22.4	189	12.1	117	45.6	324	35.9	2.2
30.10.78	30.3	24.6	199	22.3	117	73.7	376	44.2	2.2
09.11.78	34.8	25.5	232	23.1	107	37.9	363	42.5	3.1
26.11.78	43.2	28.7	254	14.6	141	54.0	403	46.0	3.8
Springbokwater									
20.12.77	28.7	23.0	289	17.7	27.8	27.2	261	96.6	3.0
01.78	27.7	22.2	303	14.8	37.1	35.9	248	113	6.7
19.01.78	142.1	110	1678	55.9	51.6	80.5	1510	314	3.4
28.02.78	25.2	20.1	244	19.0	30.7	28.7	232	88.0	3.8
25.03.78	34.9	25.6	336	18.7	39.1	42.9	304	104	3.5
24.04.78	31.0	23.1	293	18.5	31.6	35.1	256	103	7.1
04/5.78	55.6	45.0	591	36.4	37.1	61.9	545	189	6.9
04/5.78	58.5	47.7	637	37.9	52.5	76.5	561	216	4.9
07.06.78	26.2	20.1	255	16.0	33.6	22.2	223	83.0	6.4
07.78	21.0	17.9	237	18.2	28.1	26.9	203	88.0	5.6
09.11.78	27.2	21.2	265	23.1	24.2	34.8	238	101	1.9
05.12.78	37.0	24.9	284	19.5	30.7	38.3	286	98.0	2.1
Unibvlei									
02.78	31.2	24.1	293	25.6	43.3	44.7	355	35.5	2.1
22.03.78	34.2	25.1	293	26.7	47.4	68.6	372	38.5	2.4
20.04.78	34.7	24.7	293	26.7	47.4	45.6	358	33.4	1.6
05/6.78	35.0	24.5	293	19.5	42.6	46.4	354	35.2	2.4
06/7.78	32.2	24.8	296	34.9	44.7	48.3	370	38.6	1.6
08/9.78	29.7	21.0	226	31.3	39.8	43.2	303	33.3	1.8
08/9.78	34.6	25.9	291	37.4	45.9	52.1	376	13.3	1.2
09.11.78	26.1	21.1	244	30.3	39.4	39.6	299	40.6	2.2
26.11.78	30.0	19.0	210	19.5	33.5	38.7	272	32.4	2.5

* Exact month was uncertain in some cases. Thus the notation '06/7', for example, means the sample was taken in June or July.

samples, July 1969 – February 1982) are relictual estuarine lagoons. These data are used to show the effects of evaporative concentration on systems with marine influence. Barber's Pan (44 samples, March 1972 – September 1987) has the most complete record of evaporative concentration for an inland salt pan. A representative set of samples (most dilute, median, most concen-

Table 3. The data for the South African systems. Conductivity, where available, is in mS cm^{-1} at 25 °C. Total dissolved solids are in g l^{-1}: those from Seaman *et al.* (in press) have been calculated from the sum of the major ions; the others were determined analytically. Ions are in meq l^{-1}. ALK = $HCO_3 + CO_3^{2-}$. Lowercase letters refer to different samples from the same site. Localities are indicated in Fig. 3 according to the numbers in the left-hand column of this Table.

Locality	Cond.	TDS	Na$^+$	K$^+$	Ca^{2+}	Mg^{2+}	Cl$^-$	SO$_4^{2-}$	ALK	Ref[a]
1 Lake Sibayi	0.60	0.4	3.4	0.2	1.2	0.7	3.4	0.2	2.0	1
2 Abiekwasputs pan	0.61	0.5	4.5	1.5	0.1	0.6	1.0	0.9	4.7	1
3 Barber's Pan a	1.26	0.9	7.9	0.6	1.0	3.5	5.3	0.7	6.7	1
Barber's Pan b	2.91	1.9	21.4	1.6	0.4	6.1	15.9	1.4	10.5	1
Barber's Pan c	6.40	5.3	59.3	3.8	0.9	13.3	39.6	2.6	34.1	1
4 Pearly Beach Vlei	1.64	1.1	11.2	0.2	1.7	2.8	13.4	0.7	1.2	2
5 Gourikwa Vlei	1.64	1.4	9.5	0.3	4.4	2.6	14.6	0.7	4.8	2
6 Verlorevlei	2.75	1.6	18.7	0.1	3.1	5.8	24.5	1.2	1.8	1
7 Rondeberg Vlei	2.68	1.8	18.6	0.4	1.9	4.3	23.3	0.6	0.1	2
9 De Hoop Vlei a	3.44	2.2	27.1	0.5	2.2	6.5	28.7	5.4	2.0	1
De Hoop Vlei b	9.46	6.7	64.9	1.2	16.5	33.0	80.8	28.1	4.0	1
De Hoop Vlei c	22.22	9.8	214.0	2.2	7.0	39.9	241.0	13.0	12.0	1
10 Noordhoek Pan	4.17	2.9	32.5	0.9	4.0	9.5	36.2	3.8	2.7	2
11 Brakpan 1		3.0	31.1	0.5	4.1	5.4	30.4	10.4	6.2	3
12 Brakpan 3		3.1	30.4	0.1	11.5	6.9	23.9	16.7	6.6	3
13 Finch Pan		3.1	33.0	1.3	9.5	7.4	37.8	13.1	0.3	3
14 Riet Pan 1		3.1	27.0	0.1	9.5	13.2	19.2	18.8	9.0	3
15 Apex Pan		3.1	30.4	0.2	13.5	9.1	39.4	10.0	3.0	3
16 Grootwitpan		3.2	29.1	0.2	13.5	8.2	26.5	19.4	7.5	3
17 Konga Pan		3.4	39.1	0.4	12.0	5.4	36.6	13.0	4.1	3
18 Soetendalsvlei	4.45	3.6	42.4	0.6	4.6	10.1	50.1	3.1	2.4	2
19 Tete Pan		3.7	36.4	0.1	15.5	14.9	59.0	3.4	2.7	3
20 Ganna Pan 2		3.8	32.2	0.5	17.5	14.8	45.1	16.0	1.6	3
21 Rynfield Pan		3.9	60.0	0.3	0.2	0.2	29.9	14.2	12.8	3
22 Brakpan 2		4.1	55.7	0.7	0.4	5.4	25.4	15.6	17.5	3
23 Annas Pan East		4.3	59.1	1.0	1.0	8.2	35.5	23.5	6.6	3
24 Elandsfontein Pan		4.3	65.2	0.6	0.4	0.3	30.4	12.5	18.4	3
25 Nyamithi Pan		4.4	60.9	0.1	3.7	23.9	64.3	0.1	6.3	3
26 Vermont Pan	5.00	4.5	53.0	1.3	8.0	11.7	51.9	6.4	4.3	2
27 Present Pan		4.9	67.4	1.4	2.5	9.1	59.2	17.9	2.1	3
28 Modder Pan		4.9	73.9	0.3	1.7	2.6	45.1	21.7	8.5	3
29 Varkensvlei	5.66	5.1	53.9	1.6	11.1	12.9	51.8	10.8	3.3	2
31 Langklip Pan		5.4	72.2	0.2	1.9	9.1	27.3	43.8	9.2	3
32 Ratelgat Pan		5.5	54.8	1.2	23.5	14.0	84.5	9.4	2.1	3
33 Amabele Pan		5.9	60.9	0.2	22.0	11.5	33.8	52.1	4.1	3
34 Wolwepan		6.1	90.9	1.0	0.3	0.6	42.0	24.0	22.0	3
35 Ferndale Pan		6.4	95.7	1.7	1.0	0.3	58.6	9.2	26.6	3
37 Klein Aarpan		6.9	71.3	0.3	24.0	14.8	45.1	58.3	3.0	3
38 Rietpan 2		7.1	116	0.3	1.3	2.5	73.2	41.7	6.6	3
39 Kooppan Noord		8.1	104	2.7	16.5	14.0	98.6	31.0	2.5	3
40 Gemsbok Pan		8.5	83.0	0.4	11.5	34.6	39.4	89.6	3.1	3
41 Kingswood Pan		8.9	136	2.3	1.3	2.6	84.5	34.4	16.4	3
42 Rooikraal Pan		9.4	143	1.0	0.1	0.2	49.3	36.3	0.1	3
44 Toronto Pan		10.1	136	1.5	7.9	13.2	99.8	56.1	6.6	3
45 Kamfersdam		10.3	126	2.9	21.4	16.8	106	59.0	1.5	3
46 Kooppan Suid		11.6	120	3.0	45.0	24.7	130	56.3	4.1	3
47 Middlepos Pan		12.7	144	1.9	33.0	37.9	186	30.6	2.3	3
48 Nigel Pan		12.9	200	1.9	0.5	0.4	125	56.7	18.0	3
49 Flamingo Pan		16.4	245	5.1	8.2	15.8	210	42.2	11.1	3
50 Melkbos Pan	15.60	16.6	208	2.6	26.0	60.4	240	20.9	2.7	2
51 Dealesville Pan		17.5	313	0.1	0.9	7.5	276	63.0	1.4	3
52 Leeuwpan		18.1	278	2.8	0.5	0.9	149	96.9	26.6	3
53 Soutpan		18.7	198	3.7	75.0	43.6	310	17.9	1.2	3
54 Seven Springs Pan		21.5	39	3.6	2.3	4.2	256	81.3	8.2	3
55 Rooipan	42.50	26.5	325	9.0	47.0	64.2	234	40.0	2.1	2
56 Pretoria Salt Pan		31.6	504	0.1	0.1	0.1	287	63.0	156.0	3
57 Geduldpan		34.8	583	1.8	0.1	0.5	510	52.1	11.8	3
58 Teviot Pan		35.0	474	16.7	46.9	41.1	461	115	2.2	3
59 Koppieskraal Pan N		35.2	478	7.6	33.5	71.6	486	104	1.8	3
60 Swinkpan		38.2	587	8.2	12.5	10.7	409	198	8.2	3
61 Kareepan		41.8	661	1.8	0.3	0.3	482	177	15.6	3
62 Agulhas Salt Pan	48.20	42.3	532	9.6	29.5	95.2	653	24.4	1.8	2
63 Koppieskraal Pan S		48.6	652	18.0	38.0	115	718	104	4.6	3
64 Stinkpan		51.1	851	7.3	49.9	64.0	702	291	9.8	3
65 Yserfontein Pan	53.60	74.2	885	14.5	76.4	195	1020	59.0	2.8	2
66 Rensburg Pan		102	1577	2.8	56.0	41.8	1450	268	3.1	3
67 Skoppan		160	2652	0.9	3.0	82.3	2560	146	0.4	3
68 Britten Pan		182	2767	57.7	14.5	165	2680	380	1.2	3
69 Florisbad Pan		197	3391	0.6	3.4	156	3100	152	0.1	3
70 Koekiespan	97.30	276	3800	8.5	18.0	464	3780	130	10.3	2

[a] Ref 1 = Original DWA data; 2 = Silberbauer & King (1991); 3 = Seaman *et al.* (1991).

trated) is presented for each of these systems except for Lake Sibayi, the TDS of which was <1 g l⁻¹ for all of the samples. The data for Abiekwasputs (1 sample, July 1966) represent the only known record of the chemical composition of a large (several hectares when filled) freshwater pan in the arid north-west. Details of the analytical methods used by the South African DWA can be found in van Vliet *et al.* (1988). Briefly, metals were analysed by atomic absorption spectrophotometry, chloride and alkalinity titrimetrically, and TDS and sulphate gravimetrically. These samples were not preserved; the reservations recorded above for the Namibian samples apply here as well. TDS values from Silberbauer & King (1991) and from DWA are analytical and those from Seaman *et al.* (1991) are calculated from the sum of the ions.

Data for hot springs and East African waters

Analyses of data for hot springs (Table 4) are taken from Hoffmann (1979) and Ashton & Schoeman (1984) and those for East Africa from Wood & Talling (1988).

Results

Chemical composition of surface waters

Total dissolved solids and conductivity
The systems with the highest TDS values are: in the central Namib Desert, Hosabes, a tiny spring in a gypsous crust (Day & Seely, 1988); in the southern Namib Desert, a coastal salt pan at Oranjemund; in South Africa, Koekiespan in the south-western Cape and in East Africa,

Table 4. The data for the hot springs. Conductivity is in mS cm⁻¹; no temperature was indicated in the original data. Total dissolved solids are in g l⁻¹ and were calculated from the sum of the major ions. Ions are in meq l⁻¹. ALK = $HCO_3^- + CO_3^{2-}$.

Locality	Conductivity	TDS	Na^+	K^+	Ca^{2+}	Mg^{2+}	Cl^-	SO_4^{2-}	ALK
Hoffmann (1979)									
Cradock	0.044	0.03	1.8	0.0	1.1	1.0	1.5	0.4	2.4
Citrusdal	0.08	0.05	0.3	0.1	0.1	0.0	0.5	0.0	0.1
Montagu	0.09	0.06	0.3	0.1	0.2	0.2	0.3	0.0	0.6
Lekkerrus	0.20	0.20	1.1	0.1	0.9	0.2	0.2	0.2	2.0
Libertas	0.26	0.20	1.0	0.1	1.2	0.2	0.3	0.1	2.2
Rhemardo	0.50	0.24	1.7	0.1	1.1	0.1	0.7	0.2	2.2
Die Oog	0.36	0.30	1.8	0.1	1.7	0.2	0.7	0.2	3.1
Badplaas	0.54	0.38	4.7	0.1	0.3	0.1	3.5	0.4	1.1
Tshipise	1.02	0.50	6.3	0.2	0.5	0.4	4.1	1.3	2.3
Die Eiland	1.75	1.03	13.9	0.4	1.7	0.1	13.8	1.6	1.0
Aliwal North	1.70	1.29	14.8	0.1	4.3	0.5	9.4	8.5	1.6
Rob Ferreira	2.24	2.19	27.8	0.3	2.3	3.0	17.6	5.4	11.2
Ashton & Schoeman (1984)									
Stampriet	1.64	0.54	3.2	0.2	2.1	3.0	2.5	0.4	5.2
Osterode süd	1.79	0.58	4.0	0.3	2.0	3.5	2.8	0.6	6.1
Omapyu	1.50	1.12	14.3	0.7	1.4	0.7	9.6	4.6	2.7
Blydeverwacht	1.65	1.14	14.8	0.2	2.0	0.5	7.9	5.2	3.7
Gründorn	2.70	1.33	17.8	0.2	1.7	0.7	11.0	6.7	2.6
Gröss Barmen	2.60	1.36	18.3	0.5	0.3	0.1	4.2	11.0	3.3
Rehoboth	4.00	1.75	22.2	1.3	0.7	0.5	5.4	14.6	4.5
Warmbad	3.50	1.76	24.3	0.4	2.0	0.7	17.2	9.4	0.9
Ganikobis	4.20	2.38	32.0	0.2	1.5	1.5	14.6	15.6	5.4
Ai-Ais	4.95	2.40	33.5	0.7	0.5	0.4	13.8	19.6	1.7
Warmbad N.	4.00	2.61	36.5	0.4	1.0	1.0	16.6	19.4	3.1
Riemvasmaak	5.40	2.62	36.3	0.5	2.0	1.6	24.5	12.3	4.1

Lake Katwe in the western Rift Valley of Kenya.

Conductivity and TDS correlate well for the Namibian ($r = 0.92$, $n = 76$, $P < 0.05$) and East African ($r = 0.99$, $n = 23$, $P < 0.05$) systems. Too few conductivity data are available for a similar analysis of South African hot spring or surface waters.

Seasonal variations in conductivity for five Namib gypsous springs are listed in Table 2. The heaviest rains in more than 40 years fell on 24 January 1978. The lags in response to rain can be explained by the fact that these springs are all fed by underground sources, some of which are linked to underground streams far inland, where rainfall was heaviest. Conductivity values continued to fall for up to six months after the rain.

Major ions
All Namibian waters with TDS values > 3 g l^{-1} have dominance by Na$^+$ and Cl$^-$ (Fig. 4). The

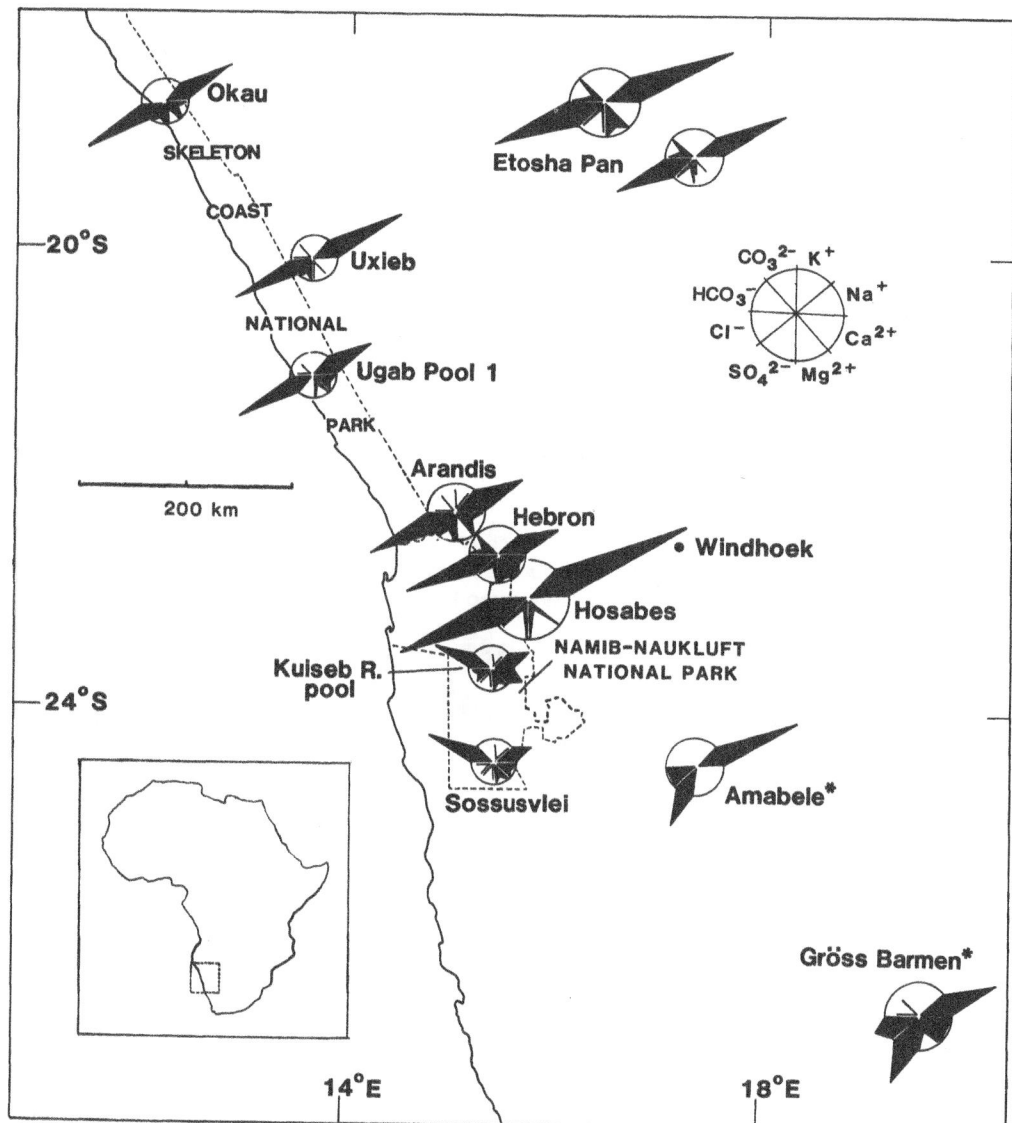

Fig. 4. Maucha ionic diagrams showing the distribution of selected waters of different ionic proportions in central and northern Namibia. Data from Table 1 and from Ashton & Schoeman (1984). For Etosha Pan, two samples are represented.

48

dominant cations of the less saline waters (from temporary pools in river beds, boreholes, waterholes and hot springs) are Na^+ or Na^+ and Ca^{2+}, and the dominant anions tend to be HCO_3^-/CO_3^{2-} or SO_4^{2-}. The same holds true for South African waters (Fig. 5). A different picture (Fig. 6) emerges for East Africa, where again the less saline waters are usually dominated by HCO_3^-/CO_3^{2-} and a variety of cations, while most saline systems are soda lakes dominated by HCO_3^-/CO_3^{2-} and Na^+ ions. Only the most saline of all (Lake Katwe in Kenya and Lakes Afrera and Assal on the Red Sea coast) are dominated by Na^+ and Cl^- ions.

Maucha diagrams (Broch & Yake, 1969) are useful for comparing waters in which the proportions of the major ions are clearly different. In order to see more subtle differences, particularly between ions other than the dominant ones, more detailed analysis is necessary. As chloride ions are conservative and are the last major ions to be precipitated during evaporative concentration, plots of the concentrations of the other ions against Cl^- concentration can provide further chemical characterisation of different types of waters. Results for these plots are shown in Fig. 7 for Na^+, Ca^{2+}, HCO_3^-/CO_3^{2-} and SO_4^{2-}, which are the most abundant ions in most systems.

As might be expected, Na^+ correlates strongly with Cl^- for all four systems, the regression lines for Namibian and South African waters being insignificantly different statistically (ANOVA: d.f.

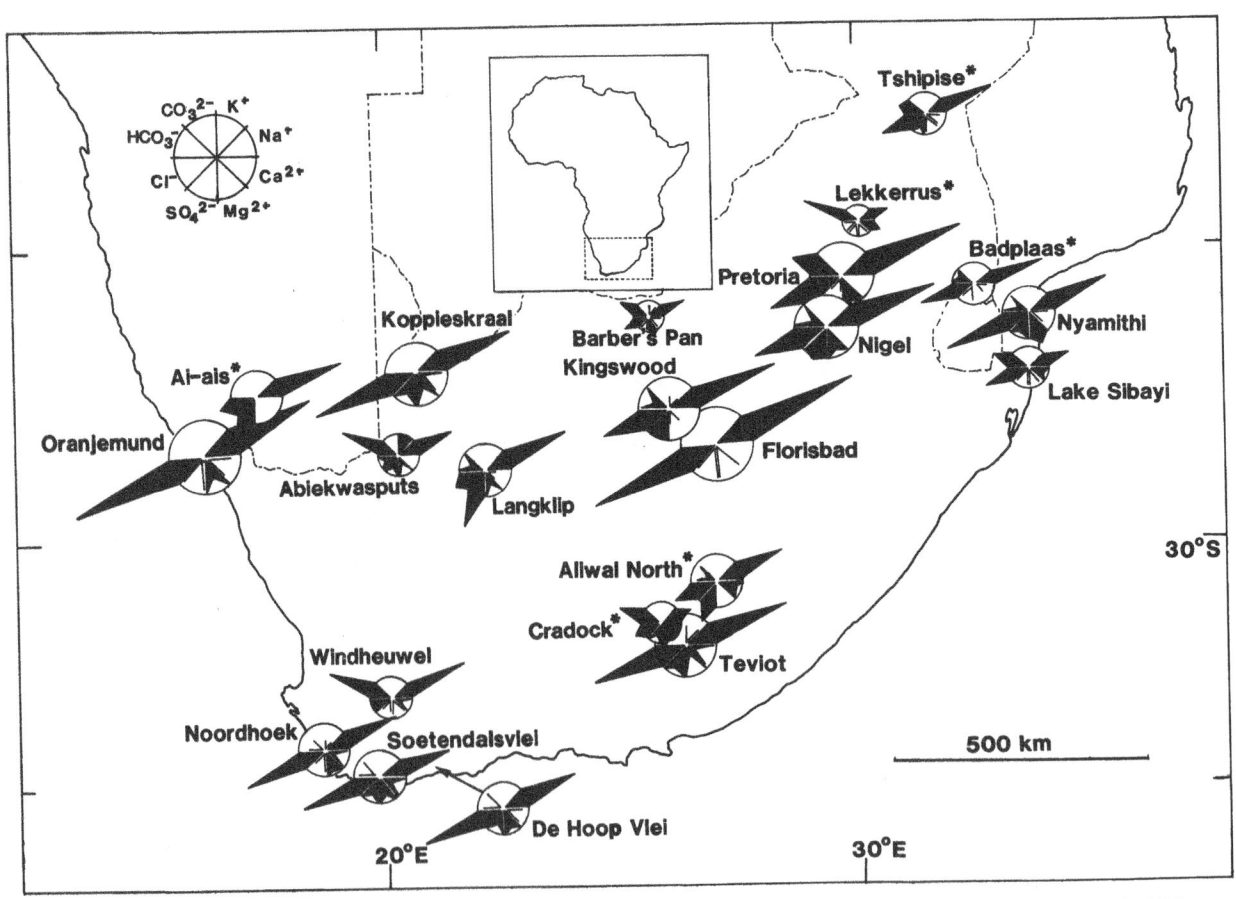

Fig. 5. Maucha ionic diagrams showing the distribution of selected waters of different ionic proportions in South Africa and southern Namibia. Data from Table 3 and from Hoffmann (1979).

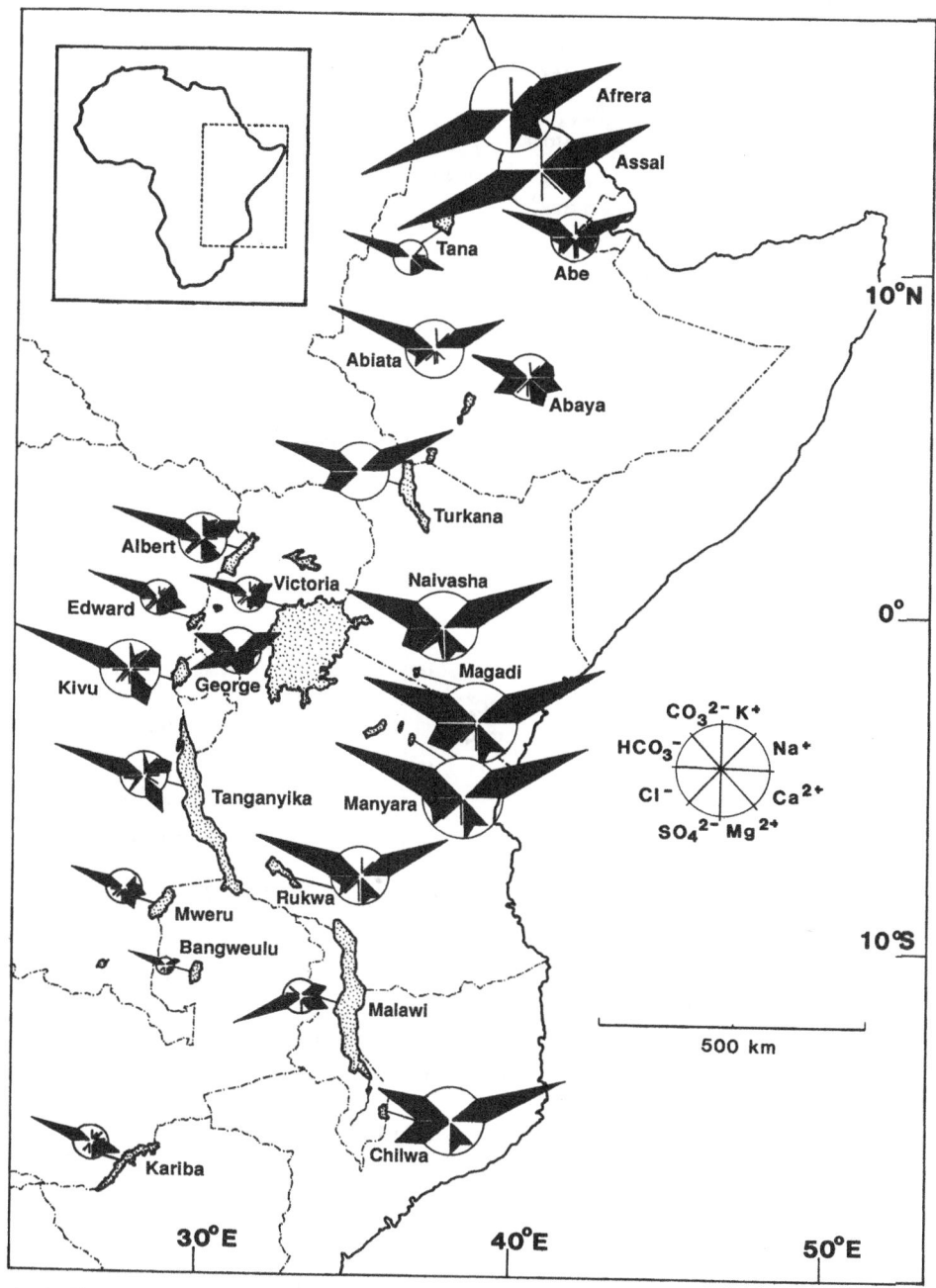

Fig. 6. Maucha ionic diagrams showing the distribution of selected waters of different ionic proportions in East Africa. Data from Talling & Talling (1965) and Wood & Talling (1988).

143, 1; $F = 2.92$; $P > 0.05$) and coinciding with the value for sea water. The curve for the hot springs is slightly steeper, indicating a relatively more rapid increase in the $Na^+:Cl^-$ ratio with concentration. The curve for the East African lakes lies above the others because of the relatively smaller amount of Cl^- in these waters. This curve closely follows the line drawn by Wood & Talling (1988) (and dashed in Fig. 7a) for passive concentration from rainwater in East Africa.

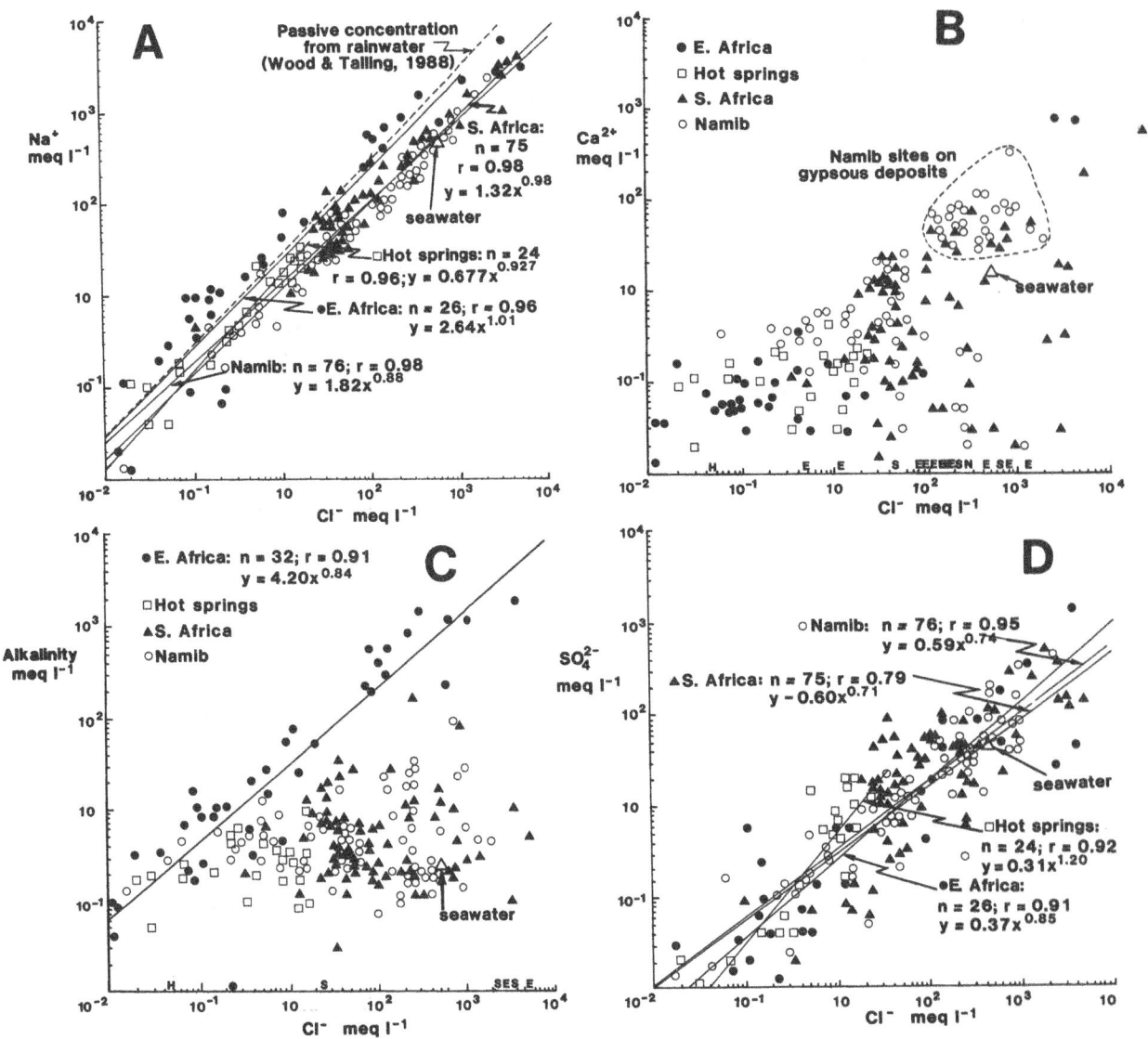

Fig. 7. The relationship between concentrations of (a) Cl^- and Na^+, (b) Cl^- and Ca^{2+}, (c) Cl^- and alkalinity and (d) Cl^- and SO_4^{2-} for the Namibian and South African salinity series and hot springs, and East African saline lakes. The dashed line is that given by Wood & Talling (1988) for passive concentration of East African rainwater.

No clear relationship exists between Cl^- and Ca^{2+} concentrations (Fig. 7b), although Ca^{2+} ions are least abundant in north-east African systems and, as might be expected, most abundant in the Namib springs on gypsous crusts.

Alkalinity and Cl^- concentrations (Fig. 7c) correlate significantly only for the East African lakes, in which the dominant ions are HCO_3^-/ CO_3^{2-}.

SO_4^{2-} and Cl^- concentrations (Fig. 7d) correlate significantly for all systems and those for the Namibian, South African and East African waters are insignificantly different from each other (ANOVA: d.f. 173, 2; $F = 3.87$; $P < 0.05$). The curves for all three fall very close to the value for sea water. The relatively high SO_4^{2-}-levels in the hot springs are reflected in the somewhat steeper slope of that curve.

The effects of evaporative concentration on ionic ratios

Figure 7 illustrates the relationships between the absolute values of Cl^- and various other ions. Changes in ionic proportions with evaporative concentration are better illustrated by plotting ionic ratios against concentration. Thus the ratios of the dominant cations (Na^+ and Ca^{2+}) and anions (Cl^- and HCO_3^-/CO_3^{2-}, and Cl^- and SO_4^{2-}) for all Namibian waters are plotted against TDS in Figs 8–10. In all cases the ratios are derived from values in mg l^{-1} to correspond with Gibbs's (1970) diagram (see Fig. 8 and below).

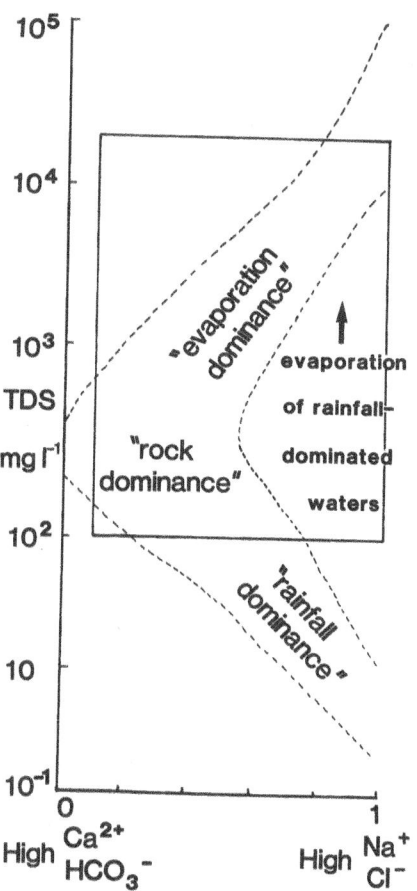

Fig. 8. The relationship between ionic ratios, TDS and the mechanisms controlling the major ion composition of natural waters. The rectangle indicates the ranges covered in Figs 9–11. Modified from Gibbs (1970).

The changing relationship between Na^+ and Ca^{2+} concentrations as TDS levels increase is shown in Fig. 9 for Namibian waters. The regression line for De Hoop Vlei (a South African coastal lake) represents the evaporative concentration of sea water and that of Barber's Pan that of a saline inland pan. The slopes of the lines are identical but the intercepts with the x axis are not, indicating that, although the rate of precipitation is the same for both, virtually all Ca^{2+} is lost from solution in Barber's Pan at a much lower concentration than in De Hoop Vlei. This is probably due to the formation of $CaCO_3$ from the proportionally greater quantities of HCO_3^- in Barber's Pan. This is substantiated by the curve for Barber's Pan, which falls within the area on the graph occupied by the East African lakes, in which HCO_3^-/CO_3^{2-} (and Na^+) predominate, and that for De Hoop Vlei falls within the area occupied by the South African lakes, in which Cl^- (and Na^+) predominate. The Namibian waters form a concentration series spanning both of these types. Generally, ground water and temporary waters are most dilute and have the greatest proportion of Ca^{2+} ions. All the springs on gypsous deposits retain measurable quantities of Ca even at extremely high TDS levels, although the ratio does not correlate with TDS because of precipitation of calcium at high concentrations. The actual ratio of $Na^+:Ca^{2+}$ seems to be characteristic for a number of gypsous springs, that for the Poacher's Springs being particularly low.

The changing relationship between Cl^- and HCO_3^- ions as TDS levels increase is shown in Fig. 10 for Namibian waters. The values for De Hoop Vlei all lie well to the right (dominance by Cl^- ions, as might be expected for sea water) but the correlation is not statistically significant. The curve for Barber's Pan lies well to the left, indicating dominance by HCO_3^- ions, a feature shown even by concentrated East African lake waters. South African saline lakes at similar concentrations are dominated by Cl^- ions and therefore fall to the right of the figure. The more dilute Namibian temporary and ground waters are dominated by HCO_3^-/CO_3^{2-}, the pattern shifting to Cl^- domination with increasing concentration.

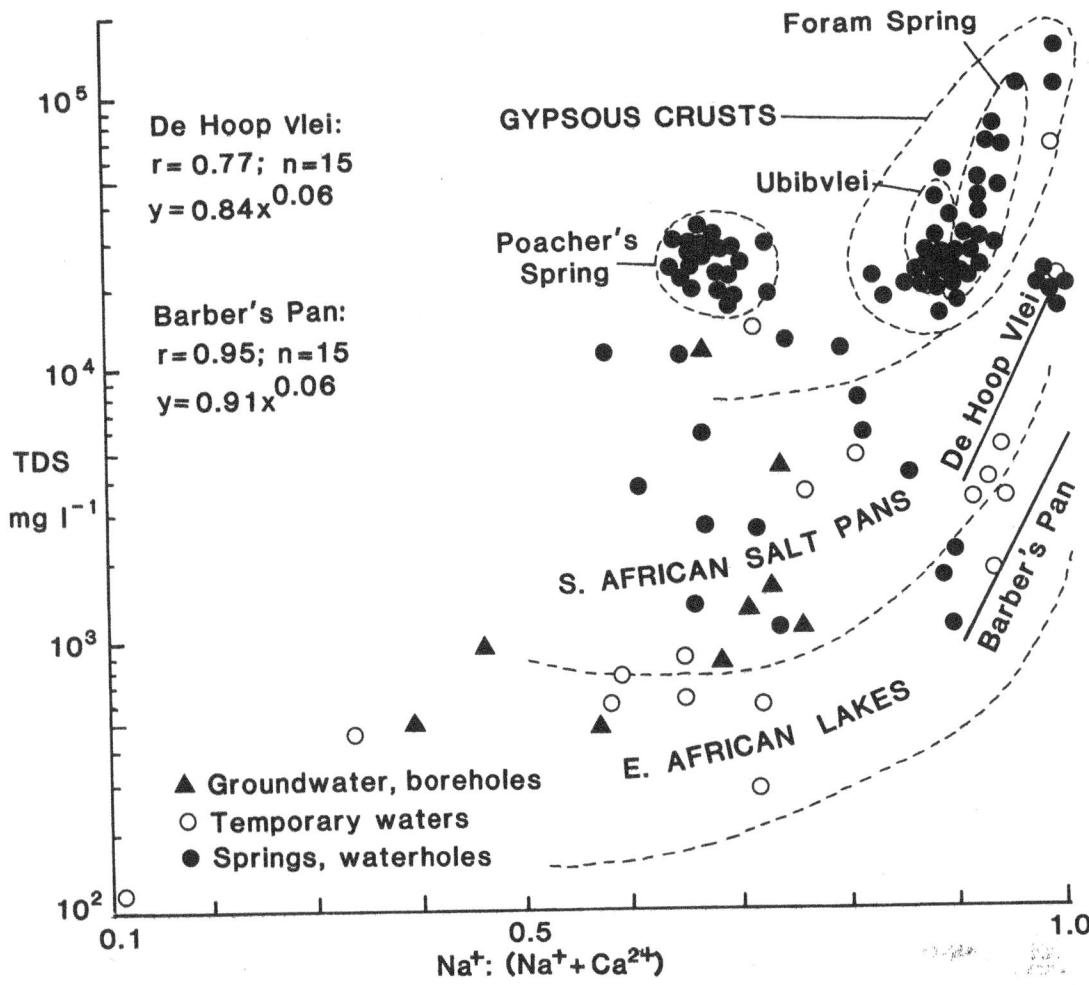

Fig. 9. The ratio of Na$^+$:(Na$^+$Ca^{2+}) plotted against TDS for the Namibian waters listed in Table 1. The areas of the graph occupied by South African and East African waters are indicated by dashed lines. The lines labelled 'De Hoop Vlei' and 'Barber's Pan' are regression lines for the concentration series for these two vleis and are given for comparative purposes.

All springs on gypsous deposits tend to cluster to the right and to follow the general trend of increasing dominance by Cl$^-$ ions as HCO$_3^-$ precipitates out.

The changing relationship between Cl$^-$ and SO$_4^{2-}$ ions as TDS levels increase is shown in Fig. 11 for Namibian waters. Most ratios are high. The slope is much steeper for the waters of Barber's Pan than for those of De Hoop Vlei. The positions of the East African and South African lakes are not indicated on the graph because no clear patterns are evident: points for both regions fall widely within the area bounded by the Namibian points. The Namibian data are also fairly

widely scattered, although once again those for gypsous pools tend to cluster and to show high proportions of SO$_4^{2-}$ ions relative to TDS.

Ionic proportions: summary

Differences in ionic proportions are summarised by means of a ternary diagram (Fig. 12). In it are plotted the relative proportions of the major anions and cations for most Namibian systems and the general positions occupied by a variety of other types of system. Perhaps the most striking features indicated by this diagram are the anoma-

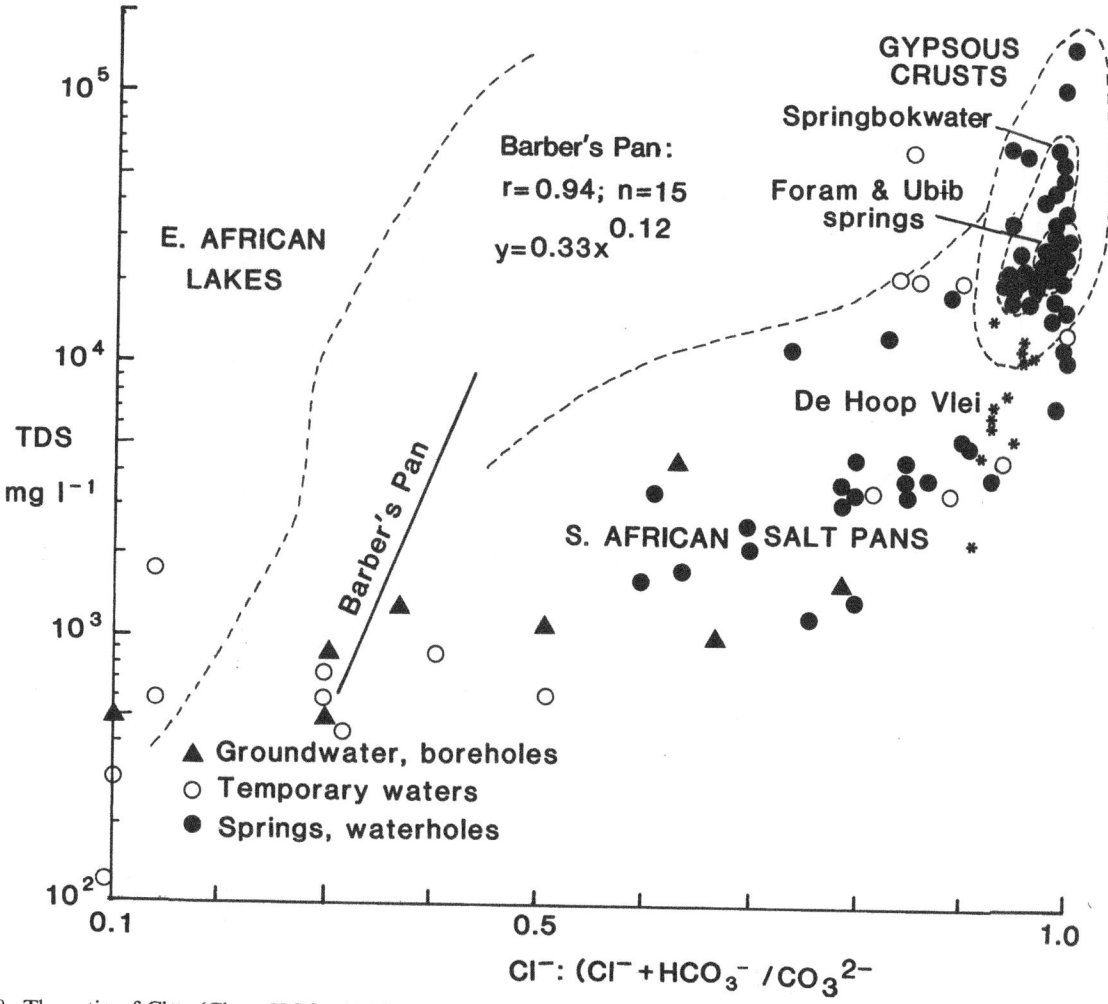

Fig. 10. The ratio of $Cl^-:(Cl^- + HCO_3^-/CO_3^{2-})$ plotted against TDS for the Namibian waters listed in Table 1. The area of the graph occupied by South African and East African waters is indicated by dashed lines. Concentration series for De Hoop Vlei (asterisks: not significant) and Barber's Pan are given for comparative purposes.

lously high proportions of Ca^{2+} and Mg^{2+} in the springs on the gypsous crusts in the Namib, the high SO_4^{2+} levels in the Namibian hot springs, and the virtual absence of Cl^- ions in East African fresh waters.

Discussion

Sources of ions

Ions dissolving in surface waters must come either from the atmosphere in the form of rain, fog or wind-blown particles (snow can be discounted here) or from the substratum or the surrounding catchment area.

Given the extreme aridity of the area under discussion, the atmosphere would seem to be a minor source of ions. This assumption may not be valid, though, given the following simple calculation. The rate of accumulation of Cl^- ions for, say, a closed-basin pan in the northern Cape Province (rainfall 400 mm y^{-1}, $[Cl^-]$ in rain 11 μeq l^{-1}: Bosman & Kempster, 1985), yields a value of 156 g m^{-2} in 1000 years. For a pan 1 m deep, this is equivalent to an increase of 4.4

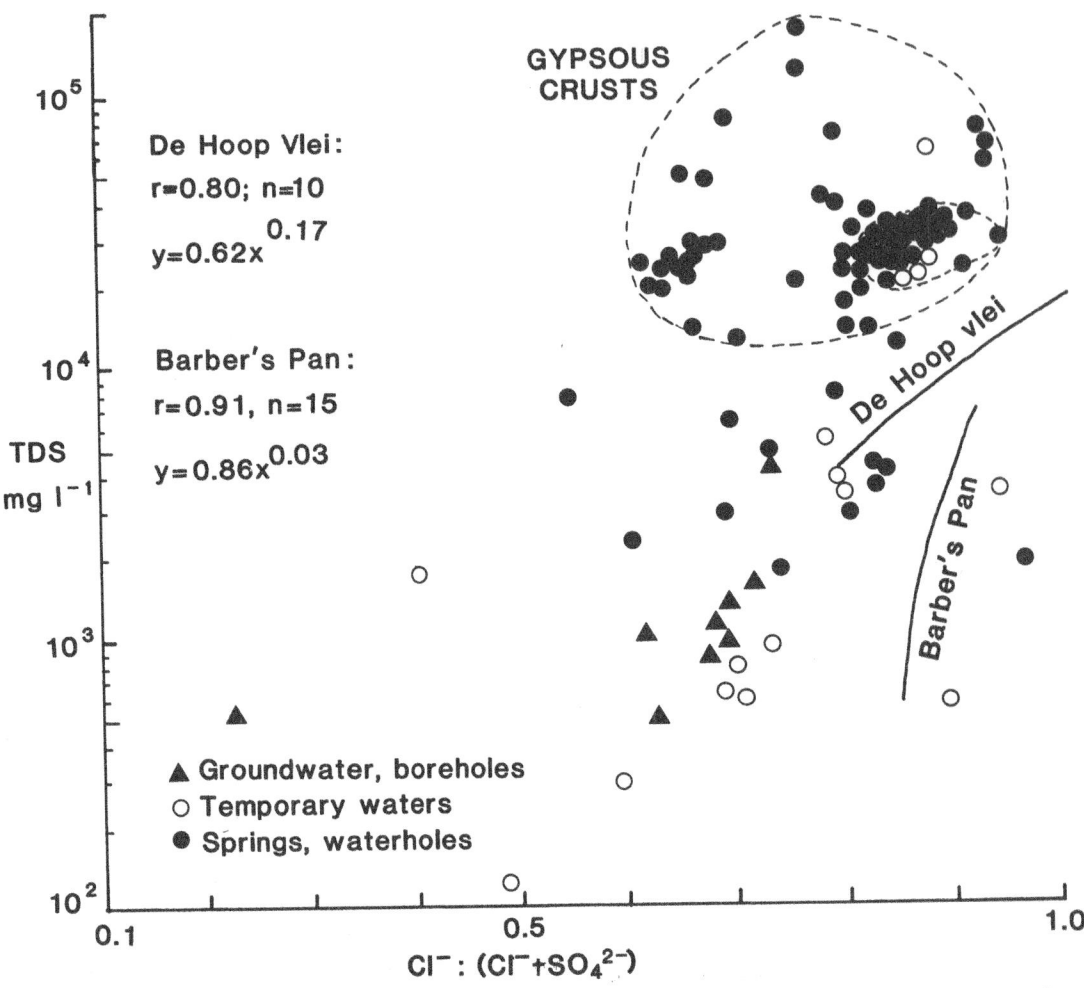

Fig. 11. The ratio of Cl^- : $(Cl^- + SO_4^{2-})$ plotted against TDS for the Namibian waters listed in Table 1. Concentration series for De Hoop Vlei and Barber's Pan are given for comparative purposes.

$\mu eq\ l^{-1}$ over 1000 years. Lancaster (1979) has reported the existence of stromatolites (an indication of salinity and therefore possibly of aridity) from saline Kalahari lakes dated at 17000–15000 B.P. If arid conditions have indeed persisted for this long, then atmospheric precipitation could account for the salinity of many pans. Clearly this is an oversimplification of the situation, in that even closed-basin pans can lose ions (by deflation and seepage, for instance) but it does indicate that atmospheric precipitation should not be ignored as a potentially significant source of ions in arid areas. In general, though, major ions

in saline waters are derived from the catchment or the substratum (e.g. Kilham, 1990).

Ionic ratios

Gibbs (1970) postulated that the ratios of Na^+ : Ca^{2+} and Cl^- : HCO_3^- plus CO_3^{2-} ($mg\ l^{-1}$) relative to salinity (measured as TDS) should indicate whether the chemical composition of river waters is determined largely by geological parent material ('rock dominance'), atmospheric precipitation ('precipitation dominance') or evaporative

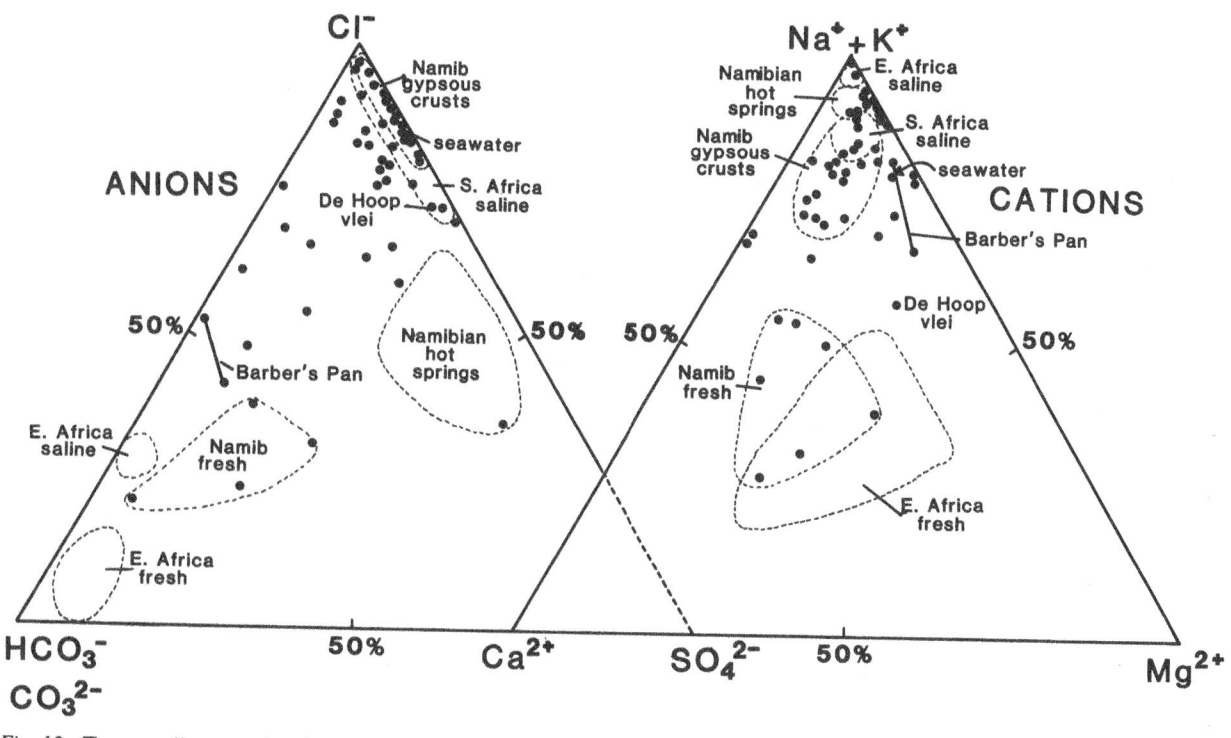

Fig. 12. Ternary diagram showing the relative ionic proportions of the Namibian waters listed in Table 1 (dots). Dashed lines outline areas occupied by East African fresh and saline lakes, South African saline systems, and Namibian gypsous springs and hot springs.

concentration ('evaporation dominance') (Fig. 8). Gibbs' schema needs to be modified in various ways for particular systems. For instance, some waters (e.g. the curves from Barber's Pan in Figs 9 and 11) fall to the right of the area outlined by Gibbs because their chemical composition results from direct evaporation of rainfall-dominated waters. Others fall to the right because of the very high Na:Ca ratio in the dominant rocks or subsoils of the catchment (e.g. much of Australia: Cornish, 1987). Kilham's modification for certain East African waters is discussed in some detail below. In general, though, Gibbs' schema is applicable for saline waters.

All of the less saline waters illustrated in Fig. 5 (Lekkerrus, Lake Sibayi, Barber's Pan, Cradock, Abiekwasputs, Windheuwel and Langklip), conform in ionic proportions to those known for fresh waters from the same areas, and the ionic proportions are largely explained by the surface geological formations in each area (e.g. Bond, 1946;

Day & King, in press). For instance, the high concentration of SO_4^{2-} ions in Langklip (and also in Gröss Barmen and Amabele in Namibia) is characteristic of waters on the basement-complex granites in the region (e.g. Bond, 1946; van der Merwe, 1962).

Little published information is available on the more dilute waters of the Namib for making similar generalizations. Some conclusions can, however, be reached from an examination of the ionic proportions exhibited by the least saline waters (Table 1 nos 1–10, Table 4, Osterode Sud and Stampriet) plotted in Figs 9–10. Ca^{2+} and HCO_3^- dominate in these waters, which are rock dominated *sensu* Gibbs (1970); all (except Sossusvlei – on sand and hardpan) occur on the igneous rocks of the Khomas series of the Damara system or on basement-complex granites. No data are available on fresh waters associated with any of the other geological formations in the area, so no further conclusions can be reached about the relationship between geology and ionic dominance in the Namib.

Nonetheless, the general tendency for the Namib salinity series to fall within the upper arm of Gibbs's diagram (Figs 9, 11) indicates that the chemical composition is a result of evaporative concentration of rock-dominated waters. The salinity series can be explained as follows. Most of the temporary waters, both in endorheic pans and in river beds, and most of the boreholes and waterholes, tend to be dilute. With relatively high proportions of divalent cations and HCO_3^-, they reflect the rock dominance typical of Gibbs' (1970) moderately dilute waters. (Etosha Pan is of a different nature, being a very large, endorheic salina.) The springs and lakelets are most saline, and all are dominated by Na^+ and Cl^- ions as a result of precipitation of $CaCO_3$. It can thus be concluded that the surface springs of the Namib Desert tend to arise as dilute, rock-dominated groundwater that becomes subject to evaporative concentration once exposed to the atmosphere.

A similar situation is reported by Jones, Eugster & Rettig (1977), who describe five 'hydrologic stages' in the Lake Magadi Basin in Kenya: dilute streamflow (in the Namib this is of an episodic nature); dilute groundwater (as seen in boreholes and waterholes in the Namib); saturated brines (Oranjemund Salt Pan); residual brines (Etosha Pan) and saline ground-water (not yet reported for the Namib). In comparison, the saline (TDS $> 3 \, g \, l^{-1}$) South African pans under discussion are temporary, filling only after rain. The chemical composition of their waters is determined by the evaporative concentration of salts of connate origin (see also Seaman et al., 1991). They also show increasing loss of Ca^{2+} and HCO_3^- ions by crystallization and thus the proportion of Na^+ and Cl^- ions increases with rising TDS.

The origin of the gypsous crusts in the Namib Desert

Gypsous deposits generally result from igneous activity (e.g. in the Lake Bengweulu region of Central Africa: Symoens, 1968) or from evapo-ration of bicarbonate-poor brines. In the case of the Namib gypsous crusts, neither of these explanations is sufficient: the area is not volcanic (and has not been for millions of years) and, although calcium is present in the parent material (and therefore both calcium and bicarbonate ions should be present in solution), sulphate levels are too low to account for the formation of gypsum. A much more unusual explanation has been offered (Martin, 1963). The coastal seas off the Namib Desert, forming part of the Benguela upwelling system (see, for instance, Shannon, 1985), are extremely productive. Occasionally, as a result of as-yet unexplained perturbations in the marine environment, massive die-backs of phytoplankton result in anoxic conditions and the release of H_2S into the water column. Fish die, rock lobsters beach themselves and sometimes irruptions of H_2S form clouds that waft inland, sometimes together with the frequent fogs. As the fog precipitates, H_2S is oxidised. These occurrences are not common and no quantitative data are available to confirm or reject this suggested explanation for the presence of sufficient sulphate to form gypsous crusts. Kilham & Cloke (1990) report on a comparable phenomenon in some Tanzanian saline lakes. The same process of evaporative concentration of dilute groundwater, as described above, occurs in the gypsous springs, but the ionic proportions of these systems are modified by the presence of gypsum-saturated soils.

Individual ions

Sodium is the dominant cation through the entire South African salinity series, in all but three of the Namibian salinity series and in all of the hot springs. That it acts here as a conservative ion is shown by the strong positive correlation with chloride (Fig. 7b).

Potassium never accounts for more than 10% of the cations except in Abiekwasputs, an episodically-filled, ephemeral pan in the northwestern Cape, where it accounts for 23% of the cations.

Calcium, the dominant cation in dilute waters in the Transvaal and parts of Natal (Day & King, in press), accounts for no more than 27% of the cations in the South African salinity series and decreases in proportion with increasing salinity. Proportions are relatively high in temporary pools in the Kuiseb River (23–50% of cations) and, although the actual proportions are low for the sites on gypsous crusts, they are high relative to concentration (Fig. 7c). Values for the hot springs vary between 1 and 24% of cations in Namibia and 5 and 47% in South Africa. The four hot springs with high Ca^{2+} proportions (36–47%) are clustered geologically on feldspars of the Rooiberg Series of the Transvaal Supergroup, presumably rich in calcium.

The proportion of *magnesium* is usually lower than that of calcium at low salinities but, as Mg^{2+} ions are more soluble than Ca^{2+} ions, the proportion reverses up the salinity series. In sea water Mg^{2+} accounts for about 18% of the cations, and Ca^{2+} about 3%. Thus one might expect, and indeed finds, similar proportions in the coastal lakes (Verlorevlei, De Hoop Vlei). Barber's Pan is unusual for an inland system in that Mg predominates over Ca (7–27% as opposed to 1–7%), even at low salinities.

Chloride is the dominant anion in all of the South African salinity series, except in Abiekwasputs and in the most dilute samples from Barber's Pan (where HCO_3^- dominates) and in four saline pans (Langklip Pan, Amabele, Klein Aar Pan and Gemsbok Pan) where the dominant anion is SO_4^{2-}. The anionic composition of six of the 12 South African hot springs is dominated by Cl^- and the other six by HCO_3^-. All but a few of the borehole and temporary waters of the Namib salinity series are dominated by Cl^- ions.

Bicarbonate (and, where pH is high, *carbonate*) ions dominate the anions only of the more dilute surface and ground waters of Namibia (Sossusvlei, Kuiseb River boreholes and temporary pools) and South Africa (Abiekwasputs, Barber's Pan). The fact that these waters contain significant quantities of HCO_3^- ions supports the contention that southern African saline waters are dominated by Na^+ and Cl^- ions because of pro-

cesses involved in evaporative concentration. When CO_2 dissolves in water, the resulting HCO_3^- ions bring Ca^{2+} (and Mg^{2+}) ions into solution from the parent material. Since $CaCO_3$ is the first salt to precipitate out as concentration increases, even waters initially high in both ions will lose them to the sediments at high salinities.

The high $HCO_3^-/CO_3^{2-}:Ca^{2+}$ ratio in Barber's Pan is less easy to explain. Calcium levels are fairly low (1–7% of cations: see Table 3) and the pH is high (7.5–9.7) and, according to Hutchinson *et al.* (1932), charophytes are present on the bottom of the pan when it has water in it. Whether or not the biota of this pan is able to sequester sufficient calcium to reduce the concentration of Ca^{2+} ions is not known.

Sulphates generally account for <20% of the anions in the pans of the south-western Cape and between 0 and 68% of the anions in the rest of South Africa. In contrast to the Namibian hot springs (SO_4^{2-} ions accounting for 20–60% of anions), those of South Africa, with one exception are much poorer in SO_4^{2-} ions (>20% of anions). The corresponding value for Namibian waters is always <25%.

The curves in Fig. 8 show an exponential >1 for $SO_4^{2-}:Cl^-$ for the hot springs, indicating increasing dissolution of sulphates at increasing concentrations, and exponentials <1 for the surface waters, indicating loss of sulphates with increasing concentration. In contrast to the other major ions, which are lost almost entirely to the sediments during evaporative concentration, this loss can be to the sediments in the form of salts such as $CaSO_4$ or to the atmosphere in the form of S-containing gases (largely H_2S), as a result of physical turnover of the water column or of biotic activities.

The very steep curve for Barber's Pan in Fig. 11 suggests biological regeneration or physical dissolution of sulphate as total concentration increases, however. Both this steep curve and the anomalously high SO_4^{2-} levels in the Namib gypsous springs are possibly the result of continuously oxidising conditions (and therefore less loss as H_2S) in these biologically very productive systems (Hutchinson *et al.*, 1932; pers. obs.).

Comparison between East and southern African lakes

Whether saline or fresh, most East African lakes are dominated by Na^+ and HCO_3^- ions (Figs 6, 9, 10) because HCO_3^- ions do not readily precipitate out in the virtual absence of Ca^{2+} ions (e.g. Fig. 9 and Talling & Talling, 1965; Wood & Talling, 1988).

In a recent paper entitled 'Mechanisms controlling the chemical composition of lakes and rivers: data from Africa', Kilham (1990) modified Gibbs' (1970) 'boomerang-shaped envelopes of data' to accommodate a large data set from 'intertropical Africa' (largely Uganda, Kenya, Tanganyika, Ethiopia and Zaire). Kilham concluded that for Na^+ and Ca^{2+} 'the envelope of data for African waters is shaped like an alchemists's retort rather than a boomerang' (*i.e.* that from TDS values of about 600 mg l^{-1} the $Na/(Na + Ca)$ ratio is essentially >0.8) and that for Cl^- and HCO_3^-, the data clustered largely at a low ratio of $Cl:Cl + HCO_3$ (<0.5) except at TDS values >8000 mg l^{-1} (*i.e.* data points seldom fell within either arm of the 'boomerang').

This is clearly not the case for the southern African data, presented above, which generally fit within Gibbs' 'boomerangs' but not within Kilham's 'retort'. The reason for Kilham modifying Gibbs' original figures was to accommodate the East African waters, many of which are particularly deficient in calcium. It is not appropriate, however, to equate East Africa, or even 'intertropical Africa', with Africa as a whole. It must be emphasised that the hydrochemistry of both the southern African saline pans and the springs of the coastal Namib Desert is very different from that of the saline lakes of East Africa. Generalizations about this fairly small area should not be applied to an entire continent. It would nonetheless be interesting to examine Central and West African data in relation to geology and climate in order to gain a wider understanding of the determinants of the chemical composition of African waters in general.

Acknowledgements

I should like to thank the Department of Nature Conservation in Namibia and the Department of Water Affairs in South Africa for providing me with data; the Department of Water Affairs in Namibia for chemical analyses; Michael Silberbauer for providing the PC programme that draws Maucha diagrams; Dr Mary Seely and the staff of the Desert Research Unit of Namibia and Dr Mike Griffin of the Department of Nature Conservation and Tourism of Namibia for their help and encouragement, and for logistical support; and Delny Britton, Belinda Day and Liz Reynolds for assistance with the illustrations.

References

Ashton, P. J. & F. R. Schoeman, 1983. Limnological studies on the Pretoria Salt Pan, a hypersaline maar lake. 1. Morphometry, physical and chemical features. Hydrobiologia 99: 61–73.

Ashton, P. J. & F. R. Schoeman, 1988. Thermal stratification and the stability of meromixis in the Pretoria Salt Pan, South Africa. Hydrobiologia 158: 253–265.

Ashton, P. J. & F. R. Schoeman, 1984. A preliminary limnological investigation of twelve southern African geothermal waters. J. limnol. Soc. sthn Afr. 10: 50–56.

Berry, H., 1972. Flamingo breeding on Etosha Pan, South West Africa, during 1971. Madoqua series 1, 5: 5–31.

Bond, G. W., 1946. A geochemical survey of the underground water supplies of the Union of South Africa with particular reference to their utilisation in power production and industry. S. A. Dept of Mines Geological Survey Memoir no. 41. Government Printer, Pretoria, 216 pp.

Bosman, H. H. & P. L. Kempster, 1985. Precipitation chemistry of Roodeplaat Dam catchment. Water S.A. 11: 157–164.

Broch, E. S. & W. Yake, 1969. Modification of Maucha's ionic diagram to include ionic concentrations. Limnol. Oceanogr. 14: 933–935.

Cornish, P. M., 1987. World streamwater chemistry and the relative importance of sodium in the Australian environment. Search 18: 89–91.

Day, J. A., 1990. Environmental correlates of aquatic faunal distribution in the Namib Desert. In: Seely, M. K. (ed.), Namib ecology: 25 years of Namib research. Transvaal Mus. Monograph 7: 99–108.

Day, J. A. & J. M. King, in press. Geographical patterns, and their origins, in the dominance of major ions in South African rivers. S. Afr. J. Sci.

Day, J. A. & M. K. Seely, 1988. Physical and chemical conditions in an hypersaline spring in the Namib Desert. Hydrobiologia 160: 141–153.

de Bruiyn, H., 1972. Pans in the western Orange Free State. Ann. geol. Surv. S. Afr. 9: 121–124.

Gevers, T. W., O. Hart & H. Martin, 1963. Thermal waters along the Swakop River, South West Africa. Trans. geol. Soc. S. Afr. 66: 157–189.

Gibbs, R. J., 1970. Mechanisms controlling world water chemistry. Science 170: 1088–1090.

Grobbelaar, J. U., 1976. Some limnological properties of an ephemeral waterbody at Sossus Vlei, Namib Desert, South West Africa. J. limn. Soc. sthn Afr. 2: 51–54.

Grobbelaar, J. U. & M. K. Seely, 1980. The composition of water collected from the Kuiseb River, Namib Desert, at Gobabeb. J. limn. Soc. sthn Afr. 6: 46–48.

Hoffmann, J. R. H., 1979. Die chemiese samestelling van warmwaterbronne in Suid- en Suidwes-Afrika. National Institute for Water Research, Special Report no. WAT 56. South African Council for Scientific and Industrial Research, Pretoria, 21 pp.

Hutchinson, G. E., G. E. Pickford & J. F. M. Schuurman, 1932. A contribution to the hydrobiology of pans and other inland waters of South Africa. Arch. Hydrobiol. 24: 1–154.

Jones, B. F., H. P. Eugster & S. L. Rettig, 1977. Hydrochemistry of the Lake Magadi basin, Kenya. Geochim. Cosmochim. Acta 41: 53–72.

Kilham, P., 1990. Mechanisms controlling the chemical composition of lakes and rivers: data from Africa. Limnol. Oceanogr. 35: 80–83.

Kilham, P. & P. L. Cloke, 1990. The evolution of saline lake waters: gradual and rapid biogeochemical pathways in the Basotu Lake District, Tanzania. Hydrobiologia 197: 35–50.

Kok, D. B. & J. U. Grobbelaar, 1980. Chemical properties of waterholes in the Kuiseb River Canyon, Namib Desert. J. limn. Soc. sthn Afr. 6: 82–84.

Kok, D. B. & J. U. Grobbelaar, 1985. Notes on the availability and chemical composition of water from the gravel plains of the Namib-Naukluft Park. J. limn. Soc. sthn Afr. 11: 66–70.

Lancaster, I. N., 1978. The pans of the southern Kalahari, Botswana. Geogr. J. 144: 80–98.

Lancaster, I. N., 1979. Evidence for a widespread late Pleistocene humic period in the Kalahari. Nature 279: 145–146.

Lancaster, J., I. N. Lancaster & M. K. Seely, 1984. Climate of the central Namib Desert. Madoqua 14: 5–61.

Martin, H., 1963. Suggested theory for the origin and a brief description of some gypsous deposits of South West Africa. Trans. Proc. Geol. Soc. S. Afr. 66: 345–350.

Martin, H., 1965. The precambrian geology of South-West Africa and Namaqualand. Precambrian Research Unit, University of Cape Town.

Mitchell, S. A. & M. T. Seaman, 1988. Observations on the coexistence of fresh and saltwater invertebrates in an inland saltworks. J. limnol. Soc. sthn Afr. 14: 121–123.

Musil, C. F., J. O. Grunow & C. H. Bornman, 1973. Classification and ordination of aquatic macrophytes in the Pongolo River pans, Natal. Bothalia 11: 181–190.

Prior, B., 1978. Properties of two halophilic bacteria from a salt pan. Water S.A. 4: 119–124.

Schoeman, F. R. & P. J. Ashton, 1982. The diatom flora of the Pretoria Salt Pan, Transvaal, Republic of South Africa. Bacillaria 5: 63–99.

Seaman, M. T., P. J. Ashton & W. D. Williams, 1991. Inland salt waters of southern Africa. Hydrobiologia 210: 75–91.

Shannon, L. V., 1985. The Benguela ecosystem. Part 1. Evolution of the Benguela, physical features and processes. Oceanogr. mar. Biol. Ann. Rev. 23: 105–182.

Shaw, P. A., 1988. Lakes and pans. In: B. P. Moon & D. F. Dardis (eds), The geomorphology of southern Africa. Southern Book Publishers, Johannesburg.

Silberbauer, M. J. & J. M. King, 1991. The water chemistry of selected wetlands in the south-western Cape Province, South Africa. sthn Afr. J. aquat. Sci. 17: 82–88

Symoens, J.-J., 1968. La minéralisation des eaux naturelles. Exploration hydrobiologique du bassin de lac Bangweolo et du Luapula. 2: 1–199.

Talling, J. F. & I. B. Talling, 1965. The chemical composition of African lake waters. Int. Revue ges. Hydrobiol. 50: 421–463.

Tankard, A. J., M. P. A. Jackson, K. A. Eriksson, D. K. Hobday, D. R. Hunter & W. E. L. Minter, 1982. Crustal evolution of South Africa. Springer-Verlag.

Truswell, J. F., 1970. An introduction to the historical geology of South Africa. Purnell.

Tyson, P. D., 1986. Climatic change and variability in southern Africa. Oxford University Press.

van der Merwe, C. R., 1962. Soil groups and subgroups of South Africa. Science Bulletin no. 356, Chemistry Series no. 165. South African Department of Agricultural Technical Services.

Van Vliet, H. R., P. L. Kempster, D. P. Sartory, F. A. Gerber & I. J. Schoonraad, 1988. Analytical methods manual. Hydrological Research Institute Report no. TR136. South African Department of Water Affairs, Pretoria.

Watson, A., 1979. Gypsum crusts in deserts. J. arid Envir. 2: 3–20.

Weir, J. S., 1966. Seasonal variation in alkalinity in pans in central Africa. Hydrobiologia 32: 69–80.

Weir, J. S., 1969. Studies on central African pans. III. Fauna and physico-chemical environment of some ephemeral pools. Hydrobiologia 33: 93–116.

Williams, W. D., 1964. A contribution to lake typology in Victoria, Australia. Verh. int. Ver. Limnol. 15: 158–163.

Wood, R. B. & J. F. Talling, 1988. Chemical and algal relationships in a salinity series of Ethiopian inland waters. Hydrobiologia 158: 29–67.

Formation of manganese oxyhydroxides on the Dead Sea coast by alteration of Mn-enriched carbonates

Aminadav Nishri[1] & Arie Nissenbaum[2]

[1] *Kinneret Limnological Laboratory, Israel Oceanographic & Limnological, Research Company, P.O.B. 345, Tiberias, Israel 14102;* [2] *Weizmann Institute of Science, Rehovot, Israel 76100*

Key words: Manganese, Dead Sea

Abstract

Manganese enriched carbonates preferentially accumulate in near-shore, shallow water sediments of the Dead Sea. These carbonates are formed by coprecipitation of Mn with authigenic aragonite, as well as by direct precipitation of Mn-carbonate from the pore water of the shallow sediments. The primary source of the Mn that accumulates as carbonates is allochthonous Mn-enriched oxides that are eroded from the nearby coasts and become buried within the near-shore shallow water sediments. Due to the decline in the level of the Dead Sea between 1960–1990, bands of sediments (parallel to the current shoreline) which were previously submerged, became exposed to air and consequently desiccated. We suggest that in order to approach new hydraulic equilibrium in some of those coastal areas, the decline in the level of the lake was followed by a lakeward advance of fresher groundwater from the shallow coastal aquifer. Those fresher waters are characterized by a higher pH than the interstitial brine, and therefore a new state of water-rock interaction is established which results in oxidative alteration of Mn-carbonates to Mn-oxides. In addition, manganese-oxidizing bacteria, shown to be active in water with lower salinity than that of the Dead Sea, may also play a part in oxidation of divalent manganese released from the sediment. As a result, some segments of the Dead Sea coast are characterized by black Mn-enriched sediments that in places form crusts over the surface.

Introduction

The gradual decline of the level of the Dead Sea between 1960 and 1990 resulted in the exposure of a supralittoral sediment pavement parallel to the shoreline. At several localities this pavement is more than two km wide and consists primarily of detrital boulders and pebbles cemented by aragonite (Garber, 1980).

Along this pavement there are several areas that are covered by black, Mn enriched crusts. The surface crusts are often the outer manifestation of a layered structure with alternating white aragonite layers and black, Mn-enriched layers.

The black laminae are usually a few mm to a few cm thick (Fig. 1). They form continuous hard crusts on pebbles and boulders as well as a laminae alternating with aragonite within accretionary boulders. No such black laminae were found in the deep basin sediments of the Dead Sea and only detrital manganese-rich particles are found in the bottom sediments of the near-shore shallow areas. Several studies have been conducted on the origin of the black layers. However, no completely satisfactory mechanism for their formation has yet been proposed.

Garber *et al.* (1980) showed, by SEM, that the manganese-enriched grains form discrete struc-

Fig. 1. A cross section through a multi-layer accretionary boulder: Sand (detrital) in center. White laminae are araongite: Black-Mn enriched material. Real size: 6 × 4 cm.

tureless aggregates that are embedded in an aragonite matrix. They also showed, by electron microprobe analysis, that the distribution of Mn correlates with that of Al and of Si, which presumably reside in clay minerals. About 10% of the black material can be leached by 0.1 M hydroxylamine in 0.01 M HNO_3 and therefore is considered to be in an easily reducible phase. Manganese (50%) and iron (10%) are the major components of this phase. About 15% of the black material is in an insoluble residue (6N-HCl) which contains mostly quartz and clays.

Infrared spectra of the black material were obtained by computer generation of the difference between the spectra of the carbonate-free black laminae leached with 0.1 M hydroxylamine in 0.01 M HNO_3 and the unleached carbonate-free black laminae (Nishri, 1983). The resulting differential IR spectrum of the black material resembles that of MnO_2.

Based on the visible resemblance of the accretionary boulders to stromatolites, Druckman (1981) proposed that the black Mn-enriched layers are formed by algae that oxidize manganese dissolved in Dead Sea water. However, in thin sections, Garber *et al.* (1981) could not identify algal textures within these layers. Ehrlich & Zapkin (1981) could not grow Mn-oxidizing bacteria from Dead Sea sediments in a culture prepared in a synthetic Dead Sea water medium and concluded that the salinity of the water is too high for these organisms to survive. In a later publication, Ehrlich & Zapkin (1984) reported the occurrence of spore-forming manganese-oxidizing bacteria in fresh waters from the Dead Sea coast and in the black laminae. They proposed that the manganiferous layers were formed, at least in part, by bacterial activity in a fresh water environment along the coast.

Nishri (1983; 1984) determined the value of the first order rate constant for the oxidation of dissolved divalent manganese, Mn^{+2}, in Dead Sea water and concluded that it is much too small to account for substantial direct removal of dissolved Mn^{+2} from the water column by oxidation followed by precipitation of the oxidation products.

The two major questions that this study addresses are: a) what are the sources of Mn for the coastal deposits, and b) what are the particular conditions on the coasts that differ from those prevailing in the lake sediments which enable the formation of the black layers?

Materials and methods

Sampling

Sediment cores and grab samples were collected during 1978–1980. The cores were sealed at both ends and transported, in an upright position, until opened in the laboratory (within 72 hours). The data given in this report are for cores that represent the different depositional environment in the Dead Sea. Cores GF and EG-1 represent deep

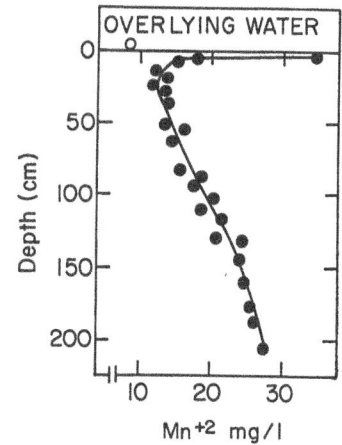

Fig. 3. Pore water-Mn profile in core GF, Deep basin.

basin and shallow basin sediments respectively. Core BS-3 was collected several hundred meters offshore from the 1980 shoreline of Wadi Tseelim (for location map of sampling sites see Fig. 2). Cores BS-1 and BS-2 were taken directly on the shoreline (May 1980), 15m apart, in the coastal area of Wadi Tseelim (Fig. 3).

Analysis

After the sediment cores were sectioned, the samples were centrifuged for 20 minutes at 20 °C and 15 000 rpm. The supernatant pore waters were then filtered through a 0.45 μm membrane filter and diluted appropriately for chemical analysis. The 'punch-in pH' was determined, before sectioning, by inserting a small glass calomel electrode into the wet sediment. The electrode used was a GK-2321 type coupled to a pH-M62 Radiometer pH-meter. After each set of three measurements, the electrode was re-calibrated with standard NBS buffers. Upon inserting the electrode into the sediment, the reading usually stabilized within less than two minutes. Pt electrode readings were obtained simultaneously by a method similar to that of Emerson (1976), and the results are expressed in an arbitrary mV scale, with reference to aerated Dead Sea water.

Due to the liquid junction potential involved, the extract meaning of pH measurements in such a highly saline brine is still somewhat controversial (Ben-Yaakov & Saas, 1977; Krumgalz &

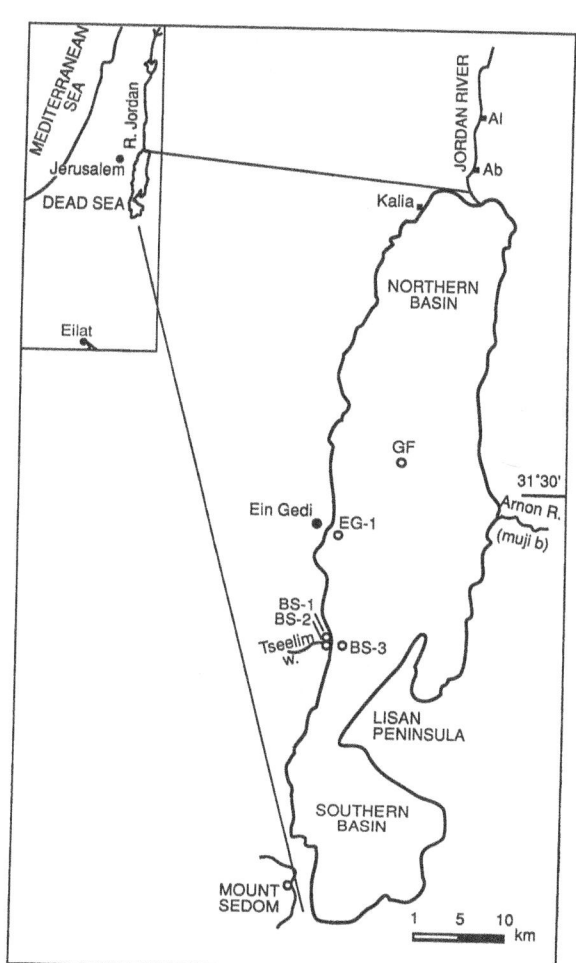

Fig. 2. Location of sampling sites.

Millero, 1982). The pH readings reported here are relative to the infinite dilution scale.

For analysis of sediments, the centrifuged sediments were first dried, ground and then leached at least four times with double distilled water in order eliminate the entrapped brines. The washed sediments were dried again, weighed and stored in a desiccator until analysed.

As shown by Nishri (1983), removal of dissolved Mn^{+2} by air oxidation during centrifugation and filtration is negligible in the pH range of Dead Sea sediment pore water, Analysis of dissolved Mn was done within several hours after core opening. Mn, Na, Ca, K and Li were measured by atomic absorption spectrometry. Li was measured on a 20-fold dilution of Dead Sea water using artificial Dead Sea water as a matrix for the Li standards. Cl was determined by $AgNO_3$ titration. Analytical uncertainties are: Li: 1% Mn, Ca: 2%, Mg, K, Cl: 3% (Dead Sea water contains several g l^{-1} Bromine that is included in Cl values) and Na: 5%. The chemical data are presented either in grams per liter or in moles per liter, as appropriate.

The distribution of manganese in different mineralogical phases was determined by sequential chemical attack on the sediments, in the following order:

(a) leaching with 1 M acetic acid – to dissolve the carbonate fraction (Chester & Hughes, 1967);

(b) leaching with 0.1 M hydrocylamine hydrochloride in 0.01 M HNO_3 to dissolve the socalled easily reducible fraction (Chaw, 1972);

(c) leaching with hydrogen peroxide in 6% HNO_3 to dissolve the organic and sulfide (oxidizable) phases (Jackson, 1958);

(d) leaching with 0.04 M hydroxylamine hydrochloride in 25% acetic acid to dissolve the socalled moderately reducible phases (Chester & Hughes, 1967).

Between each leaching step the sediment samples were rinsed with deionized water. The concentration of total, non-silicate associated manganese in the sediment was determined by aqua regia dissolution.

Results and discussion

The chemistry of the pore waters and the Mn profiles in the sediments depend on the conditions that prevail under the different sedimentary environments of the Dead Sea bottom sediments. The data are presented here according to these environments.

Subaqueous lake sediments

Deep basin: Core GF (320 m water depth)

Pore water and sedimentary Mn profiles shown in Fig. 3 are typical for the deep basin sediments. There is a relatively steady increase with depth in the concentration of dissolved manganese in the pore water and no maxima appear below the sediment-water interface. The 'punch-in pH' is about 5.65 (± 0.2 pH units) which typical for Dead Sea bottom sediments.

Most of the Mn in the sedimentary column is associated with the carbonate phase. Less than 5% of the Mn is associated with the easily reducible phase (Table 1).

Shallow basin: Core EG-1 (25 m water depth)

This core was sampled in February 1978, about 1.5 km offshore, opposite Ein-Gedi (Fig. 2), where the surface of the sediments was oxidized. The sediments consisted of 50–60% carbonates, of which 10 to 80% is aragonite with some calcite (which is probably detrital: Garber, 1980). Most of the Mn in the sediments (about 80%, Table 2) is associated with the carbonate fraction. The distribution of iron in these mineralogical phases differs from that of manganese. In many layers most of the iron is associated with a non-carbonate phase. Local rates of sedimentation were estimated by Stiller & Chung (1984) to be about 0.1 cm y^{-1}.

Pore water profiles of Mn, Fe^{+2}, titration alkalinity and calculated $CO_3^{=}$, as well as Pt electrode readings and 'punch in' pH in core EG-1 are given in Fig. 5.

The concentration of dissolved CO_3 was cal-

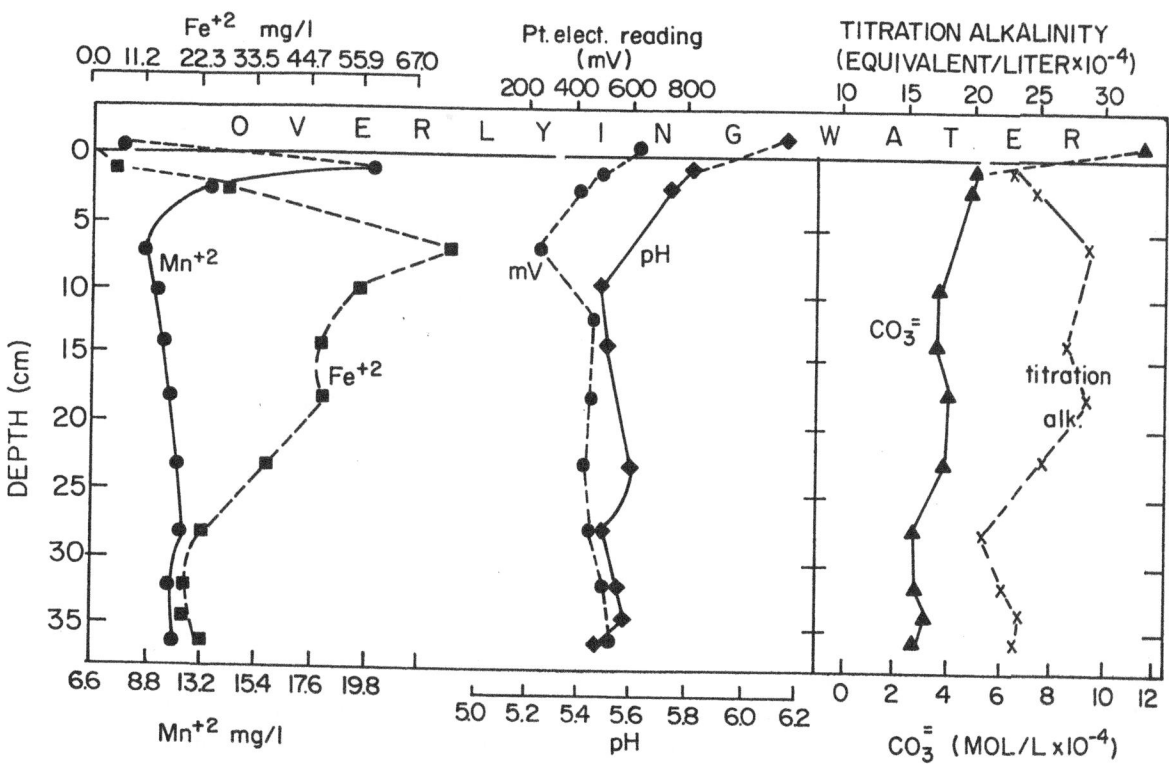

Fig. 4. Pore water profile in core EG-1: Mn^{+2}, Fe^{+2}, 'punch-in' pH, Pt electrode reading, titration alkalinity and calculated CO$_3$ = for (for details text).

culated by using the apparent solubility constants of the carbonate system in Dead Sea water (Sass & Ben-Yaakov, 1977) and by assuming that the

predominant protolytic system in the pore water, as well as in the overlying Dead Sea water, is the carbonate system. As yet this assumption cannot

Table 1. Manganese (Mn) concentration in different mineralogical fractions along core GF. C – carbonate fraction; E – easily reducible phase; S – oxidizable phase; M – moderately reducible phase.

Sediment layer depth (cm)	Mn concentration in (ppm)				Total* C + E + S + M	Total**	Ratio of total* over total** (in %)
	C	E	S	M			
0.0–7.0	102	7	28	29	166	158	105
7.0–8.0	–	–	–	–	–	88	–
8.0–12.5	150	8	33	32	222	254	88
12.5–14.0	–	–	–	–	–	248	–
14.0–17.0	159	4	44	38	245	–	–
17.0–20.4	–	–	–	–	–	208	–
20.4–26.0	132	10	69	30	242	248	98
26.0–30.0	–	–	–	–	–	221	–
30.0–41.0	192	8	33	28	281	280	100
41.0–49.2	129	9	36	32	227	233	97

* Calculated by summation of the concentration of Mn in the four phases.
** Measured after 'aqua-regia' dissolution.

Table 2. Managanese (Mn) concentration in different mineralogical fractions along core EG-1. C-carbonate fraction; E – easily reducible fraction; S – sulphide-oxidizable fraction; M – moderately reducible fraction.

Depth of layer (cm)	Mn concentration (ppm)			
	C	E	S	M
0.0–1.7	305	7	20	32
1.7–4.7	150	5	4	27
4.7–9.4	143	4	10	17
9.4–10.7	194	9	10	28
10.7–16.2	189	21	18	7
16.2–20.9	186	19	11	7
20.9–27.9	116	39	8	–
27.9–30.5	300	17	40	–
30.5–33.0	380	79	22	12
33.0–35.8	160	16	13	19
35.8–39.5	105	7	5	10

be completely proven. It should be noted that in core EG-1 the calculated concentration of carbonate in the profile is almost constant, at about 3×10^{-4} M l^{-1}. The corresponding concentration is also constant and identical to that of the overlying water. Since local sediments are 5–50% aragonite, we suggest that the pore waters are saturated with respect to aragonite. This implies, in turn, that the overlying lake waters are highly supersaturated with regard to aragonite (pH of Dead Sea water: ~ 6.23, titration alkalinity = 38.5×10^{-4} M l^{-1}). Occasional 'whitening' events in the Dead Sea, during which aragonite is massively deposited from the water column (Neev & Emery, 1967), support this suggestion. It seems possible therefore that diagenetic precipitation of aragonite takes place within Dead Sea bottom sediments, due to the downward diffusion of dissolved carbonate species.

Sedimentary aragonite is particularly abundant in coastal shallow water areas, near to sources of fresh water influx to the Dead Sea, such as valleys that carry water during the rainy season and perennial springs.

Coprecipitation of manganese with aragonite may explain the association between Mn and the carbonate phase in Dead Sea bottom sediments. The value of the apparent distribution coefficient of Mn in aragonite precipitated from Dead Sea water (*i.e.* the Mn/Ca ratio in The precipitate divided by the corresponding ratio in solution) was determined experimentally by Nishri (1983) at 25 °C to be quite high and equal to 0.23. It is expected therefore that aragonite precipitating from Dead Sea water with a Ca concentration 17.2 g l^{-1} and a Mn concentration between 6 (Dead Sea water) and 160 mg l^{-1} (pore water) would contain between 37 ppm Mn (aragonite precipitated in the water column) and about 1000 ppm Mn (aragonite precipitated in the sediments).

As indicated above, Mn concentrations in pore water extracted from samples collected opposite Wadi Tseelim (Fig. 2) may reach several hundred mg l^{-1}. In such a regime, coprecipitation of Mn with aragonite may yield Mn-enriched aragonite. Data presented in Table 3 shows that the concentration of Mn associated with the carbonate phase in near-coast sediments may exceed 1000 ppm.

Under suitable conditions Mn may precipitate as a pure Mn-carbonate mineral. A plausible mineral is rhodochrosite, $MnCO_3$. In the following section we examine the possibility that this mineral may precipitate from Dead Sea water, by evaluating the ion activity product (IAP) of Mn

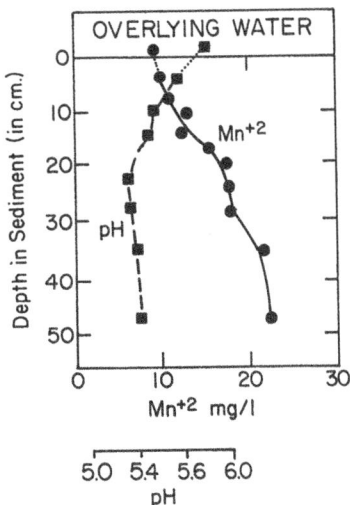

Fig. 5. Pore water-Mn^{+2} and pH profile in core BS-3, shallow basin.

Table 3. Manganese (Mn) concentration in various mineralogical phases of discrete sediment samples. C-carbonate fraction, E – easily reducible fraction, S – sulphide-oxidizable fraction, M – moderately reducible fraction.

Sample site and number	Mn concentration (ppm)			
	C	E	S	M
Stations along the River Jordan				
Ab	301	109	12	63
Al.	289	53	37	13
Coastal sediments (without MnOx enriched black layers)				
Kalia	163	12	7	23
C-1	1166	5	78	12
C-2	383	14	171	55
C-3	1414	243	171	51
Near shore sediments (south of TSE zone)				
S-1	709	30	142	73
S-2	279	28	48	104
S-6	683	83	105	112
P-7	791	104	227	44

and CO_3 in Dead Sea water and comparing it to the thermodynamic solubility of this mineral.

Krumgalz (personal communication), using the Pitzer approach, has calculated several thermodynamic parameters of Dead Sea waters. He found that the total activity coefficients of Ca^{+2} and Mn^{+2} ($yT\ Ca^{+2}$, and $yT\ Mn^{+2}$) in Dead Sea water are 0.879 and 0.517 respectively. The chemical composition of Dead Sea water selected by Krumgalz for modeling was Cl-226 g l^{-1}; SO_4-0.45 g l^{-1}; Mg-44.5 g l^{-1}; Na-41.0 g l^{-1}; K-7.9 g l^{-1}; Ca-17.2 g l^{-1}; density-1.233 g cm^3; temp.-25 °C; pH-6.25 and titration alkalinity of 4×10^{-3} M l^{-1}. Krumgalz also calculated the degree of saturation (DOS) for aragonite in Dead Sea water (the ion activity product (IAP)/the thermodynamic solubility constant ($K_{SO, aragonite}$)), at 25 °C, to be 33.9. However, we are aware that in this calculation the possible contribution of borate species to the titration alkalinity was ignored and that the carbonate alkalinity therefore may be somewhat lower than the titration alkalinity. So we refer to this value of 33.9 as the maximum possible value for $DOS_{aragonite}$ in Dead Sea water. We suggest that $DOS_{aragonite}$ in Dead Sea water can vary between 1.0 and 33.9. On the other hand, the supersaturation of aragonite in the pore water of Dead Sea sediments results in $DOS_{aragonite}$ values ≥ 1.

Since the titration alkalinity in Dead Sea water is negligible compared with the total ionic strength of the solution, we feel that it is safe to assume that the activity coefficients for Ca^{+2} and Mn^{+2}, as mentioned above, are properly evaluated by the Pitzer approach. Assuming saturation with respect to aragonite within the pore waters, the a_{CO_3} ($K_{SO, aragonite} = 10^{-8.21}/a_{Ca^{+2}}$) $= 10^{-7.789}$. At saturation with respect to rhodochrosite ($K_{SO} = 10^{-10.4}$, Li *et al.* 1969), the concentration should be equal to $(10^{-10.4}/10^{-7.789})/$ $\sim 5 \times 10^{-3}$ M l^{-1}) $= 275$ mg l^{-1}. Such high values of Mn are rarely detected in the pore waters of coastal sediments. In the overlying lake water: $10^{-7.789} < a_{CO_3} = < 10^{-6.259}$. At saturation with respect to rhodochrosite and in the extreme case of $a_{CO_3} = 10^{-6.259}$, $a_{Mn^{+2}}$ should be 7.063×10^{-5}, a value that is quite close to a concentration of about 7.5 mg l^{-1}. This is slightly above the concentration of Mn in the Dead Sea (6–7 mg l^{-1}) and it may imply that Dead Sea water is more or less at saturation with respect rhodochrosite.

Near shore and coastal sediments

Manganese enriched incrustations, covering recessional coastal sediment surfaces, are often found several meters landwards from the shoreline, at 10 to 15 cm above the lake level. Between 1960 and 1989, lake level declined by an average of 30 cm per year. This implies that exposed coastal sediments at an elevation of 10 to 50 cm above lake level were submerged two years before collection of samples. We therefore examined the possibility that near shore shallow water sediments are the source of Mn contained within the Mn-oxide incrustations, and that the formation Mn oxide on the exposed coastline occurs by a relatively fast process that is completed within less than a year.

Near shore shallow water sediments

The results of the sequential chemical attack performed on several individual samples taken from near-shore shallow water (up to 1 m water depth), coastal sediments (without black layers) and Jordan River bed sediments are given in Table 3. The sampling sites are shown in Fig. 2.

In all of the near shore samples Mn is preferentially associated with the carbonate phase and its concentration there approaches about 1000 ppm. The Jordan River bottom sediments, shown for comparison, are relatively more enriched with Mn that is associated with the easily reducible phase.

Core BS-3, which is 2 meters long, was sampled about 1 km from the Wadi Tseelim shoreline (Fig. 3) at a water depth of 17 m. Its pore water-Mn profile, shown in Fig. 5, reveals a prominent maximum within 20 mm of the top of the sediment column, probably indicating the local reduction of fresh, allochthonous Mn-oxides to a soluble species.

On-shore littoral cores

Core BS-2

Core BS-2 was collected at the May 1980 shoreline. Its surface was covered by a layer of coarse gravel. Below this layer the sediment is typical of Dead Sea bottom sediments; *i.e.* it consists mainly of carbonates with dark grey and black color, suggesting the presence of iron sulphide.

Pore water profiles of Ca, Li, Mn, Fe^{+2} and pH are given in Fig. 6. Both Li and Ca are depleted relative to Dead Sea water (about 14.5 mg l^{-1} and 17.2 g l^{-1}, respectively). This may be due to the occurrence of a subsurface fresh water reservoir that is mixed with the entrapped brines. The punch-in pH profile along this core is relatively constant at about 5.5 and is typical of Dead Sea bottom sediments. The pore water salinity profile prevailing here probably represents that of the local subaqueous surface sediments just before the decline in lake level, before the now exposed surface began to dry out.

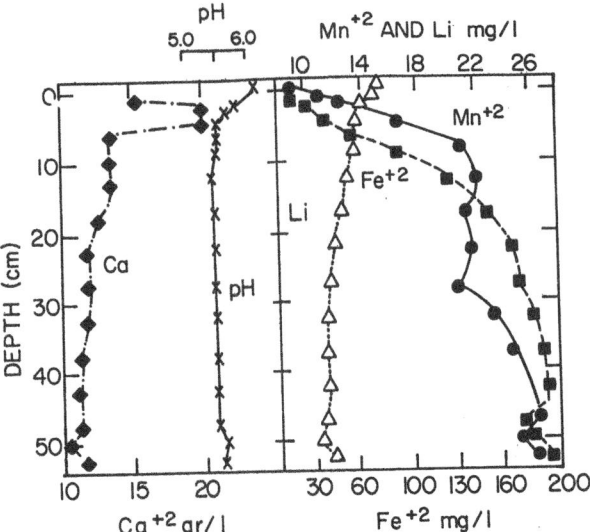

Fig. 6. Pore water profile of Core BS-2, sampled at the shoreline (May, 1980): Ca, Li, Mn, Fe^{+2} and 'punch-in' pH.

Core BS-1

Core BS-1 was taken 15 m west of the May 1980 shoreline where the local coastal sediment surface was 22 cm above the lake level Coring was done several weeks after the entire area was inundated by floods originating from the mountains of Judea on the west side of the lake. Remnants of this fresh flood water could still be found in nearby ponds. The sediment in the upper 1–2 cm of the core had a reddish-brown color, due to oxidation of iron sulfides. Between 2 and 60 cm most of the sediment had a jet black color, indicating the occurrence of amorphous acid volatile iron sulfides. Well developed halite crystals were found between 10 and 22 cm. Between 60 and 120 cm the sediment color was mostly grey with several white laminae of aragonite.

Pore water data are presented in Fig. 7. The concentration profiles are the composite result of several processes, among which is diffusion along the pore water sediment column. In the profiles two major zones can be recognized: a) An upper zone, between 0 and 30 cm, and b) the lower zone, below 30 cm.

The upper zone is characterized by steep concentration gradients of the various ions and pH. The uppermost zone of the sediment is charac-

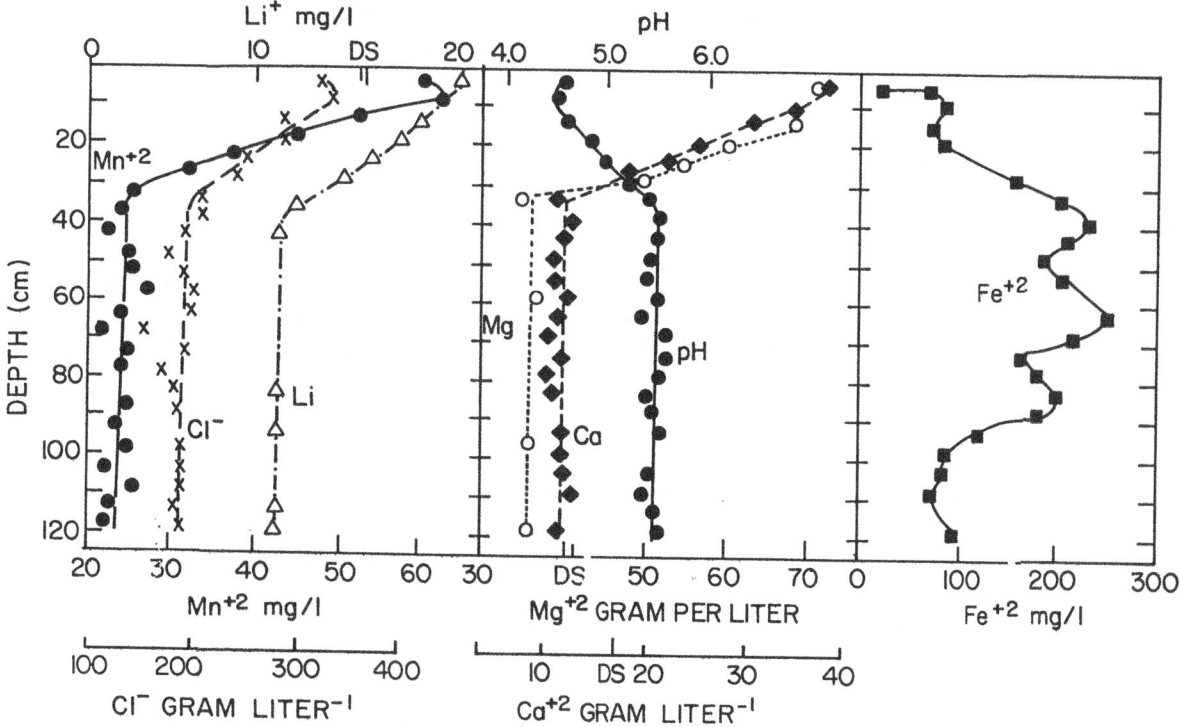

Fig. 7. Pore water profile of core BS-1, sampled on the Dead sea coast: Mn, Fe^{+2}, Li, Ca, Mg, Cl and 'punch-in' pH profile. Dead Sea denotes, average Dead Sea composition.

terized by higher concentrations of Mn, Li, Ca, Mg and Cl than are found in Dead Sea (Fig. 7) water, implying that the sediments have started to dry out. The occurrence of halite crystals in this zone supports this idea. In this upper layer, the 'punch-in' pH is relatively low, about 4.5. Such decrease in pH was observed during evaporation of Dead Sea brine (Nishri, 1983).

The lower part of core BS-1, below 30 cm, is characterized by relatively constant pore water concentrations of Li, Mn^{+2}, Ca, Mg, Cl and pH. The salinity here is about 15% lower than that of Dead Sea water. Again, such relatively low salinity within coastal cores supports the existence of a fresher body of water underlying the sediments that mixes with the residual brines occluded within them.

Katz *et al.* (1977) found that only trace amounts of lithium coprecipitate with NaCl (halite) during evaporation of Dead Sea water. This was also found for coprecipitation of Li and carnallite (Nishri *et al.*, 1988). Thus, Li is considered here

as a conservative ion in the sediments of core BS-1, *i.e.* it is not being removed from solution to the solid phase due to evaporation of Dead Sea water. We therefore, use the Li concentration profile to show the extent of evaporation/dilution of pore water in core BS-1.

Assuming that vertical migration is negligible, the ratio between dissolved Li concentration in the upper part of the pore water column to that of the lower part $((Li)_{up}/(Li)_{low})$ indicates the degree of evaporation in the upper zone. It is now possible to evaluate the criteria for determining whether a certain dissolved element (x) in pore water is relatively enriched or depleted above or below that produced by evaporation, by defining the parameter β:

$$\beta = \frac{(x)_{up}/(x)_{low}}{(Li)_{up}/(Li)_{low}} \; .$$

If at a certain depth $\beta > 1$, then element x is enriched as compared with Li and must have been

70

contributed from the solid phase. If $\beta < 1$ then element x is depleted, probably by removal into the solid phase. The β values, presented in Fig. 8, indicate that:

(a) chloride is slightly depleted in the upper part of the pore water column, probably due to the precipitation of halite:

(b) dissolved Ca, Mn and, to a lesser extent, Mg are enriched in the upper layer, suggesting that these elements are probably leached from a solid phase possibly (Mn enriched carbonates).

One of the working hypothesis in the present study is that Mn-oxide formation takes place on the dry land along the coast. If true, then this process must be completed within less than a year, because these incrustations are found in areas that several months earlier were still submerged below Dead Sea water. The increase in salinity of the residual evaporating brine is accompanied by a decrease in pH and thus reduces even more the chances for oxidation of Mn^{+2} in the drying coastal areas.

In Dead Sea water diluted by fresh water the rate of chemical oxidation of Mn^{+2} is faster than

in undiluted Dead Sea water (Nishri, 1983). This effect is due to an increase in pH of the diluted brine, as compared to the original brine. Such a mixture may also be favorable for bacterially mediated oxidation of Mn^{+2}, an idea supported by Ehrlich & Zapkin (1984) who succeeded in growing spore-forming Mn^{+2} oxidizing bacteria from Dead Sea sediments in a culture prepared in a fresh water media, and therefore suggested that these bacteria may contribute to the formation of manganese incrustations along the Dead Sea shores.

In either case, whether the oxidation is bacterially or chemically mediated, two conditions must be fulfilled before substantial oxidation of Mn^{+2} can take place. First, at the site of oxidation there must be a large enough source of dissolved Mn^{+2} for the rather thick incrustations found, and, second, the oxidation process must take place in diluted Dead Sea water that has a relatively high pH.

Fresh water floods along the Dead Sea coast are not very frequent, whereas 'fresh' Mn incrustations are common in many coastal localities. For example, during 1989–1990 while the recession of the Dead Sea continued, there were no freshwater floods along the coastal belt and yet 'fresh' incrustations formed there. We therefore propose that the medium that facilitates oxidation of Mn^{+2} is not necessarily a mixture of Dead Sea water and fresh flood water that infiltrates from above but rather a mixture of residual brine with fresh water that occurs at some depth within the sediments.

The possible occurrence of relatively fresher water within shallow aquifers along the Dead Sea coasts is evident from cores BS-1 and BS-2. Also, Yechieli *et al.* (1992) showed the penetration of fresh groundwater and flood water into shallow perched horizons in the Tseelim area. It is expected that a further fall in the level of the Dead Sea would be accompanied by a change in the hydraulic regime, resulting in a lakeward advance of the fresher water within these aquifers. A schematic model is presented in Fig. 9. Under these conditions there would be competition between the rate at which the sediment surface is dried by

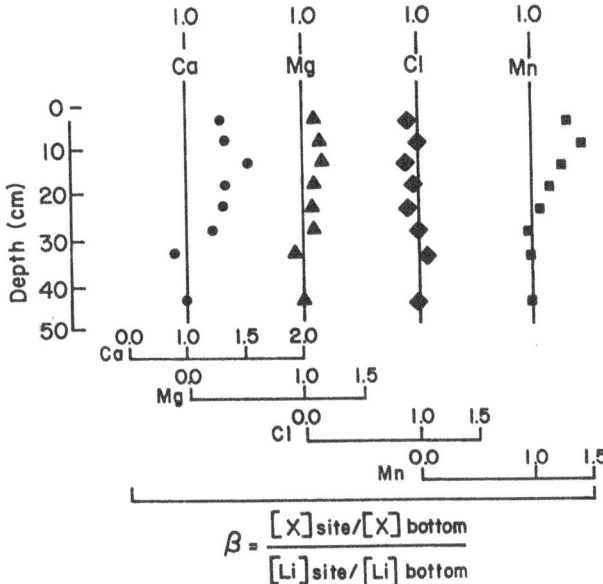

$$\beta = \frac{[X]site/[X]bottom}{[Li]site/[Li]bottom}$$

Fig. 8. Beta (β) values for core BS-1, sampled on the Dead Sea coast.

← LAKE   COAST →

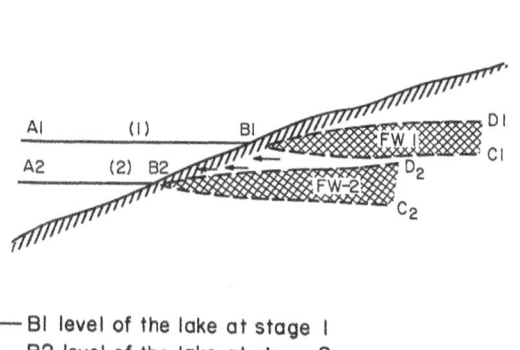

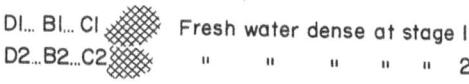

A1——B1 level of the lake at stage 1

A2——B2 level of the lake at stage 2

B1---C1 Interface between fresh and saline ground water
         at stage 1

B1   D1 Phreatic interface at stage 1

B2   D2 Phreatic interface at stage 2

D1...B1...C1 Fresh water dense at stage 1
D2...B2...C2 "   "   "   "   " 2

B1   C1 Interface between fresh and saline ground water
         at stage 1

B1   B2 Coastal surface exposed due to fall in lake level
         at stage 1

◀▬▬ Migration of ground water due to fall in lake level

Fig. 9. A schematic model presenting the advance towards the Dead Sea of fresher water from shallow aquifers due to the decline in the level of the lake.

evaporation and the rate of introduction of the less saline ground water. In our specific research area evaporation dominates; in other localities the advance of fresh water may dominate. If the second mechanism dominates there will be less saline brine in contact with a sediment that formed previously under the more saline conditions of the Dead Sea.

It has been shown that the natural redox chemistry of manganese can be influenced by sunlight. Photochemical reactions (Sunda and Huntsman, 1990) increase the rate of reductive dissolution of Mn oxides and decrease the rate of oxidation of divalent Mn. We are not able to evaluate properly the role of photochemical processes in the Dead sea coast, except by noting that it is unlikely that such processes can influence reactions that occur inside the accretionary boulders, although they might affect surface reactions.

The alteration model

Hager (1980) examined the interaction between dissolved manganese (as highly concentrated $MnSO_4$), and silica, clay minerals and carbonates. She found that aragonite was altered to manganese carbonate within 6 months at 25 °C, and within one week at 70 °C. The addition of smectite served as a substrate for manganese carbonate precipitation in the presence of carbonate. The manganese oxy-hydroxides hausmannite (Mn_3O_4) and manganite (MnOOH) were formed by direct nucleation on smectite through oxidation of manganese carbonate previously formed at a similar temperature-dependent rate. Hager (1980) suggests that (a) the mechanism of conversion of aragonite to manganese carbonate involved the initial adsorption of manganese onto the surface of the aragonite, followed by dissolution of this phase and (b) oxidation of the Mn-carbonate into the Mn-oxyhydroxide is to be expected, reflecting an approach to chemical equilibrium predicted by Eh-pH diagrams.

There are several reasons to suggest that Hager's (1980) alteration model may also explain the formation of Mn oxide along parts of the Dead Sea coast. First, as shown above, most of the manganese that is contained in Dead Sea nearshore sediments is associated with the carbonate phase whereas the black Mn-enriched coastal layers are associated with the easily reducible phase. Second, the mineral assemblage composing the Dead Sea coastal incrustations of Mn-oxides is associated with clay minerals which are embedded in an aragonite matrix. This resembles the conditions used during Hager's alteration experiments. Third, the time scale obtained in Hager's experiments of one week to several months for oxidation of Mn-carbonates is similar to that assumed here for formation of Mn-oxides on the Dead Sea coast.

In summary, we suggest the following mechanisms: After the fall in the Dead Sea water level the interstitial brine is replaced by fresher water from very shallow aquifers. The mixture of fresher water with Dead Sea residual brine comes into contact with Mn enriched carbonates that are common in near-shore sediments. These fresher waters may have higher pH, thus facilitating direct oxidation of Mn-carbonates to manganese oxyhydroxides, as suggested in Hager's model.

The concentration of manganese in the black layers exceeds by an order of magnitude the concentration of manganese in near-shore carbonates. This raises the question of whether there is enough manganese in the carbonates to form the black, manganese-rich layers. This may be due to our analytical procedures that compare manganese in separated black layers to that of the total manganese in the bulk sediments. It may well be that manganese-rich layers exist in the sediments, but they are mixed with detrital, manganese-poor components.

Conclusions

The following model is proposed: Mn-enriched carbonates are formed in near shore sediments of the Dead Sea. The source of Mn is allochthonous Mn-oxides that either have been transported from the neighboring highlands or are erosional products of the coastal sediments. Once buried within the hostile (lower Eh-pH) environment of the Dead Sea sediments these oxides are reduced and the dissolved manganese goes into the pore water. This Mn^{+2} may diffuse upwards to the overlying lake water due to the concentration gradient. If Mn^{+2} concentration in the pore water exceeds about 270 mg l^{-1} it will be in equilibrium with rhodochrosite that could be formed diagenetically. Aragonite precipitates at the sediment-water interface due to introduction of HCO_3^- bearing fresh water within near coast sediments, by floods issuing from the outlets of dry valleys during the winter rainy season. Part of the Mn^{+2} that diffuses upward may coprecipitate with this aragonite. Both processes lead to the preferential accumulation of Mn carbonates.

The decline in the level of the lake, between 1960–1989, caused a lakeward advance of relatively fresher ground water. The Mn-enriched sediments may, therefore, be in contact with these fresher media. Alteration through direct oxidation of Mn-carbonates to Mn-oxides, as suggested by Hager (1980), may facilitate the formation of the black Mn-oxide layers. Further decline of the level of the lake provides a 'fresh' source of Mn-oxides that may be subsequently re-eroded and transported to the 'newly formed' nearshore bottom sediments, and so, for as long as the lake level declines, the process continues to accumulate additional Mn in the incrustations. A complementary pathway could be the bacterial oxidation of manganese. As shown by Ehrlich & Zapkin (1984) these bacteria are active at only salinities lower than that of the Dead Sea. Thus, the lake level decline and the encroachment of fresh waters (Fig. 9) provide the appropriate environmental conditions for oxidizing the divalent manganese that is released from the sediments. As Fig. 8 shows, pore waters from core BS-1 are substantially enriched in both manganese and calcium, suggesting that Mn-carbonates are unstable in salinities lower than that of the Dead Sea and thus provide the source of reduced manganese that is later oxidized either biologically or non-biologically or by both processes.

Therefore, we suggest that the formation of Mn-oxide enriched incrustations may take place at a certain locality only while the following conditions are fulfilled:

(a) The localized composite result of the two processes of drying out of the sediment surface on one hand, and lakeward advance of less saline water from shallow aquifers.
(b) There is a source of Mn-oxides and means of transportation of these oxides to the lake, such as intensive surface erosion.
(c) Mn-enriched carbonates were previously formed subaqueously in local near-shore sediments

Coastal areas that are situated at the inlets of major dry valleys on the Dead Sea may satisfy these three requirements. Other coastal areas, such as those where cores BS-1, BS-2 were taken, do not satisfy these requirements and no manganese oxides were found there.

Acknowledgements

We thank Dr B. Krumgalz (IOLR) for sharing with us unpublished data. The Dead Sea Works Ltd. are warmly thanked for their financial support.

References

Ben Yaakov, S. & A. Saas, 1977. Independent estimate of the pH of Dead Sea brine. Limnol. Oceanogr. 22: 374–376.

Chaw, T. T., 1972. Selective dissolution of manganese oxide from soil and sediments with acidified hydroxylamine hydrochloride. Soil Sci. Soc. Proc. 36: 764–768.

Chester, R. & M. J. Hughes, 1967. A chemical technique for the separation of ferro-manganese minerals, carbonate minerals and adsorbed trace elements from pelagic sediments. Chem. Geol. 2: 249–262.

Druckman, Y., 1981. Subrecent manganese-bearing stromtolites along the shorelines of the Dead Sea, In: Phanerozoic Stromatolites. C. Monthy, (ed.) Springer Verlag, Berlin: 197–208.

Ehrlich, H. L. & M. A. Zapkin, 1981. Mn^{+2} oxidizing bacteria from the Dead Sea region of Israel. Abstr. Amm. Mtg. Amer. Soc. Microbiol. MGO, p. 183.

Emerson, S., 1976. Early diagenesis in anaerobic lake sediments: chemical equilibria in interstitial waters. Geochim. Cosmochim. Acta 40: 925–934.

Garber, R., 1980. The sedimentology of the Dead Sea. unpublished PhD thesis, Rensselaer Polytechnic Institute, Troy, N.Y.

Garber, R., A. Nishri, A. Nissenbaum & G. M. Friedman, 1980. Modern deposition of manganese along the Dead Sea shore. Sed. Geol.: 257–274.

Hager, L. H., 1980. Sorption of manganese and silica by clay and carbonate. Mar. Chem. 9: 194–209.

Jackson, M. L., 1958. Soil Chemical Analysis. Prentice Hall, Englewood Cliffs, N.J.

Katz, A., A. Starinsky & N. Teitel-Goldman, 1979. The solubility of Gypsum in the Dead Sea. Limnol. Oceanogr. 26: 709–716.

Krumgalz, S. B. & F. J. Millero, 1982. Physico-chemical study of Dead Sea waters. I Activity coefficients of major ions in the Dead Sea. Mar. Chem. 11: 209–222.

Levy, Y., 1981. Suspended matter in the Dead Sea. Report No. MG/1981. The Geological Survey of Israel. Jerusalem, Israel.

Li, Y. H., J. Bischoff & G. Mathieu, 1969. The migration of manganese in the arctic basin sediments. Mar. Geol. 3: 457–474.

Neev, D. & K. Emery, 1967. The Dead Sea depositional processes and environment of evaporites. Geol. Survey Israel. Report. No. MG/41/87.

Nishri, A. 1983. The geochemistry of manganese and iron in the Dead Sea. Unpublished. PhD thesis, The Weizmann Inst. of Science, Israel.

Nishri, A., 1984. The geochemistry of manganese in the Dead Sea. Earth and Planet Sci. Lett. 71: 415–426.

Nishri, A., H. J. Herbert, N. Jockwer & W. Stichler, 1988. The geochemistry of brines and minerals from the Asse salt mine, Germany. Appl. Geochem. 3: 317–332.

Saas, A. & S. Ben Yaakov, 1977. The carbonate system in hypersaline solutions: Dead Sea brines. Mar. Chem. 5: 183–199.

Starinsky, A., 1974. Relationship between Ca-chloride brine and sedimentary rocks in Israel. Unpublished PhD thesis, the Hebrew University of Jerusalem, Israel (in Hebrew).

Stiller, M. & Y. C. Chung, 1984. Radium in the Dead Sea: a possible tracer for the duration of meromixis. Limn. Ocean. 29: 574–586.

Sunda, W. G. & S. Huntsman, 1990. Diel cycles in microbial manganese oxidation and manganese redox speciation in coastal waters of the Bahama Islands. Limn. Ocea., 35: 325–338.

Yechieli, Y., Magaritz, M., Shatkay, M., Ronen, D. & I. Carmi, 1992. Processes affecting interstitial water in the unsaturated zone at the newly exposed shore of the Dead Sea, Israel. Isotope Geoscience (in Press).

Morphology, distribution, and preservation potential of microbial mats in the hydromagnesite-magnesite playas of the Cariboo Plateau, British Columbia, Canada

Robin W. Renaut

Department of Geological Sciences, University of Saskatchewan, Saskatoon, Saskatchewan, S7N 0W0 Canada

Key words: Saline lakes, playas, mudflats, microbial mats, stromatolites, microbolites, hydromagnesite, magnesite, dolomite

Abstract

Benthic microbial mats are common in the alkaline hydromagnesite-magnesite playa lakes of Interior British Columbia. Four main zones are recognized based on mat morphology that can be related to the type and duration of wetting. From the basin margin toward the playa centre they are: (i) vegetated hummocky ground; (ii) polygonal hummocky ground; (iii) low-domal and stratiform mats, and (iv) laterally continuous and pustular mats. Mats in peripheral mudflats are commonly mineralized by hydromagnesite, mostly precipitated by capillary evaporation of shallow groundwaters. Mats forming in the ephemeral lake tend to have lower carbonate content.

Although widespread, the mats are poorly preserved in the Holocene sedimentary record. Underlying sediments are commonly weakly bedded, disturbed or massive. Desiccation, dehydration, wetting-drying cycles, and grazing by invertebrates cause fragmentation of mats at the surface, facilitating erosion. Cryogranulation, interstitial mineral precipitation, vesiculation, bioturbation, compaction, and volume changes associated with diagenesis, disrupt and destroy lamination in the upper few centimetres. Most surviving organic matter is lost by early microbial degradation.

Introduction

Microbial mats, constructed by cyanobacteria, bacteria and algae are commonly found in saline lake basins (e.g. Anderson, 1958; Moss & Moss, 1969; Walter *et al.*, 1973; Halley, 1976; von der Borch *et al.*, 1977; Pueyo-Mur, 1978; Bauld, 1981a; Osborne *et al.* 1982; De Deckker, 1983; Casanova, 1986; Hammer, 1986; Last & De Deckker, 1990; Kempe *et al.*, 1991). They occupy a wide range of subenvironments, including zones of groundwater discharge, springs and their outflow channels, streams, lake-marginal mudflats, and are found as various benthic and biohermal forms within the saline lake. The mats may be preserved in the geological record as microbolites (Riding, 1991), defined by Burne & Moore (1987) as 'organosedimentary deposits that have accreted as a result of a benthic microbial community trapping and binding detrital sediment and/or forming the locus of mineral precipitation'. Stromatolites are microbolites with internal lamination (Walter, 1976; Riding, 1991) and are commonly associated with filamentous cyanobacteria. Thrombolites are microbolites with a clotted internal structure and are commonly produced by coccoid forms (Kennard & James, 1986).

Stromatolites have been described from many

ancient saline lake deposits. In many formations they are associated with perennial saline pale-olakes or periods when brine was normally present. They are also preserved in the deposits of ancient ephemeral lakes and playas that were subject to periodic desiccation. Examples are recorded from the Eocene Green River Formation of Wyoming, (Surdam & Wolfbauer, 1975; Surdam & Wray, 1976), the Cambrian of South Australia (White & Youngs, 1980; Southgate *et al.*, 1989), the Jurassic of New England (Demicco & Gierlowski-Kordesch, 1986), the Tertiary of France (Truc, 1978), and many other paleolakes. In some basins, stromatolites are associated with periods of brine freshening; in others they appear to have formed under a wide range of salinities. Morphology has commonly been used to infer water depth and proximity to paleoshorelines (e.g. Smith & Mason, 1991; Surdam & Wolfbauer, 1975).

Modern laminar microbial mats are very common in the playas and saline ephemeral lakes of Interior British Columbia, Canada. Some are partially mineralized by carbonates (both precipitated *in situ* and accretionary) and/or other salts. However, in common with ephemeral lakes elsewhere, most shallow (<1 m) cores recovered from the playa basins show very poor preservation of the microbial lamination. Sediments underlying modern mats are typically crudely bedded, disturbed or massive. Although clots and detrital mineralized fragments of mat (e.g. laminite intraclasts) are occasionally found, only rarely are the mats well preserved as stromatolites.

As part of a broader study of Canadian playas and ephemeral lake basins, an examination is being made of the microbial mats and their role in sedimentation. The main aims of this paper are: (i) to present a preliminary account of the distribution and morphology of the modern mats in one group of saline lakes in British Columbia – the Mg-carbonate playas, and (ii) to outline some processes that limit their potential for preservation in the paleolimnological and geological records.

Environmental setting

Saline lakes are found across much of the inter-montane south-central Interior Plateau region of British Columbia, which lies between the Coast Mountains and the Columbia-Rocky Mountain ranges. They are most numerous on the southern Cariboo Plateau, near the villages of Clinton and 70 Mile House (Fig. 1). The plateau, which lies at an altitude of 1050–1250 m, is a gently undulating surface covered by coniferous forest, broken in places by open grassy meadows.

The Cariboo Plateau is underlain by Neogene basalts (Mathews, 1989) and has a thin (0–5 m) veneer of till and/or glaciofluvial sediments (Campbell & Tipper, 1971). Deglaciation occurred about 10 000 years ago (Fulton, 1984), resulting in extensive meltwater channels that cut through the till and, in places, incised the lava bedrock. The adjacent Marble Range is composed of marine sediments, basic lavas and ultramafic rocks of Permian to Jurassic age (Monger, 1989).

The climate is semi-arid to sub-humid with a mean annual precipitation of 300–400 mm, a total similar to the mean annual moisture deficit (Valentine & Schori, 1980). Mean July temperatures range from 13 to 17 °C, compared to −9 to −11 °C in January. Temperatures can fall to below −40 °C in winter and can exceed 35 °C in summer. Less than 90 days annually are frost-free. Snow and ice usually cover the plateau from November until late March. Winds are predominantly from the west and north.

Drainage on the plateau is highly disordered with few streams, abundant marshy ground and several thousand lakes, ranging from fresh to hypersaline (<1 to >350 g l^{-1}). The saline lakes include playas, small saline pans precipitating natron, epsomite or mirabilite, and perennial saline lakes, several of which are meromictic. Most lie in small, topographically closed basins between linear mounds of till or eskers. Many are clustered along the paleomeltwater channels. The playas and ephemeral lakes are fed directly by groundwater discharge, lake marginal springs and

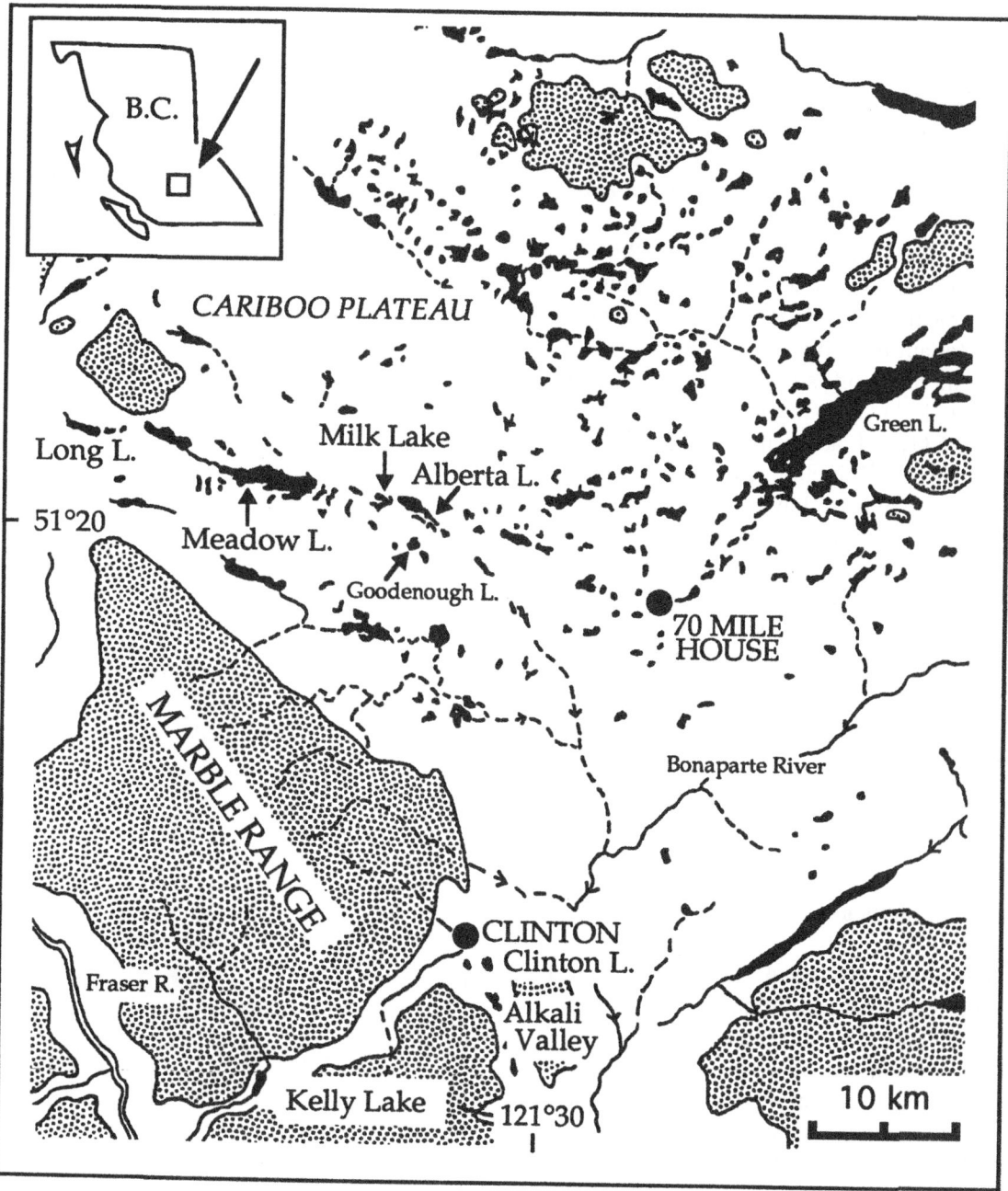

Fig. 1. The southern Cariboo Plateau, showing location of the lakes studied. Shaded area: land more than 1200 m in altitude.

seepages, snowmelt and unchannelled wash. Channelled inflow is rare.

Three main groups of playas and ephemeral lakes that desiccate annually or every few years are found: *Siliciclastic playas* are dry mudflats that are seasonally flooded and are floored by clastic detritus derived mostly from slopewash and small slope failures. Most of the *carbonate playas* are shallow alkaline lakes that desiccate to produce hard, dry mudflats composed predominantly of hydromagnesite and magnesite. The *saline mudflat – ephemeral lake complexes* have peripheral,

78

brine-soaked mudflats composed of siliciclastics and carbonates, including abundant dolomite, that surround a shallow (< 1 m) saline lake or a saline pan. Most playas and ephemeral lakes are small, ranging from < 100 m to a few km in length, and up to ± 1 km² in area. Further details are given in Renaut & Long (1989) and Renaut et al. (in press).

More than 200 analyses of Cariboo Plateau waters have shown a very wide range of salinity, pH and chemical composition (Topping & Scudder, 1977; Renaut & Long, 1987; Renaut, 1990). The main ions in runoff, spring waters and fresh lakes (< 3 g l^{-1}) are usually Mg^{2+}, Na^+, HCO_3^-, and SO_4^{2-}. Lakes with intermediate salinities (3 to 50 g l^{-1}), including most hydromagnesite-magnesite playas, have similar compositions to runoff, but with very high molar Mg/Ca ratios (from 0.7 to > 300). There are three main types of hypersaline brine (50 to > 350 g l^{-1}): (i) highly alkaline brines (pH: 8.5 to 10.5), poor in calcium and magnesium, with $Na-CO_3-(SO_4)-Cl$ composition; (ii) more neutral brines (pH: 7.5 to 8.8), poor in HCO_3 and CO_3, with $Mg-Na-SO_4$ composition; and (iii) $Na-Mg-SO_4-CO_3$ brines (pH: 8.0–9.5) with somewhat lower salinities (20–70 g l^{-1}).

Materials and methods

The saline lakes discussed were visited and sampled four times during different seasons between 1988 and 1991. Water and sediment samples were collected on each occasion. Sediments, including the microbial mats, were collected as grab samples at the surface, from shallow (< 1.2 m) pit and trench sections, and from short cores (3.8 cm diameter, < 1 m long). Mineralogy was determined by X-ray diffraction (XRD) and petrographic methods. Sixteen samples of sediments and mats were examined by scanning electron microscopy (SEM), using a JEOL JXA 8600 electron microprobe, equipped with wavelength and energy-dispersive spectrometers. For determination of percentages of carbonate and organic carbon (TOC), samples of

mud and mat were air dried, ground to < 100 mesh, then digested with hot HCl. Then followed combustion in a Leco induction furnace and measurement of organic carbon as CO_2. Details of other analytical methods are given in Renaut (1990). Determination of the microbiological composition of the mats has not yet been undertaken.

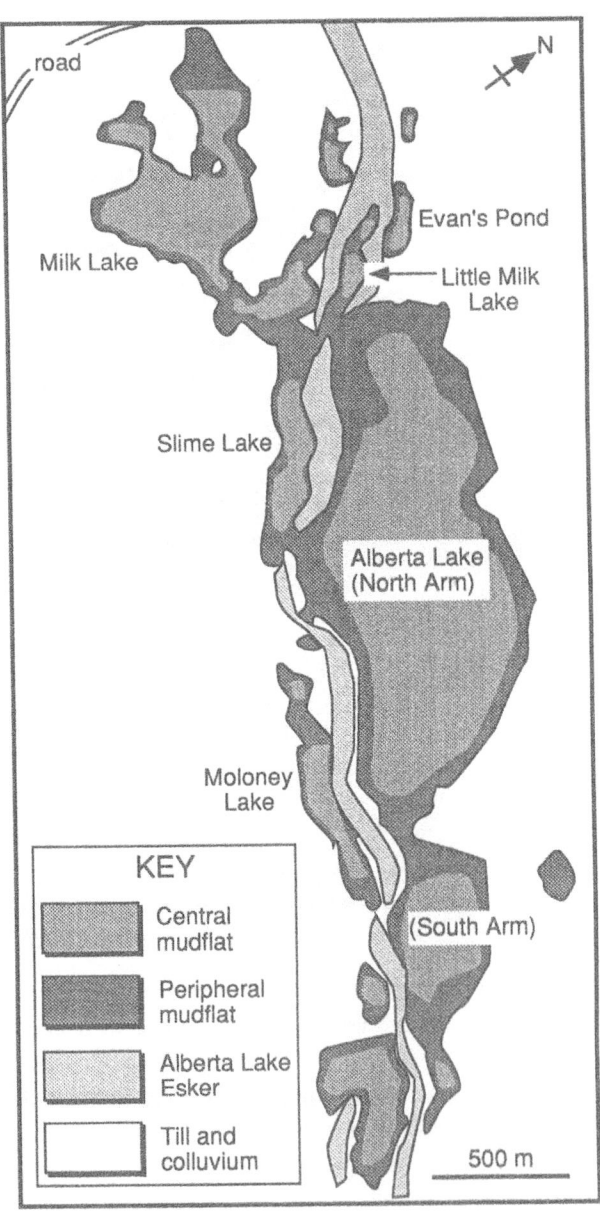

Fig. 2. Map showing the Alberta Lake group and the general setting.

The Mg-carbonate playa lakes of the Cariboo Plateau

Carbonate playas occur across much of the Cariboo Plateau. General descriptions were presented by Reinecke (1920) and Cummings (1940), and were summarized recently by Grant (1987). A large group is found along a NW-SE paleomeltwater channel system that extends from Long Lake, along the axis of the Alberta Lake esker (Fig. 1). Three of these playas, Milk Lake, Alberta Lake and Slime Lake (Fig. 2), have been examined and sampled. The lakes lie at an altitude of ± 1095 m.

Milk Lake (Fig. 6A), which has a surface area of 0.3 km^2, consists of three elongate lobes, and a broad central playa flat. It is confined by low rounded hills of till. Along the eastern edge of the basin, the Alberta Lake Esker stands 10–15 m above the valley floor. There is no channelled inflow, but several areas of spring seepage occur around the shoreline, commonly at the base of headlands composed of sandy till. During desiccation, the eastern lobe becomes isolated from the rest of the lake. Several smaller ephemeral carbonate lakes lie to the northeast, separated by carbonate mudflats with grass or reed cover.

Alberta Lake is a larger (0.95 km^2), elongate playa, 3 km long by up to 750 m wide. The esker forms a steep western edge. The eastern shore is formed by till ridges broken with small slope failures, and with many spring seepages along the base. Slime Lake (Fig. 6C) is a small (0.08 km^2) elongate playa-lake, 500 m long. It is separated from Alberta Lake by the esker, and from Milk Lake by a mudflat. At various stages in their history, all three lakes, together with others to the southeast, likely combined as a single fresher water body.

The lakes are highly alkaline (recorded pH range: 7.4–9.8), and in early summer following winter snowmelt range in salinity from 2 to > 10 g

Table 1. Chemical analyses of waters from the Milk, Alberta and Slime lake group.

Sample[a]	Date	pH[b]	Concentration (mg l^{-1})									
			Na	K	Ca	Mg	HCO$_3$	CO$_3$	Cl	SO$_4$	F	SiO$_2$
Milk lake												
ML-1-88	5–88	7.4	22	15	18	32	270	0	1	1	0.1	21
ML-11-90	6–90	8.4	67	12	5	210	1140	0	5	57	0.6	21
ML-3-89	8–89	7.9	143	3	20	222	1235	0	6	284	0.5	17
ML-2-90	6–90	9.0	4140	385	1	55	2150	3600	275	1250	5.4	33
ML-5-90	6–90	8.6	805	80	2	10	1300	450	31	170	0.4	17
ML-7-88	5–88	8.3	380	42	2	89	1285	110	12	45	0.5	8
ML-16-90	7–90	8.8	1850	225	1	81	990	1920	110	470	4.8	33
ML-5-89	8–89	9.5	3350	385	0	42	2890	2400	195	970	4.3	14
Other lakes												
Little Milk	6–90	9.4	1550	160	<0.5	10	900	1650	62	180	4.5	14
Slime	9–89	9.4	1700	205	2	78	2250	1145	40	285	3.7	14
Alberta 1	7–90	8.7	1785	250	0.5	41	1020	1740	78	310	3.7	21
Alberta 2	7–88	9.1	9880	1160	0	15	11500	7200	450	1180	14.7	42

[a] *Index to samples*: ML-1-88: Marsh, northwest edge of Milk Lake (Zone 1); ML-11-90: Groundwater at 60 cm, peripheral mudflat, south-central shore (Zone 3); ML-3-89: Groundwater at 50 cm, peripheral mudflat, southwest shore (Zone 1–3 boundary); ML-2-90: Groundwater at 50 cm, peripheral mudflat, northwest embayment (Zone 3); ML-5-90: Lake water, 3 m off southern shore; ML-7-88: Lake water, northwest shore; ML-16-90: Lake water, 5 m off northeastern shore; ML-5-89: Lake water, 3 m off southern shore; Little Milk: Lake water, southern shore; Slime: Lake water, northern shore; Alberta 1: Lake water, northwestern shore; Alberta 2: Shallow groundwater (perched?) at depth of 10 cm, north-central playa.
[b] Field pH is given.

1^{-1} (Table 1). Before complete desiccation, salinities may exceed 25 g 1^{-1}. In composition, they are dominated by Na^+, CO_3^{2-} and HCO_3^-. Notable are the very high Mg/Ca ratios (commonly >50:1). These may reflect (i) contact of the inflow waters with subsurface volcanic rocks, and (ii) extensive early precipitation of calcite cements in till and as coatings on gravels (Renaut, 1990). When tested using the PC version of the WATEQF program (Rollins, 1988), most lake waters in spring and summer are theoretically supersaturated with respect to magnesite and dolomite, but undersaturated with respect to hydromagnesite. Calcite and aragonite vary from slightly undersaturated to just saturated.

Depositional subenvironments, stratigraphy and mineralogy

Each playa has three main depositional subenvironments. The centre of the basin is occupied by a *central mudflat* that for a few months annually is covered by a shallow ephemeral lake. This passes transitionally into a *peripheral mudflat* that lies above the average annual maximum lake level. Groundwater below the peripheral mudflat is shallow (<1 m depth) and is locally discharged at the surface as seepage or is lost by capillary evaporation and evapotranspiration. The mudflats extend toward the vegetated *hillslopes* of glacial till or eskers that define the margins of most basins.

Cores and pits suggest that the playas have similar stratigraphy (Fig. 3). XRD analyses of the sediments have shown that the recent muds are predominantly hydromagnesite ($Mg_5(CO_3)_4(OH)_2\,4H_2O$) and magnesite ($MgCO_3$), the latter mineral increasing in relative abundance with depth and toward the playa centres (Renaut & Stead, 1991a and unpublished data). Below about 30 cm depth, non-stoichiometric dolomite, aragonite, Mg-calcite and calcite are locally found in

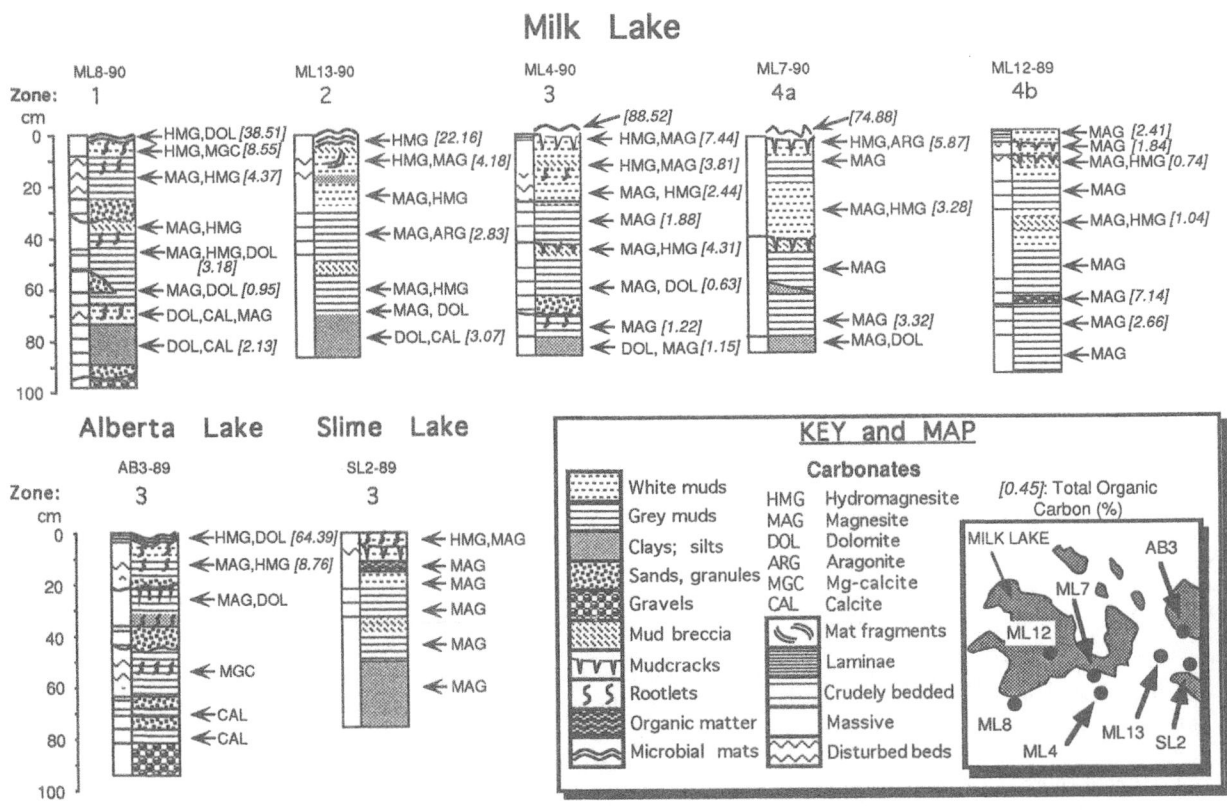

Fig. 3. Stratigraphy and carbonate mineralogy of selected core and pit profiles from Milk, Alberta and Slime lakes.

the muds. Dolomite also occurs in muds near sites of dilute groundwater inflow. Huntite $(CaMg(CO_3)_4)$ and nesquehonite $(MgCO_3 3H_2O)$ are also locally present in highly alkaline carbonate playas. Sepiolite $(Mg_4Si_6O_{15}(OH)_2 6H_2O)$, palygorskite $(MgAlSi_4O_{10}(OH) 4H_2O)$ and opaline silica $(SiO_2 nH_2O)$ are found in several Mg-carbonate playas on the plateau (e.g., south of Meadow Lake). Their formation, together with diatoms, may contribute to the low silica concentration of many alkaline waters (Table 1).

In many sections, carbonates lie with sharp contact on a dense ochrous clay a few dm thick that, in turn, rests on till or glaciofluvial sands and gravels. Until paleolimnological and geochemical analyses are undertaken, interpretation of the mineral stratigraphy must remain open. The overall vertical sequence from calcite to Mg-carbonates suggests a progressively higher Mg/Ca ratio and a higher Mg concentration of the lake or pore waters through time. This might reflect a relative increase in aridity and the early fractionation of Ca^{2+} from groundwaters by precipitation of calcite in soils and till (Renaut, 1990). Similar changes in carbonate mineralogy are recorded from other local lake basins.

Annual cycle of sedimentation in the Mg-carbonate playas

Magnesium carbonates are precipitated annually in the carbonate playas. Throughout winter, the playas are covered by a layer of snow and ice from several decimetres to more than a metre thick. Following melting during late March to mid-April, they are flooded to depths of a few decimetres in Milk Lake, and up to a metre in Alberta Lake and Slime Lake. Between late April and June the lakes normally attain their maximum annual level as the air temperature rises and early summer rains fall. The lakes appear highly productive at this time. Benthic and floating microbial mats, and green and yellow gelatinous microbial masses, are found extensively in the littoral zone and extend outward into the central playa lake.

Gradually, the lake waters during summer become white and turbid with a milky appearance. Water samples were collected from Alberta Lake and Milk Lake during July 1990 and filtered on site. SEM observations revealed suspended fine (< 0.5–$3 \mu m$) aggregates of calcium-free (EDS analyses) Mg-carbonates. Some from the sublittoral zone of Alberta Lake have the platy morphology of hydromagnesite. Samples from Milk Lake are anhedral aggregates and could not be identified with confidence, although many particles have vague rhombic faces suggestive of magnesite. Although individual filter samples provided insufficient material for reliable XRD identification, a composite sample showed a dominant reflection at 2.735Å, also indicating magnesite. At least some of these suspended crystals represent annual carbonate nucleation and crystallization in the waters, although a proportion may be stirred up from shallow bottom sediments. Carbonate precipitation may result largely from evaporative concentration and photosynthetic removal of CO_2, but some nucleation and precipitation may be microbially induced, as shown elsewhere by Thompson & Ferris (1990).

The bottom sediments during summer are a carbonate soup a few centimetres thick, locally developing a yoghurt-like consistency, resting on a harder carbonate mud layer. Although some crystallization may have occurred after collection, XRD analyses and SEM observations (Fig. 7C and D) of Milk Lake samples show the bottom sediment ooze to be a mixture of very fine magnesite and hydromagnesite crystal aggregates, the former increasing toward the centre of the playa-lake. During late June and July, the lakes gradually desiccate producing a dry white mudflat that develops small (3–10 cm wide; 1–2 cm deep) mudcracks. For a few weeks following desiccation, the upper 1.0–1.5 cm of the playa-lake muds show fine grey and white laminae 1–2 mm thick. These suggest active precipitation and settling from suspension, but laminae are often rapidly destroyed by a range of processes, some of which are described below.

After the mudflats dry out completely in late summer, thin (1–2 mm), white crinkly crusts and soft powdery efflorescences of hydromagnesite may develop by capillary evaporation. Locally, aragonite and dolomite crusts are found, usually near calcium-bearing freshwater spring seepages. Depending on the pattern of rainfall, mudflat surfaces may remain dry and hard until snowfall in October–November, or may become partially reflooded by autumnal rains. When dry in late summer, the water-tables may be withdrawn to a depth of a few decimetres (10–50 cm). The salinity of the groundwater below the playas may increase (to >25 g 1^{-1}) and they become weak sodium carbonate-bicarbonate brines (Table 1: Alberta 2). Minor highly soluble, evaporite salts (trona?) were found as puffy efflorescent crusts along mudcracks at Alberta Lake during July 1988, but their formation is rare. Complete desiccation does not occur in each basin every year. Slime Lake only dries out completely in exceptionally dry summers.

In each basin, hydromagnesite is usually the dominant carbonate in the youngest muds of the peripheral mudflats (Fig. 3), whereas surficial muds collected from the central mudflats of the dry playas are commonly nearly pure magnesite, or a mixture of magnesite and hydromagnesite. The magnesite/hydromagnesite ratio also increases down-profile (Fig. 3). In the littoral zone, both magnesite and hydromagnesite may be found together at the surface.

Little is known about the factors controlling precipitation of Mg-carbonates in lakes (Kelts & Hsü, 1978). Precipitation of metastable hydromagnesite is commonly favoured over magnesite in lakes because of the strong hydration of Mg^{2+} (Christ & Hostetler, 1970). However, most evidence to date suggests that magnesite is currently precipitating from Milk Lake waters. When tested using the PCWATEQ program, most analysed lake waters during early summer are theoretically supersaturated with respect to magnesite and dolomite by a factor of 1 to 3, but are undersaturated with respect to hydromagnesite. As the lake begins to desiccate, the salinity increases with a consequent decrease in the activity of water in the solution. The proportion of less-strongly hydrated Mg^{2+} should increase significantly, perhaps al-

lowing magnesite to form. Periodic dilution by fresh runoff or groundwater, or a decrease in the amount of available Mg^{2+} as a consequence of magnesite precipitation, may favour hydromagnesite (cf. Rosen et al., 1988, p. 120), accounting for mixed carbonate muds, especially in the littoral zone. Alternatively, (i) metastable hydromagnesite may have altered rapidly to diagenetic magnesite, (ii) thin crusts of hydromagnesite may have been removed by deflation (unlikely), (iii) precipitation of magnesite is biologically influenced (cf. Thompson & Ferris, 1990). Further analyses and seasonal monitoring are needed to resolve the controls of precipitation.

Distribution and morphology of the microbial mats

Microbial mats are found in all the carbonate playas examined, and in each, the morphological zonation and distribution are similar. They are most extensive and abundant in the peripheral mudflats, but can be found across much of the playa surface. Four main zones can be recognized, each with its characteristic forms, although not all are necessarily present or equally developed at each playa (Figs 4 and 5):

Zone 1: Vegetated hummocky ground

The contact between the hillslope and peripheral mudflat is commonly a damp zone of groundwater seepage, marked in places by dense reed beds or standing water (e.g., E. Slime Lake). This may pass lakeward into a zone of vegetated hummocky ground (Fig. 6A), characterized by irregular, subcircular mounds that rise up to 40 cm above the intervening hollows. The hummocks range from 30 to 100 cm in diameter and are commonly encrusted by pink, brown and orange, microbial mats (Fig. 6B). The interhummock depressions (hollows) and some hummock surfaces are vegetated discontinuously by coarse grasses. Where present, this zone ranges from a few metres to >20 m wide. During spring (May–June), inter-

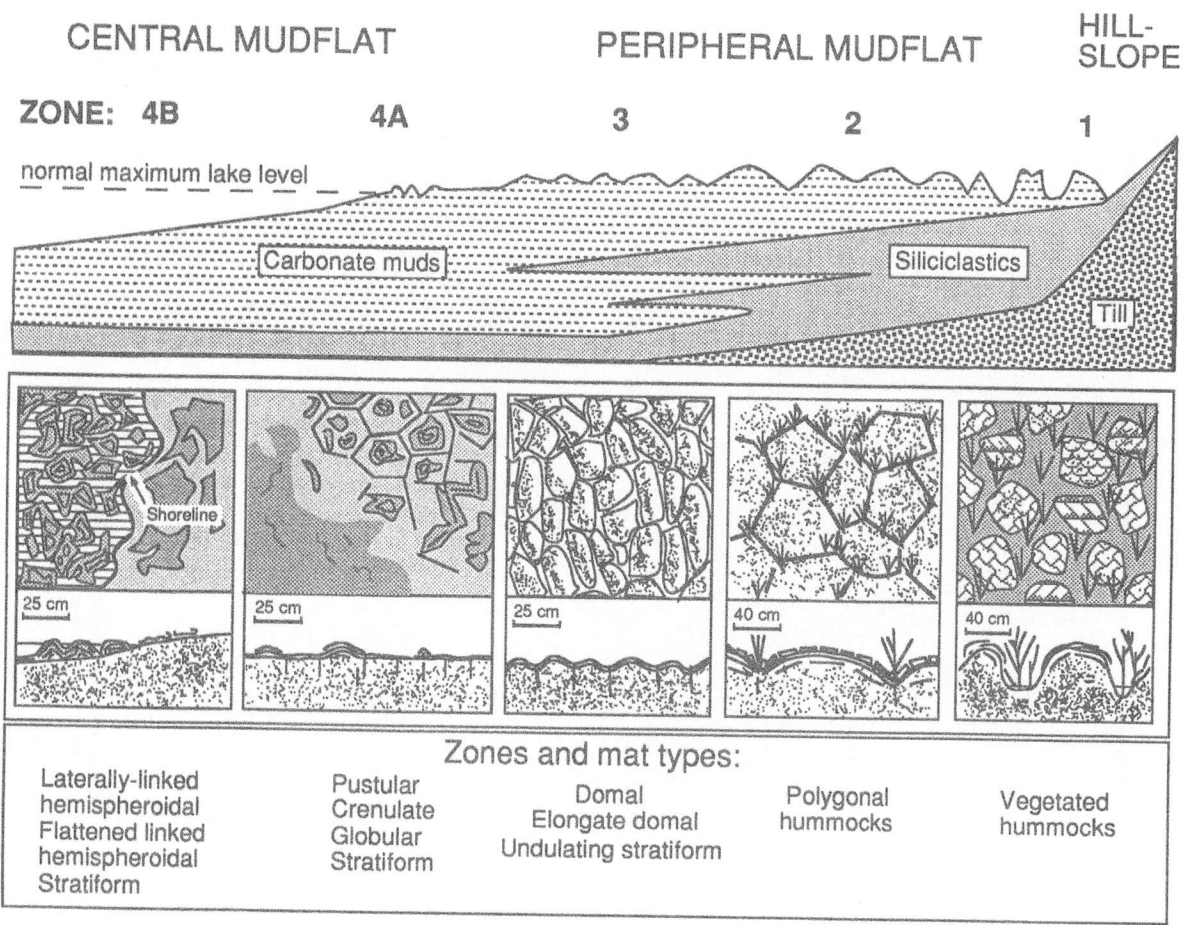

CENTRAL MUDFLAT   PERIPHERAL MUDFLAT   HILL-SLOPE

ZONE: 4B   4A   3   2   1

normal maximum lake level

Carbonate muds   Siliciclastics

Till

Shoreline

25 cm   25 cm   25 cm   40 cm   40 cm

Zones and mat types:

Laterally-linked hemispheroidal
Flattened linked hemispheroidal
Stratiform

Pustular
Crenulate
Globular
Stratiform

Domal
Elongate domal
Undulating stratiform

Polygonal hummocks

Vegetated hummocks

Fig. 4. Generalized cross-section from central playa (left) to the basin margin (right) showing typical zonation of microbial mats. Not all mat zones shown are necessarily present in each playa basin.

hummock depressions may be flooded. Ground-water is found at depths of 5–40 cm in late summer (August–September).

The mats are typically from 1 to 3 cm thick, leathery and, except during early spring, are extensively cracked into downward-curling fragments from a few cm to about 20 cm long. They have a well defined lamination (0.5–2 mm) with filaments normal to the mat surface. In most carbonate playas, the mats are extensively mineralized, usually by hydromagnesite, but locally, by aragonite. If present, aragonite may constitute all the carbonate or may be subsidiary to hydromagnesite. Analyses reveal up to 43 weight % carbonate. Observations by SEM (Fig. 7A) show platy, anhedral to subhedral crystal aggregates of

hydromagnesite, or acicular aragonite with stubby crystal terminations, encrusting cyanobacterial filaments and mucilage. Other minerals that may be present in the mats include anhedral to subhedral dolomite (0.5–1.5 μm), calcite, magnesite, and detrital siliciclastic grains – mainly plagioclase, quartz, volcanic rock fragments (basalt) and clay minerals. Pollen grains and epiphytic pennate diatoms are also present.

Cores and pits from below these mounds reveal massive or weakly bedded carbonate mud with intercalated lenticular units of brown detrital sand and silt, locally derived by wash from adjacent slopes (Fig. 3). Except in the upper 2–3 cm, there is little evidence for microbial lamination. Most of the carbonate mud is grey and white, and is ex-

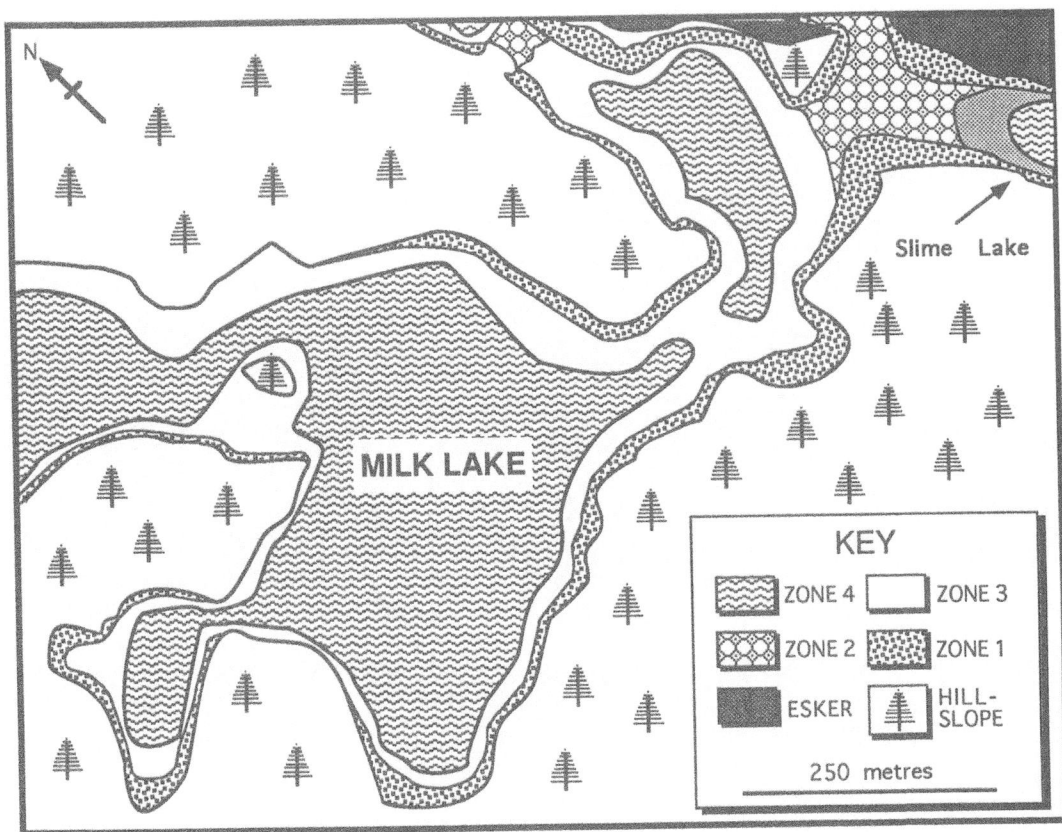

Fig. 5. Simplified sketch map of mat zonation for part of Milk Lake.

tensively mottled. Rootlet horizons, burrows (?), and deformed mudcracks are common near the surface. The carbonate in the upper 10–20 cm is mostly white hydromagnesite. At a few decimetres depth, this is accompanied by grey magnesite and, locally (e.g. southeastern margins of Milk Lake), dolomite.

Zone 2: Polygonal hummocky ground

Zone 2 is represented by white, polygonal hummocky ground. This distinctive surface resembles that of Zone 1, but the hummocks are usually larger with a well defined polygonal pattern (Fig. 6, C to F). The hummocks are typically from 50 to 150 cm across, rising 5–20 cm above adjacent hollows. Most hummocks lack macrovegetation but intervening hollows still have grasses

that emphasize the polygonal pattern. The surficial sediment is typically > 90% hydromagnesite. Groundwater below hummocks is found at depths of 10–40 cm in spring (May), and from 50–80 cm in late summer (August–September). This zone is discontinuous around most lakes, but is common at the extreme ends of some elongate lakes (Fig. 5). Southeast of Meadow Lake (Fig. 1), this surface type is continuous for more than 200 m.

Microbial mats, in places, may cover up to ± 80% of the surface in this zone (Fig. 6F), but are not always present. Some hummocks remain free of living mats throughout the year. When dry, mats are usually fragmented, giving the hummocks the appearance of large cauliflowers (Fig. 6D; also cf. Plate IV in Reinecke, 1920; Cummings, 1940). Mats are 1–3 cm thick, white or grey and highly friable when dry, crumbling readily between the fingers. They are heavily min-

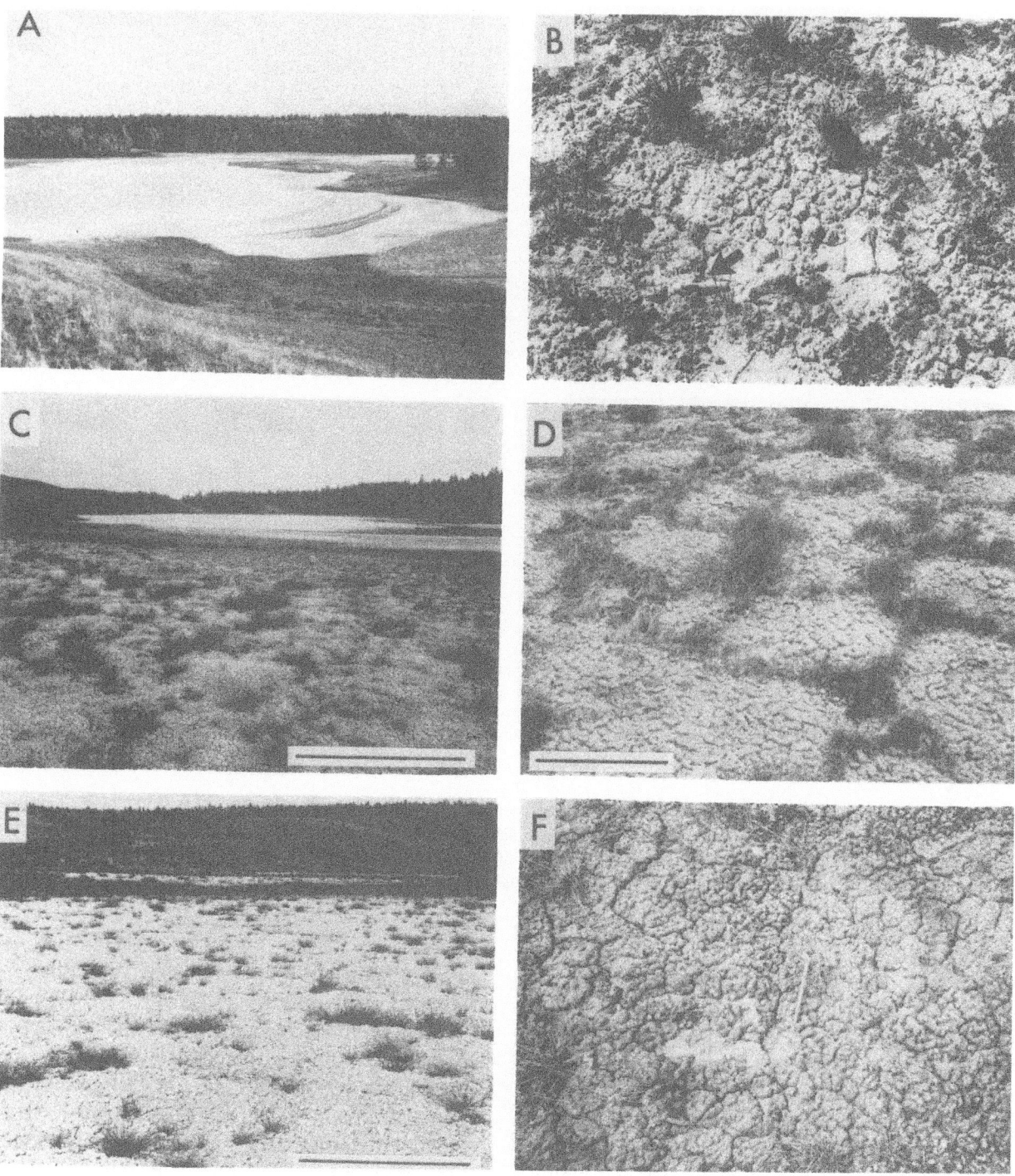

Fig. 6. Mat morphologies from Zones 1 and 2. A: View toward the southwest across the eastern arm of Milk Lake (dry). Middle foreground shows vegetated hummocks of Zone 1. B: Mat-covered hummocky ground of Zone 1, southern shore of Milk Lake. Mats here cover 90% of the surface, which is atypical for much of Zone 1. Marker pen (13 cm) for scale. C: Slime Lake viewed southward across polygonal hummocks of Zone 2. Scale bar: 1 m. D: Mat-encrusted polygonal hummocks (hydromagnesite) north of Slime Lake. Hummocks are 70–100 cm in diameter. Most mat fragments are curled downward. Scale bar: 50 cm. E: White polygonal hummocks (hydromagnesite) east of Milk Lake. Alberta Lake Esker in background. Scale bar: 80 cm. F: Detail of mats on polygonal hummocks shown in D. Pen (15 cm) for scale.

86

Fig. 7. Scanning electron photomicrographs of Milk Lake sediments. A: Cyanobacterial (?) filament encrusted by hydromagnesite plates, surficial mat in Zone 1, southern shore. B: Filament moulds in hydromagnesite mat, Zone 2, eastern shore. Two different sizes can be seen. The larger moulds have a diameter of 40–50 μm; smaller moulds in the matrix are 4–10 μm. C: Aggregates of very fine magnesite crystals from modern soupy surface muds of Milk Lake. D: Mixed magnesite-hydromagnesite mud from northern shore (Zone 4a).

eralized (up to 85 weight % carbonates) and are composed almost entirely of hydromagnesite (Fig. 7B), with subsidiary magnesite and siliciclastic silt. Although erect filaments are present, much of the hydromagnesite occurs as ovoid and spherical peloids, 1–3 mm in diameter. Lamination is generally less well defined than in Zone 1, especially when mats are dry and highly peloidal.

Cores from below the hummocks reveal white and cream massive carbonate muds that are commonly mottled (Fig. 3). Interfingering siliciclastics are rare to absent below most mounds in the upper 30 cm, but quartz-feldspar silt (eolian?) is found in some hollows. A few lava pebbles were found in hollows. Texturally, the muds range from loose and granular to weakly cohesive at and near the surface when dry, and are commonly composed of cemented aggregates of subspherical peloids identical with those found in some modern mats. Recognizable fragments of mineralized microbial mat < 2 cm long are present near the surface, but are uncommon. Where found, they may be oriented subvertically, especially toward the hollows. Mudcracks and rootlets are present locally. Beneath the loose friable layer (10–30 cm depth), the muds become more compacted and the hydromagnesite is mixed with magnesite. Below 40 cm, the muds are mostly massive magnesite, with some interfingering siliciclastics. No visible evidence for microbial mats (intraclasts or laminites) was seen below about 15 cm depth.

Zone 3: Low domal and stratiform mats

Transitionally toward the lake, the coarse hummocks of Zone 2 are locally replaced by a flatter, gently undulating surface with only about 5–15 cm of relief, characterized by the development of low hemispheroidal stratiform mats or circular and elongate domal mounds from ± 10 cm to 30 cm in diameter (Fig. 8A). Mats may cover most of the surface during late spring and early summer, broken in places by tufts of grass between the domes and by shallow rills. With desiccation, the surface cracks into low domal polygons (Plate 3-3-3 in Renaut & Stead, 1991a). Groundwater is found at depths of ± 20–40 cm in spring (May–June), and at 40–60 cm in late summer (August–September).

Most mats are thin (2–5 mm) across the domes and may line the walls of intervening cracks to a depth of 2 cm. The mats are variably mineralized by hydromagnesite and magnesite (4–24 weight % total carbonate; 4 samples). They are filamentous and generally less peloidal than in Zone 2. The underlying sediments show a pattern similar to Zone 2: massive white hydromagnesite muds rest on grey mottled muds containing increasing amounts of magnesite (Fig. 3). At ± 1 m depth the muds become increasingly calcareous, and locally contain dolomite, Mg-calcite and calcite. Rare mat fragments are found in the subsurface sediments.

Zone 4: Laterally continuous and pustular mats of supralittoral and sublittoral zones

As the late spring–early summer (May–June) shoreline is approached, the exposed supralittoral mudflats become essentially flat, but develop considerable microtopography associated with mat development, mudcracks and shallow rills (Fig. 8B). Two subzones are recognized based on duration of submergence:

Subzone 4a. As the lakes fall from their maximum level, a thin skin of microbial mats, only 1–3 mm thick, locally covers >80% of the supralittoral surface. Its presence is not everywhere obvious, but is revealed by a wide range of elongate wrinkles, pustular growths, blisters, and crenulations (Fig. 8B to E), the latter forming along the edges of polygons and rills (cf. Walter et al., 1973). Small (1–3 cm diameter) isolated globular mounds and domes with thicker (5–10 mm), erect filamentous mats are also found. Most of these forms are coreless but isolated gravel clasts are locally colonized. This subzone has the most morphological variety.

Subzone 4b. The mats of Zone 4a may continue below the lake as laterally continuous benthic mats, showing pustular and flattened, laterally-linked hemispheroidal morphology for several metres offshore. Observations over three years have shown that the morphology, mat thickness and width of this subzone depend mostly on the rate of lake recession. If the lake desiccates rapidly, the mats are essentially the same as in Zone 4a. If, on the other hand, the lake waters remain throughout most of the summer (as happened in 1990 and 1991), hemispheroidal mats, 5–15 mm thick, may develop in the shallow littoral zone for a distance of several metres offshore (Fig. 8F). About 5 to 10 m from the spring shoreline, the benthic mats in Milk and Alberta Lakes lose continuity and are partially replaced by a yellowish green microbial scum that periodically covers much of the lake bottom. With increasing distance from the littoral zone, the lake floor substrate becomes soft and soupy during early summer, which together with the turbid waters, may inhibit colonization by benthic mats. However, as the lake dries up and the substrate hardens thin mats like those in Zone 4a may colonize the retreating littoral zone. In Slime Lake, discontinuous benthic mats and gelatinous mi-

88

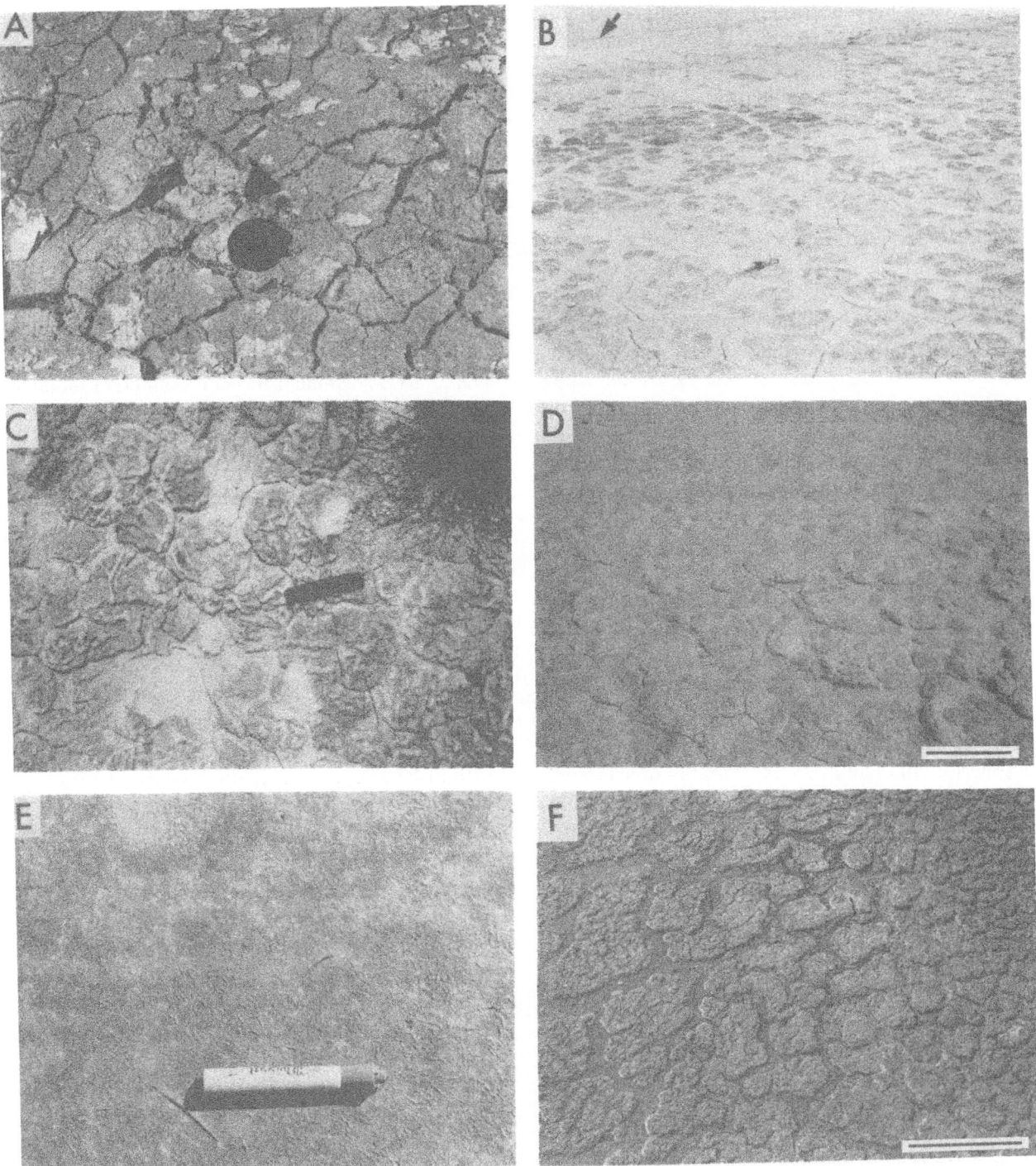

Fig. 8. Characteristic mat morphologies from Zones 3 and 4. A: Desiccating stratiform mats, southeastern shore of Milk Lake. Mats range from very low amplitude, gentle undulations shown here to well developed linear domes. Mats in this zone are mineralized both by magnesite (?detrital) and hydromagnesite. White efflorescent patches forming at the surface are nearly pure hydromagnesite. Lens cap is 5 cm in diameter. B: Littoral zone (4a) of Alberta Lake showing irregular pustular mounds in the centres of developing desiccation polygons. Trowel is 23 cm long. Arrow shows shoreline. C: Laterally continuous mats (Zone 3–

crobial masses appear to cover much of the lake floor.

The mats of Zone 4 differ from those of Zones 1–3 principally by being more weakly laminated, and having lower carbonate mineralization (<2 to 14 weight % total carbonate; 5 samples). A few pennate diatoms may be found. The sediments below the mats, like those of Zone 3, are mainly white massive hydromagnesite-magnesite muds, becoming mottled and darker with depth (Fig. 3).

Significance of zonation and environments of the carbonate sedimentation

The morphological variation in the mats and their distribution is mainly a reflection of the type, depth and duration of wetting. Zones 1 and 2 (peripheral mudflat) normally lie *above* the recent annual maximum lake level and today receive almost all their moisture from runoff (including snow melt) and capillary rise of shallow groundwater. Zones 3 to 4 (central mudflat) are regularly submerged, but except in unusually wet years, mats are periodically emergent, during which times they may also derive some capillary moisture from shallow groundwaters.

The peripheral mudflat is a zone where shallow groundwater with a high molar Mg/Ca ratio (from 5 to >50) is moving toward the playa centre. Three lithologies appear to act as the main aquifers: (i) sands and gravels, probably of early postglacial age, that underlie the carbonate sediments, (ii) brecciated, partially lithified, carbonate mudstones that represent former exposure surfaces, and (iii) interfingering lenticular and sheetlike units of sand and fine gravel, representing colluvial detritus washed into the basin from adjacent hillslopes. Groundwaters are drawn upward by capillary evaporation and evapotranspiration, and hydromagnesite, aragonite or dolomite is precipitated in the upper part of the profile, on and within the microbial mats, and as surficial crusts. Although many muds have low permeability, crack networks (some expanded by ephemeral ice), fenestrae (some representing decayed mats), vesicles, and the granular nature of many carbonates appear to permit some upward fluid migration.

The carbonate mudflats have characteristics similar to the 'dry mudflat' subenvironment of Hardie *et al.* (1978) and Smoot & Lowenstein (1991), given the abundance of desiccation features and relative paucity of interstitial soluble salts. With present data, it is not possible to determine the origin(s) of the carbonate sediments of the peripheral mudflats. Much of the carbonate probably was deposited subaqueously during former periods of lake expansion and has since been modified by mudflat processes.

The precise origins of the hummocky surfaces of Zones 1 and 2 are unclear. In both zones groundwater lies at shallow depth. Hummocky and self-rising ground are commonly associated with (i) intrasediment mineral precipitation in porous, permeable sediments (e.g. Motts, 1970), (ii) differential expansion and contraction associated with wetting and drying (gilgai) (e.g. Hallsworth *et al.*, 1955; Verger, 1964), or (iii) ephemeral ground ice (e.g. Tufnell, 1975; Mackay, 1980). The morphology of Zone 1 in places resembles that produced by cryoturbation, but the other two processes may have also contributed to development. Similar morphology is found in the same setting at local siliciclastic playas. Although the polygonal pattern may have resulted from desiccation (?), the hummocks of Zone 2 may have developed by intrasediment precipitation of carbonate from shallow groundwaters (Cummings, 1940). In support of this interpretation, this morphology is only found in the carbonate mudflats and is absent in local siliciclastic mudflats.

4a), composed of hydromagnesite, northwestern margin of Alberta Lake. Comb is 12 cm long. D: Pustular and crenulate stromatolites, littoral zone (4a) of Alberta Lake. Scale bar: 25 cm. E: Laterally continuous mat with small elongate wrinkles, Zone 4a, northeastern shore of Milk Lake. Marker pen is 13 cm long. F. Partially-submerged, flat-topped laterally continuous mats in the littoral zone of Milk Lake (SW corner). Scale Bar: 25 cm.

Mats that form in Zones 3 and 4 are regularly submerged by the playa lake. During spring and early summer the surface is moist and the mats appear to have their maximum phase of growth. By late summer, the surface is broken extensively by mudcracks, and thin efflorescences of hydromagnesite form on and within the mats.

Origin of the carbonate sediment in the mats

At present, the role played by microorganisms in Mg-carbonate precipitation and in influencing mat morphology is unknown. In Zones 3 and 4, the benthic mats are usually poorly mineralized and may contain, besides hydromagnesite, magnesite, aragonite, clay minerals, quartz, and feldspars. The fine siliciclastic grains are clearly washed or blown onto the mat surface, adhering to the mucilage. Suspended Mg-carbonates also may adhere to mat surfaces, but biologically-influenced carbonate precipitation is possible, especially in Zone 3, where shallow groundwaters are prone to capillary evaporation.

In contrast, in Zones 1 and 2, where mats derive much of their moisture from capillary rise of groundwater, most filaments are heavily encrusted by nearly pure hydromagnesite (Fig. 7A) or fine acicular aragonite (1–2 μm), with very little siliciclastic debris. Zone 1 is commonly the least prone to complete desiccation. Because they are elevated above normal lacustrine flooding, hummock surfaces derive little carbonate sediment from suspension, except eolian dust. Therefore, most carbonate sediment is probably precipitated *in situ* following capillary evaporation. Whether this hydromagnesite precipitation is biologically induced ('biomineralization', in the sense of Riding, 1991) or the filaments are merely acting as substrates or templates for external precipitation ('mineralization', in the sense of Riding, 1991), remains to be shown, but the peloidal microfabrics in Zone 2 suggest at least some microbial influence.

Preservation potential of microbial mats and lamination in mudflats

Although the microbial mats are common and well developed, cores, pits and trenches cut into the sediments rarely reveal well developed lamination. The sediments of most mudflats examined are massive and mottled. There are two possible explanations – either microbial mats have only recently colonized the substrate, or else the mats and lamination are being destroyed. The *localized* preservation of mat fragments at depths of down to a metre shows that mats have probably existed for several thousand years of sedimentation. Therefore, it is most likely that mats are being destroyed and some of their mineralized remains are incorporated into the sediment.

As in the marine environment (e.g. Park, 1977) many processes inhibit preservation of microbial mats as stromatolites. Several are essentially physical; others are chemical and/or biological. Some processes are effective in destroying the mats at the surface; others destroy microbial lamination in the sediment. In this section, the causes and consequences of these surficial and early diagenetic processes (Fig. 9) will be briefly outlined.

(i) Desiccation, and wetting and drying

Desiccation and dehydration, following fall in lake level is a major factor in mat destruction, especially in Zones 3 and 4. The spring-early summer shoreline is littered with brittle, poorly mineralized, brown, yellow or black mat fragments, that shrivel and curl both upward and downward (Fig. 10). These become detached from the substrate and may be reworked by wind, water and ice. This is perhaps the most commonly cited, physical agent of microbial mat destruction.

In Zones 1 and 2, repeated wetting and drying of the mudflat is effective in sediment disruption. Surficial mats, fragmented by desiccation, are reworked on the edges of the hummocks and may be reincorporated into the sediment when it is next wetted. Some fragments are washed into the

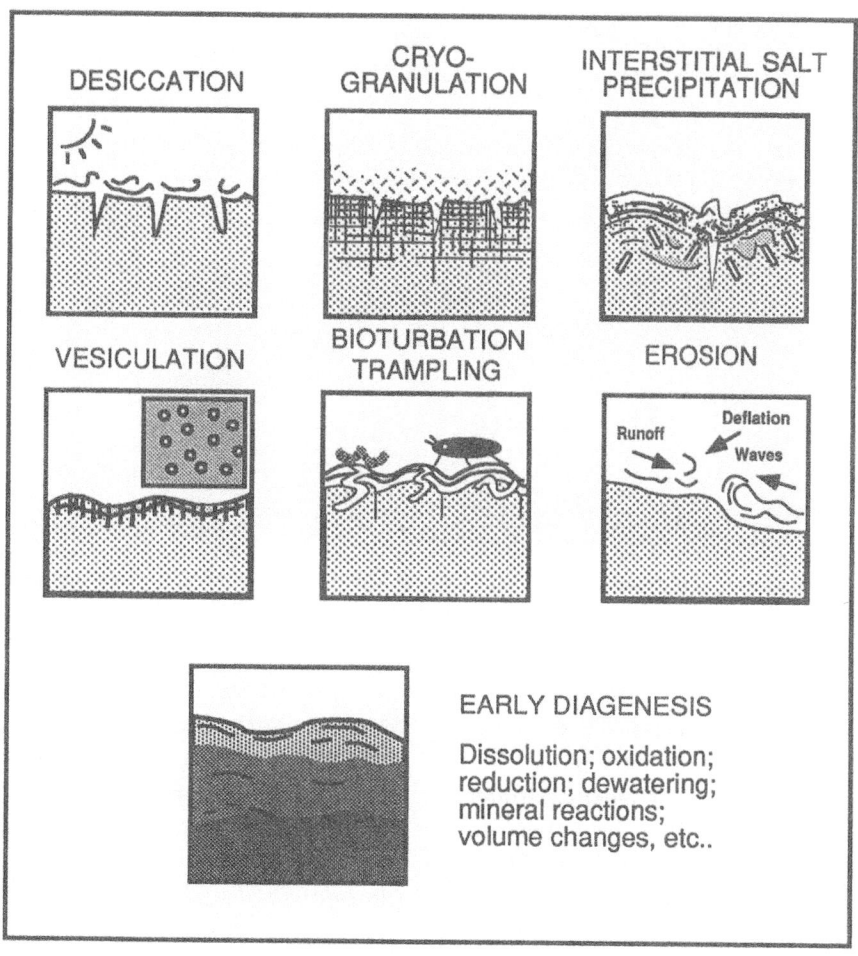

Fig. 9. Processes that destroy mats and lamination in the carbonate playas.

gaps between the hummocks. Others disintegrate *in situ* and become incorporated in the underlying sediment, including peloids produced within the mats. New mats may recolonize the surface following surface wetting or groundwater discharge, and the process is repeated. Observations over several years have shown that many hummocky surfaces of Zone 2 are colonized only intermittently, probably when groundwater levels are high. Thus destructive processes may continue several years without any mat renewal. Repeated wetting and drying also leads to brecciation of the playa muds in Zones 3 and 4, producing 'crumb fabrics' (Smoot & Olsen, 1985) that disrupt lamination.

(ii) Cryogranulation (ice action)

During winter, the Cariboo Plateau receives up to 2 m of snow. The mean daily temperatures from November until March are below 0 °C and the minimum temperature can fall below − 45 °C. Consequently, most lakes begin to freeze over in October–November and many remain frozen until April. Even Last Chance Lake, which has a salinity of > 350 g l^{-1}, develops an ice cover of several decimetres during winter.

Within the upper 20–40 cm of the saline mud-flats, segregation ice can form. The effects of ephemeral freezing on the sediment structure and microbial mats can be severe, especially where

Fig. 10. Desiccated microbial mats in the littoral zone (Zone 3–4a) of Milk Lake (SW margin). Note sharp contact with vegetated hummocky ground of Zone 1, top left corner. Line of desiccated mats is about 70–100 cm wide.

snow cover is thin. Excavated mudflats in winter have revealed that most ground ice occurs as fine (1–5 mm) clear laminar layers or lenses parallel to the surface, and subvertical or reticulate sheets and veins normal to the surface. Close to the surface many ice layers are <1 cm apart, but the spacing increases to a few centimetres with increasing depth. Many of the subvertical sheets form within desiccation cracks produced during the preceding summer. Some cracks may also result from water migration in the sub-surface to form segregation ice (Williams & Smith, 1989).

The net effect of the ice is to fracture the playa muds into small subrectangular (1–10 mm) blocks. When the ice melts, the muds take on a granular appearance and much of the original stratification, including microbial lamination, is severely disrupted or destroyed. When the sediment dries out it may disaggregate into small hard granules. Some of these may be reworked in the littoral zone during spring, becoming subrounded and producing carbonate peloids. In a few places (e.g. NW Alberta Lake), these are subsequently coated by cyanobacterial films. Carbonate (hydromagnesite, aragonite) is precipitated on erect filaments in the coating, producing microooncoids with thin cortices. As yet, these have not been found in the underlying sediments.

(iii) Interstitial carbonate and salt crystallization

Interstitial carbonate and efflorescent salts are precipitated from shallow groundwater drawn upward by capillary action across much of the playa mudflats, wherever the water table is shallow. Observations suggest that the process is most effective where the latter is less than ± 70–80 cm below the surface. In the carbonate playas, hydromagnesite (or aragonite) is precipitated as surficial crusts, and magnesium carbonates are precipitated interstitially within mudcracks, fenestrae and other pores in the upper sediment profile, locally raising the surface. Other salts, such as sodium carbonates, are only of minor consequence in the carbonate playas, but are common interstitial precipitates in the hypersaline playa basins. Whether significant phreatic precipitation of carbonate takes place is unknown.

The effects of interstitial mineral precipitation

and crystallization on surficial mats are difficult to assess. In the porous, open microfabrics of mats, filaments commonly serve as substrates for precipitation. In contrast, interstitial precipitation is known to destroy sedimentary structures in saline mudflats (Hardie *et al.*, 1978; Van Houten, 1980; Renaut & Long, 1989) and marine sabkhas (Park, 1977), and is thus likely to disrupt any surviving subsurface lamination.

(iv) Vesiculation

The upper 1–2 cm of the peripheral mudflat sediments commonly show well developed vesicular structure, characterized by dense concentrations of elongate, circular vesicles from 1 to 4 mm in diameter, giving the appearance of a honeycomb. Vesicles occur in damp carbonate muds at the surface, particularly in the littoral zone, but are commonly seen below the stratiform mats near the shoreline. They are also well developed below soil crusts in siliciclastic silts and sands of small washes draining the hillslopes, and in muds lacking mats. Vesiculation is most obvious in spring and early summer, before desiccation and cracking of the mudflat.

Vesiculation results from upward-escaping gases and is very common in semi-arid soils. Vesicles may form below superficial soil crusts (McIntyre, 1958) where trapped air is heated and expands on drying, or they may develop when air escapes during drying from a muddy slurry deposited by floods (Springer, 1958; Cooke & Warren, 1973). The latter process can explain those found in washes. On the margins of the hydromagnesite playas some have been observed forming as gases (air and/or CO_2?) escape from the soft soupy muds while the lakes desiccate. Trapped gases, resulting from bacterial degradation of organic matter (including old mats) in the sediment, also might contribute to vesiculation. Whatever the process, they are common in these and other playas and can result in disruption of lamination, and the production of birdseye and fenestral porosity, often to be later modified by compaction.

(v) Activities of organisms

Both invertebrates and vertebrates contribute to destruction of the mats. Invertebrates play a dual role. Some consume mats directly at the surface (Walter *et al.*, 1973; De Deckker, 1987); others burrow in the sediments and mats destroying lamination (e.g. Gerdes & Krumbein, 1987). Scudder (1969) described the fauna of several Cariboo saline lakes. Mat-grazing ephydrids are abundant in the local sodium carbonate lakes, but although present, are less common in the Mg-carbonate playas. Living gastropods, which are common mat grazers in marine-marginal environments, have not yet been found, but gastropod shell fragments are recorded from core sediments near spring seepages west of Milk Lake.

Burrows occur in some of the thicker mats of Zones 3 and 4. The muds of the littoral and supralittoral zones are commonly riddled by horizontal and subvertical burrows, generally 3–5 mm in diameter, with oxidized rims. Insect larvae (chironimids, mosquitos, beetles) are commonly found in the burrows and are important locally in bioturbation (Renaut and Sarjeant, 1991). Microscale burrows are seen in some thin sections.

Wading birds that feed on insects and crustaceans, particularly the killdeer (*Charadrius vociferus*), and visitors from local fresh lakes (e.g. common snipe, phalaropes and sandpipers) are common both in the littoral zone and across the dry mudflats. The shore zone, including mats, is often covered by their footprints. Birds have been observed to dislodge mats from the substrate, both in feeding and in take off. Trampling by large mammals (e.g. elk, deer, bear and domestic cattle) is locally significant. Many animals cross the mudflats to drink from freshwater springs and seepages, some of which flow even in midwinter.

(vi) Other diagenetic processes

Many other physical, chemical and biochemical processes may contribute to mat destruction, but have yet to be studied. Physical compaction, associated with early dewatering and shrinkage of the carbonate muds, may lead to some loss of structure as the platy grains of hydromagnesite compact like clays. During early burial, hydromagnesite alters to magnesite (Müller *et al.*, 1972; Christ & Hostetler, 1970), accompanied by expulsion of water and perhaps some loss of structure.

Organic diagenesis, within and below the mats, is probably a major factor in their destruction. Many processes, involving activities of aerobic and anaerobic bacteria are known to decompose organic mats (e.g. Doemel & Brock, 1977; Bauld, 1981b; several papers in Cohen *et al.*, 1984). Their role in these playas is unknown. Organic matter is more likely to be preserved and fossilized anaerobically (Golubic, 1991). Reducing conditions are evident in muds below mats in the littoral zone in the spring and in some marsh (Zone 1), but most of the near-surface muds are oxidizing in the central playa following desiccation.

Although the structure of the mats is not well preserved, some organic matter does remain in the sediment. The total organic matter (TOC) content for a series of pits in each zone at Milk Lake is shown in Fig. 3. This shows that most of the organic matter is lost within the upper 3–5 cm in the carbonate playas. Not all the organic matter is microbial. High values recorded from Zones 1 and 2 may reflect the preservation of some macrovegetal detritus, including wood fragments.

(vii) Erosion

Debris from mat destruction in Zones 1 and 2 is mostly reworked *in situ*. In Zones 3 and 4, poorly mineralized mats that have been physically disaggregated by desiccation, ice and trampling, are commonly reworked by surface wash, wind or waves in the littoral zone. Although waves are of very low energy compared to most lakes, accumulations of mat detritus, usually mixed with other organic debris (e.g. invertebrate egg cases, sheaths, etc.), form recessional strandlines that show some littoral sorting can occur. Surface

wash may entrain dried mat fragments on exposed playa flats following desiccation and transport them a short distance. Small brush heaps of fine (< 1 cm) mat fragments have been observed in Alberta Lake in shallow rills. Wind may deflate fine (silt) fragments, but does not appear significant in mat removal. Very little organic matter leaves the basin by deflation: the vegetation forms an effective wind-break for small playas and dust storms are uncommon.

Discussion

Few modern lakes with hydromagnesite-magnesite deposits have been described in detail, those from the Coorong region of South Australia being an exception (e.g. Aldermann, 1965; von der Borch, 1965, 1976; Walter *et al.*, 1973; Rosen *et al.*, 1988; Warren, 1990). Other examples have been reported from East and Central Europe (Irion & Müller, 1968; Müller *et al.*, 1972; Molnar, 1990), Spain (Pueyo-Mur & Ingles-Urpinell, 1987) and Uzbekistan (Popov & Sadykov, 1987). In the Coorong region, Walter *et al.* (1973) found stratiform, crenulate, and globular stromatolites similar to those in Zones 3 and 4 and were able to correlate morphology with microbial assemblage. However, they attributed stromatolite growth to trapping and binding of resuspended carbonate ('agglutinated stromatolites' of Riding, 1991), rather than chemical or bio-induced mineralization. They also noted their poor preservation potential.

Clearly, the lack of stromatolites and laminites in the sedimentary record of carbonate playas and ephemeral lakes does not preclude their former presence. In the lakes studied microbial mats are present at most sites of modern carbonate precipitation. Many factors known to favour their development are present. Most substrates in the carbonate playas are stable, hardening regularly upon desiccation. Environmental factors, including elevated salinities and the climatic extremes, exclude many competing and grazing organisms. The low siliciclastic sediment input (Renaut & Long, 1989) results in relatively pure mineralogical compositions. The possible roles of microbial mats in sedimentation must, therefore, be considered in interpreting the record. The low percentages of organic matter in the underlying sediments suggest that most mats that survive early physical destruction are lost to microbial degradation and oxidation. Although microbolites are occasionally preserved, muds, peloids, intraclasts and a poorly-defined organo-mineral residue account for most of the surviving evidence (Fig. 11).

Park (1977) noted that marine-marginal mats are most likely to be preserved as stromatolites in two main settings: (i) where there are high sedimentation rates and rapid burial, and preferably near-surface anoxia, which usually preserve horizontal laminites, and (ii) where they undergo early lithification. These conditions are also applicable to non-marine microbial mats. Although few lakes have yet been cored, former mats are only commonly preserved in two situations in the Cariboo – as microbial laminites below ephemeral and perennial lakes with anoxic sediments, and associated with calcareous springs.

At Clinton Lake (Fig. 1), an epsomite-precipi-tating saline pan (Reinecke, 1920; Renaut & Stead, 1991b), black and greenish-grey microbial laminites, with alternating fine (1–2 mm) laminae of carbonate and organic matter, are preserved in carbonate muds of the peripheral saline mudflats. The black muds are variably composed of hydromagnesite (surface only), magnesite and non-stoichiometric dolomite, but unlike those described, they are strongly reducing at about 1 cm depth and yield H_2S. Mats commonly cover the mudflat surface. Similar laminites, composed largely of carbonates (including dolomite) and organic matter derived from benthic mats, underlie many sodium carbonate lakes on the Cariboo Plateau. Although sedimentation rates are low, anoxic conditions in very shallow sediments help preserve the organic lamination.

The best preserved stromatolites in the Cariboo saline lakes are those composed of calcite. Though rare, they are found as crusts 1–2 cm thick on gravel around some lake-marginal and sublacustrine spring orifices, as for example at Goodenough Lake (Fig. 1). Even there the lithified carbonate laminae are shattered by ice crystallization.

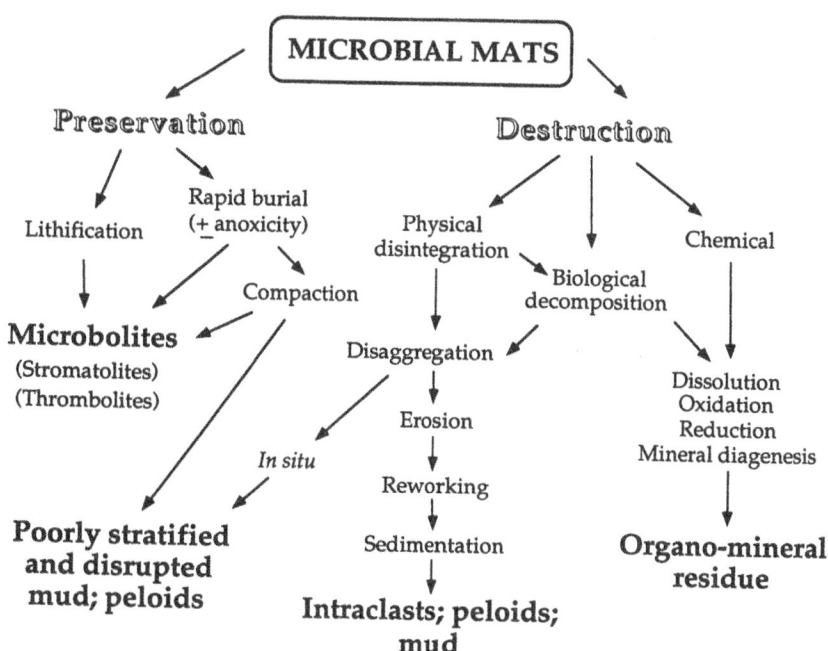

Fig. 11. Flow chart to summarize the fate of microbial mats in the carbonate playas.

An aspect not yet addressed is the effect of mineralogy on preservation. Hydromagnesite is metastable (Langmuir, 1965; Christ & Hostetler, 1970) and usually does not preserve in the geological record. It may alter diagenetically to magnesite or may dissolve to provide a source of Mg^{2+} for dolomitization (Rosen *et al.*, 1988). It is somewhat analogous to the diagenetic alteration of modern aragonitic mats to calcite or dolomite microbolites. However, the effects of burial diagenesis on stromatolites with hydromagnesite mineralization remain unknown.

Conclusions

Many playas and ephemeral lakes in the Interior of British Columbia are actively precipitating magnesium carbonates. Hydromagnesite, accompanied locally by aragonite, is the dominant mineral in the peripheral mudflats, and much is precipitated by capillary evaporation. Both magnesite and hydromagnesite may be forming at different times in the ephemeral lakes. Microbial mats are common and form both subaqueously in the ephemeral lakes and in peripheral zones where moisture is provided by shallow groundwater. Mats that develop on hummocky, polygonal ground on the margins of the playas are commonly mineralized by *in situ* hydromagnesite precipitation, some of which may be bio-induced. These mats are laterally continuous and leathery, but break up with desiccation. Mats formed in the playa lakes tend to be less mineralized and more ephemeral, but have greater morphological variation, with low domal, crenulate, globular, pustular, stratiform, and laterally-linked forms present.

The mats have low preservation potential as stromatolites. A range of processes, including desiccation, wetting-drying cycles, vesiculation, cryogranulation, interstitial mineral precipitation, and organic and inorganic diagenesis, lead to their destruction. Rare mat fragments do survive, but sediments below the mats generally have relatively low contents of organic matter that decrease with depth in the upper few decimetres.

Acknowledgements

This work was supported by grants from the Natural Sciences and Engineering Research Council (Canada) and the British Columbia Geoscience Research Grant program. I thank Douglas Stead and Cherdsak Utha-aroon for their assistance during fieldwork, Chris Boys for analyzing many sediment samples, and Brian Jones and Michael Rosen for their helpful reviews of the manuscript.

References

Aldermann, 1965. Dolomitic sediments and their environment in the South-East of South Australia. Geochim. Cosmochim. Acta 29: 1355–1365.

Anderson, G. C., 1958. Some limnological features of a shallow saline meromictic lake. Limnol. Oceanogr. 3: 259–270.

Bauld, J., 1981a. Occurrence of benthic microbial mats in saline lakes. In W. D. Williams (ed.), Salt Lakes. Developments in Hydrobiology 5. Dr W. Junk Publishers, The Hague: 87–111. Reprinted from Hydrobiologia 81/82.

Bauld, J., 1981b. Geological role of cyanobacterial mats in sedimentary environments: a production and preservation of organic matter. J. Aust. Geol. Geophys. 6: 307–317.

Burne, R. V. & L. S. Moore, 1987. Microbialites: Organosedimentary deposits of benthic microbial communities. Palaios 2: 241–254.

Campbell, R. B. & H. W. Tipper, 1971. Geology of the Bonaparte Lake map area, British Columbia. Mem. Geol. Surv. Can. 363, 100 pp.

Casanova, J., 1986. East African Rift stromatolites. In: L. E. Frostick, R. W. Renaut, I. Reid & J. J. Tiercelin (eds), Sedimentation in the African Rifts. Spec. Publ. Geol. Soc. Lond. 25: 201–210.

Christ, C. L. & P. B. Hostetler, 1970. Studies in the system $MgO\text{-}SiO_2\text{-}CO_2\text{-}H_2O$ (II): The activity product constant of magnesite. Am. J. Sci. 286: 439–453.

Cook, R. U. & A. Warren, 1973. Geomorphology in deserts. Batsford, London, 394 pp.

Cohen, Y., R. W. Castenholz & H. O. Halvorson (eds), 1984. Microbial mats: Stromatolites. MBL Lectures in Biology Vol. 3, Liss, New York, 498 pp.

Cummings, J. M., 1940. Saline and hydromagnesite deposits of British Columbia. Bull. B. C. Dept. Mines 4, 160 pp.

De Deckker, P., 1983. Australian salt lakes: their history, chemistry and biota – a review. In U. T. Hammer (ed.), Saline Lakes Developments in Hydrobiology 16. Dr W. Junk Publishers, The Hague: 231–244. Reprinted from Hydrobiologia 105.

De Deckker, P., 1987. Biological and sedimentary facies of

Australian salt lakes. Palaeogeogr. Palaeoclimatol. Palaeo-ecol. 62: 237–270.

Demicco, R. V. & E. Gierlowski-Kordesch, 1986. Facies sequences of a semi-arid closed basin: the Lower Jurassic East Berlin Formation of the Hartford Basin, New England, USA. Sedimentology 33: 107–118.

Fulton, R. J., 1984. Quaternary glaciation, Canadian Cordillera. In: R. J. Fulton (ed.), Quaternary stratigraphy of Canada – A Canadian contribution to I.G.C.P. Project 24. Pap. Geol. Surv. Can. 84–10: 39–47.

Gerdes, G. & W. E. Krumbein, 1987. Biolaminated deposits. Lectures Notes in Earth Sciences 9. Springer, Berlin, 183 pp.

Golubic, S., 1991. Modern stromatolites: A review. In R. Riding (ed.), Calcareous algae and stromatolites. Springer, Berlin: 541–561.

Grant, B., 1987. Magnesite, brucite and hydromagnesite occurrences in British Columbia. Open File Rep. B.C. Geol. Surv. Branch 1987–13, 68 pp.

Halley, R. B., 1976. Textural variation within Great Salt Lake algal mounds. In: M. R. Walter (ed.), Stromatolites. Elsevier, Amsterdam: 435–446.

Hallsworth, E. G., G. K. Robertson & F. R. Gibbons, 1955. Studies in pedogenesis in New South Wales, VII: The 'gilgai' soils. J. Soil Sci. 6: 1–31.

Hammer, U. T., 1986. Saline lake ecosystems of the world. Dr W. Junk Publishers, Dordrecht, 616 pp.

Hardie, L. A., J. P. Smoot & H. P. Eugster, 1978. Saline lakes and their deposits: a sedimentological approach. In: A. Matter & M. E. Tucker (eds), Modern and ancient lake sediments. Blackwell, Oxford: 7–41.

Irion, G. & G. Müller, 1968. Huntite, dolomite, magnesite and polyhalite of Recent age from Toz Gölü, Turkey. Nature 220: 130–131.

Kelts, K. & K. J. Hsü, 1978. Freshwater carbonate sedimentation. In A. Lerman (ed.), Lakes: chemistry, geology, physics. Springer, New York: 295–323.

Kempe, S., J. Kazmierczak, G. Landmann, T. Konuk, A. Reimer & A. Lipp, 1991. Largest known microbialites discovered in Lake Van, Turkey. Nature 349: 605–608.

Kennard, J. M. & N. P. James, 1986. Thrombolites and stromatolites: two distinct types of microbial structures. Palaios 1: 492–503.

Langmuir, D., 1965. Stability of carbonates in the system MgO-CO_2-H_2O. J. Geol. 73: 730–754.

Last, W. M. & P. De Deckker, 1990. Modern and Holocene carbonate sedimentology of two saline volcanic maar lakes., southern Australia. Sedimentology 37: 967–981.

Mackay, J. R., 1980. The origins of hummocks, western Arctic coast, Canada. Can. J. Earth Sci. 17: 996–1006.

Mathews, W. H., 1989. Neogene Chilcotin basalts in south-central British Columbia: Geology, ages and geomorphic history. Can. J. Earth Sci. 26: 969–982.

McIntyre, D. S., 1958. Soil splash and the formation of surface crusts by raindrop impact. Soil Sci. 85: 261–266.

Molnar, B., 1990. Modern lacustrine carbonate (calcite, dolomite, magnesite) formation and environments in the Hungary. Abstr. (Pap.) 13 Int. Sedimentol. Congr., Nottingham, U.K.: 363–364.

Monger, J. W. H., 1989. Overview of Cordilleran geology. In: B. D. Rickets (ed.), Western Canada Sedimentary Basin – a case history. Canadian Society of Petroleum Geologists, Calgary: 9–32.

Moss, B. & J. Moss, 1969. Aspects of the limnology of an endorheic African lake (Lake Chilwa, Malawi). Ecology 50: 109–118.

Motts, W. (ed.), 1970. Geology and hydrology of selected playas in Western United States. U.S. Air Force Cambridge Research Laboratories, Bedford, Massachussetts, Final Scientific Report (Part II), AFCRL-69-0214, 288 pp.

Müller, G., G. Irion & U. Förstner, 1972. Formation and diagenesis of inorganic Ca-Mg carbonates in lacustrine environments. Naturwissenschaften 59: 158–164.

Osborne, R. H, G. R. Licari & M. H. Link, 1982. Modern lacustrine stromatolites, Walker Lake, Nevada. Sed. Geol. 32: 39–61.

Park, R. K., 1977. The preservation potential of some recent stromatolites. Sedimentology 24: 485–506.

Popov, V. S. & Sadykov, T. S., 1987. Magnesium carbonate deposits of the Lake Beshkod region (Western Uzbekistan). Lithol. Mineral. Res. 21: 394–400.

Pueyo-Mur, J. J., 1978. La precipitación evaporitica actual en las lagunas saladas del área: Bujaraloz, Sástago, Caspe, Alcañiz y Calanda (provincias de Zaragoza y Teruel). Rev. Inst. Inv. Geol. Disputación Prov. Barcelona 33: 5–56.

Pueyo-Mur, J. J. & M. Ingles-Urpinell, 1987. Magnesite formation in recent playa lakes, Los Menegros, Spain. In: J. D. Marshall (ed.), Diagenesis of sedimentary sequences. Spec. Publ. Geol. Soc. Lond. 36: 119–122.

Reinecke, L., 1920. Mineral deposits between Lillooet and Prince George, British Columbia. Mem. Geol. Surv. Can., 118.

Renaut, R. W., 1990. Recent carbonate sedimentation and brine evolution in the saline lake basins of the Cariboo Plateau, British Columbia, Canada. In F. A. Comín & T. G. Northcote (eds), Saline Lakes. Developments in Hydrobiology 59. Kluwer Academic Publishers, Dordrecht: 67–81. Reprinted from Hydrobiologia 197.

Renaut, R. W. & P. R. Long, 1987. Freeze-out precipitation of salts in saline lakes – examples from Western Canada. In: G. L. Strathdee, M. O. Klein & L. A. Melis (eds), Crystallization and precipitation. Pergamon, Oxford: 33–42.

Renaut, R. W. & P. R. Long, 1989. Sedimentology of the saline lakes of the Cariboo Plateau, Interior British Columbia. Sed. Geol. 64: 239–264.

Renaut, R. W. & W. A. S. Sarjeant, 1991. Salt tracks – evidence for animal activities in saline mudflats and their paleolimnological implications. Prog. Abstr., Sedimentary and Paleolimnological Records of Saline Lakes Conf., Saskatoon, Canada: 40.

Renaut, R. W. & D. Stead, 1991a. Recent magnesite-hydromagnesite sedimentation in playa basins of the Cari-

98

boo Plateau, British Columbia. Pap. B.C. Geol. Surv. Branch 1991-1: 279–288.

Renaut, R. W. & D. Stead, 1991b. Carbonate-evaporite sedimentation at Clinton Lake, British Columbia. Prog. Abstr., Sedimentary and Paleolimnological Records of Saline Lakes Conf., Saskatoon, Canada: 41.

Renaut, R. W., D. Stead & R. B. Owen, in press. The saline lakes of the Fraser Plateau, British Columbia, Canada. In E. Gierlowski-Kordesch & K. Kelts (eds), Global geological record of lake basins, Vol. 1. Cambridge University Press.

Riding, R., 1991. Classification of microbial carbonates. In R. Riding (ed.), Calcareous algae and stromatolites. Springer, Berlin: 21–51.

Rollins, L., 1988. PCWATEQ: A simple, interactive PC version of the water chemistry analysis program WATEQF.

Rosen, M. R., D. E. Miser & J. K. Warren, 1988. Sedimentology, mineralogy and isotopic analysis of Pellet Lake, Coorong region, South Australia. Sedimentology 35: 105–122.

Scudder, G. G. E., 1969. The fauna of saline lakes on the Fraser Plateau in British Columbia. Verh. int. Ver. Limnol. 17: 430–439.

Smith, A. M. & T. R. Mason, 1991. Pleistocene, multiple-growth, lacustrine oncoids from the Poacher's Point Formation, Etosha Pan, northern Namibia. Sedimentology 38: 591–600.

Smoot, J. P. & T. K. Lowenstein, 1991. Depositional environments of non-marine evaporites. In J. L. Melvin (ed.), Evaporites, petroleum and mineral resources. Elsevier, Amsterdam: 189–347.

Smoot, J. P. & P. E. Olsen, 1985. Massive mudstones in basin analysis and paleoclimatic interpretation of the Newark Supergroup. Circ. U.S. Geol. Surv. 946: 4–10.

Southgate, P. N., I. B. Lambert, T. H. Donnelly, R. Henry, H. Etminan & G. Weste, 1989. Depositional environments and diagenesis in Lake Parakeelya: a Cambrian alkaline playa from the Officer Basin, South Australia. Sedimentology 36: 1091–1112.

Springer, M. E., 1958. Desert pavement and vesicular layer of some desert soils in the desert of the Lahontan Basin, Nevada. Proc. Soil. Sci. Soc. Am. 22: 63–66.

Surdam, R. C. & C. A. Wolfbauer, 1975. The Green River Formation – a playa-lake complex. Bull. Geol. Soc. Am. 86: 335–345.

Surdam, R. C. & J. L. Wray, 1976. Lacustrine stromatolites, Eocene Green River Formation, Wyoming. In: M. R. Walter (ed.), Stromatolites. Elsevier, Amsterdam: 535–541.

Thompson, J. B. & F. G. Ferris, 1990. Cyanobacterial pre-cipitation of gypsum, calcite and magnesite from natural lake water. Geology 18: 995–998.

Topping, M. S. & C. G. E. Scudder, 1977. Some physical and chemical features of saline lakes in central British Columbia. Syesis 10: 145–166.

Truc, G., 1978. Lacustrine sedimentation in an evaporitic environment: the Ludian (Palaeogene) of the Mormoiron basin, southeastern France. In: A. Matter & M. E. Tucker (eds), Modern and ancient lake sediments. Blackwell, Oxford: 189–203.

Tufnell, L., 1975. Hummocky microrelief in the Moor House area of the Northern Pennines, England. Biul. Perygl. 24: 353–368.

Valentine, K. W. G. & A. Schori, 1980. Soils of the Lac La Hache-Clinton area, British Columbia. B.C. Soil Surv. Rep. 25, 118 pp.

Van Houten, F. B., 1980. Late Triassic part of Newark Supergroup, Delaware River section, west-central New Jersey. In: W. Manspeizer (ed.), Field studies of New Jersey geology and guide to field trips. 52nd Annual Meeting, New York State Geological Association, New York: 264–276.

Verger, F., 1964. Mottureaux et gilgais. Annal. Géogr. 73: 413–430.

von der Borch, C. C., 1965. The distribution and preliminary geochemistry of modern carbonate sediments of the Coorong area, South Australia. Geochim. Cosmochim. Acta 29: 781–799.

von der Borch, C. C., 1976. Stratigraphy of stromatolite occurrences in carbonate lakes of the Coorong Lagoon area, South Australia. In: M. R. Walter (ed.), Stromatolites. Elsevier, Amsterdam: 413–420.

von der Borch, C. C., B. Bolton & J. K. Warren, 1977. Environmental setting and microstructure of subfossil lithified stromatolites associated with evaporites, Marion Lake, South Australia. Sedimentology 24: 693–708.

Walter, M. R. (ed.), 1976. Stromatolites. Elsevier, Amsterdam, 790 pp.

Walter, M. R., S. Golubic & W. V. Preiss, 1973. Recent stromatolites from hydromagnesite and aragonite depositing lakes near Coorong Lagoon, South Australia. J. Sed. Petrol. 43: 1021–1030.

Warren, J. K., 1990. Sedimentology and mineralogy of dolomitic Coorong lakes, South Australia. J. Sed. Petrol. 60: 843–858.

White, A. H. & B. C. Youngs, 1980. Cambrian alkali playa-lacustrine sequences in the northeastern Officer Basin, South Australia. J. Sed. Petrol. 50: 1279–1286.

Williams, P. J. & M. W. Smith, 1989. The frozen earth: Fundamentals of geocryology. Cambridge University Press, Cambridge, 306 pp.

Effects of microbial activity on the hydrochemistry and sedimentology of Lake Logipi, Kenya

Sabine Castanier[1], Marie-Claire Bernet-Rollande[2], André Maurin[2] & Jean-Pierre Perthuisot[3]

[1] *Faculté des Sciences, Université d'Angers, 2 Bd. Lavoisier, Belle Beille, F-49045 Angers cedex; Service of Microbiogeology of the Laboratory of Biogeology, University of Nantes; IGPC 252;* [2] *TOTAL Compagnie Française des Pétroles, cedex 47, F-92069 Paris la Défense;* [3] *Laboratory of Biogeology, Faculté des Sciences, Université de Nantes, 2 rue de la Houssinière, F-44072 Nantes cedex 03. IGPC 252*

Key words: soda lake, bacteria, sediment, recent, Kenya

Abstract

Lake Logipi is a saline soda and alkaline lake which marks the northern termination of the Suguta River drainage system. It also receives waters from streams, possible seepage from Lake Turkana, and hot springs. Present hydrochemistry and sedimentology is controlled by numerous factors including seasonal variations, composition of incoming waters, water depth and, above all, bacterial activity. Given the scarcity of Ca^{2+} and Mg^{2+} in the lake waters, bacterial activity seems to intensify the alkalinization of the waters which inhibits the deposition of organic matter and leads to the genesis of a poorly organic, zeolitic mud that reaches 1.5 m in tickness in the deepest part of the lake. This black layer may be overlaid with thin crusts of trona and halite which prograde over the basin from its southern bank when the lake is drying out and which are dissolved in the lake waters during the rainy season.

Introduction

Lake Logipi is the northernmost saline lake of the Kenyan Eastern Rift (Fig. 1). Within the limits of a former Pleistocene lake, it marks the northern termination of the Suguta River endorheic drainage system. This system is isolated from Lake Turkana by a transverse volcanic barrier (the Barrier) joining the southern Turkana and Nyiru breakaways or distensive faults (Fig. 2). The position of this barrier coincides with an 'accommodation zone', *i.e.* a zone where the terranes accomodate to the displacements of both faults (Bosworth, 1989; Lambiase & Bosworth, in press).

Lake Logipi is surrounded on the west, north and east sides by volcanic formations, mainly ba-

salts, and superficial detrital covers. The alluvial plain of the Suguta River extends southward with delta fans along its eastern flank (Fig. 3).

The region may be considered as a semi-desert caused partly by foehn effects on the rift depression (the foehn is a dry and hot south wind blowing down from the Alps to the upper valley of the Rhône River). The rainfall is less than 300 mm per year (Gwynne, 1969) but is very irregularly distributed with generally two monsoon periods, spring and fall.

The present paper addresses some biogeodynamical aspects of the behavior of Lake Logipi by taking into account observations and studies made during the course of the Logipi Project, primarily during a few days of field work in November, 1988. This project was aimed at provid-

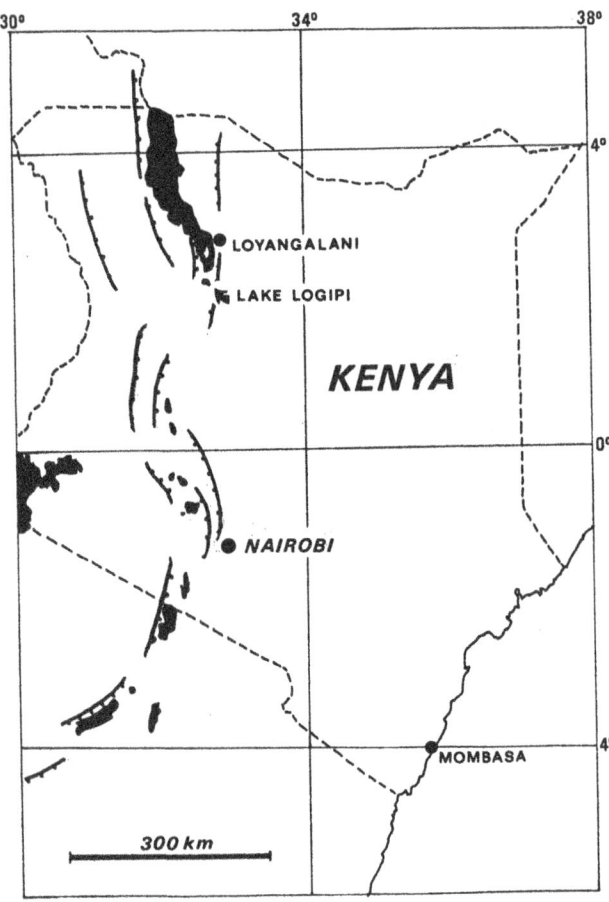

Fig. 1. Location of Lake Logipi.

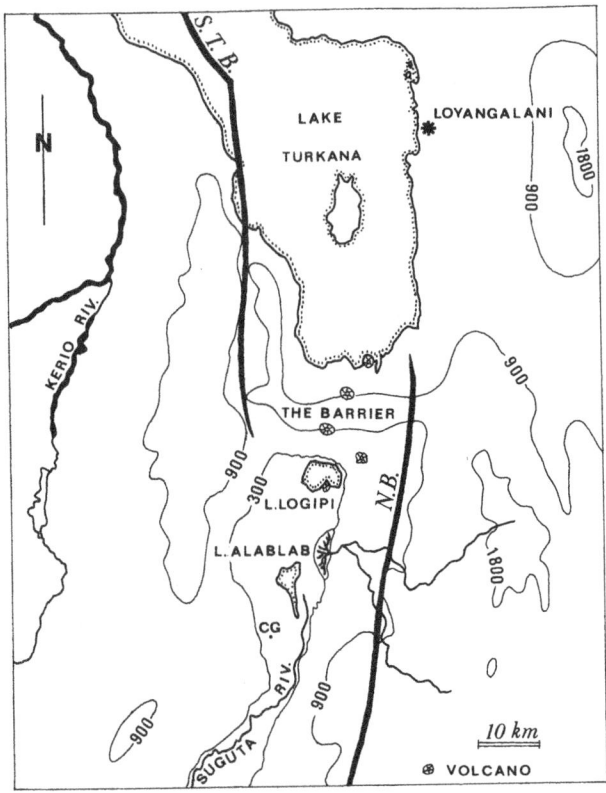

Fig. 2. Schematical map of South Turkana region (Modified after Lambiase and Bosworth, in press). S.T.B.: South Turkana breakaway; N. B.: Nyiru breakaway; CG: Crescent Geyser.

ing a modern model in order to improve understanding of the Kenyan fossil rift basins in which oil exploration had been undertaken. The preparation of the project included a short helicopter trip in November, 1985. A few samples of water and sediments were also collected in 1990 during a short additional expedition. The present paper is, therefore, not an exhaustive study, but aims at offering most of collected data and discussions on a highly inaccessible lacustrine system.

Hydrology

The water level of Lake Logipi depends upon water discharge and evaporation. The main water discharge comes from the Suguta River and other less important tributaries so that it varies widely as a function of rainfall in the river basin. The floods induce lacustrine transgressions when water invades the whole of the bottom of the rift and forms a single lake, inundating Lake Logipi and Lake Alablab, as in September 1975 (Fig. 4). During dry periods the water body progressively shrinks and may be reduced to small salt pans restricted to the deepest parts of both lakes. At its maximum extension we estimate the lake has a maximum depth of 3–5 m. The lake is also fed with freshwater by a few wadis (Pool) and by aquifers that emerge on the southern slope of the Barrier notably at Pelicans Bar (Fig. 5). These waters could come *via* a subterranean route from Lake Turkana although this has yet to be proven.

Lake Logipi is also fed by hot springs emerging on Central Island (Cathedral Rocks) and on the southern slope of the Barrier.

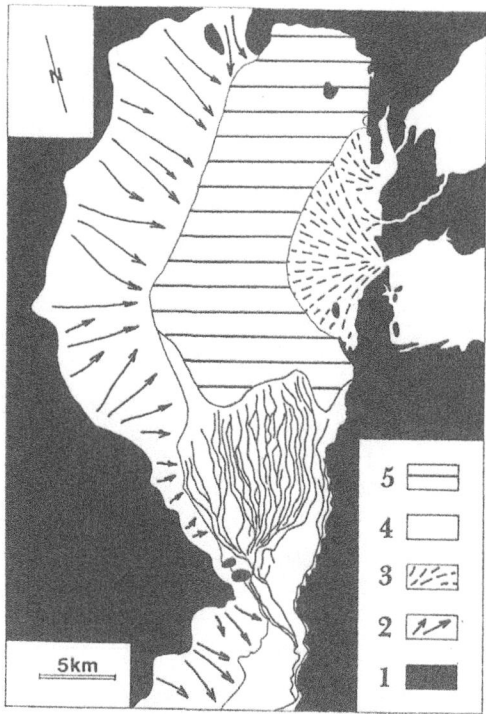

Fig. 3. Geological sketch map of Lake Logipi region. 1. Basalts; 2. Detritical coverspreadings; 3. Delta fans; 4. River alluvium; 5. Lacustrine deposits.

At the end of the dry periods, Lake Logipi and Lake Alablab remain partly flooded because of these inflows and represent windows in the water table of the alluvial plain. When Lake Logipi reaches its minimum level, three ponds remain. A large one is situated along the northern shore and two tiny ones, north and south of Central Island, correspond to hot springs discharges (Fig. 4, February 1987).

Hydrochemistry

Materials and methods

The October 1988 expedition took place during an intermediate stage of the lake, similar to the March 1975 situation. On October 20, 21, 22 and 26 several variables were measured directly in the field: pH and Eh (with a pHmeter CG 837 F Schott Gerate), dissolved oxygen content (with a DOmeter HI 8543 Bioblock) at 11 stations, and salinity at 59 stations (Table 1; Fig. 5). Salinity was measured by means of a refractometer calibrated with pure NaCl solutions (g l^{-1}). This is, of course, a very approximate method, especially since the various dissolved salts do not necessarily have the same effect as NaCl on the refractive index of water. Dissolved organic compounds may furthermore be present in the water. It is, nevertheless, a very easy method which gives rapid information in the field.

Lake water samples were collected in the morning (October 20, 21, 22 and 26) at 12 stations and analyzed in the evening in a field laboratory carried from France and arranged in a room of the Oasis Lodge in Loyangalani. Analyses were performed using spectrophotometric methods for SO_4^{2-} (Rodier, 1984), NH_4^+ (Solorzano, 1969), NO_2^- (Rodier, 1984); we had no mean to measure the NO_3^- concentrations) and total alkalinity by acidification (Rodier, 1984, modified after Castanier, 1987). Alkalinity is due to carbonate, bicarbonate, borate and hydroxyl ion concentrations. Hydroxyl concentration is negligible even at pH 9.5. Borate concentrations in the lake and surrounding water sources have not been measured but they are probably low so that alkalinity is essentially due to carbonate and bicarbonate ions concentrations. Data are given Table 1.

A brine sample (LL) was also collected in February 1990 when the lake was near its lowest level. Analyses were made back in the laboratory two months later for major ions (Na$^+$ and K$^+$ by flame emission spectrometry, Ca^{2+} and Mg^{2+} by flame atomic absorption spectrometry, Cl$^-$ by titrimetry (Rodier, 1984), SO_4^{2-} by turbidimetry (Rodier, 1984), silica and phosphate by colorimetry (Rodier, 1984), and trace metals by electrothermal atomic absorption spectrometry. As the sample was stored in a refrigerator and there were no deposit in the bottle we assume the obtained values are representative for the above components. Ammonium and nitrate (after reduction to nitrite) were also analysed but they might have varied during storage and the results are possibly questionable (Table 2).

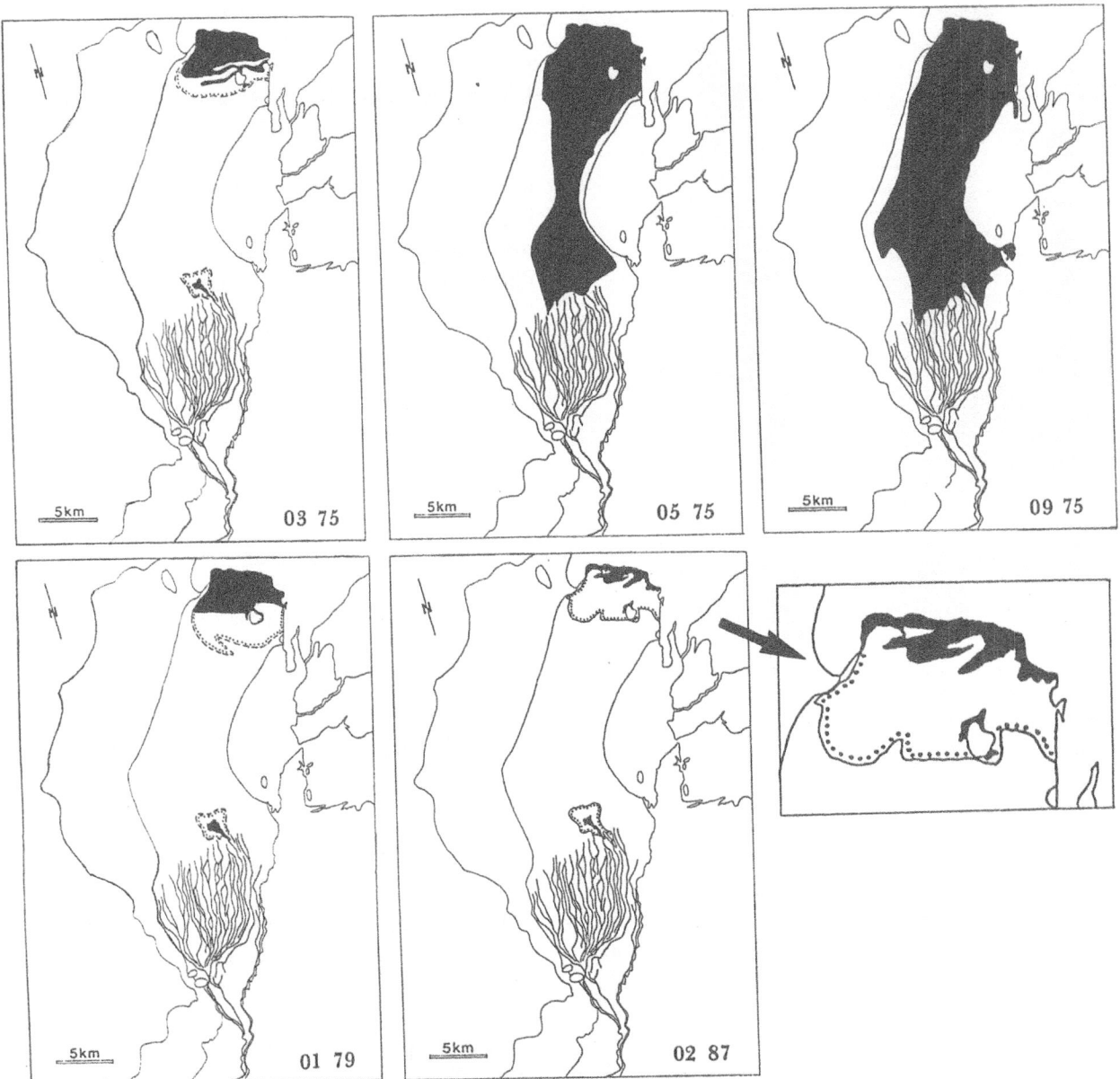

Fig. 4. Several stages of the extension of Lake Logipi, 1975–1987. Dotted line indicates the southern margin of salt crusts.

Several inflows were sampled in October 1988 (Fig. 5). These were: inlet of a small stream east of the lake and pool a few hundred meters upstream (PL 1, PL 2); interstitial water in oxygenated and reduced sediment, respectively, at Pelicans Bar (PB 1, PB 2); waters sampled directly in the vent of the hot spring (70 °C) in the southern part of Central Island, in the effluent channel and in the inundated sand downstream (SV 1, SV 2, SV 3); and water from the vent of the hot spring (54 °C) in the northern part of Central Island and interstitial water in the inundated sand downstream (NV 1, NV 2). They were analysed in the field laboratory following the same methods as mentioned above. Data are given Table 3.

. Two more samples were collected in February

Table 1. Hydrochemical data on Lake Logipi waters, October 1988. For location of stations see Fig. 5.

Station no.	Salinity ($g\,l^{-1}$)	Eh mV	DO ppm	SO_4^{2-} ($G\,l^{-1}$)	NH_4^+ ($mg\,l^{-1}$)	NO_2^- ($\mu g\,l^{-1}$)	Alkalinity ($g\,l^{-1}\,HCO_3^-$)	pH
0				0.900	1.718	22.3	2.722	9.54
1	12	+89	12.8	0.938	2.440	30.3	2.573	9.53
2	5	+87	12.0				2.486	9.53
3	10	+86	11.6				2.560	9.55
4	2	+86	13.1	0.750	2.637	43.7	2.593	9.54
6	13	+90	8.9				2.269	9.52
8	10						2.595	9.52
9	12	+87	7.7				2.542	9.61
18	10	+86	7.5	0.900	2.232	85.0	2.625	9.53
25	5			0.788	1.400	78.3	2.734	9.54
29	15	+87	7.2	0.375	1.390	71.9	2.584	9.55
43	17			0.412	2.341	118.3	2.678	9.56

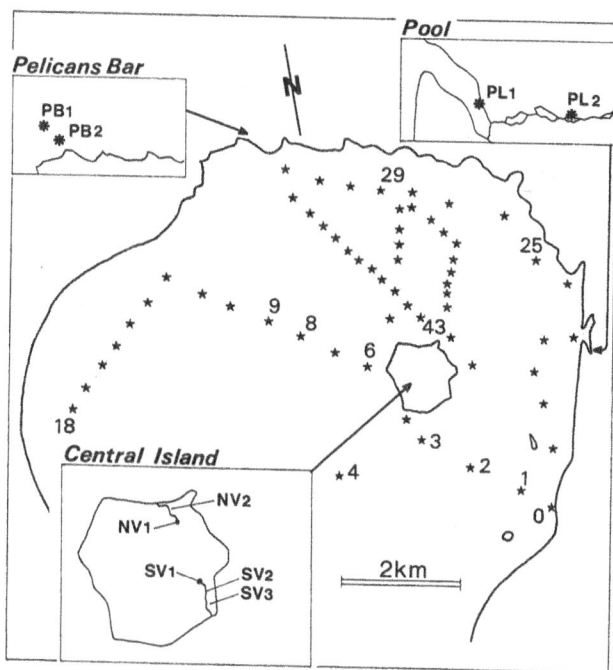

Fig. 5. Location of hydrochemical measurements and sampling stations, October, 1988. Stars: lake waters; dots and asterisks: surrounding inflows.

Table 2. Hydrochemical data of samples collected in February 1990. Results are expressed in $mg\,l^{-1}$, n.a.: not analysed.

Variable	Lake Logipi brine (LL)	Central Island Pond (SV)	Crescent Geyser (CG)
pH	10.5	9.5	9.5
Ca^{+2}	0.63	0.14	0.02
Mg^{2+}	0.09	0.09	0.03
Na^+	17500	2650	4650
K^+	382	188	194
Cl^-	n.a.	5.0	n.a.
SO_4^{2-}	1600	50	1150
HCO_3^-	n.a.	n.a.	n.a.
SiO_2	83.8	140.5	86.7
Al	12.5	23.0	25.5
Fe	4.4	1.4	0.6
Mn	0.06	0.02	<0.02
Zn	0.36	0.20	0.14
Pb	<0.01	<0.01	<0.01
Cu	<0.01	<0.01	<0.01
NO_3^-	1.45	1.70	0.70
NH_4^-	0.06	n.a.	0.04
PO_4^-	0.25	51.7	0.25

1990. The first one was collected in the Crescent Geyser (CG), SW of Lake Alablab (Fig. 2), and the second one was collected in the pond receiving the water from the hot spring in the southern part of Central Island (SV). They were analysed in the same ways as sample LL (Table 2).

Results and discussion

The isohaline map of lake waters indicates probable clockwise water movements as fresh water inputs mainly come from the south of the basin and from the northern bank of the lake (Fig. 6). The central position of the most concentrated

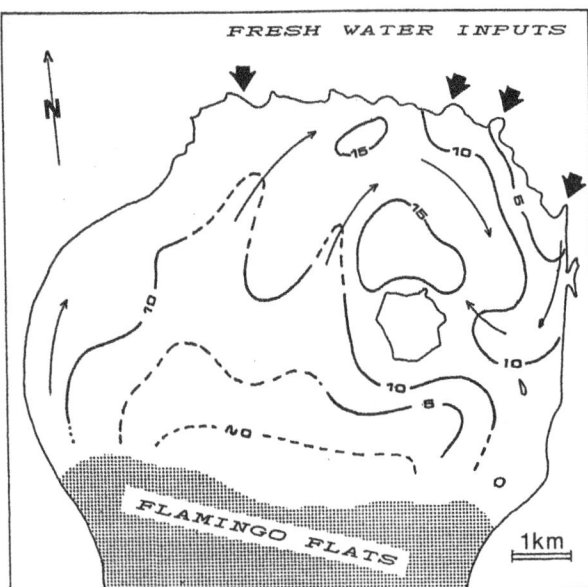

Fig. 6. Salinity map of Lake Logipi surficial waters, October, 1988, Salinity expressed as optical equivalent of $g\,l^{-1}$ NaCl.

waters must be interpreted partly in terms of the recycling of brines and evaporitic salts deposited in the central part of the basin at the end of previous dry period.

Measured pH and Eh values were rather constant all over the basin (respectively around 9.6 and $+90$ mV). The dissolved oxygen content in surface waters varied from 11–12 ppm south of Central Island to 6–7 ppm in the central part of the lake. Values rapidly decreased with depth especially in the central part of the basin where

Table 3. Hydrochemical data of surrounding inflows, October 1988. For location of stations see text and Fig. 5.

Station	SO_4^{2-} $(g\,l^{-1})$	NH_4^+ $(\mu g\,l^{-1})$	NO_2^- $(\mu g\,l^{-1})$	Alkalinity $(g\,l^{-1}\ HCO_3^-)$	pH
PL 1	0.825	1.335	18.7	2.585	9.51
PL 2	1.312	0.208	0.0	4.519	9.52
PB 1	3.337	1.696	62.3	4.084	9.39
PB 2	1.425	2.605	87.5	3.953	9.34
SV 1	2.475	18.186	17.7	6.972	9.44
SV 2	2.437	17.879	4.0	7.043	9.45
SV 3	2.700	31.405	27.8	7.430	9.40
NV 1	1.313	0.000	2.8	1.067	9.40
NV 2	0.750	7.254	175.5	2.104	9.40

nearly anoxic conditions were reached around 1.5 m below the surface.

The lake waters total alkalinity due to carbonate and bicarbonate ions is very high which corresponds to the high pH. Sulfate concentration decreases towards the saltier central part of the basin whereas nitrite concentration increases. This might be related to higher bacterial reducing activities in the deepest part of the basin. Besides, the constant and high ammonium concentrations together with the high nitrite and, presumably, nitrate (Table 2) concentrations in the surface water reflect the efficiency of the bacterial nitrogen cycle and the accumulation of soluble nitrogen compounds in the lake.

The brine collected in 1990 is very poor in Ca^{2+} and Mg^{2+} which is rather surprising in such a mafic geological environment (Table 2). This will be discussed later. On the other hand the high Fe, Mn and Zn concentrations logically result from this environment. Lastly, the high pH accounts for the high Al and silica content of the lake brine.

The inflow waters, especially the hot springs, are generally richer in sulfate ions, in nitrogen compounds and in carbonate and bicarbonate ions than the lake water itself. It is possible that some of the inputs from hot springs comprises recycled, infiltrated brines. More generally, all water sources seem to reflect a common hydrochemical background reflecting the regional geological and bio-climatic conditions.

Lake Logipi acts as a concentrator for most of dissolved ions and compounds coming from the environment *via* streams, aquifers and hot springs. Additionally, variations in the flow of the Suguta River induce changes in the salinity and composition of the lake waters. For instance, each flooding of the river induces a dilution of the lake brines and a dissolution of the previously deposited salt crusts.

Sedimentology

Materials and methods

In October 1988 no salts deposits were evident on the lake bottom. On October 23 and 24, surficial

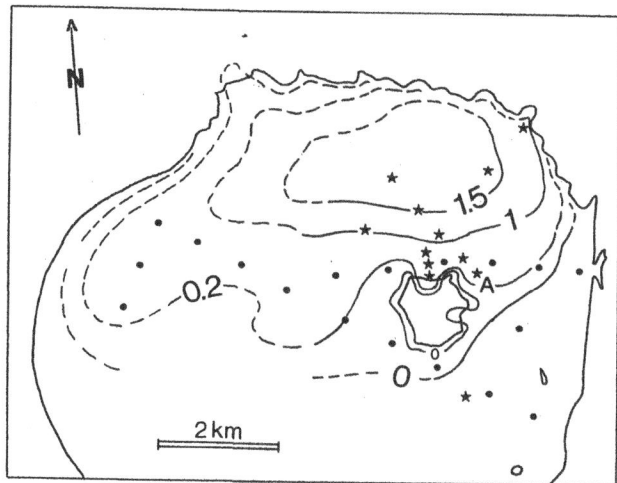

Fig. 7. Location of sediment sampling stations (October, 1988) and schematical map of the black layer thickness (Isopachs in m). Dots: samples directly observed in the field,; stars: collected cores.

lake sediments were cored at 17 stations and examined directly on the boat. A few samples were collected. On October 24 and 25, 12 cores were collected (Fig. 7). They were stored in cool boxes with carbonic ice (dry ice) and sent to France within a few days. Core A (60 cm long) was the only one to be studied in the laboratory by X-Ray diffractometry, SEM imagery and microprobe analysis. Three samples taken from this core (top, middle, and base of the black layer) were analysed for interstitial water content (weight loss at 60 °C after 24 h), organic matter and linked water content (weight loss at 450 °C after 24 h of dessicated sample at 60 °C), organic carbon content following the Anne's method by Ottman (1960), organic nitrogen and N-NH$_4^+$ with a distillation unit (Büchi 320) and a mineralisation ramp

(Büchi 425), and carbonate content (expressed as CO$_3^{2-}$ % of dry weight) by acid attack (HCl N/10). Results are given in Table 4.

Two samples of salt crusts a few millimeters thick were collected in January 1990 and analysed through X-Ray diffractometry.

Results and discussion

There was no sharp interface between water and sediment, but under the very turbid water, there was a zone of black mud that became denser with depth and contained fish (*Tilapia?*) remains. The very thin superficial sediments were smooth and black. They lay on a hardened reddish layer which prevented the penetration of coring tubes. A map of the approximate thickness of the black layer is given Fig. 7. It reaches a maximum thickness (around 1.5 m) in the deepest part of the basin. The black layer disappears southwards and the reddish layer is directly overlain by microbial mats on which flamingos feed.

X-Ray diffractograms of the black sediments were very difficult to interpret because of the interference of many mineral species and because of the relatively low peaks heights with respect to the background. Most of them must have been zeolites with clay minerals and, probably, sodium silicates. Neither pyrite nor carbonates were detected. SEM imagery (Fig. 8A, B, C) and microprobe analyses (Fig. 8A', B', C') tended to confirm the above interpretation. Such minerals would be expected from an environment of this kind (Eugster, 1967, 1970; Maglione, 1974; Zins-Pawlas, 1988). The sedimentary particles include clay minerals (Fig. 8A), several forms of zeolites

Table 4. Analyses of core A (for location see Fig. 7).

Level	Weight loss at 450 °C %	Org. C %	Org. N %	NH$_4^+$ μg g^{-1}	C/N	Water content %	Carbonates as CO$_3^{2-}$ %
Top (mud)	10.62	0.30	0.05	0.040	5.8	54.9	7.30
Middle (30 cm)	9.52	0.34	0.05	0.026	7.1	52.6	6.50
Base (60 cm)	10.13	0.24	0.04	0.068	5.9	51.7	7.41

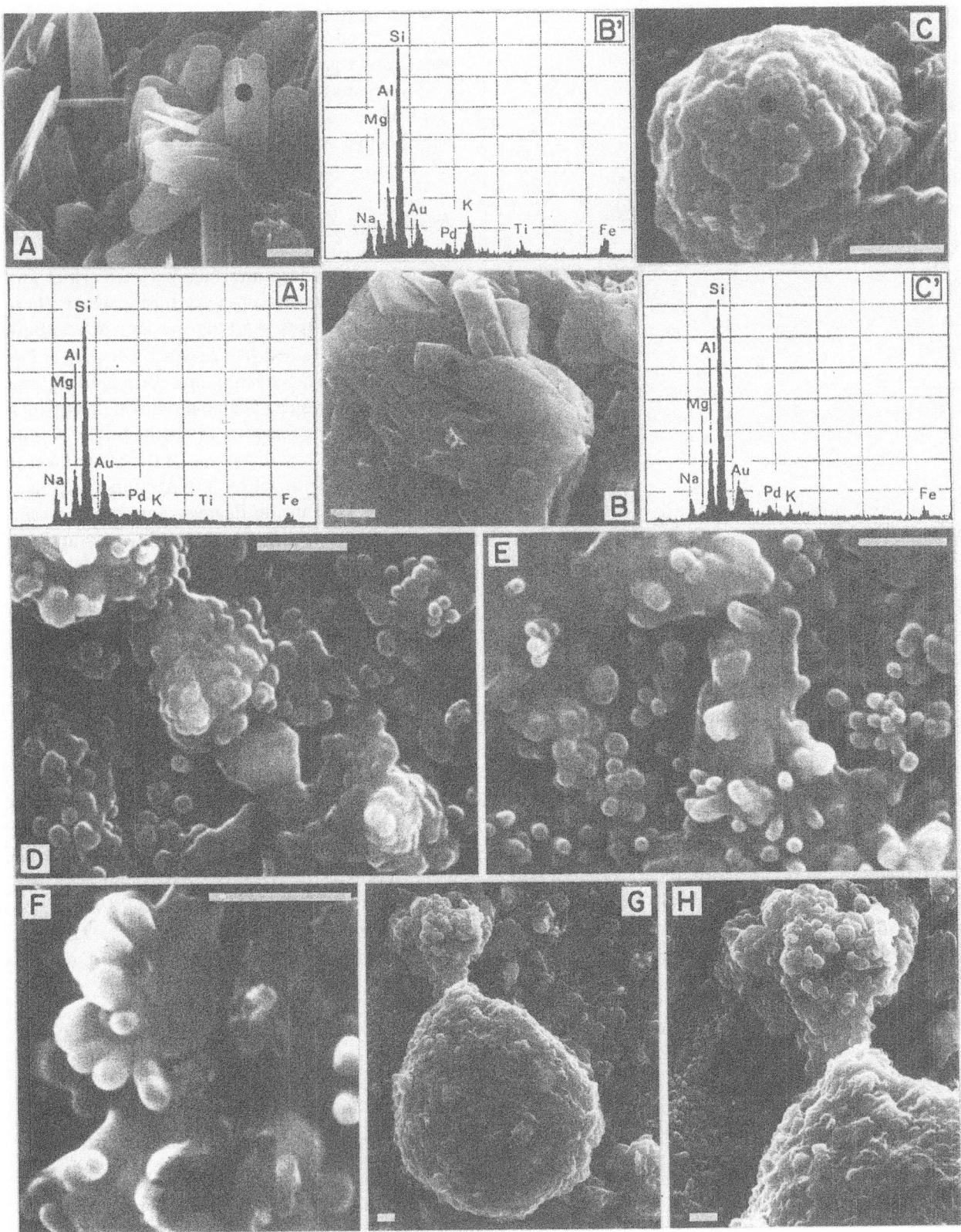

among which probable clinoptilolite (Fig. 8B). No pyrite, which is easily recognizable through SEM, was detected.

The analysed aluminosilicates display noticeable amounts of Na and variable amounts of Mg and K but are devoid of Ca. The scarcity of Ca^{2+} in the water and absence of Ca-carbonates in the lake sediments suggest that most of the calcium released by weathering in the hydrological basin is trapped before reaching the lake. In numerous neighboring areas the pebbles and granules lying on the soil are discretely coated on their lower side with a thin carbonate crust, probably of bacterial origin. Thus, calcium must be retained upstream from the lake by the inconspicuous but widespread precipitation of carbonates in and at the surface of the soils. Some of the magnesium released from the mafic formations of the basin may behave in the same way.

It is worth noting that most of the aluminosilicates display noticeable Fe contents. Given the absence of sedimentary pyrite, in such an alkaline environment Fe seems to be preferentially linked to aluminosilicates rather than to sulfides even though the lake water contains large amounts of Fe and bacterial activity produces hydrogen sulfide.

The black sediments are very poor in organic matter (Table 4). It is even possible that a large part of organic matter remains in solution in the pore water of sediments. Thus, the black color of sediments is probably due to the presence of Fe in the lattices of silicates or aluminosilicates.

Caution should be taken in interpreting C/N ratios (Table 4) because the concentration of nitrogen is low and any small errors in its measurement will considerably change this ratio. Nevertheless, the three values obtained are low which tends to indicate a microbial origin for the organic matter in sediment and/or a preferential early biodiagenesis of carbohydrates that produces large quantities of carbonate and bicarbonate ions (Zajic, 1969; Castanier, 1987). This is in accordance with the high alkalinity of the lake waters.

A large number of sedimentary particles of the black layer seem to be composed of bacterial cells coated with cocoon-like mineral matrix (Fig. 8F) which often form more or less spherical (Fig. 8C, D), plate-like or urchin-like (Fig. 8E) bio-mineral assemblages which seem to be usually composed of aluminosilicates (Fig. 8C'). These could be amorphous compounds such as those that accumulate in the early stages of the formation of bacterially-produced carbonate bio-mineral assemblages (Castanier, 1987; Castanier et al., 1988).

During the course of SEM exploration a very strange structure was discovered (Fig. 8G). It is composed of an avoid body and a cone-shaped appendix which are separated on the micrograph but seem to have been linked together. A close-up (Fig. 8H) shows that the appendix is hollow and its wall is made of several layers of mineralized bacterial cells. Here is not the place to discuss further this obviously biological structure. In our opinion, it has to be placed, with similar structures we have found in other modern environments (Castanier, 1987, and unpublished studies) and with fossil counterparts known from Precambrian times (Schopf & Walter, 1983), among a group that might be termed 'metaprokaryotes'.

The reddish layer overlain by the black sediment has a similar mineralogical composition as shown by X-Ray diffractometry but it is likely that the iron has been released from silicates or aluminosilicates to form oxides or hydroxides.

Fig. 8. Some nannofacies of sediments in core A and microprobe analyses. Black dots indicate analyses points. Scale bars are 1 μm. A: Probable clay minerals; B: Probable zeolite (clinoptilolite?). Notice the round body being included in crystal (possible bacterial body); C: Mineral assemblages. The round body could be of biological origin and constituted of amorphous compounds; A', B' & C': microprobe analyses of mineral particles; D: Globular or more or less tetrahedral bio-mineral assemblages. E: First stage of bio-mineral build-up. Numerous bacterial cells are trapped in mineral assemblages which are probably composed of amorphous compounds; F: Some bacterial cells have been separated from tiny bio-mineral assemblages during SEM preparation which lets appear they are coated by a rigid mineral matrix or cocoon; G: *Incertae sedis* biological structure with an ovoïd body and an appendix. H: Close up of preceding micrograph showing detail of the appendix and structure of its wall.

This probably occurred in a period when the lake dried out completely for a long time.

The salt crust which covered the lake on January 1990 comprised two parts. The outer one was composed of halite, the inner one was trona (Na_2CO_3, $NaHCO_3$, $2H_2O$). These crusts are dissolved when floods occur and, according to satellite imagery, they cover the basin northward starting from the southern bank.

Microbiology

Materials and methods

On October 20–26, 1988, samples for counting living bacteria were collected aseptically in the field early each morning. The culture media were inoculated in the field laboratory within 3 h in order to obtain reliable counts.

Water samples were collected at the following points (Fig. 5 & 7): station 43 in the lake; PL 1 and PL 2; PB 1 and PB 2; SV 1, SV 2 and SV 3; NV 1 and NV 2. Sediment samples were: Core A

(mud and sediment); SV 1 and SV 3; NV 1 and NV 2; and a piece of silicified wood from Central Island.

The media inoculated for counts were: for heterotrophic strict aerobic bacteria, the solid medium 2216E (Oppenheimer & Zobell, 1952); for heterotrophic aerobic bacteria able to precipitate carbonates (here after called carbonate-precipitating, bacteria), the solid medium of Castanier (1987); for heterotrophic strict and facultative anaerobic bacteria a liquid culture medium described by Marty (1981) prepared following the technique of Hungate (1969) and inoculated following the method of the most probable number (MPN) (McCrady, 1918); and for the sulfate-reducing bacteria the liquid medium of Marty & Garcin (1987) prepared following the technique of Hungate (1969) and inoculated following the method of the most probable number (MPN) (McCrady, 1918). After 15 days incubation at room temperature counts were performed by the Service of Microbiogeology in Nantes. Results are given Table 5.

As bacterial carbonate precipitation may be

Table 5. Bacterial numerations on water (W) and sediment (S) of lake samples and surrounding inflows. Counts are numbers of living cells in 1 ml water or 1 g sediment. For location of samples see text and Figs 5 & 7.

Sample	Heterotrophic aerobic bacteria		Heterotrophic anaerobic bacteria	
	Total	Carbonate-precipitating	Total	Sulfate-reducing
St. 43 (W)	1 300 000	675 000 (52%)	600	6 (1%)
Core A (mud)	2 320 000 000	39 500 000 (2%)	20 000 000	1 300 000 (7%)
Cora A (S)	3 860 000 000	52 100 000 (1%)	600 000	6 000 (1%)
PL 1 (W)	36 700 000	6 540 000 (18%)	700 000	6 000 (1%)
PL 2 (W)	385 000 000	50 000 000 (13%)	20 000	2 500 (13%)
PB 1 (W)	76 700 000	16 500 000 (22%)	5 000	6 (<1%)
PB 2 (W)	1 080 000 000	600 000 000 (56%)	700 000	250 (<1%)
SV 1 (W)	32 900 000	4 400 000 (13%)	1 300	0 (0%)
SV 1 (S)	72 000 000	14 500 000 (20%)	12 000	25 (<1%)
SV 2 (W)	140 000 000	17 500 000 (13%)	25 000	0 (0%)
SV 3 (W)	395 000 000	68 800 000 (17%)	1 100 000	60 (<1%)
SV 3 (S)	333 000	269 000 (81%)	20 000	25 (<1%)
NV 1 (W)	30 000 000	4 360 000 (14%)	1 100 000	600 (<1%)
NV 1 (S)	388 000 000	105 800 000 (27%)	250 000	20 000 (8%)
NV 2 (W)	161 000	63 700 (40%)	200 000	250 (<1%)
NV 2 (S)	3 790 000	2 100 000 (55%)	60 000	0 (0%)
Silicified wood	24 800 000	17 500 000 (71%)	700 000	2 500 (<1%)

unfamiliar to some readers, it is worth offering a few statements about it. Bacteria precipitate carbonates in two ways. First, there is passive precipitation resulting from an increase in the pH of the medium produced by the end-products of bacterial metabolism. This can be brought out by three metabolic pathways. These are: the anaerobic ammonification of amino-acids; the anaerobic and microaerophilic dissimilatory reduction of nitrate; and the anaerobic reduction of sulfate accompagnied by hydrogen sulfide production. Second, there is the active precipitation of carbonates by still poorly known membrane processes. This has been demonstrated under aerobic conditions but still remains conjectural for anaerobic ones (Castanier, 1987). From several experiments carbonate precipitation appears to be the reaction of bacterial populations to an increase in metabolisable organic substrates. When this occurs, the bacterial populations first increase rapidly, causing gradual increases in the concentrations of carbonate and bicarbonate ions and of other end-products of metabolism, and in pH. A steady state is reached within 24 to 48 hours. In these conditions and given the presence of Ca^{2+} and Mg^{2+} in the medium, active carbonate precipitation occurs first and is followed by passive precipitation (Castanier, 1987). The first carbonate particles seem to be generally composed of amorphous, perhaps hydrated, compounds (Castanier et al., 1988). If Ca^{2+} and Mg^{2+} are lacking, the precipitation of carbonates does not occur and carbonate and bicarbonate ions simply accumulate in the medium.

Results and discussions

Primary producers
A large part of microbial primary production seemed to occur on the shoals south of Central Island which were covered by extensive cyanobacterial mats. Cyanobacteria also colonized the hot springs vents, their effluent channels and the beds of peripheral streams. Bacterial populations in the lake itself include methane producing bacteria, the activity of which was demonstrated by

methane bubbles as assayed by the flame of a lighter. The color of the lake waters varied widely. In October, 1988, they were dirty pink whereas in November 1985, when the level was a little lower, they were deep green with pink patches. This could have represented purple and green photosynthetic bacteria populations changing in relation to oxygen availability. The surface oxygen content of the water was high for sulphur bacteria but let us recall that the dissolved oxygen content of the lake waters rapidly decreased with depth. Moreover, the microscopic observation of water showed the absence of algal phytoplankters that could have been responsible for the color of the lake water. Lastly, the lake sediments did not contain diatom frustules that might have lived during high stand periods, either because no diatom population developed in the lake or because their siliceous tests were rapidly dissolved. Thus, in Lake Logipi the bulk of primary producers must be prokaryotes.

Heterotropic bacteria
The populations of heterotrophic bacteria were much denser in the inflowing waters, notably the hot spring environments and the seepage zone of the Barrier (Pelicans Bar), than in the lake water itself, presumably because the primary production is higher there. Contrarily they are denser in the lake sediments than in sediments of the surrounding water inflows.

As far as heterotrophic aerobic bacteria populations are concerned, there is a clear increase in the proportion of carbonate-precipitating bacteria downstream, towards the lake where they dominate in water (Table 5). The high percentage of such bacteria in the 'silicified wood' collected on the beach near NV 2 is very typical. This means that the organic substrates in peripheral environments are colonized by carbonate precipitating bacteria and at least partly transformed into carbonate and bicarbonate ions. The same processes occur in the surface waters of the lake itself. In both cases, because of the scarcity of Ca^{2+} and Mg^{2+} in water the carbonate-precipitating bacteria are unable to precipitate carbonates, so that the carbonate and bicarbonate ions contents tend to increase in the inflowing waters as well as in the lake which accounts partly for their high alkalinity and pH.

The bacterial populations of the lake sediments are very dense with up to nearly 4.10^9 living cells per gram. This contrasts sharply with the low organic matter content of these sediments. In addition, even though carbonate-precipitating bacteria are numerous in sediments, they fall to a few percent of the total bacterial counts (Table 5). The organic matter produced in the lake or coming from the outside apparently is partly recycled into the bacterial populations themselves and partly transformed, through bacterial metabolism, into carbon dioxide and carbonate and bicarbonate ions with very little accumulation of particulate organic matter in the sediments. Lake Logipi appears to be a terminal pool of natural detergent, unable to sediment large quantities of organic matter at its present stage of evolution.

In such conditions of alkalinity and pH, the carbon dioxide produced by heterotrophic bacteria must be mostly converted to carbonate and bicarbonate ions. Additionally, a part of it is possibly recycled by primary producers so that probably very little carbon escapes the system as gaseous CO_2 or CH_4.

Anaerobic heterotrophic bacteria are much less numerous than aerobic ones, even in the few restricted environments studied and in the mud at the bottom of the lake. In this mud sulfate-reducing bacteria are fairly numerous, which explains the relatively low sulfate concentration with regard to the salinity in the water of the central part of the lake (Table 1). Bacterial sulfide oxidation probably occurs near the top of the water column but does not prevent a part of the produced hydrogen sulfide from escaping towards the atmosphere as attested by typical odors when travelling over the lake.

Many hydrochemical and sedimentological features of Lake Logipi are related to prokaryotic activity. This is true for all ecosystems but is particularly evident here because the prokaryotic activity is not camouflaged by other biological activity.

Higher organisms

None of the expedition members was a zoologist. Nevertheless, Lake Logipi is such an isolated place that we feel it might be useful to offer our observations on higher organisms.

In October 1988, birds flocks were distributed in two zones. On the southern shoals of the lake, hundreds of flamingos were feeding on microbial mats. The birds were trampling on the mats perhaps in order to break it up into small edible pieces but perhaps also to draw out small larvae or worms. Some insects (Coleoptera) were visible in the very shallow (up to 5 cm depth) shoreline waters. Elsewere observations were impossible because of the very high turbidity of water. We tried to collect zooplankton by means of a zooplankton net pulled by the boat. Under the microscope the collected material revealed no recognizable metazoans, only plant detritus and feathers. The feeding behaviour of flamingos produces round and linear tracks on the surface sediment. In the northern deeper part of the lake a few tens of pelicans were gathered, probably to feed. Indeed there were fish in the lake. Live ones were seen at the water surface and dead ones were occasionally brought to the surface by the boat's turbulence, but none was collected. They were approximately 20 cm in length. In November 1985, the lake was lower, smaller, saltier and probably less oxygenated. Part of the fish population was dead and floating on the surface especially in the northeastern corner of the lake, which attracted vultures. Again, no fish were collected but from photographs taken then, it appears they may be *Tilapia* sp.

The wastes of the lake's numerous birds enrich the lake water with phosphate which favours bacterial activities and with uric acid the bacterial degradation of which gives carbon dioxide and ammonia. These bacterial processes act in the same way as the ammonification of amino-acids which increases pH and enhances the carbonate and bicarbonate ions production. It is difficult to know if such processes are quantitatively significant. Besides, the disturbance and oxygenation of sediments caused by birds activities could be even more important but birds probably play an important role in the hydrochemical behavior of Lake Logipi.

Conclusions

From its hydrological situation Lake Logipi appears to be an accumulator for mineral ions and dissolved organic and inorganic compounds, even though there could be some recycling through hot springs.

At its present stage of evolution, and given the hydrochemical background, *i.e.* mainly the lack of Ca^{2+} and Mg^{2+} in incoming waters, the microbial processes (organic matter production and consumption) maintain a high alkalinity in the lake waters because they are chiefly responsible for carbonate and bicarbonate ions production. This induces:

- the precipitation of sodium aluminosilicates (perhaps with sodium silicates) and clay minerals from the silica and alumina saturated brines;
- the crystallization of ephemeral salt crusts (mainly Na-carbonates) at the last stage of brine evaporation;
- the retention in solution and biological recycling of most of the organic matter produced by the system itself or imported.

Both kinds of mineral deposits have to be interpreted in terms of concentration of solutions and in terms of biological activity of the sedimentary environment. As biological processes account at least partly for their genesis, they might be termed bio-evaporites even though, under other conditions, they may form in absence or without the intervention of organisms.

Lastly, it is necessary to keep in mind that carbonate production and/or precipitation are usual and widespread metabolic processes for bacterial populations in natural environments. Then, let us consider the whole basin of the Suguta River. Calcium and magnesium which are abundant in such a geological environment must be massively trapped in carbonates precipitated

at the surface or inside the soils of the slopes because the general bacterial activity is favored by the high temperatures. On the other hand, as rainfall is low, dissolution of carbonates is minor and very little Ca and Mg reach the terminal lake. This allows microbial activity to intensify the accumulation of carbonate and bicarbonate ions in the lake. Weathering processes within the basins (which are partly controlled by bacterial activity) and evaporative concentration are surely causes of the high pH and alkalinity of Lake Logipi and of most soda lakes but the influence of microbial processes on these characteristics must not be underestimated.

Acknowledgements

The authors are indebted to MARATHON, MOBIL and TOTAL petroleum companies which financially supported the mission of October 1988 and made possible the transfer to the field of a whole microbiogeological laboratory of microbiogeology, which was probably the first experience of that kind in Kenya. They also thank: Frances Westall for her careful clearing up the first english version; Stuart Hurlbert for his useful comments and suggestions, and final clearing up of the text; A. Nissenbaum and J. Melack who considerably helped us to precise statements and discussions; Maryvonne Piron-Frenet, Nelly Margerel, Alain Barreau and Alain Cossard for their diverse contributions to the present paper.

References

Bosworth, W., 1989. Basin and range style tectonics in east Africa. J. Afr. Earth Sci. 8: 191–201.

Castanier, S., 1987. Microbiogéologie: processus et modalités de la carbonatogenèse microbienne. State Doctorate Thesis, University of Nantes, France, 541 pp.

Castanier, S., A. Maurin & J.-P. Perthuisot, 1988. Les Cugnites: carbonates amorphes de Ca et Mg, précurseurs possibles de la dolomite. C. r. Acad. Sci., Paris, 306, II: 1231–1235.

Eugster, H. P., 1967. Hydrous sodium silicates from Lake Magadi, Kenya. Contr. Mineral. Petrol. 22: 1–31.

112

Eugster, H. P., 1970. Chemistry and origin of the brines of Lake Magadi, Kenya. Mineral. Soc. Amer. Spec. Paper 3: 215–235.

Gwynne, M. D., 1969. The South Turkana Expedition. Scientific Papers I. Preliminary report on the 1968 season. Geogr. J. 135: 331–342.

Hungate, R. E., 1969. A roll tube method for cultivation of strict anaerobes. In Norris and Ribbons (eds), Methods in Microbiology, Vol. 3B, Academic Press, London and New York: 117–132.

Lambiase, J. J. & W. Bosworth, 1991. Structural Controls on Sedimentation in Continental rifts. Geol. Soc. Amer. Bull., in press.

Maglione, G., 1974. Géochimie des évaporites et silicates néoformés en milieu continental confiné. Les dépressions interdunaires du Tchad, Afrique. State Doctorate Thesis, University P. & M. Curie, Paris, 334 pp.

Marty, D., 1981. Distribution of different anaerobic bacteria in Arabian Sea sediments. Mar Biol. 63: 277–281.

Marty, D. & J. E. Garcin, 1987. Présence de bactéries méthanogènes méthylotropes dans les sédiments profonds du détroit de Makassar (Indonésie). Oceanol. Acta 10: 249–253.

McCrady, M. H., 1918. Tables for rapid interpretation of fermentative tube results. Can. J. Publi. Health, 9: 201–216.

Oppenheimer, C. H. & C. E. Zobell, 1952. The growth and viability of sixty three species of marine bacteria as influenced by hydrostatic pressure. J. mar. Res. 11: 10–18.

Ottman, J. M., 1960. Essai de détermination qualitative et quantitative de quelques constituants de la matière organique dans un sédiment marin. Revue Geogr. Phys. Geol. Dynam., Paris, 3, 1: 49–52.

Rodier, J., 1984. L'analyse de l'eau, eaux naturelles, eaux résiduaires, eau de mer. Dunod, Paris: 564 pp.

Schopf, J. W. & M. R. Walter, 1983. Archean Microfossils: New Evidence of Ancient Microbes. In Schoph (ed.), Earth's Earliest Biosphere. Its origin and evolution. Princeton University Press, Princeton: 214–289.

Solorzano, L., 1969. Determination of ammonia in natural waters by the phenylhypochlorite method. Limnol. Oceanogr.: 14–779.

Zajic, J. E., 1969. Microbial biogeochemistry. Academic Press, New York & London, 247 pp.

Zins-Pawlas, M.-P., 1988. Géochimie de la silice dans les saumures et les milieux évaporitiques. Doctorate Thesis, University of Strasbourg, France, 200 pp.

Physico-chemical characteristics of a permanent Spanish hypersaline lake: La Salada de Chiprana (NE Spain)

B. Vidondo, B. Martínez, C. Montes & M. C. Guerrero *
Departamento Interuniversitario de Ecología, Facultad de Ciencias, Universidad Autónoma de Madrid, 28049 Cantoblanco, Madrid, Spain (author for correspondence)*

Key words: Water chemistry, anaerobic deep layer, planktonic green sulphur bacteria, microbial mats, paleochannels

Abstract

La Salada de Chiprana Lake, located in the Ebro River basin, northeastern Spain, is the only permanent and deep water hypersaline ecosystem in all of western Europe. With a total surface of 31 ha and a maximum depth of 5.6 m, it has several basins bounded by elongated sandstone-bodies or *ribbons* which are paleochannels of Miocene age. Its salinity varied from 30 to 73 g l^{-1} during the 1989 hydrological cycle and the most abundant ions were magnesium and sulphate. Depth-time distributions of major physico-chemical variables demonstrated that the lake was stratified in two distinctive layers during most of the year. The chemocline disappeared only in October, with the complete overturn of the water column. In the deep water, three conditions occurred which allowed development of green sulphur bacteria populations: (1) oxygen depletion, (2) presence of hydrogen sulphide and (3) presence of light. Benthic microbial mats covered the sediments of shallow shores of moderate slope.

Introduction

Hypersaline environments are known since Precambrian times. They present many environmental, social, economic and scientific values (Hammer, 1986; Williams, 1986). The present state of knowledge about the geochemistry and microbiology of the hypersaline systems is reviewed by Javor (1989).

Such ecosystems constitute a feature of the landscape of the arid and semiarid regions of Spain. The majority of saline lakes in Spain are shallow and ephemeral. These lakes have salinities ranging from 10 g l^{-1} to 400 g l^{-1} and depths of 0.05 to 0.70 m. They have mixed or sodium-chloride dominated ionic composition, and they show very high annual and interannual environ-

mental fluctuations (Montes & Martino, 1987; Comín & Alonso, 1988; Baltanas *et al.*, 1990).

In this context, La Salada de Chiprana Lake has great environmental value because it is a unique ecosystem in Western Europe. The basin of La Salada is a depression bounded by paleochannels (elongated sandstone-bodies), while all other Spanish saline lakes have basins of karstic, hydroeolic or tectonic origin. Its maximum depth is about 5.6 m, in contrast with the shallow character of other lakes. Because of this depth and of ground water inflow, La Salada is a permanent lake, showing very slight water level fluctuations. Its magnesium-sulphate dominated ionic composition also stands out from the sodium-chloride or mixed composition of other Iberian saline lakes.

The aim of this study is to describe the special

environmental characteristics of La Salada, the whole of which forms a suitable habitat for the development of two singular communities of phototrophic bacteria: planktonic green sulphur bacteria in the anoxic deeper water and benthic microbial mats which cover the flat sediments not deeper than 1.5 m.

Study area

La Salada (41° 14′ W, 0° 10′ N) is located in the south-central region of the Ebro River basin in northeastern Spain (Fig. 1). The lake is at 150 m elevation, and about 3.5 km from the Ebro River which here is at 130 m.a.s.l.

The Ebro River basin has a semiarid climate with 330 mm of annual precipitation. Annual mean temperature is 16 °C. The high moisture deficit during the summer season is intensified by dry winds from the northwest.

The natural shrub vegetation of the region has been replaced in most areas by agricultural fields, an increasing proportion of which are irrigated.

The substrate of the La Salada area is characterized by Miocene fluvial sediments that present lithological formations known as *ribbons* which are very singular due to their size, extent and state of conservation in this region (Riba *et al.*, 1967). Ribbons are sinuous old river beds (paleochannels) that were filled with sand and fossilized when the basin was later a site of sediment deposition. Afterwards, erosion of softer surrounding material left these elongated sandstone bodies as emergent structures (Friend *et al.*, 1986). This type of landscape is referred to as 'inverted relief'.

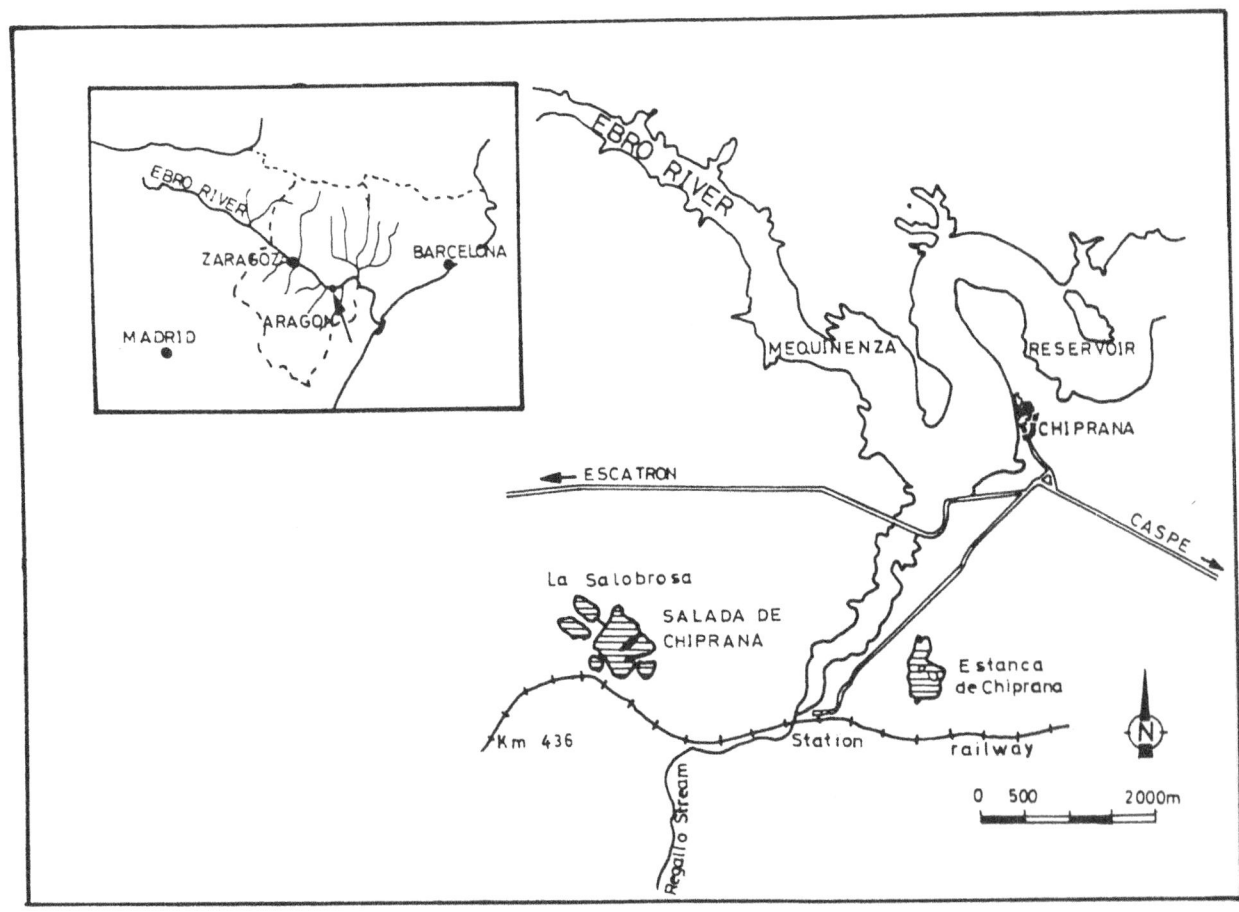

Fig. 1. Location of the study area.

The ribbons sometimes intersect, thereby delimiting depressions and playing an important role in the origin of some lakes in this area. In some cases, the ribbons can also act as conduits for ground water flow. Most of these depressions do not contain water because of the high moisture deficit of the region on the one hand, and losses by drainage on the other. Only those basins that receive ground water inflow ever contain temporary or permanent water bodies (Bernaldez, 1987). This probably explains the permanent nature of La Salada (Pueyo, 1978/79).

Apart from La Salada, three small former saline lakes exist in the Chiprana-Caspe region that are now freshwater (Fig. 1). The first one, Estanca de Chiprana (13.24 ha), is a deep and permanent freshwater lake as a result of receiving water from the Guadalope river by a drainage canal. Adjacent to La Salada, the second one, Prado del Farol (1.51 ha) is very shallow, almost completely filled with sediments and covered by solid bed of reeds (*Phragmites australis* (Cav.) Trin.). The third one, La Salobrosa or Roces Lake (2.8 ha), also receives by a drainage canal, freshwater rich in nutrients from agricultural lands and, in turns, drains through a canal into La Salada.

Methods

The depth-time distribution of major physicochemical variables was studied during the 1989 hydrological cycle. Every month vertical profiles of temperature, conductivity, pH and redox potential were measured *in situ*, Secchi depth was determined, and water samples were taken for chemical analysis, at midday, at the sampling station located at one of the deepest points in the larger basin (Fig. 2). Samples were taken at different depths with the purpose of observing physical, chemical or biological gradients throughout the water column.

Conductivity at 25 °C and temperature were measured using an Instrand-10 compensating conductivity meter with a thermistor. The pH and redox potential were measured with a Crison-506

portable pH/redox meter equipped with a Metrohn platinum combination electrode. Because of the slight vertical variation in salinity and ionic proportions, it was considered unnecessary to determine separate temperature-conductivity adjustment curves for the different strata of the water column as Hall & Northcote (1986) recommend.

Dissolved major ion, nutrient (phosphate, nitrate and ammonia), oxygen, hydrogen sulphide and chlorophyll *a* concentrations were determined following the methodology proposed by A.P.H.A. (1985) with some modifications (Bernués *et al.*, 1990). Bacteriochlorophyll *a* concentration was calculated following Clayton (1966).

The percentage of error of chemical analysis was calculated as [(cations − anions) / (cations + anions)] × 100. All values of our analysis gave an error under 5–10 percent, which is considered admissible for saline waters (Custodio & Llamas, 1976). All absorbance values were obtained using a Hitachi U-2000 Spectrophotometer.

One sampling station for microbial mats was set up at 1 m depth off a sandy shoreline on the north side of the lake (Fig. 2).

Results and discussion

Basin morphometry and origin

La Salada de Chiprana lake is found on a landscape of 'inverted relief' as discussed earlier. The lake was formed when a Quaternary alluvial fan dammed the drainage into the Ebro river. This fact confirms the recent age of the lake.

This lake has several basins separated by ribbons which, above the water surface, are represented by narrow peninsulas and linear islands (Fig. 2). Its main morphometric parameters are given in Table 1. Its total surface is about 31.5 ha, of which 0.4 ha correspond to islands, giving an insulosity value of 1.3%. The bottom drops off very steeply in some areas and very gradually in others. The phytobenthos is composed of multi-

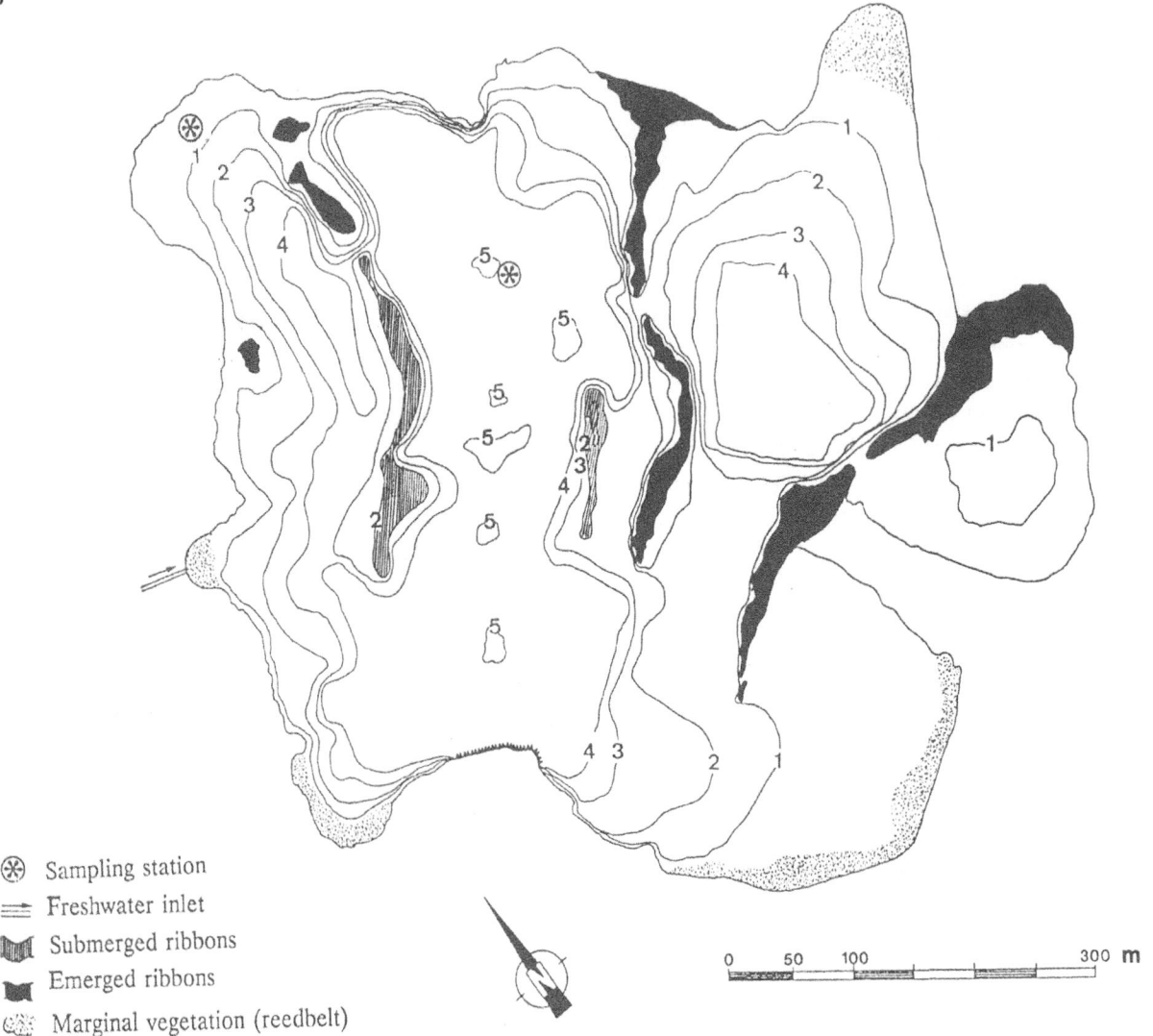

Fig. 2. Bathymetric map of La Salada de Chiprana Lake (May 1988), showing sampling stations (*) for vertical profiles and microbial mat communities.

laminated communities known as microbial mats which cover horizontal sediments not deeper than 1.5 m. Some stands of the submerged macrophyte *Ruppia maritima* L. var *maritima* are also present in these shallow areas. At greater depths, up to 3 m, sediments are covered by the charophyte *Lamprothamium papulosum* (Wallr.) J. Groves.

Maximum depths of 5.6 m occur in some areas of the central and largest basin. This basin is characterized by steep shorelines owing to the disposition of the ribbons, and a flat bottom con-

sisting of fine organically rich sediments up to 1 m thick in some areas. This morphometry play an important role in inhibiting the mixing of deep waters which are anoxic during most of the year.

Water regime

The original water regime of La Salada was characterized by the absence of outflows as well as by natural surface inflows (Reyes Prosper, 1915). Its

Table 1. Morphometric parameters of Salada de Chiprana Lake during the 1987/88 hydrological cycle. Parameter definitions follow Hakanson (1981).

Parameter	Value
Area (ha)	31.5
Volume (hm^3)	7.7
Maximum depth (m)	5.6
Mean depth (m)	2.3
Relative depth (m)	0.86
Shore development	1.89
Volume development	1.27
Mean slope (%)	3.4
Insulosity (%)	1.3
Bottom roughness	5.67
Shoreline length (m)	3750
Maximum length (m)	792
Maximum effective length (m)	792
Maximum width (m)	693
Maximum effective width (m)	523

permanence and slightly fluctuating level are related to ground water discharges. Main losses occur via evaporation and evapotranspiration. Precipitation is responsible for the small seasonal variations in water level and salinity. In 1989, the water level twice (in March and December) rose about 20 cm in response to the precipitation maxima in February and November (40 and 90 mm precipitation, monthly totals, respectively).

Nowadays, the water regime is altered by continuous freshwater input via a small canal coming from La Salobrosa lake. Several other canals in the vicinity of La Salada drain surrounding agricultural lands and supply water to La Salada intermittently from March to October, causing year to year fluctuations in water level and salinity. In May 1989, the salinity was about 64 g l^{-1} while values of 84 g l^{-1} (May 1987) and 76 g l^{-1} (May 1988) have been obtained previously (Guerrero et al., 1991). These freshwater inflows, as well as the natural runoff, are difficult to quantify because of their irregularity and diffuseness.

Nevertheless, the potential of man's activities to reduce the salinity of La Salada is great, as demonstrated by the histories of the three small lakes near La Salada, mentioned earlier.

Mixing and stratification regime

This regime at La Salada was strongly under the influence of the hydrological factors and variations in salinity. Temperature was a factor of secondary importance.

Salinity

There was a strong stratification during most of the year, with surface salinity lower than deep salinity by more than 10 g l^{-1}. The steepest salinity gradient or chemocline was located at ca. 2 m in January, and gradually descended to about 4.5 m by August and September, when upper mixed layer had a salinity of 56 g l^{-1} and was about 4 m thick (Fig. 5a). This gradient can be much more clearly appreciated in the depth-time distribution of conductivity (Fig. 5b). In October, strong winds caused mixing of the water column rendering it isosaline at ca. 53 g l^{-1} and isothermal at ca. 19–20 °C. Salinity stratification was almost immediately reestablished by freshwater inflow and dilution of surface waters as a consequence of heavy rains in November.

Temperature

Thermal stratification was generally weak, being strongest in August when there was a 6 °C difference between surface and bottom waters (Fig. 5c). During midwinter, inverse stratification was present, with surface waters colder than deeper waters. Both phenomena mentioned above were possible in this shallow lake because of the stabilizing effect of the vertical salinity gradient. Seasonal temperature variation was greater in the surface oxic waters (7.1 °C to 24 °C) than in the deeper anoxic waters (11.8 °C to 21.4 °C).

Long term behavior: meromictic or holomictic?

One of the most interesting features of the lake is its apparent position at the holomictic/meromictic boundary. It clearly was holomictic in 1989; but as the dynamic of the lake depends on the weather conditions, there may be years when the lake behaves like a meromictic one.

Precipitation and temperature data for 1989 and the previous twenty years (1969 through 1988) are represented in Fig. 3. These data were registered at Torre Los Baños (Chiprana) meteorological station, located at 177 m a.s.l. Monthly mean temperatures in 1989 were somewhat higher than the twenty-year average (1969 through 1988)

values, especially in mid-summer. From March to September of 1989, the chemocline was progressively displaced downward by wind-induced mixing. Because of higher than normal summer air temperatures, however, the rate of this downward displacement might have been slower than in cooler years. That is, meromixis or at least a delay of complete overturn may have been favored.

On the other hand, precipitation maxima in 1989 occurred earlier in spring (February) and later in autumn (November) than is typical, and total August-October rainfall was only about fifty percent of the twenty-year average for that period. This may have made it more likely that the water column would mix completely in 1989 than in a more typical year. If August-October precipitation in 1989 had been average or above average, the freshening of surface waters might have created a salinity gradient sufficient to have prevented complete mixing during that autumn. In previous years, when salinities were even higher (as noted earlier), it would have been even more likely for La Salada to have behaved as a meromictic lake if heavy rains occurred in early autumn.

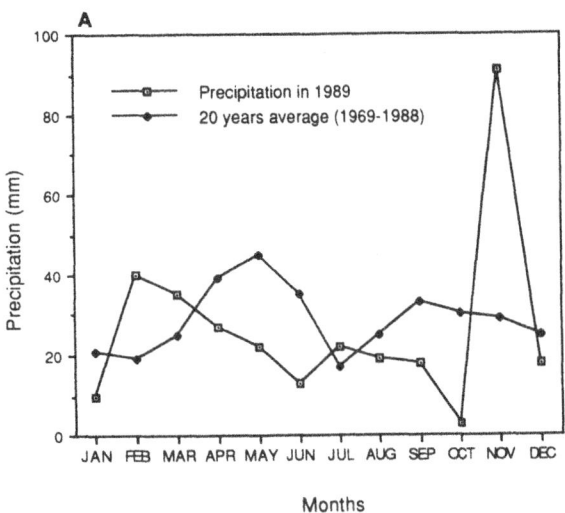

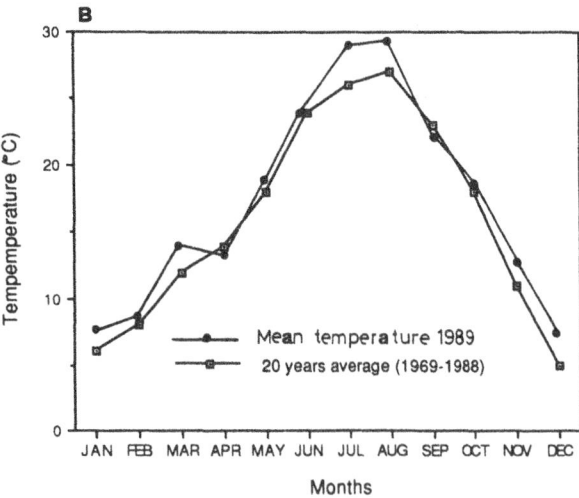

Fig. 3. Meteorological data for Torre Los Baños (Chiprana) meteorological station. A – Monthly Precipitation (mm). B – Mean monthly Temperatures (°C).

Biological processes and water chemistry

Ionic composition

The ionic composition showed some variation with both depth and season. The most abundant ions were always magnesium and sulphate as a consequence of dissolution of evaporitic salts (gypsum, dolomite, carnallite) in the soil by the ground and pluvial waters that feed the lake. All samples taken in 1989 and during some years before (Pueyo 1978/79; Mingarro *et al.*, 1981) belong to SO_4-Cl-Mg-Na type of Eugster & Hardie's (1978) ionic series.

In March, calcium decreased throughout the water column, reflecting precipitation of calcium carbonate (Fig. 4a). As chlorophyll did not increase at this time, suggesting that photosynthetic activity did not either, it seems likely that calcium carbonate precipitation was caused by chemical

119

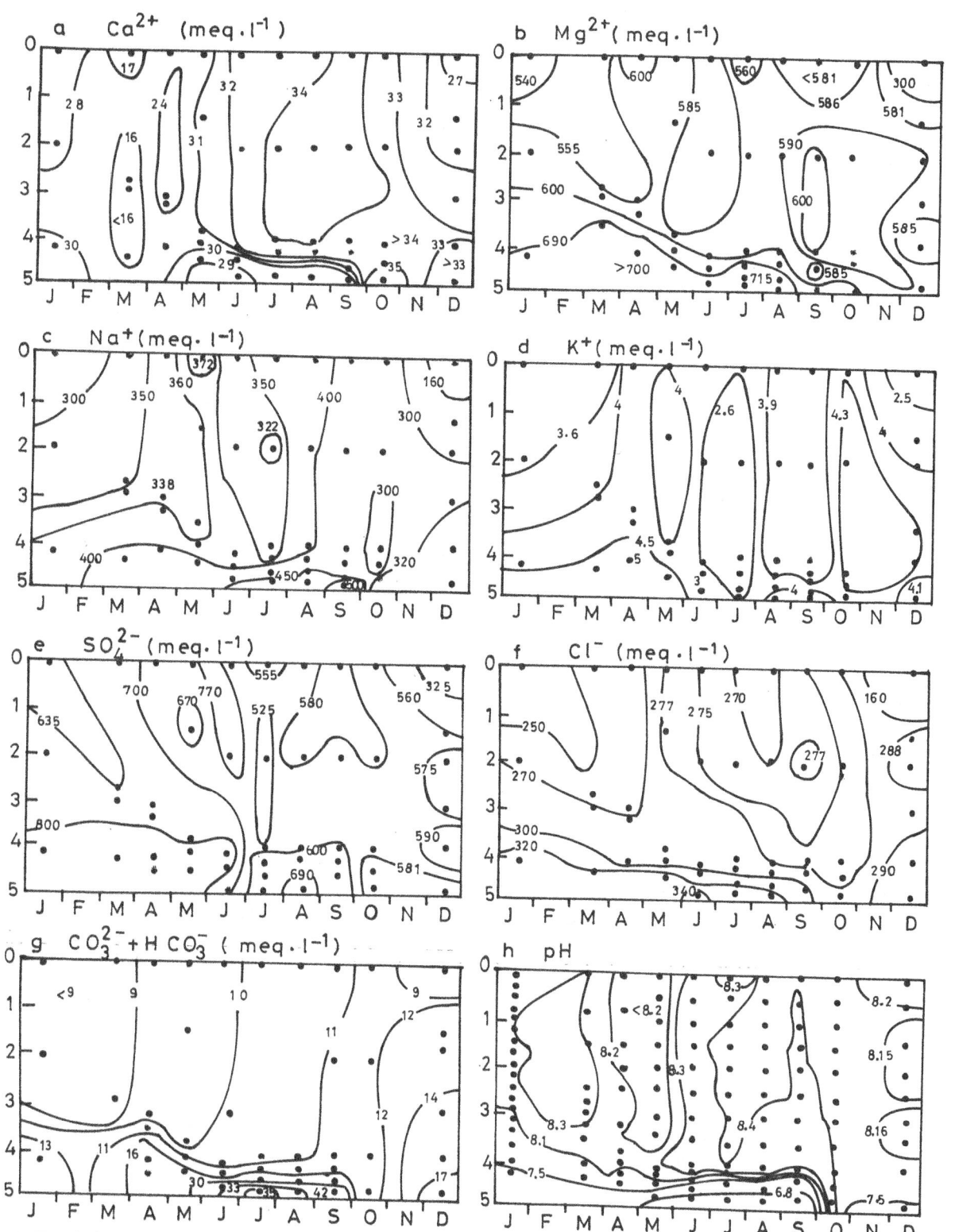

Fig. 4. Depth-time distribution of five major ions (meq l⁻¹) and pH in La Salada Lake during annual cycle 1989. a- Calcium, b- Magnesium, c- Sodium, d- Potassium, e- Sulphate, f- Chloride, g- Total alkalinity, h- pH.

processes. The rainfall (and runoff) maximum in February caused an increased input of clay and humic acids and these easily could have formed complex ions with calcium carbonate that would have precipitated at the basic pH of La Salada waters. The slight springtime temperature increase may also have favored calcium carbonate precipitation.

In summer, sulphate concentration tended to decrease slightly (Fig. 4e). Sulphate-reducing activity associated with organic matter decomposition at the higher temperatures of this period probably was responsible.

Total alkalinity was highly correlated with pH ($r = -0.94$, $N = 49$, $P = 0.0001$). In surface waters, alkalinity varied between $8.47 \, \text{meq} \, l^{-1}$ and $11.06 \, \text{meq} \, l^{-1}$. In bottom waters, a maximum bicarbonate concentration of $42.49 \, \text{meq} \, l^{-1}$ coincided with a pH minimum at the end of summer (Fig. 4g, h). These variations reflected two processes. First, the higher water temperatures of summer season stimulated microbial production of CO_2 and a consequent lowering of the pH; and second, carbonate precipitated from surface waters during spring redissolved readily in these summertime low pH conditions.

Oxycline, nutrients and metabolic influences

The variables which best defined the chemical gradient of the water column were oxygen and hydrogen sulphide which presented complementary distributions (Fig. 5d, e). Bottom waters were anoxic during most of the hydrological cycle. Only in October, when mixing of the water column occurred due to strong winds, was oxygen detectable (up to $1.3 \, \text{mg} \, l^{-1}$) in the deepest water.

There were several factors which led to this oxycline. One of them was the morphology of the basin which favours the isolation of the deep water and diminishes the stirring effect of wind. Another factor was the salinity gradient formed as a consequence of freshwater inflow from precipitation and runoff. Finally, the development of a temperature gradient also caused density differences that inhibited mixing. In the stagnant and

progressively anoxic deep water, sulphide accumulated. This was produced by the anaerobic respiration of sulphate-reducing bacteria and the decomposition of organic matter in the bottom sediments. Thus, it was not unusual that hydrogen sulphide levels as high as $220 \, \text{mg} \, l^{-1}$ were found near the bottom in late summer.

The reducing power of the deep water was indicated by a redox potential of $-436 \, \text{mV}$ as well as by a decrease in pH values down to 7.5–6.8 (in contrast with pH values of 8.1–8.4 in oxic surface waters).

With regard to nutrients, phosphate-phosphorus concentration was very low ($0-0.7 \, \mu\text{g at} \, l^{-1}$), even in anoxic waters, showed no detectable spatial or temporal patterns. Nitrate-nitrogen was only present in oxic waters and exhibited a maximum of $10-30 \, \mu\text{g at} \, l^{-1}$ from January to April (Fig. 5f). This was associated with the rainfall and runoff maximum in February, which delivered nitrate from the surrounding agricultural fields. A nitrate minimum of $3.93 \, \mu\text{g at} \, l^{-1}$ was recorded in May, coinciding with increasing chlorophyll *a* and bacteriochlorophyll *a* concentrations. The decrease in nitrate concentration with depth throughout the year probably was due to denitrification in the anoxic strata. Maximal ammonia concentrations were about $40 \, \mu\text{g at} \, l^{-1}$ in surface waters, and up to $600 \, \mu\text{g at} \, l^{-1}$ in the deep water in late summer near the end of the long period of stratification.

Planktonic photosynthetic organisms

Distribution, abundance and activity of planktonic photosynthetic organisms were inferred from pigment concentrations and Secchi disk measurements (Fig. 5g, h). The lake is highly eutrophic. In oxic waters chlorophyll *a* concentration varied from 1 to $30 \, \mu\text{g} \, l^{-1}$, with a minimum in March and April and maxima were during summer. From January to May, the Secchi disk was visible only to 1.0–1.5 m when chlorophyll concentrations in the oxic surface waters were low. The clearest water occurred in May. In June, chlorophyll *a* concentrations begin to in-

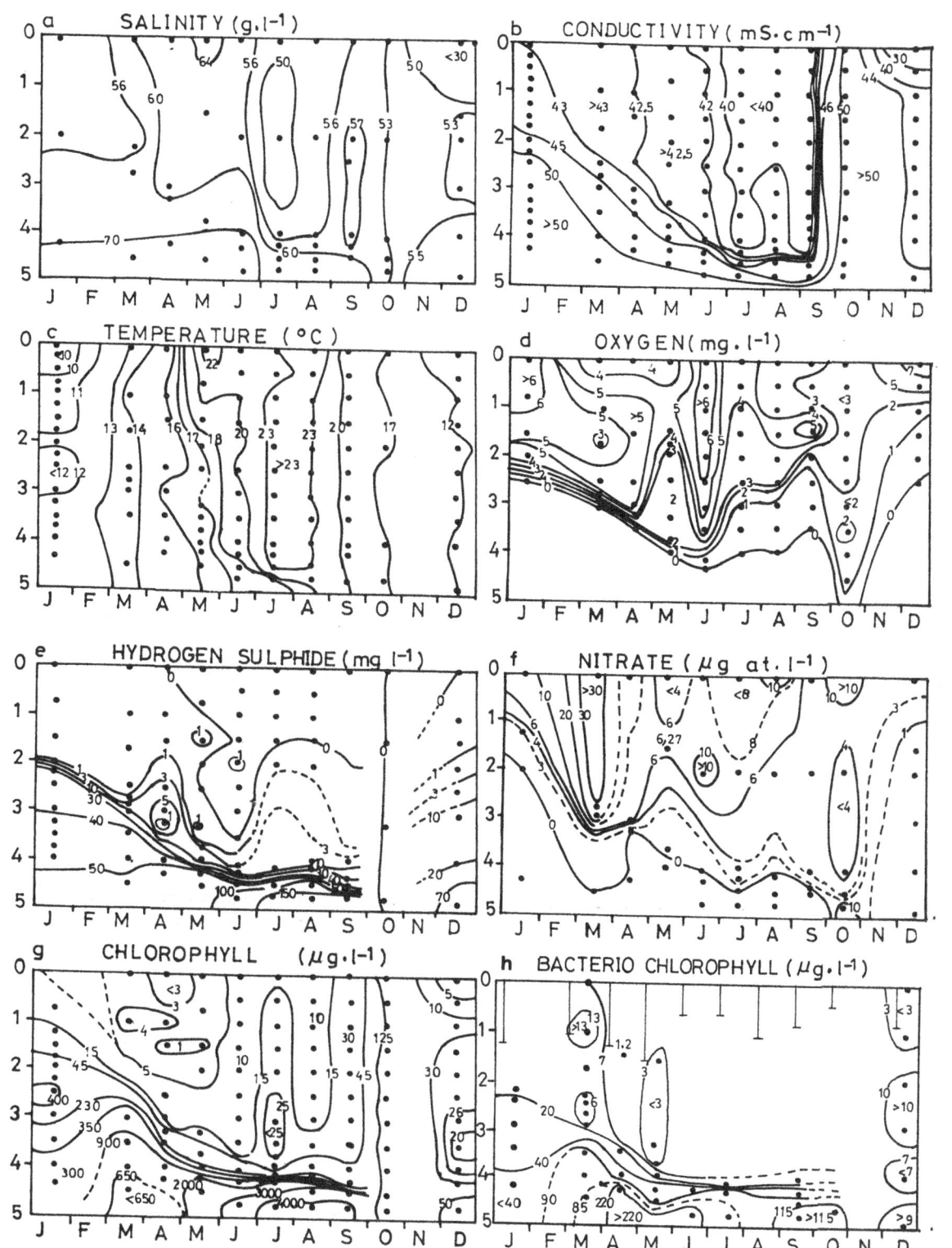

Fig. 5. Depth-time distribution of main limnological variables in La Salada lake during annual cycle 1989. a- Salinity (g l⁻¹), b-
Conductivity at 25 °C (mS cm⁻¹), c- Temperature (°C), d- Oxygen (mg l⁻¹), e- Hydrogen sulphide (mg l⁻¹), f- Nitrate (μg at l⁻¹),
g- Chlorophyll *a* (μg l⁻¹), h- Bacteriochlorophyll *a* (μg l⁻¹) and Secchi disk depth (⊥).

122

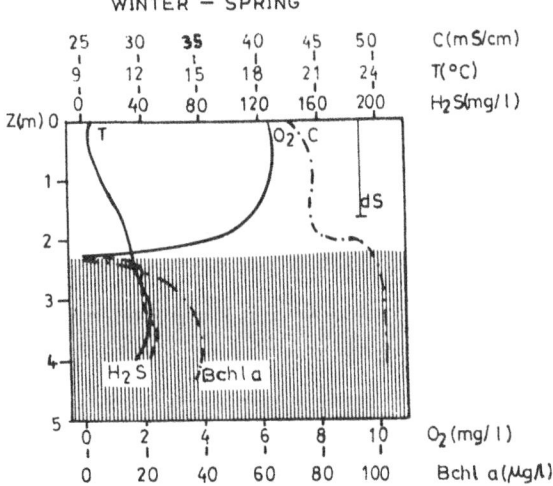

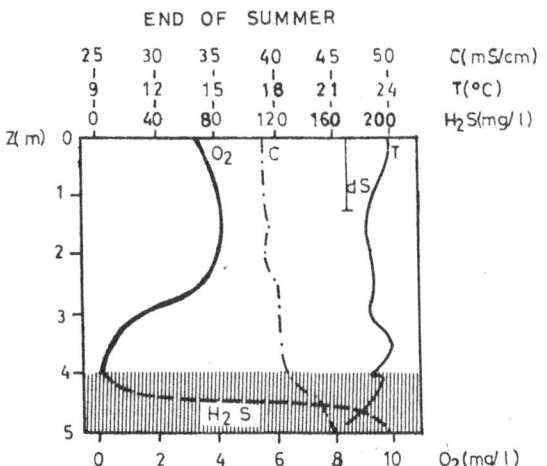

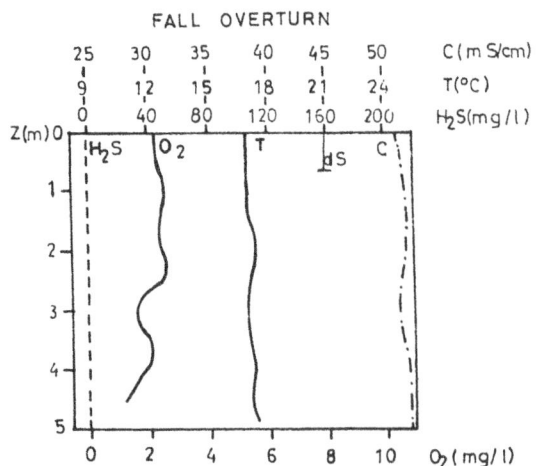

crease, perhaps because of higher solar radiation and temperature. Algal blooms from July to October caused Secchi disk depth to decrease to a minimum. During the fall overturn, there was a large increase in chlorophyll *a* concentration, reflecting a plankton bloom perhaps stimulated by erosion of the chemocline and upward mixing of nutrients.

Chlorophyll *a* concentration apparently increased with depth at all times except during the period of mixing. Highest concentrations were at or below the oxic/anoxic interface, with a maximum of 4000 $\mu g l^{-1}$ at the end of the summer. These high concentrations could not have corresponded to living algal populations because no algal cells were observed at these depths. Two explanations seem possible. (1) Chlorophyll *a* accumulated in anoxic waters where decomposition rates were low because of high salinity and absence of oxygen. (2) We overestimated chlorophyll *a* concentration because of the presence of bacteriochlorophyll *d*, which has a similar absorption maximum. Spectrophotometry may not provide an unequivocal method to distinguish and quantify these pigments. Spectrophotometric data must be confirmed by the chromatographic identification of the extracted bacteriochlorophylls as Gloe *et al.* (1975) recommend.

Depth-time distribution of bacteriochlorophyll *a* is shown in Fig. 5h. From April to June, bacteriochlorophyll *a* was found in small amounts in surface waters, but higher values, up to 220 $\mu g l^{-1}$, were recorded in deeper water. In summer, bacteriochlorophyll *a* concentrations decreased to 85 or 100 $\mu g l^{-1}$.

The presence of this pigment revealed anoxygenic phototrophs growing in the deep anaerobic water of Salada de Chiprana Lake. Absence of oxygen, presence of sulphide, and sufficient solar radiation permitted the development of green sulphur bacteria. Absorption spectra obtained for

Fig. 6. Seasonal variation of the structure of the water column of Salada de Chiprana Lake during annual cycle 1989. (T = temperature, C = conductivity, dS = Secchi disk depth). The shaded area shows the distribution of photosynthetic bacteria.

raw water samples indicated that the dominant bacterial population was the same throughout 1989. Biochemical and morphological characteristics of the isolated dominant population identified this species as *Chlorobium vibrioforme* Pelsh, possibly with some physiological adaptations to the environmental conditions of this lake (Guerrero *et al.*, 1991).

The seasonal variation of the structure of water column at La Salada is summarized in Fig. 6, where profiles of significant variables defining the depth gradient are represented. We have distinguished three periods:

(a) In winter and spring, the oxygen-sulphide interface was located high in the water column,

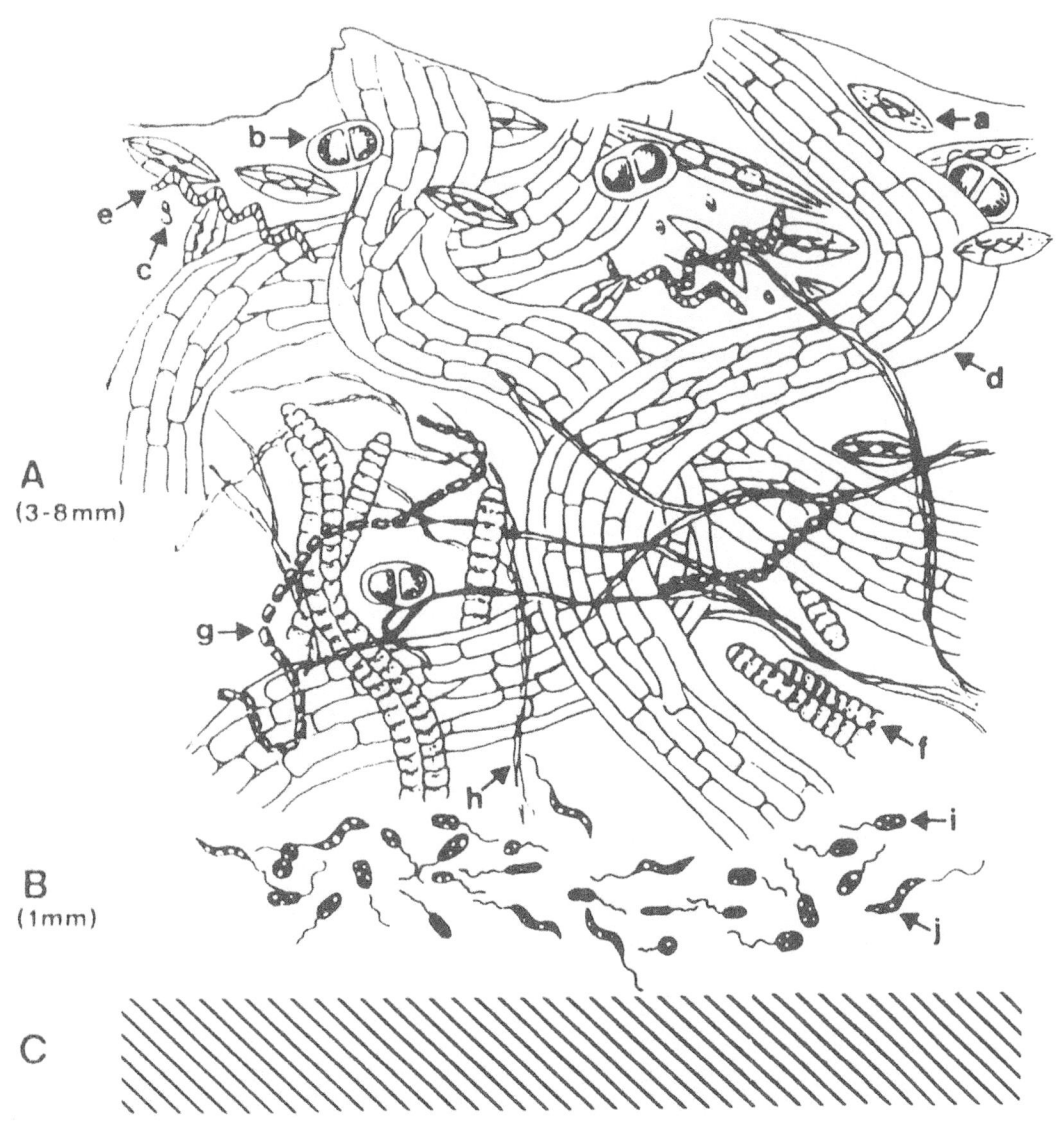

Fig. 7. Schematic drawing representation of vertical section through a summer microbial mat from Salada de Chiprana Lake showing the different constituent layers of the green and red layers. The letters on the drawing refer to the following: a, diatoms; b, *Gloeocapsa* sp.; c, *Synechococcus* sp.; d, *Microcoleus chthonoplastes*; e, *Spirulina* sp.; f, *Oscillatoria* spp.; g, *Pseudoanabaena* sp.; h, *flexibacteria*, i, *Chromatium* sp.; j, *Thiospirillum* sp.; A = green layer; B = red layer; C = black sediment.

124

and solar radiation reached the anoxic layer.

(b) At the end of summer, the chemocline was located nearer the bottom, and the anoxic water was confined to the last meter of the water column.

(c) In autumn, complete oxygenation of the water column occurred because of strong winds. By December, if not earlier, strong stratification had been reestablished again.

During the first period (January to May), there was a large increase in *Chlorobium* density in bottom waters as the bacteriochlorophyll *a* indicate (Fig. 5h). Towards the end of the summer, light reduction in the deep water probably caused a decrease in this population. In autumn, oxygenation of bottom waters and elimination of sulphide would have been expected to cause a major decline in the *Chlorobium* population. High bacteriochlorophyll *a* levels at the bottom of the water column in October may have been due in part to non-metabolizing cells accumulated in this oxic and poorly lit region. In December, the distribution of the *Chlorobium* population had again expanded upward in concert with the thickening of the anoxic layer and perhaps aided by vertical mixing during the fall overturn.

Benthic microbial mats

The special physico-chemical characteristics of Chiprana Lake make it also suitable for the development of microbial mat communities. These benthic formations covered the sediments of shallow areas of moderate slope. The thickness and cohesion of this microbial formation was considerable and was due to the presence of a filamentous cyanobacterium *Microcoleus chthonoplastes* Thuret which was the principal component of the mats. The stratification of microbial populations was related to the gradient of oxygen, sulphide and light through the mat (Fig. 7). The surface layer was composed primarily of diatoms and unicellular cyanobacteria. Several green layers of filamentous cyanobacteria were present underneath. The deepest layer was red and constituted

by unicellular purple sulphur bacteria in contact with the black sediments, rich in sulphide produced by sulphate-reducing bacteria.

Conclusions

The extreme character of La Salada de Chiprana Lake causes a low species diversity and some unusual limnological and microbial features. These justify a rigorous protection of this ecosystem. A decrease of salinity, such as might result from the irrigation plans currently being considered for the surrounding agricultural area, could cause the disappearence of the microbial communities now present.

Acknowledgements

We thank J. Balsa, M. Pascual and F. Moya, for their help in sampling the lake as well as in analyzing samples in the laboratory, and Dr S. H. Hurlbert, for the inestimable help of his comments, suggestions and enthusiasm during the revising of this article.

References

American Public Health Association, 1985. Standard methods for the examination of water and waste-water (16 edn). APHA. AWWH. Washington, D.C.

Baltanás, A., C. Montes & P. Martino, 1990. Distribution patterns of ostracods in Iberian saline lakes. Influence of ecological factors. Hydrobiologia 81: 87–111.

Bernaldez, F. G., 1987. Las zonas encharcables españolas. El marco conceptual. In: Bases científicas para la protección de los humedales en España: 9–30. Real Academia de Ciencias de Madrid.

Bernués, M., F. Moya, A. G. Besteiro & C. Montes, 1990. Análisis químicos de aguas dulces y saladas. Dpto. Interuniversitario de Ecología. Universidad Autónoma de Madrid.

Clayton, R. J., 1966. Spectroscopy of bacteriochlorophylls in vivo and in vitro. Photochem. Photobiol. 5: 669–677.

Custodio, E. R. & M. R. Llamas, 1983. Hidrología subterránea. 2nd edition. Ed. Omega, S.A., Barcelona.

Eugster, H. P. & L. A. Hardie, 1978. Saline lakes. In A. Lernion (ed.), Lakes, chemistry, geology, physics, Springer-Verlag, New York: 237–293.

Friend, P. F., J. P. P. Hirst & G. J. Nichols, 1986. Sandstone-body structure and river process in the Ebro basin of Aragón, Spain. Cuadernos Geología Ibérica 10: 9–30.

Gloe, A., H. Pfennig, H. Brockmann & W. Trowitzsch, 1975. A new bacteriochlorophyll from brown-colored chlorobiaceae. Arch. Microbiol. 102: 103–109.

Guerrero, M. C., J. Balsa, M. Pascual, B. Martínez & C. Montes, 1991. Caracterización limnológica de la laguna Salada de Chiprana (Zaragoza) y sus comunidades de bacterias fototróficas. Limnetica 7: 83–96.

Hakanson, L., 1981. Manual of lake morphometry. Springer-Verlag, New York, 78 pp.

Hall, K. J. & T. G. Northcote, 1986. Conductivity-temperature standardization and dissolved solids estimation in a meromictic saline lake. Can. J. Fish. aquat. Sci. 43: 2450–2454.

Hammer, V. I., 1986. Saline lakes ecosystems of the world. Dr W. Junk Publ., Boston, 616 pp.

Javor, B. J., 1989. Microbiology and biogeochemistry of hypersaline environments. Brock/Springer. Berlin, 328 pp.

Margalef, R., 1983. Limnología, Ed Omega. Barcelona, 951 pp.

Mingarro, F., S. Ordóñez, M. C. López & M. A. García del Cura, 1981. Sedimentología de las lagunas de los Monegros y su entorno geológico. Bol. Geol. Min. XCII-III: 171–195.

Montes, C. & P. Martino, 1987. Las lagunas salinas españolas. In: Seminario sobre bases científicas para la protección de los humedales en España. Real Academia de Ciencias Exactas, Físicas y Naturales de Madrid, Madrid: 95–145.

Pueyo, J. J., 1978/79. La precipitación evaporítica actual en lagunas saladas, Diputación Provincial, Universidad de Barcelona. Rev. Inst. Invest. Geol. 33: 5–56.

Reyes Prosper, E., 1915. Las estepas de España y su vegetación. Madrid.

Riba, O., J. Villena & J. Quirantes, 1967. Nota preliminar sobre la sedimentación en paleocanales terciarios de la zona de Caspe-Chiprana (Zaragoza). Anal. Edaf. Agrob. 26: 617–634.

Ward, P. P. B., K. J. Hall, T. G. Northcote, W. Gheung & T. Murphy, 1990. Autumnal mixing in Mahoney Lake, British Columbia. Hydrobiologia 197: 129–138.

Williams, W. D., 1986. Limnology, the study of inland waters: a comment on perception of studies of salt lakes, past and present. In P. D. Deckker & W. D. Williams (eds), Limnology in Australia, Dr W. Junk Publ., Boston: 471–496.

The Dead Sea – an economic resource for 10000 years

Arie Nissenbaum
Weizmann Institute of Science, Rehovot 76100, Israel

Key words: Dead Sea, economic resources, history, potash, bromine, asphalt, medicine, mummies, salt, therapeutics

Abstract

The archaeological and historical record of the Dead Sea as an economic resource is longer than that of any other hypersaline lake. Although it is completely devoid of life, except for a few bacteria and algae, the climatic and geological conditions in the Dead Sea basin have produced circumstances which made this lake important for the economy of the area. The salt which was produced by evaporation of the water, or by quarrying from the salt diapir of Mt. Sodom, on the Dead Sea coast, is referred to in the Bible and in the Talmud. It was harvested until the 1930's. Potash has been extracted from the brine, by solar processes, since 1931 and today the Dead Sea is a major source of potash and bromine. The asphalt, which is found in seepages along the shores and in large blocks, occasionally found floating on the lake, has been used by the inhabitants of the area for waterproofing baskets and for decorative purpose, since the Pre-ceramic Neolithic Period, 10000 years ago. Later, the asphalt became a major export item to Egypt. During the Early Bronze age, 4000 years ago, it was used mostly to glue flint implements to wooden handles and in the Graeco-Roman period it was used as one of the components in the embalming of Egyptian mummies. The area around the Dead Sea was the only source of balsam, perhaps the most important incense and medication of the Ancient World. Remains of a 7th century B.C. perfume factory, were found in Ein Gedi. During later periods, until the Arab conquest in the 7th century A.D., the growing of balsam was an imperial monopoly. The area of the Dead Sea was famous, for over 2000 years, for its dates and sugar. The therapeutical and medicinal properties of Dead Sea water and the hypersaline hot springs on its shore, were famous throughout the Ancient World. For example, King Herod the Great, 2000 years ago, used to visit the area to cure his many diseases. This practice continues today, and the lakes has become a major center for treatment of psoriasis. There is pictorial, archaeological and historical evidence to support the Dead Sea's importance as a trade artery for over 2300 years.

Introduction

Hypersaline lakes are usually a negligible component of the regional economic scenario. it is true, that since the advent of the industrial revolution, the need for chemicals such as soda, borax etc., several salt lakes have become an important natural resource. However, during most of man's history, salt lakes did make a negligible contribution to human welfare. The dominant reason for this is perhaps the location of such lakes in zones which are marginal to human habitation. The only possible deviation from this is the utilization of this environment for production of salt which was traded over large distances. The Dead Sea, however, is an exception. Its location in an area which is close to centers of civilization, its climate, the geological phenomena associated with it, the pe-

128

culiar properties of its water and the occurrence of several well-watered oases near its shores, resulted in this lake being a major economic resource. Yet, this might have gone unheeded if the historical and archaeological records of this area would have been as meagre as those for other lakes. The Dead Sea is outstanding in this respect, not only in comparison with other salt lakes but also in comparison with fresh water lakes. No other part of the world has been historically as well documented as the Middle East. Thus, it is possible to garner from the literature a comprehensive picture of the role that the Dead Sea played in the economy and geopolitics of this area. In modern times, the importance of this lake as a natural resource increased even further. Today, the products harvested from the Dead Sea constitute the largest profit generating operation from natural resources in Israel and are major contributors to the economy of Jordan.

The Dead Sea – background

The Dead Sea is a land-locked hypersaline lake located in the deepest part of the Dead Sea-Jordan Rift Valley (Fig. 1). About 80 km long and 13 km wide, its water level, currently at −407 m. below mean sea level, makes it the lowermost surface on the face of the earth. For many years the Dead Sea has been physiographically divided into a northern basin, 330 m deep, and a shallow southern basin with water depth not exceeding 10 m. The two basins were separated by the Lynch Strait, only 3 m deep (Neev & Emery, 1967). The recent drop in lake level by about 15 m in as many years, might have been expected to result in the drying up of the southern basin. However, since conversion of the southern basin into huge evaporation ponds for production of potash and bromine, water has been pumped into it from the northern basin (Fig. 1).

The Dead Sea has attracted the attention of mankind for more than 2000 years (Nissenbaum, 1979). The most thorough investigation of the lake was carried out by Neev & Emery (1967) between 1959 and 1963. In 1979 the stratification

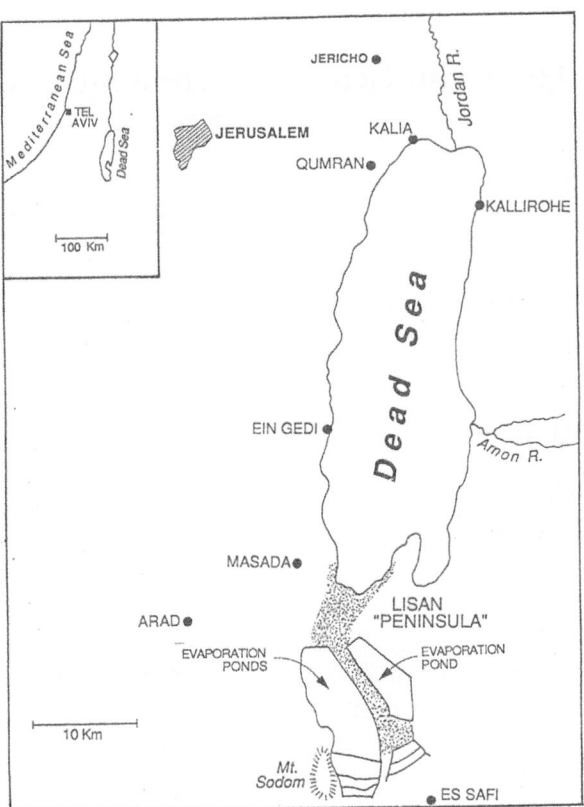

Fig. 1. Location map for the Dead Sea and its environs.

of the water column was destroyed by an overturn for the first time in almost 300 years (see references in Nissenbaum *et al.*, 1990b). The average chemical composition (in 1977) according to Beyth (1980) was (in $G l^{-1}$): $Mg = 44$; $Ca = 17.2$; $Na = 40.1$; $K = 7.7$; $Cl = 224.9$; $Br = 5.3$; $So_4 = 0.45$; $HCO_3 = 0.1$; Total dissolved salts $= 339.6$ gr l^{-1}; Density $= 1.232$. Today, due to a further drop in lake level, the salt concentration is somewhat higher although the chemical composition is basically the same. The climate of the region is dry and hot. The average rainfall is, at Sodom, 47 mm, with the average temperature in the summer ranging between 27 and 40 °C and in the winter between 11 and 21 °C. Fresh water springs are found only in a few localities along the shores such as in Ein Gedi, the Arnon River and at Es Safi (Fig. 1) where they form natural oases.

Salt

There are very few, if any, natural commodities which have been as important to mankind as salt (Bloch, 1976). Salt has been produced either by quarrying or evaporation of water. In the Dead-Sea, both sources have been used. Solid rock salt is found in Mt. Sodom, in the southwestern corner of the Dead Sea (Fig. 1). Mt. Sodom is a salt body with the structure of a salt wall, originating from the Pliocene Sedom member of the Dead-Sea Group (Zak, 1967). This salt body is a member of salt dome family which can be found in the subsurface of the Dead Sea Basin (Neev & Hall, 1979). Mt. Sodom is the one case where the salt penetrated the sediment cover to form a low lying hill, 11 km long and up to 2 km wide, about 100 to 150 m above the Dead Sea level (Zak, 1967). The lower part of the mountain is built of pure rock salt. The paucity of historical and archaeological evidence does not enable the reconstruction of the quarrying of salt from Mt. Sodom in ancient times since it is not always possible to attribute references to Dead Sea salt to this particular source. For example the Biblical reference to 'Mikhre Melah' in Zephania 2:9, means in modern Hebrew a salt mine. However, ancient translators assumed that it means salt heaps or salt pits (Braslavsky, 1943). Neither is it clear if the term 'Salt of Sodom', which is used by the Talmud, in the 2nd to 4th centuries A. D., and by Galen in the 2nd century A. D. (Walsh, 1924), refers to Mt. Sodom salt or to salt extracted from the brine, since the Dead Sea was frequently called the Sea of Sodom. The 6th century A. D. Madaba mosaic map shows boats on the Dead-Sea, one of which is carrying a red heap and the other a gray one. Bloch (1976) interpreted this as representing two sources of salt. The red salt coming from evaporation ponds, with the color being due to presence of halophilic algae, and the grey heap being salt quarried from Mt. Sodom. In 1919 an Arab from Bethlehem, obtained a concession for quarrying commercial quantities of salt from Mt. Sodom (Braslavsky, 1943). The concession later passed to a Mr S. Deib from Hebron. The quarried salt was sent by boats to

Kalia in the north (Fig. 1) and from there to Jerusalem and other inland locations. The amounts involved were rather small and between 1930 and 1944 a total of 14,872 tonnes were mined. Following the Second World War, just a small amount of salt −1,571 tonnes in 1946- was quarried in Mt. Sodom, which represented only 7% of the total amount of salt produced that year in Palestine (Picard, 1954). Following the establishment of the State of Israel in 1948, attempts were made in the early 1950's to renew commercial mining of salt from Mt. Sodom. In 1953 and 1954 about 15000 tonnes of salt were produced, which represented about 50% of the total amount produced in the country. However, the production and transportation costs from the Dead Sea area to central Israel were too high in comparison with the extraction of salt from Mediterranean salt pans, and within a short time this endeavor was abandoned (Almog & Eshel, 1957).

Salt was also obtained by evaporation of Dead Sea brine. Indeed, the Hebrew name for the Dead Sea is the 'Salt Sea'. It is quite probable that evaporation of Dead Sea water in salt pans near the shore was practiced for a long time. The Bible mentions a location called the City of Salt (Joshua 15:62). Although its exact location is unknown modern archaeologists placed it near the northwestern shore of the take, where the only source of salt is the lake itself. The Book of Ezekiel (47:11) describes how the Dead Sea will sweeten in the future but adds 'and the marshes thereof, shall not be healed; they shall be given for salt'. Obviously, the requirement for salt shall not be abated in the Millennia. During the Hellenistic regime, a tax was levied on 'salt pans' (Book of Maccabees) which may refer to the Dead Sea area. Several medieval Arab historians, also refer to salt from the Dead Sea, but no indication as to its source is given. Perhaps the reference by Mukkadasi (10th century A.D.) to powdered salt, indicates an origin from evaporation ponds rather than from the massive salt of Mt. Sodom. Many of the travellers to the Dead Sea mention the occurrence of salt along the shore. For example, the Russian pilgrim, Archimandrite Agrefenii, writes in the 1370's about salt being found along the

Dead Sea shore. Robinson (1841) describes how the inhabitants of the area manufacture salt from a salt pan near Ein Gedi (Fig. 1) as well as in the north end of the lake. Similar descriptions can be found in almost all 19th century travelogues on the Dead Sea area. In 1862, the Turkish Ottoman government, took over the monopoly on salt production, and the use of Dead Sea brine for this purpose was forbidden. Undoubtedly, illegal production of salt by the local Beduins, not notorious observers of the law, continued. During the Second World War, a shortage of salt developed in the Middle East, and salt which was precipitated in the potash ponds in the north end of the lake was exported to the British Army. Today, 45 000 tonnes of salt are annually extracted from the evaporation ponds which are used primarily for potash production in the southern basin of the lake.

Asphalt

Seeps of asphalt are very common in the southern part of the Dead Sea basin (Nissenbaum & Goldberg, 1980). The origin of the asphalts has been debated in the literature. Amit and Bein (1979) proposed that the asphalts were generated by biodegradation of oils. Spiro *et al.* (1983) and Rullkotter *et al.* (1985) suggested that the asphalts are immature and have been produced in an early maturation stage from the Upper Cretaceous bituminous chalks of the Ghareb formation. Tannenbaum & Aizenshtat (1985) supported a dual source with some of the asphalts being immature and others being alteration products of crude oils.

Two major types of asphalt occurrences are known: (a) as veins and seeps in the Upper Cretaceous rocks which comprise the western escarpment of the Rift valley and as cement of recent or sub-recent wadi gravels and (b) as large masses of extremely pure asphalt which can occasionally be found floating on the Dead Sea (Nissenbaum, 1978; Nissenbaum *et al.*, 1980). Both types of asphalts can be collected quite readily. The quantity of pure asphalt which can be gathered from the seeps and cements, is however

quite small, since it is intimately mixed with the stony matrix. On the other hand, the floating blocks are totally devoid of any mineral matter but their appearance is highly irregular. The asphalt blocks easily float on the lake's surface since their density is much lower than that of the brine (1.1 vs 1.23) and within a few days they are swept to the shore by winds and currents. This asphalt could thus be collected either when it appeared on the surface of the lake as was described by Diodorus Siculus in the 1st century A. D. or collected on the shore as described by Robinson (1841) for the asphalt which was released by the large earthquake of 1837 (Nissenbaum, 1978). In any case, this raw material was easily available since very early times and it should not come as any surprise to see that mankind learned to use this natural resource as long as 10 000 years ago.

Decoration

A lesser known use of asphalt in the Fertile Crescent was for artistic purposes. This material was used for decoration, for making small decorated utensils such as bowls, for coating, partly or completely, statues, for making ornaments and so on. (Forbes, 1954; Abraham, 1960). Recently, Connan *et al.* (1990) conducted a comprehensive geochemical analysis of asphalts from Ugarit (Ras Shamra), Syria including material which was used to coat orthostats and a statue of the God El. The earliest use of Dead Sea asphalt (and perhaps of any other asphalt) for decorative purposes dates to about 9000 B.P. Bar Yosef (1985) described art objects from a Pre-Pottery Neolithic B cave from Nahal Hemar (= Asphalt Valley), not far from the Dead Sea. Some of them were coated with asphalt. One of the most interesting finds was of skulls of adult males which were coated with asphalt on which a net pattern was embossed. It has been proposed that this cave was used as a store for cult objects used in a Neolithic ritual (Bar Yosef, 1985). The asphalt must have been an important component of the ritualistic objects. Kelso & Powell (1944) described asphalt in the Tel beit Mirsim excavation, in the mountains of Judea, of Early Bronze age, and suggested that it was used for making paints or varnishes.

Cement

Asphalt was of great importance in the building trade of ancient Mesopotamia. The scarcity of suitable rock material to produce cement, resulted in the widespread use of asphalt as mortar. The Bible, when describing the building of the Tower of Babel says: 'and slime (= asphalt) they had for mortar' (Genesis 11:3). Indeed, excavations in Babylon, Accad and Summer showed the wide use of asphalt, often mixed with straw to prevent flow of asphalt under pressure, to cement burnt clay bricks in buildings and walls. The excavations of the Babylonian Great Wall of Nebuchadnezzar (6th century B.C.), showed that it was constructed of bricks cemented with asphalt (Abraham, 1960). This very important mode of utilization of asphalt was probably not very common in ancient Canaan since stony building materials are very abundant. Abraham (1960) writes that walls of the city of Jericho, built at about 2500–2100 B.C., were made of brick cemented with asphalt. If indeed so, then the asphalt was very probably of Dead Sea origin.

Adhesive

An underappreciated use of asphalt in the ancient world was its utilization as an adhesive. The asphalt softens and becomes a very viscous liquid at a temperature of 50 to 60 °C and on cooling it re-solidifies. Thus, it can be used to glue together various utensils. It can also be poured, after heating, into a mould, and while still warm and sticky, gems, decorative stones and other materials can be embedded in it (Forbes, 1954). In the excavations of Tel Arad and Small Tel Malhata, dated to the Early Bronze Age, Nissenbaum et al. (1984) found Dead Sea asphalt which was used to glue a copper awl into a bone handle. Many flint implements showed smears of asphalt on one side of their surface, indicating that the asphalt was used to glue them to since-decayed wooden handles. Forbes (1954) provides a picture of a flint sickle set in bitumen, from Fayum, Egypt dated to 3000 B.C. Although this sample was not analyzed, Connan et al. (1992) showed that during this period, Early Bronze Age, Dead-Sea asphalt was extensively exported to Egypt and remains of it were found in the Ma'adi excavation not far from Fayum, Egypt.

Agriculture

One of the major uses of Dead Sea asphalt was in agriculture. Although this usage probably goes back to antiquity, most of the references in the literature are from the Arab period. The Persian traveller, Nasir-i-Khusrau who visited the area in 1047 A.D. describes how in the Dead Sea region asphalt is used to coat the roots of fruit trees and thus protect them against 'worms and things that creep below the surface' (Khusrau, 1896). This explorer adds, however, a word of caution by saying that this information is only what he has been told. To this hearsay evidence he also adds that he was told that druggists use asphalt to preserve their drugs against the worm 'nuktah'. Istakhari (951 A.D.) and Ibl Haukal (978 A.D.) write that asphalt was used for fertilization of vines (in Le Strange, 1890). Abu-I Fidaa writes in 1321 A.D. that asphalt to is used to coat vines and fig trees (Schiller, 1988). Another traveller, Abu Mansur Muwaffak, wrote in about 1000 A.D. that asphalt is an excellent preservative and can be used as a disinfectant (Abraham, 1960). The agricultural application of asphalt probably continued until at least the 17th century. Posniakov, writing in 1559–1560 A.D., says that 'black pitch' coming out of the Dead Sea is used to smear branches of vines in order to kill grubs (Raba, 1986). It is possible that the high sulphur content of the asphalt, which on exposure to air may oxidize and emit SO_2, an excellent fumigant, may be the reason for this effect. It is also possible that the aromatics in the asphalt also act as preservatives in a manner similar to that by which creosote protects wooden piles. During the *Phylloxera* scourge which devasted the vineyards of France at the end of the 19th century, it was proposed that Dead Sea asphalt might be used to combat the disease and a small amount was sent to France for testing (Delachanal, 1883).

Armament

A mixture of sulphur, resins, naphtha and quick-lime known as 'Greek Fire' and which self ignited on contact with moisture, was one of the major weapons in the Medieval age (Forbes, 1954). This mixture was used to make fire-bombs and was widely used by the Byzantines and the Crusaders. According to Prawer (1975), grenades of Greek-Fire, from the Crusader period in the Holy Land, were made with Dead Sea asphalt. However, it is not clear what the source for this information is.

Boatcraft

The use of asphalt for caulking boats was widespread in ancient Mesopotamia. The lack of trees forced the inhabitants to construct their vessels from reeds growing in the marshes of southern Iraq. Those boats were then caulked with asphalt. This practice continued until recent times (Abraham, 1960). Indeed, in the description of Noah's Ark is written: 'and shalt pitch it within and without with pitch' (Genesis 5:14). The Hebrew word for pitch is 'Kopher' which is probably of the same origin as the Accadian word 'Kaparu'. According to Forbes (1964) Kaparu of Kupru is native impure asphalt which was widely used for boat building in Mesopotamia. For similar reasons, namely lack of suitable wood, asphalt was used in ancient Egypt as well. This is reflected in the story of Moses and the Ark of Bulrushes (Exodus 2:3) which was coated with 'hemar' which is usually taken as to be asphalt. Josephus Flavius (1st century A.D.) in his description of the Dead Sea specifically writes that its asphalt was used to caulk boats (Josephus, 1959). This must have been an important application since

the Dead Sea region is short of trees suitable for boat construction, but it is rich in reeds growing in the areas where fresh-water springs exists. Diodorus od Sicily (1st century A.D.) describes the collection of asphalt from the Dead Sea, by the use of reed rafts (Diodorus, 1947). Theoderich, who visited the Holy Land in 1172 A.D. writes that the asphalt, also called Jew's pitch, is of great use to sailors (Theoderich, 1986). The Russian Vassily Posniakov, travelling to the area in 1560–1559 writes that the Dead Sea exudes 'hot sulphur' which is sold to merchants and which is used to caulk boats sailing in the Red Sea. The term 'hot sulphur' very probably refer to asphalt which is sulphur rich, and when heated does produce a strong sulphurous reek. He says that the ships sailing in the Red Sea are built without iron nails, and that the wooden planks are bound by palm fibres which are then coated with 'hot sulphur' which is imported from the Dead Sea (Raba, 1986).

Medicine

Asphalt was widely used as a medicine in the ancient Middle East. Pliny the Elder in the 1st century A.D. describes the use of asphalt for curing boils, eye inflammation, cataracts and leucoma, skin diseases and in checking bleeding; mixed with soda it is effective against aching teeth, and in stopping diarrhoea; mixed with vinegar it is potent against lombago and rheumatism and together with barley flour it healed torn muscles. Together with mint and myrrh it is effective against certain types of fever, and together with other ingredients it was potent in the treatment of certain gynecological disorders (Pliny, 1948). According to Dioscorides (2nd century A.D.) asphaltos is almost a panacea, being effective against a large number of diseases; the list of which corresponds quite closely to that of Pliny, with a few additions, such as for the relief of snake bite (Dioscorides, 1968). The asphalt from the Dead Sea is mentioned specifically by several authors as having potent medicinal properties. Josephus Flavius, in the 1st century A.D., wrote that is was included in many medical prescriptions (Josephus, 1959). Dioscorides, writes in his book on pharmacology that 'Judaicum Bitumen' is better than others. Galen, 2nd century A.D., the most influential doctor of medicine in the Roman and Medieval world, travelled to the Dead Sea area in order to procure Dead Sea asphalt for medical use (Walsh, 1924). In the Middle Ages, and particularly in Arab medicinal practices. Dead Sea asphalt became very important. This may be partly ascribed to its association with 'mummia', the shavings of black wrappings of Egyptian mummies, which were considered to have magic properties. The power of the 'mummia' is ascribed by Pliny to its ability to preserve the dead for many centuries (Abraham, 1960). In the 12th century, Fulk, the Crusader King of Jerusalem, granted to the inhabitants of the village of Tekoa, the monopoly on the collection of asphalt from the Dead Sea. In his charter, the medicinal properties of Dead-Sea asphalt are specifically mentioned. In the late 1930's a team of dermatologists, led by Prof. A. Dostrovsky from the Hadassa Medical School in Jerusalem, distilled Dead Sea asphalts and showed that the distillate is a valuable medication against certain skin diseases. This material was named 'bitupal' (*bitu*men from *Pal*estine) and is still in use, being produced today by the Teva Pharmaceutical Company in Petah-Tiqwa, Israel. The active ingredient of bitupal is not known, although it probably relates to a sulfur containing compound or perhaps even to elemental sulfur produced by the distillation of the asphalt.

Mummification

One of the most intriguing, and controversial, uses of Dead Sea asphalt is its presumed use in ancient Egypt as one of the ingredients in the process of mummification. The term mummy itself originates in the Persian word moumiah which means bitumen (El Mahdi, 1989). The mummies were usually covered with a black, pitch-like, substance which was believed to be bitumen and hence the name. However, the presence of asphalt in mummies was often disputed, and the

black material was assumed to originate from the degradation of plant oils and resins which were used in the embalming process. Ancient Egyptian descriptions of preparations of mummies are scarce, probably due to a combination of the sanctity of the process and the wish to preserve trade secrets by the Guild of Embalmers. Detailed descriptions are given by two Greek historians. Herodotus in the 6th century B.C. does not mention asphalt as a component of mummies. On the other hand, Diodorus of Sicily, writing in the 1st century A.D., who also describes how mummies were made, describes the collection of asphalt from the Dead Sea by the local inhabitants and says: 'The barbarians who enjoy this source of income take the asphalt to Egypt to sell it for the embalming of the dead; for unless this is mixed with other aromatic ingredients, the preservation of the body can not be permanent' (Diodorus, 1947). Recent organic geochemical studies by Rullkotter & Nissenbaum (1988) and Connan & Dessort (1989) showed that Dead Sea asphalt was present in some of the investigated mummies. The criteria for the identification were based on biomarker analysis of Dead Sea asphalt (Rullkotter *et al.*, 1985) and on five mummies ranging in age from 900 B.C. (the 22nd dynasty) to the Ptolemaic period (300 B.C.–30 A.D.). Evidence for Dead Sea asphalt was found in the younger mummies but not in the 900 B.C. mummy. Connan & Dessort (1989) also found evidence of Dead Sea asphalt in Ptolemaic mummies. Rullkotter & Nissenbaum (1988) proposed that the Egyptians used Dead Sea asphalt mainly after the 4th century B.C. when the geopolitical situation in the Middle East was such that the Land of Canaan was incorporated into the economic sphere of Egypt. Thus, during the times of Herodotus, Dead Sea asphalt was not used, but during the times of Diodorus it became one of the ingredients of mummies.

Photography

In 1822 Joseph Nicephore Niepce (1765–1833) produced the first permanent heliographic copy of an engraving of the Cardinal d'Amboise by means of bitumen of Judea smeared on glass. Bitumen of Judea was used as a synonym for Dead Sea asphalt. Niepce spent several years in searching for light sensitive substances (Gernsheim, 1969) and became acquainted with this asphalt which was used in engraving because of its resistance to etching liquids. It was known that the asphalt hardens when exposed to light and thus could be used to differentiate between illuminated parts of a picture which are represented by hardened asphalts from dark parts where they asphalt can be dissolved away by solvents. The process that Niepce used was replicated by Marignier (1990) who used, however, a commercial product of bitumen from a unspecified source. In 1822, Niepce, using Dead Sea asphalt, took the first photograph ('heliograph') of nature by exposing to the sunlight a metal plate covered by bitumen on which the image of his yard was captured. Niepce's technique was used for obtaining heliographs on a variety of matrices coated by asphalt – zinc plates, pewter, and so on. The poor light sensitivity of the asphalt resulted in the rapid replacement of heliography by other techniques.

Waterproofing

The malleability of warm asphalt combined with its imperviousness to water made it a very valuable component of any water carrying system. Thus, for example, it was widely used throughout ancient Mesopotamia and the Indus valley to seal water drains, pipes and reservoirs (Forbes, 1954). The asphalt was of particular importance in waterproofing vessels which were used to contain liquids. The earliest human use of asphalt is probably from the Neolithic preceramic excavation of Gilgal, near Jericho, north of the Dead Sea (Fig. 1). At this site, dated 10 000 B.P., asphalt was used to coat woven baskets made of harvested flax. Similar remains of the same, or perhaps slightly younger age, were also found the Nahal Hemar cave, near the Dead Sea (Bar Yosef, 1985).

Oil exploration

The recognition of the possible commercial value of Dead Sea asphalt led to several exploration projects aiming at the geological evaluation of the asphalt deposits. Blanckenhorn conducted in the beginning of the 20th century a survey of the mineral resources of Palestine with special emphasis on the Dead Sea area (Blanckenhorn, 1903). According to his estimate, the asphalt deposit of Wadi Muhawat (= 'Nahal Hemar'), west of Mt Sodom, contains about 20 000 cubic meters of ore and with an average asphalt content of 11%, or about 4000 tonnes of bitumen worth about 2 000 000 German marks in 1904. Following Blanckenhorn's expedition the interest in Dead Sea asphalt continued, but from a different angle. Asphalt was not of interest in itself, but rather as an indicator for the existence of buried oil fields (Nissenbaum, 1992 and references therein). In the late 1920's G.S. Blake, the chief geologist of the British Mandatory Government of Palestine explored the geology of Dead Sea asphalt from the point of view of its potential as a natural resource (Blake, 1930). Blake's estimate of the quantities of asphalt in Wadi Muhawat were much higher than that of Blanckenhorn. The material was analyzed by the Haifa refineries as well as in England and was found to be similar to the asphalt from Lattakia, Syria or from Hasbeya, Lebanon. Blake (1930) wrote that the Dead Sea asphalt is of suitable quality for making paints and varnishes as well as asphalt emulsions. Since the asphalt is exposed on the surface it could also be quarried for use in road paving. During the Second World War, asphalt and ozocerite from the Dead Sea area were collected in order to assess their industrial usefulness and several tonnes of material were sent to England for testing. The quantities however, were far too small for commercial exploitation.

Magnesium

The Dead Sea has always been considered to be one of the world's largest potential sources of magnesium. The high concentration of magnesium in the brine has attracted the attention of many. However, the production of metallic magnesium is energy intensive, and the lack of a viable source of electricity in the Dead Sea area made this untenable. Some of the residual liquor which is left after the extraction of potash from carnallite, and which contains high concentrations of magnesium chloride, is piped to the 'Periclase' factory on the highlands on the western side the Dead Sea (Fig. 1) and is used to produce high quality magnesia (magnesium oxide) refractory bricks. The chemical process used by this factory also furnishes HCl as a side product.

Potash

The possibility of utilization of the Dead Sea as a source for chemicals has been on the agenda since the 1890's. Theodore Herzl, the founder of the Zionist movement, records in his 1896 diary a conservation relating to this subject (Braslavsky, 1943). Herzl's book 'Altneuland' which was written in 1902 and laid the foundation of the Zionist movement, devotes much space to the future utilization of the Dead Sea. In this he was deeply influenced by the German commercial attache in Constantinopole, Elschner, who proposed to German capitalists, in the beginning of the century, the establishment of a chemical plant in the north end of the Dead Sea (Blanckenhorn, 1928, as quoted by Picard, 1954). In 1904 the Zionist Organization sent a geological party to the Dead Sea area, headed by the German geologist, Max Blanckenhorn, in order to provide a detailed account of the natural resources of the Dead Sea area. In 1906, a mining engineer from Siberia, M. Novomeyski, established a company to utilize the natural resources of the Dead Sea (Braslavsky, 1953). Novomeyski visited the Dead Sea in 1911 and reached the conclusion that the best way to exploit the Dead-Sea brine is by using solar evaporation. In 1918, before the end of the First World war, a Major Tulloch submitted to the British Government a proposal to extract potash from the Dead Sea,

the reason being the importance of potash for production of explosives. Winston Churchill, the then Minister of Armament, rejected this proposal. Other proposals to extract potash from the lake, submitted between 1918 and 1920 were rejected as well (Aran, 1984). In 1922, after the British mandate on Palestine was ratified, both Novomeyski and Tulloch submitted bids to obtain the concession on Dead Sea potash (Aran, 1984). After considerable haggling, bickering and intriguing, the concession was awarded, on January 1, 1930, to Novomeyski in partnership with Tulloch. In 1929 Novomeyski established the Palestine Potash Company, and a small experimental evaporation pond was built at Kalia, in the north end of the lake (Figs 1 and 2). In 1932, the plant was finished, and bromine and a few hundreds tonnes of potash appeared on the market. The process employed for potash production involved solar evaporation in shallow ponds in

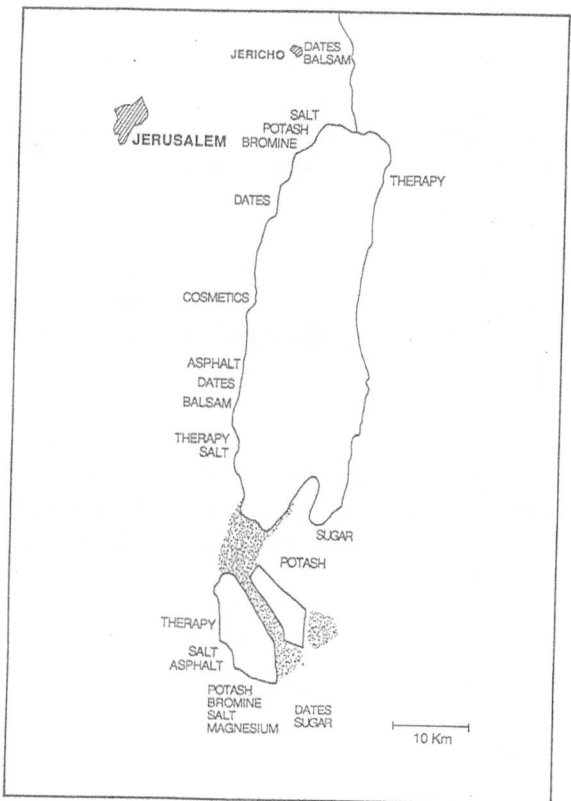

Fig. 2. Geographical distribution of Dead Sea natural resources.

order to precipitate halite, and removal of the residual liquor to another pond where carnallite ($KMgCl_3$) was precipitated. The carnallite was washed with water, and by utilization of the preferential solubility of the $MgCl_2$ component pure KCl was obtained. It was soon realized that the demand for large evaporation ponds could not be met at the north end of the lake, so in 1934 a new plant was established near Mt. Sodom at the south end of the lake (Figs 1 and 2). This enterprise was beset with many difficulties due to the lack of water, absence of roads and the harsh climate. The plant was completed in 1938 and its potash was sent by boats to Kalia (Nissenbaum, 1991). Between 1930 and 1948 the total production of potash was 1 040 000 tonnes.

In 1948, during Israel's War of Independence, both the northern and southern plants were destroyed with the Kalia plant being part of the area taken over by Jordan. In 1949 the southern plant was revived, but its output was quite small. In the early 1960's the Dead Sea Potash Company embarked on an ambitious program to convert the entire Israeli half of the southern Dead Sea basin into huge evaporation pond with concomitant large increase in the potash production capacity. With the completion of this project, the potash output increased tremendously and in 1974 the annual production exceeded 1 000 000 tonnes. In 1979 a second plant was erected in Sodom utilizing a new technique for potash production, the so called 'cold cristallization' process. In 1990, 2 185 000 tonnes of potash were produced, which were sold for 274 million dollars. Together with income from its daughter companies (including the Bromine Compounds Ltd), the Dead Sea Works are today the major generators of income from natural resources in Israel. In parallel with the Israeli plant, the Jordanian government established a similar plant in the Jordanian portion of the southern Dead Sea basin (Fig. 2).

Bromine

The Dead Sea today is one of the world's major sources of bromine. The water of this lake shown an exceptional enrichment of in bromine which is the second most abundant anion in the brine. The bromine concentration (*ca* 5.4 g l^{-1}) as well as the low chlorine to bromine ratio (*ca* 44), makes these waters unique among other surface hypersaline waters. Although bromine was discovered in 1826 by the French chemist Ballard, it is interesting to note that the German chemist Hermbstadt, who analyzed Dead Sea water in 1822 reported 'free hydrochloric acid' in the salts in Dead Sea brine which may be related to the bromine in the water (Nissenbaum, 1970). According to Picard (1954), who quotes Blanckenhorn (1928), Ballard already had in mind a large plant in the Dead Sea for the production of bromine. In 1827 Gmelin reported the first analysis of bromine in the Dead Sea. The value reported was only 25% higher than the present value (Nissenbaum, 1970). Since then almost every chemical analysis of Dead Sea water included information on bromine. According to Picard (1954), the German commercial attaché in Turkey, had proposed at the turn of this century, to several German capitalists, the establishment of a chemical plant in the north end of the Dead Sea. With the growing industrial importance of bromine, particularly with its use in preventing the deposition of lead oxide in engines through the use of tetra-ethyl-lead as an anti-knocking agent in gasoline, the German Potash Company which produced bromine from the Zechstein evaporites, sent in 1924 at the behest of General Motors, a mission to the Dead Sea in order to explore the possibility of utilizing the brine for production of bromine. However, in view of the remoteness of the area at that time, as well as the lack of proper infrastructure for industry, this project never came to fruition. In 1931 the Palestine Potash Company established a potash production plant at Kalia (Fig. 1). near the north end of the Dead Sea. After harvesting of the carnallite and separating out of the potash the residual brines, the so called 'end brines', contained as much as 10–14 g l^{-1} bromine. A small plant was established on site in 1931 which was expanded in 1932. The bromine was extracted by heating and bubbling of chlorine gas. The small area available for construction of larger evaporation ponds in the north as well as the scarcity of fresh water in the region resulted in the establishment, in 1934, of a larger potash plant in the south, near Mt. Sodom which resulted in increased production of bromine. Between 1939 and 1947 approximately 3.5 million dollars worth of bromine were exported. In 1948, during Israel's War of Independence, both the northern and southern plants were destroyed. In 1957 a new bromine plant was constructed near Mt. Sodom and it was completed in 1962. The new plant, which is owned by the Dead Sea Bromine group, is today one of the world's largest producers of bromine.

Therapeutics and cosmetics

The use of the Dead Sea and its environs for medicinal and therapeutic practices is probably rooted in ancient times. The unusual properties of the water, combined with the occurrence of hot springs, frequently rich in hydrogen sulfide, along its shores, led to the belief that those waters could be used to treat various diseases. Direct evidence on the use of the Dead Sea for therapeutics in the ancient world is rather sparse. Galen, the most famous physician of the Ancient World, visited the Dead Sea in 166 A.D. After discussing several of the properties of Dead Sea brine, he says that any water may be made similar to the brine by the addition of sufficient salt, and ridicules the rich people who take cisterns of Dead Sea water back to Italy (Walsh, 1924). The Talmud also record a discussion between two learned rabbis, in the 4th century A.D., regarding bathing in the Dead Sea on the day of Sabbath for therapeutical purposes, and whether opening the eyes while being immersed in the water is permissible (Braslavsky, 1943). This indicates the notion that Dead Sea waters are a remedy against eye diseases. In modern times, the belief in the curative properties of the brine has been revived and supported by scientific and clinical evidence. Psoriasis is a skin disease which afflicts 80 million people around the globe. This disfiguring malady is usually treated by potent steroid drugs or by combination of drugs and artificial UV irradiation. The Dead Sea provides a natural remedy against this disease, which can either work by itself, or has to be augmented by mild drugs. The treatment consists only of exposure to the sun interrupted by immersion in Dead Sea water. The reason for the effectiveness of this treatment is not entirely clear. Possibly it is due to combination of environmental factors. The UV spectrum of the sunlight in the Dead Sea area is depleted in short wave rays (290–330 nm) due to filtering by the dense atmosphere. This allows longer exposure to the sun without burning (Kushelevsky & Slifkin, 1975). In addition, the brine itself, and in particular the dissolved bromide, may have some effect on this skin disease since elevated serum bromine levels were found in patients being treated in the Dead-Sea (Shani et al., 1989). Clinical studies has indicated that after four weeks of stay in the Dead-Sea, 35% of the patients were completely cleared and additional 40% showed marked improvements. The Dead Sea today is a world recognized center for psoriasis treatment (Abels et al., 1989). The Dead Sea beneficial effect on other skin diseases, such as atopic dermatitis etc. is also recognized, but it had not been thoroughly investigated. Sediments that constituted the lake bed before the recession of the water and that are now found on-shore are used for dermatological treatments. The sediments are rich in clays and fine grained aragonite and are colored deep black due to the presence of poorly crystallized iron sulfides. In addition to the possible direct effect of the sulfide on the skin, the mud absorbs the sun's heat, which may have also an effect on the skin. The medicinal properties ascribed to Dead Sea asphalt have been reviewed in a previous section.

The hydrogen sulfide containing hot springs around the Dead Sea, and in particular those of Kallirohe, in the northeastern sector of the lake (Fig. 1), were a major balneological attraction for a very long time. The area near the springs, now known as Zerka Ma'in, was inhabited since Paleolithic times (Khouri et al., 1984). The historian Josephus Flavius, in the 1st century A.D., describes how King Herod the Great, used to go there seeking remedy for his various ailments, and possibly tranquility for his tormented soul (Josephus, 1959). Pliny, in his encyclopaedic tome on Natural History, also says 'there is a hot spring possessing medicinal value. The name of which, Kallirohe (= good spring), itself proclaims the celebrity of its water' (Pliny, 1948). The importance of this resort is indicated by the finding of ruins of a Roman harbor on the Dead Sea shore which was probably used to transport people from and to this resort (Nissenbaum, 1991). According to a quotation in Braslavsky (1943), the priest Petrus, visited these springs in 430 A.D. in order to cure a severe malady. The mosaic map of the Holy Land, from the 6th century A.D., discovered in the monastery of Madaba, overlooking the Dead Sea, also depicts these hot springs as prominent feature, attesting to their importance. In recent years, the water and muds of the Dead-Sea have became the basis for thriving manufacture of cosmetics and beauty aids. The salts which are obtains by evaporation of the brine and dried muds are sold for making beauty masks and as bath salts. Soaps and other cosmetics are produced which contain some of the components of the muds.

Agriculture

Due to the combination of climatic factors such as heat, and the existence of several perennial water sources, the Dead Sea area has become famous for its specialty agricultural products, of which some of them were not known elsewhere. The main agricultural areas in ancient times were at Ein Gedi, an oasis on the western shore and at Es-Safi south of the Dead Sea (Fig. 1). The most unique product was balsam. This incense was considered to be the most precious incense and medicine in the Ancient World and is frequently and extensively mentioned by early historians. It was used as a remedy against various diseases, such as migraines, ophthalmological disorders and other maladies. The nature of the plant from which it was extracted is unknown. The most likely candidate is *Commiphora opobalsamus* (Burseraceae), although it was suggested that it

may have been a group of plants (Porath, 1986). Chemical analysis of what was considered to be balsam oil, very recently found in a juglet found in the Judean Desert, did not provide definite clues to its plant source (Aizenshtat & Ashengrau, 1990). Balsam was produced only in the Dead Sea area, at Ein Gedi and near Jericho (Fig. 1). *C. opobalsamus* probably originated in southwestern Arabia, the land of frankinsence, and it is not known when it was introduced into the Dead Sea area, but Josephus Flavius (1st century B.C.) repeated a tradition that it was brought to Canaan by the Queen of Sheba during the reign of King Solomon. Remains of what is assumed to be a perfume factory, dated to the 7th century B.C., were excavated in Ein Gedi. The earliest historical mention is by Theophrastos, in 287 B.C., who said that two balsam groves existed in the valley of Jericho. One produced 20 litres of balsam oil annually and the other only 3.3 litres (Braslavsky, 1943). This small quantity is not surprising if we consider that one full working day was necessary in order to collect 6 ml. (Porath, 1986). According to Pliny, the balsam was already grown in the Dead Sea area at the time of the conquest of Judea by Alexander the Great in 332 B.C. The importance of the balsam as a money making resource is indicated by Pliny who tells how during the Roman conquest of Judea in the 1st century A.D., Jewish zealots attempted to destroy the balsam plantations which the Romans strongly defended by fighting over every single plant. Following the Roman victory, the balsam became an imperial monopoly. According to Pliny, the income from the balsam in the five years following the conversion into imperial monopoly, was 200 000 dinars. This should compare with the annual net income of a farmer in this period which was 150 dinars (Porath, 1986). The famous physician Galen, visited the Dead Sea area in the 2nd century A.D. in order to seek high quality, unadulterated, balsam. The Talmud also refers several times to Dead Sea balsam. Both Eusebius and Heronimus, in the 4th and 5th century A.D. respectively, mention the balsam groves of Ein Gedi. It is not known when the art of growing balsam died out. Possibly, the economic deterioration of the area following the Arab conquest in the 7th century A.D. put a stop to this endeavor, and with it the secret of the balsam disappeared.

Other plants which were used for the preparation of perfumes were grown in Ein Gedi as is indicated by the Bible: 'My beloved is unto me as a cluster of henna in the vineyards of Ein Gedi' (Song of Songs, 1:14). Later writers, such as Josephus and Eusebius also refer to the production of incense at Ein Gedi. The Dead Sea basin was famous for the high quality of its dates. The Bible refer to Jericho as the City of Palms (Deuteronomy 34:3). Ein Gedi was also known for its palm trees as is evidenced by several ancient historians such as Josephus and Pliny. The area around Es Safi (Fig. 1), also called Zoar or Segor, was known all over the world for its palms. The Talmud, in the first centuries A.D., frequently refers to Zoar dates, and they were exported overseas as well, and in particular to Rome. It is recorded that King Herod the Great sent them to Augustus Caesar. Pliny, Galen, Strabo and others often mentions the quality and variety of dates from the Dead Sea area. During the Middle ages, the dates were the most important source of income in the Dead Sea region. The Arab historians of the 9th to 11st centuries (among them El-Mukkaddasi, Ibn Haukal and El Istakhari) pay tribute to the Zoar dates, and El Istakhari says that their quality exceeds that of Iraqi dates, which were the world standard. During the Crusader period, in the 12th century, the name Zoar was replaced by Villa Palmarum (in Latin) and Paumiers (in French). Today, dates are being grown in several areas near Ein Gedi and Kalia (Fig. 2) and are re-emerging as an important source of income. Sugar was grown during the Arab and Crusader periods in several water rich areas in Jericho and in the southern part of the Basin (Fig. 2). Several locations are today still called in Arabic. Tauakhin-es-Shugar (the sugar mills). Another specialty crop was indigo which was grown in Es Safi, although its quality was not as good as the indigo grown in the northern part of Israel.

Transportation

The image of the Dead Sea was for many years that of a forbiddingly barren and hostile region embedded in an area of cardinal importance to western civilization. Integral to that image was the impression, only recently dispelled, that the lake's importance as a shipping artery was negligible. This view was intensified in the Middle-Ages by the belief that the Dead Sea was so inherently inimical to life that no ship could possibly ply its water. However, historical and archaeological evidence totally refutes this concept, pointing instead to the Dead Sea as a much used trade artery for at least 2500 years (Nissenbaum, 1991b). Archaeological evidence from the first century B.C. and from the Roman period shows the existence of small harbors and landing places in the northeastern and northwestern sectors of the lake. The historical literature is filled with references to the use of boats on the Dead Sea. Diodorus of Sicily describes a naval battle on the lake in 312 B.C.; Josephus Flavius and Strabo, in the first century B.C., describe the boats used to collect the floating asphalt blocks found in the Dead Sea. Pictorial evidence includes a scratching of a sail boat found in the Hasmonean palace of Masada dated to the second century B.C. and in the Madaba map (6th century A.D.). Recently, the first direct physical evidence for the use of the Dead Sea as a shipping artery during Antiquity was provided by Nissenbaum et al. (1990a) who dated ropes attached to stone anchors which were found near Ein Gedi (Fig. 1) on what was the lake bottom, prior to the recession of lake shoe, to the 3rd century B.C. During the Middle Ages, wheat and dates were regularly transported by boats across the lake from Es Safi to the north. Modern exploration of the Dead Sea began in the 1830's with the ill fated expeditions of Costigan and Mollineux and culminated in the 1847 expedition by Lynch which laid the basis for the study of the geology, chemistry and bathymetry of the lake (Nissenbaum, 1991). By the end of the 19th century the lake was used as a major water highway, and during World War I the German navy had gunboats on the lake. In the following years, boats were extensively used to transport potash from the south to the north. Tourist agencies also used the lake for sightseeing cruises. All this ended in 1948, during Israel's War of Independence, and since then commercial shipping on the Dead-Sea has completely ceased.

Acknowledgements

The author thanks Mr Dekel Golan (Bromine compunds Ltd.) and Mr Jacob Shatzky (Dead-Sea Works Ltd) for graciously providing material on the activities of their companies.

References

Abels, D. & J. Kattan-Byron, 1989. Psoriasis natural treatment at the Dead Sea: A natural selective ultraviolet phototherapy. Natural Therapy at the Dead Sea. Unpublished Report by the Dead Sea Regional Tourist Organization, 15 pp.

Abraham, H., 1960. Asphalts and allied substances. Volume 1. Van Nostrand, Princeton, N. J.

Aizenshtat, Z., & D. Aschengrau, 1990. Analyses of oil contained in Herodian juglet from Qumran (Israel). Israel Exploration J. 39: 55–59.

Almog, Y. & B.-Z. Eshel, 1957. The Dead Sea Region. Am. Oved. Pub., Tel Aviv. (in Hebrew).

Amit, O. & A. Bein, 1979. The genesis of asphalts in the Dead Sea area. J. Geochem. Expl. 11: 211–225.

Aran, N., 1984. The concessions for oil and potash exploration in Eretz-Israel. Catedra 31: 135–158. (in Hebrew).

Bar Yosef, O., 1985. A cave in the desert-Nahal Hemar. Israel Museum, Jerusalem.

Beyth, M., 1980. Recent evolution and present stage of Dead-Sea brines. Pages 155–166 in A. Nissenbaum, editor. Developments in Sedimentology, 28: Hypersaline Brines and Evaporitic Environments. Elsevier Sci. Pub., Amsterdam.

Blake, G. S., 1930. The Mineral resources of Palestine and Transjordan. Government of Palestine, Jerusalem.

Blanckenhorn, M., 1903. Ueber der vorkommen von phosphaten, asphaltkalk, asphalt und petroleum in Palästina und Aegypten. Z. Praktische Geologie 2: 294–298.

Blanckenhorn, M., 1928. Kali im Toten Meer. Palestina 11: 496.

Bloch, M. R., 1976. Salt in human history. Interdisciplinary Science Reviews 1: 336–352.

Braslavsky, Y., 1943. Do you Know the Land (in Hebrew). Volume 3: Around the Dead Sea. Hotzaat Hakibbutz Hameuhad, Tel Aviv.

Connan, J. & D. Dessort, 1989. Du bitume de la Mer Morte dans les baumes d'une momie égyptienne: identification par critères moléculaires. C. r. Acad. Sci. Paris 309: 1665–1672.

Connan, J., A. Nissenbaum & D. Dessort, 1992. Molecular archaeology: EXport of Dead Sea asphalt to Canaan and Egypt in the Chalcolithic-Early Bronze age (4th–3rd millenium B.C.). Geoch. Comoch. Acta 56: 2743–2759.

Delachanal, B. M., 1883. Sur la composition de l'asphalte ou bitume de Judée. C. r. Acad. Sci. Paris 97: 491–494.

Diodorus Siculus, 1947. English translation in 12 volumes by C. H. Oldfather. William Heineman Ltd, London.

Dioscorides, 1968. The Greek herbals: Englished by J. Goodyear, 1655. Hafner, New York. (Facsimile).

El Mahdi, C., 1989. Mummies-myth and magic. Thames and Hudson, London.

Forbes, R. J., 1964. Studies in ancient technology. V.1. E. J. Brill, Leiden.

Gernsheim, H., 1969. The History of Photography, Revised Edition. Thames and Hudson, London.

Hamond, P.C., 1959. The Nabatean bitumen industry at the Dead Sea. The Biblical Archaeologist 22: 40–48.

Hermbstadt, S. F., 1822. Chemisches Zergleiderung des Wassers aus dem Todten Meer. Journal fur Chemie und Physik 35: 153–195.

Josephus Flavius, 1959. The Jewish Wars: English translation by G. A. Williamson. Penguin Books, New York.

Kelso, J. L. & A. R. Powell, 1944. Glance pitch from Tell Beit Mirsim. Bull. Am. School Oriental Res. 95: 14–18.

Khoury, H., E. Salameh & P. Udluft, 1984. On the Zerka Ma'in Travertine/Dead Sea (hydrochemistry, geochemistry and isotopic composition). N. Jb. Geol. Palaeont. Monats. 8: 472–484.

Kushelevsky, A. P. & M. A. Slifkin, 1975. UV measurements at the Dead Sea and at Beer-Sheba. Internal Report to the Health Resorts Authority, Jerusalem, 6 pp.

Lucas, A., 1962. Ancient Egyptian Materials and Industries, 4th (revised and enlarged by J. R. Harris) Edition. Edward Arnold Ltd, London.

Marignier, J. L., 1990. Historical light on photography. Nature 346: 115.

Neev, D. & K. O. Emery, 1967. The Dead Sea: depositional processes and environment of evaporites. Geological Survey of Israel, Jerusalem, Bull. No. 41.

Nissenbaum, A., 1970. Chemical analyses of Dead Sea and Jordan River water, 1778–1830. Israel J. Chem. 8: 281–289.

Nissenbaum, A., 1978. Dead Sea asphalt-historical aspects. Bull. Amer. Assoc. Petrol. geol. 62: 837–844.

Nissenbaum, A., 1979. Life in a Dead Sea-fables, allegories and scientific search. Bioscience 24: 153–157.

Nissenbaum, A., 1991. The shipping lanes of the Dead Sea. Rehovot, 11: 19–24.

Nissenbaum, A., 1992. Oil exploration in the Holy Land, 1884–1955. Israel J. Earth Sci. 40: 245–250.

Nissenbaum, A. & M. Goldberg, 1980. Asphalts, heavy oils, ozocerite and gases in the Dead Sea Basin. Organic Geochemistry 2: 167–180.

Nissenbaum, A., Z. Aizenshtat & M. Goldberg, 1980. The floating asphalt blocks of the Dead Sea. Pages 157–161 in A. G. Douglas & J. R. Maxwell, editors. Advances in Organic Geochemistry 1979. Pergamon Press, London.

Nissenbaum, A., A. Seban, R. Amiran & O.Ilan, 1984. Dead-Sea asphalt from the excavations in Tel Arad and Small Tel Malhata. Paleorient 10: 157–161.

Nissenbaum, A., I. Carmi & G. Hadas, 1990a. Dating of ancient anchors from the Dead Sea. Naturwissenschaften 77: 228–229.

Nissenbaum, A., M. Miller & A. Nishri, 1990b. Nutrients in pore waters from Dead Sea sediments. Hydrobiologia 197: 83–89.

Picard, L. Y., 1954. History of mineral research in Israel. Econ. Forum 11: 10–38.

Pliny, 1948. Natural History: English transl. by H. Rackham. Harvard Univ. Press, Cambridge, Mass.

Porath, J., 1986. Aspects of development of Ancient Irrigation Agriculture in Jericho and Ein Gedi. Pages 254 pp. in A. Kasher, A. Oppenheim and U. Rappaport, editors. Man and Land in Eretz Israel in Antiquity. Yad Izhak Ben Zvi, Jerusalem, (in Hebrew).

Prawer, J., 1975. The Crusaders-a colonial society. Bialik Inst., Jerusalem, (in Hebrew).

Raba, J., 1986. Russian travel accounts of Palestine. Yad I. Ben Zvi, Jerusalem, (in Hebrew).

Robinson, E., 1841. Biblical Researches in Palestine, Mt. Sinai and Arabia Petrae. Volume 2. Crocker & Brewster, Boston.

Rullkotter, J. & A. Nissenbaum, 1988. Dead Sea asphalt in Egyptian mummies: molecular evidence. Naturwissenschaften 75: 618–621.

Rullkotter, J., B. Spiro & A. Nissenbaum, 1985. Biological marker characteristics of oils and asphalts from carbonade rocks in a rapidly subsiding graben, Dead Sea, Israel. Geochimica Comochimica Acta 49: 1357–1370.

Shani, J., S. Barak, W. Avrach, H. Robbercht & R. Van Grieken, 1989. Natural Therapy at the Dead Sea. Unpublished report by the Dead Sea Regional Tourist Organization, 15 pp.

Schiller, E., 1988. Eretz Israel and its sites in descriptions by Moslem pilgrims. Ariel Pub., Jerusalem. (in Hebrew).

Spiro, B., D. H. Welte, J. Rullkotter & R. G. Schaefer, 1983. Asphalts, oils and bitumonous rocks from the Dead Sea area-a geochemical correlation study. Bull. Am. Ass. Petroleum Geologists 67: 1163–1175.

Tannenbum, E. & Z. Aizenshtat, 1985. Formation of immature asphalt from organic rich carbonate rocks-Geochemical correlation. Organic Geochem. 8: 181–192.

Theoderich., 1986. Guide to the Holy Land (trans. by A. Stewart), 2nd Edition. Italica Press, New York.

Walsh, J., 1924. Galen visits the Dead Sea and the copper mines of Cyprus (166 A.D.). Geograph. Club Philadelphia Bull. 25: 92–110.

Zak, I., 1967. The Geology of Mt. Sodom. Ph.D. Thesis. Hebrew University of Jerusalem. (in Hebrew. English summary).

Quaternary and recent *Lamprothamnium* groves (Charophyta) from Argentina

Adriana García
Facultad de Ciencias Naturales y Museo Paseo del Bosque, 1900 La Plata, Argentina

Key words: Lamprothamnium, Charophyta, gyrogonite morphology, ecology, paleoecology, Argentina, South America

Abstract

Only recently have extant *Lamprothamnium* species been reported from the American continent. *L. succinctum* (A. Br. in Asch.) R. D. W. was found in Lago Titicaca, Bolivia and *L. haesseliae* Dont. en Laguna Luro and Laguna La Salada, Argentina.

Fossil gyrogonites of *L. succinctum* and *L. haesseliae*, however, are here reported to be widely distributed in Quaternary sediments of Argentina, associated with other charophytes, ostracods and foraminiferans. Localities include Laguna del Siasgo, Laguna Salada Grande, Laguna Mar Chiquita, Laguna La Amarga and Salina del Bebedero.

Lamprothamnium is a genus that prefers shallow, alkaline, hyposaline to mesosaline environments. Analysis of the characteristics of Laguna La Salada contributes information on the factors that influence the distribution of the genus.

The morphology characteristics of the gyrogonites and oospores of *L. haesseliae* and the gyrogonites of *L. succinctum* are described. The use of *Lamprothamnium* as a biomarker and its application in the reconstruction of Quaternary saline environments are also discussed.

Introduction

Extant forms of *Lamprothamnium* have been found recently in South America. Guerlesquin (1981) described *L. succinctum* (A. Br. in Asch.) R. D. W. from Lago Titicaca Menor, Bolivia. Donterberg (1984) established *L. haesseliae* as a new species from Laguna Luro, Argentina; and specimens found by García (1980) in Laguna La Salada, Argentina, are also assignable to this species.

These are the only two species of *Lamprothamnium* currently known from the Americas, and both are restricted to South America. Daily (1967)

proposed to include in this genus the following species: *Chara hornemannii* Wallman, *C. longifolia* Robinson, *C. bulbillifera* (Dont.) García and *C. buckellii* Allen. As analyzed by García (1990a), morphological and biological evidence suggests, however, that they may be separated as a genus distinct from *Lamprothamnium*.

In fossil material, the only preserved part of a charophyte is usually the gyrogonite which is the calcified remains of the female fruiting body. The characterization of this structure in living species thus is indispensable for the study of fossil forms. In Argentina the first fossil *Lamprothamnium* was recorded in the eo-Tertiary of the El Carrizo Formation by Musacchio & Moroni (1983), but no paleoecological interpretation was made.

This study characterizes the oospores and gyrogonites of *L. haesseliae* using a large amount of gyrogonite-bearing living material from Laguna La Salada, Argentina. It also describes ecological characteristics of Laguna La Salada so as to facilitate the use of *Lamprothamnium* as biomarker. Finally, this study reports the finding of *L. haesseliae* and *L.succinctum* (the latter not recorded living in Argentina) in five Quaternary localities of this country, which indicates a greater diversity and distribution of the genus in the recent past.

Study sites and methods

Provenance of samples

Living charophytes were obtained from Laguna La Salada in southern Buenos Aires Province and fossil forms from five other localities in Argentina (Fig. 1).

Laguna La Salada

Laguna La Salada lies North of the Colorado River in the area near its mouth and about 7 km N of the town of Pedro Luro, at sea level. The climate in this zone is semiarid with a negative hydric balance throughout the year and diminished precipitation in winter. Air temperatures are not extreme. The daily mean is about 22 °C in January and 8 °C in July. Predominant winds blow from the N–NW.

The pond is approximately 6? m deep, with gently sloping shorelines, a dark sandy bottom with intercalated gypsum crystals within the sediments. The water is clear, lacking suspended sediments, though the water surface is often disturbed by wind.

The ionic composition of the water at the moment of collecting the *Lamprothamnium* (1980, January) was dominated by sodium and chloride followed by sulphate. Salinity (TDS) was 23 g l^{-1} and pH was 7.8. According to V. Conzonno

(pers. comm.) similar conditions obtained in June 1968, but a lower salinity in June 1970 (10.5 g l^{-1}). Following Hammer (1986), this water body can be classified as hyposaline (3–20 g l^{-1}) to mesosaline (20–50 g l^{-1}), or following Ringuelet *et al.* (1967), as mesohaline (5–18 g l^{-1}) to polyhaline (18–40 g l^{-1}). The terminology of Hammer (1986) is followed in the remainder of this paper.

Associated with *L. haesseliae* were found *Chara* cf. *buckellii* G. O. Allen, *Chara* nov. sp. (Charophyta); *Cyprideis* sp., *Limnocythere* sp., and, in fewer numbers, *Cypridopsis* sp. and *Ilyocypris* sp. (Ostracoda); `Ammonia* sp., *Elphidium* sp. and *Quinqueloculina* sp. (Foraminiferida) and *Littoridina* sp. (Mollusca). All the taxa were determined by the author.

Fossil material was collected in the five localities described below:

Laguna del Siasgo

Laguna del Siasgo is part of the Salado River basin. It lies at the intersection of Ruta Provincial (State Road) 29 with the Salado River (Monte and General Paz Districts, Province of Buenos Aires).

It can be considered an 'overflow pond' and occupies a small deflation basin that is dry most of the year. The pond bed lies at 12.5 m above sea level and it is above the mean level of the Salado River. It has an area of *ca* 1250 ha, a maximum depth of *ca* 2 m, salinity (TDS) of 13 g l^{-1} (one sample), and is in a basin surrounded by river banks 4 to 10 m high that become a gently sloped beach in some sectors.

Geologic and stratigraphic studies in the area were made by Dangavs & Merlo (1980), who also collected sediments for micropaleontological analysis from the center of the basin, from a borehole 5 m deep which represents the full thickness of the lacustrine sediments at that point. The basin was formed in the Late Pleistocene, as suggested by the presence of Pampean mammals within the riverbank sediments. The lacustrine sediments are Holocene in age. The surface sediments are

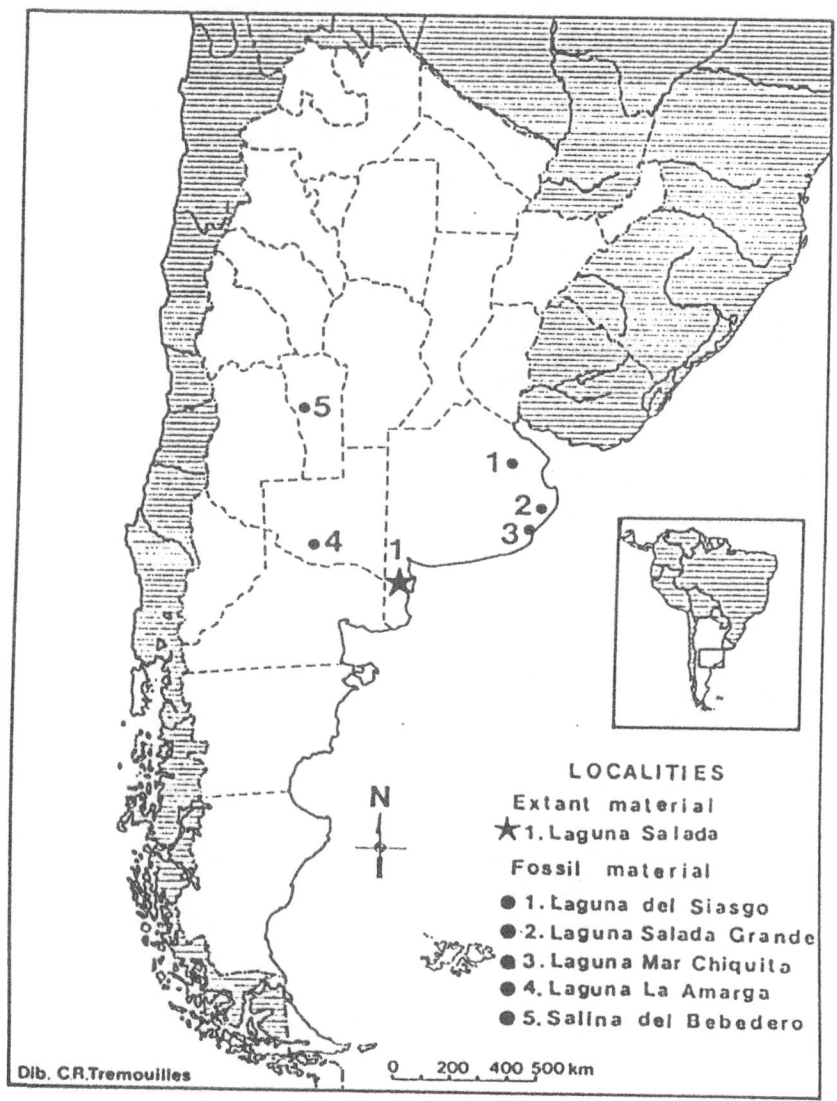

Fig. 1. Map of Argentina showing the localities where living and fossil *Lamprothamnium* were collected.

greenish silty-sands. From depths of 0.5 to 1.5 m in the core a sandy fraction predominates. Between 1.5 and 2.5 m there are sandy silty sediments, with frequent gypsum crystals. At 3.5–5 m the sediments are greyish, finer, with a clayish-sandy silt texture, and are highly saline, with many soluble salts (Dangavs & Merlo, 1980).

Between 1.5–5 m gyrogonites of *Lamprothamnium haesseliae* and *L. succinctum* were found. In the uppermost levels (1.5–2.5 m) there were also subsaline to hyposaline charophytes (*Chara lep-*

tosperma A. Br. and *C. bulbillifera*, as reported in García, 1987), foraminifera (*Ammonia* sp., *Elphidium* sp.) and ostracods (*Cyprideis* sp., *Limnocythere* sp., *Cyprinotus* sp., *Ilyocypris* sp. and *Cypridopsis* sp). Still deeper there was a large number of *Lamprothamnium* gyrogonites, a few aberrant ones of *Chara bulbillifera*, the same foraminifera and ostracods as listed above, and *Pampacythere* sp. These biotas indicate subsaline to hyposaline conditions in the uppermost sediments (0–2.5 m), and hypo-mesosaline conditions in the bottom ones (3.5–5 m) representing the Early Holocene. According to Dangavs & Merlo (1980) these associations can be referred to mixohaline environments and correlated to the Holocene Platense transgression.

Laguna Salada Grande

Laguna Salada Grande pond lies in the east-central sector of the Province of Buenos Aires (General Madariaga and General Lavalle Districts), about 30 km inland from the city of Pinamar.

At present it is a typical waterbody of the Pampas, with dark, shallow water, an area of ca 6078 ha, and a maximum depth of ca 1.30 m.

The idea that this pond represents a coastal lagoon (Pascual et al., 1965; Ringuelet, 1962) was revised by Dangavs (1983, 1988) who suggested that the basin is a result of eolic action.

A detailed geologic description comprising a larger area that includes Laguna Salada Grande can be found in Dangavs (1988), so only brief remarks are presented here.

In the lowest part of the basin there are outcrops of older sediments assigned to the upper Pleistocene, and younger ones that correspond to the Holocene. Overlying all of them are modern deposits.

Dangavs (1988) described five formations.

General Madariaga Formation is a continental series, Pleistocene in age. Salada Grande Formation (with two members), Los Zorzales Formation, Las Chilcas Formation and Resguardo Pesquero Formation are Holocene (Table 1).

Salada Grande Formation can be found in the subsurface. It has two lithofacies, the silty-clayish sandy member and the sandy-silty member. The first is thickest in the West, becoming thinner to the East where it is in contact with the other member. As to the biological content, the silty-clayish sandy member contains organisms of fresh and brackish waters, being richer in marine elements toward the East. At the boundary with the sandy-silt member the fauna changes abruptly to one representing a marine coastal environment.

Los Zorzales Formation corresponds to muddy gravel beds. The environment is mixohaline, without charophytes, and Dangavs (1988) considered it to represent a regressive phase of the Salada Grande ingression.

Las Chilcas Formation is exposed along the pond banks, and represents the bed of former ponds and the filling of older river channels. These sediments are subaqueous and in part subaerial. They contain mixohaline microfossils, with freshwater species in the South and mixohaline to marine forms in the North.

Resguardo Pesquero Formation is composed of eolic sediments (argillaceous dune, sandy dune and loess deposits).

Table 1. Schema of stratigraphy at 'Laguna Salada Grande complex' (after Dangavs 1988).

Age	Formation	Sea level
Actual	*Recent sediments*: Beach, bed, sandy dune deposits	?
Holocene	*Resguardo Pesquero*: Eolic sediments	Post-Platense regression
	· · · · · · · · · · · · · erosive discontinuity · · · · · · · · · · · · ·	
	Las Chilcas: Subaqueous and subaerial sediments	Platense ingression + 3 m (?)
	· · · · · · · · · · · · · erosive discontinuity · · · · · · · · · · · · ·	
	Los Zorzales: Mudy gravel bed	Post Salada Grande regression
	· · · · · · · · · · · · · erosive discontinuity · · · · · · · · · · · · ·	
	Salada Grande Silty-clayish sandy member Sandy-silty member	Salada Grande ingression + 2 m (?) (= Querandinense ingression)
	· · · · · · · · · · · · · erosive discontinuity · · · · · · · · · · · · ·	
Pleistocene	*General Madariaqa*: continental sediments	

Recent sediments consist of beach deposits, lake and pond beds, sandy non-consolidated and non-littoral dunes.

The charophyte *L. succinctum* was found in the silty-clayish sandy member of the Salada Grande Formation, Las Chilcas Formation, Resguardo Pesquero Formation and Recent sediments. This species is associated in the silty-clayish sandy member with *Chara contraria* s.l. (A. Br. ex Kütz.) (found living in both fresh and low saline environments) and a few foraminifera (*Ammonia* sp. and *Elphidium* sp.) and ostracods (*Cyprideis* sp.). In the sandy silty member more than 30 species of foraminifera are found. In Las Chilcas Formation the associated taxa are *Ammonia* sp., *Elphidium* sp. and *Cyprideis* sp. All these species indicate saline conditions though they can tolerate fresher water as well.

Laguna Mar Chiquita

Laguna Mar Chiquita is a coastal lagoon located almost 29 km northeast of the city of Mar del Plata on the Atlantic coast (Mar Chiquita District, Province of Buenos Aires). It has a maximum length of 25 km, a maximum width of 5 km and a maximum depth of 1.5 m. The salinity is very variable depending on the varying influence of continental and marine waters (Fasano, 1980).

The three samples studied were taken from the upper 15 cm of the lagoon bed. The sediments are dark sands. Charophytes were present only in one sample and consisted of a few gyrogonites of *L.* cf. *succinctum* that showed evidence of transport. Possibly they were carried to the lagoon by its affluent streams. The associated foraminifera and ostracods were not identified.

Laguna La Amarga

Laguna La Amarga is a pond connected to the Desaguadero River where this joins the Colorado River (Curacó Department, Province of La Pampa). The pond has a variable water level, as it is located in a semiarid area and is subject to dry periods.

The surface sediment samples were collected when the pond was almost dry. The sediments are recent dark sands. The charophyte material present included *L. haesseliae* and *Chara* nov. sp. The first is also present in Laguna La Salina, a small pond near Laguna La Amarga.

Associated microfossils were *Limnocythere* sp. and *Pampacythere* sp. (Ostracoda), identified in both ponds, and *Discorbis* sp. (Foraminiferida) found in Laguna La Amarga.

Salina del Bebedero

Salina del Bebedero is located almost 51 km southwest of the city of San Luis (La Capital Department, Province of San Luis).

This commercially exploited salt deposit occupies an area of 3650 ha and was a full pond 100 years ago. Presently it is dry and only ocasionally the Arroyo Bebedero carries water into it from the Desaguadero River.

The geology of this area was studied by González *et al.* (1980, 1981) and González (1981). Two lacustrine periods were recognized. One is known as 'Período Lacustre Mayor' (PLMa: Major Lacustrine Period) and is represented by the three uppermost shorelines. The other, represented by several lower shorelines, is known as 'Período Lacustre Menor' (PLMe: Minor Lacustrine Period). Between PLMa and PLMe there was a pluvial period, as indicated by drainage channels that cut through the PLMa shorelines but not the PLMe shorelines. The PLMa sediments, assigned to the Late Pleistocene, were identified at 60 m below the pond and extended even further down. The PLMe sediments are of Holocene age.

The following paleoecological interpretation is based on García (1990b). Charophytes found in the PLMa samples included: *Chara hornemannii* s.l. Wallman, *Chara* nov. sp. and *L. haesseliae*, associated with *Cyprideis* sp. (Ostracoda) and *Ammonia* sp. In PLMe sediments were *L. haesseliae*, *C. hornemannii* s.l., *Chara* nov. sp., plus a few specimens of *C. bulbillifera* and *C. contraria*

148

Fig. 2. Lamprothamnium haesseliae Donterberg, gyrogonites. A-D: Laguna La Salada living material. A. Lateral view. B. Basal view. C. Apical view. D. Basal plug. E–G: Laguna La Amarga subrecent material. E. Apical view. F. Basal view. G. Lateral view. Scale bar: 100 μm.

s.l., in association with *Cyprideis* sp., *Limnocythere* sp. (Ostracoda), *Elphidium* sp. and *Ammonia* sp. (Foraminiferida). In the older PLMe deposits, however, specimens of *C. bulbillifera* and *C. contraria* were more abundant, ostracods were additionally represented by *Darwinula* sp., *Ilyocypris* sp., *Cypridopsis* sp. and *Cyprinotus* sp., and the foraminiferan *Elphidium* sp. was deformed. In the lowermost PLMe shorelines the only charophytes found were *L. haesseliae* and *L. succinctum*. These were also the only levels where *Discorbis* sp. (Foraminiferida) was found.

This biota indicates a non-marine saline environment, predominantly hypo-mesosaline. The PLMe deposit probably begins when the forms identified suggest a more clearly subsaline to hyposaline environment, such as would have resulted from freshwater inflow to the pond after the dry episode at the end of PLMe.

Processing of samples

Analysis of both living and fossil material was based on the gyrogonites. They were taken directly from living material or, for fossil forms, were obtained by washing sediments samples in a screen. Standard measurements – largest polar axis (LPA), largest equatorial diameter (LED), number of spiral cells visible in lateral view(n) and the isopolarity index (ISI:100 LPA/LED), were made on 100 gyrogonites of *L. haesseliae* and 15 of *L. succinctum*. Scanning electron microscope photographs were prepared by D. Giménez at the University of Buenos Aires and P. Sarmiento, at the University of La Plata.

The material is housed at the Cátedra de Micropaleontología, Museo de La Plata, University of La Plata.

Results

Systematic description

The specimens of *L. haesseliae* collected in Laguna Salada do not show differences from the type material described by Donterberg (1984). The main vegetative characteristics are the opposite stipulodes, one for each branchlet, the uppermost whorls forming 'fox-tails', the female gametangia situated generally above the antheridia and both gametangia being borne only at branchlet nodes. For full descriptions and illustrations see Donterberg (1984).

The gyrogonite morphology of *L. haesseliae* (extant and fossil material) and *L. succinctum* (only fossil) is characterized as follows. The numbers given in parentheses are the ranges for the different dimensions.

Division Charophyta Migula 1897

Family Characeae Richard 1815 Genus Lamprothamnium *J. Groves 1916*

L. haesseliae *Donterberg 1984 (Fig. 2, Fig. 3: A-B, Fig. 4)* – L. haesseliae *Donterberg. Com. Museo arg. Bernardino Rivadavia, Bot. 2: 92–102. 1984* The following analysis is based on 100 gyrogonites from Laguna La Salada extant material. Frequency distributions of LPA, LED and n are given in Fig. 5.

Gyrogonites prolate, rarely perprolate to subprolate, apex truncate or somewhat prominent in some individuals, basal outline gently and continuously rounded. Diameter maximal near mid-line of gyrogonite (LED), diminishing very gently toward apical zone and more sharply toward basal zone. Key dimensions as follows: LPA, $\bar{x} = 716\ \mu m$ (530–850); LED, $\bar{x} = 439\ \mu m$ (310–550); ISI, $\bar{x} = 175$ (128–203); n, $\bar{x} = 9$ (8–11).

Spiral cells generally well calcified, convex or concave, in the letter case the basal outline more protruding. Equatorial zone spiral cell with ribs 70–100 μm in width.

Spiral cells slightly narrower (40–90 μm) in apical periphery widening to 100–150 μm when joining at apex. Thickness of spiral cells diminishes conspicuously toward apex, where only a thin layer can be observed. Through this layer the dark-coloured ectosporostine can be seen. Dehis-

150

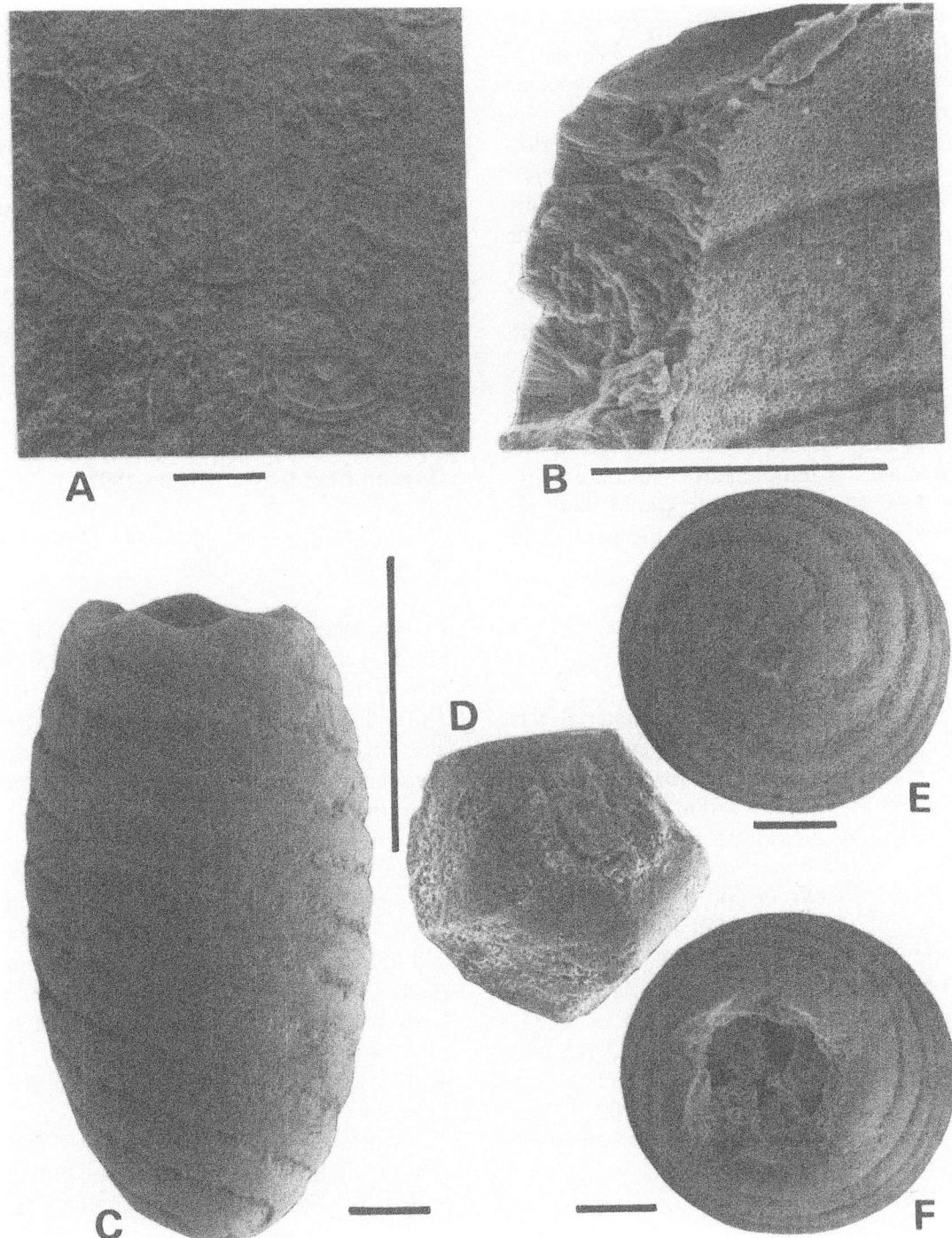

Fig. 3. A–B: *Lamprothamnium haesseliae* Donterberg, gyrogonite from Laguna La Salada. A. External detail of the wall with diatoms 'incorporated'. Scale bar: 10 μm. B. Internal view showing the undulated impression of the intercellular ridges. Scale bar: 100 μm. C–F: *Lamprothamnium succinctum* (A. Br. in Asch.) R. D. W., fossil gyrogonite from Laguna Salada Grande. C. Lateral view. D. Basal plug. E. Basal view. F. Apical view. Scale bar: 100 μm.

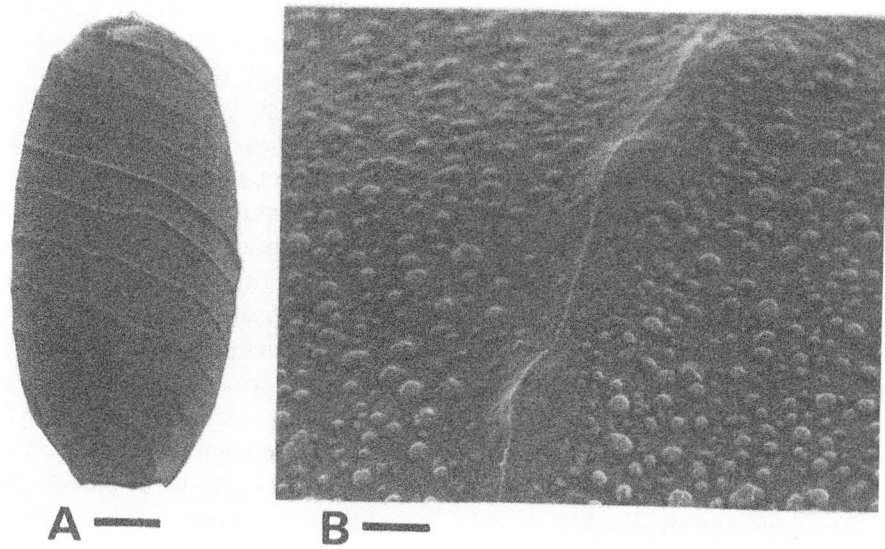

Fig. 4. *Lamprothamnium haesseliae* Donterberg, oospore from Laguna La Salada living material. A. Lateral view. Scale bar: 100 μm. B. Texture with papilliform to conical granules and undulated intercellular ridge. Scale bar: 10 μm.

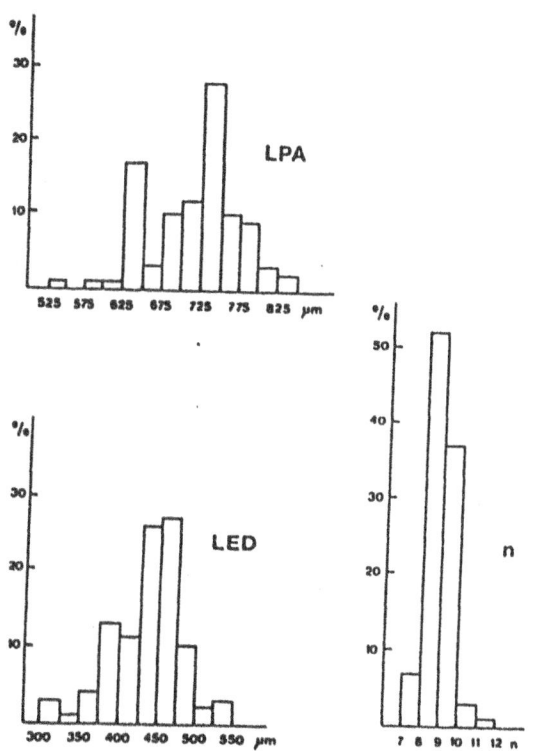

Fig. 5. *Lamprothamnium haesseliae* Donterberg: Variation of LPA (largest polar axis), LED (largest equatorial diameter) and *n* (number of spiral cells in lateral view) measured on 100 gyrogonites from Laguna La Salada living population.

cent specimens with apical opening of 150–230 μm.

Spiral cells increase in width in basal zone periphery, diminishing to half the width around basal plug, which is shown on the surface. Thickness constant. Basal pore is 50–80 μm wide. Basal plug pentagonal-pyramidal.

Oospores dark brown, 500–720 μm long, 260–330 μm wide, ten turns of spirals with granular surface texture, granules papilliform to cone-shaped and with undulated intercellular crests (as shown by Soulié-Märsche, 1979).

Material studied: MLP/Mi 475.

L. succinctum *(A. Br. in Asch.)R. D. W. (Fig. 3: C–F)* – L. succinctum *(A. Br. in Asch.) R. D. Wood. Taxon 11: 15.1962*

The following analysis is based on 15 gyrogonites from Holocene sediments in Laguna Salada Grande. Extant forms were not found, so the identification was based on the closest illustrations of this material of Soulié-Märsche (1979, plate XXXIV). In Argentinian material gyrogonites are longer.

Gyrogonites prolate to perprolate, apical outline truncate and basal outline rounded to subtruncate. Greatest diameter in the middle or below, diminishing very slightly toward apical periphery and more sharply toward base. Key dimensions as follows: LPA, $\bar{x}$ = 756 μm (680–850), all specimens dehiscent; LED, $\bar{x}$ = 418 μm (370–510); ISI (158–229); n (8–11).

Spiral cells strongly calcified, 60–90 μm wide at equator, equally wide in the subapical area, diminishing sharply in thickness toward apex, all specimens dehiscent with apical aperture 120–190 μm wide.

Basal zone with spiral cells equally wide, becoming slightly narrower around basal pore, basal pore 50–70 μm wide, basal plug on surface or slightly sunken (in this case basal pore is over 80 μm wide). Basal plug pentagonal-pyramidal, relatively high. Oospores not observed.

Material studied: MLP/MI 449

Discussion and conclusions

Gyrogonite morphology

The characteristics found in gyrogonites and oospores of Argentinian *L. haesseliae* and *L. succinctum* correspond closely to characteristics of the earliest *Lamprothamnium* species known from the Tertiary and Cretaceous (Castel & Grambast, 1969; Soulié-Märsche, 1979). These include:

– General shape cylindroid, with truncate to subtruncate apical zone and rounded basal zone.
– Apical zone weakly calcified. The spiral cells are joined at the apex, but in the subapical area there is only a thin transparent layer of calcite, which allows the observation, in extant or sub-recent forms, of the dark colour of the ectosporostine.
– Basal plug nearly as high as wide.
– The oospore texture under SEM shows granules (palliform to cone-shaped in *L. haesseliae*) and undulating intercellular ridge reflected on the inner surface of the gyrogonite.

X-ray analyses of *L. papulosum* (Wallr.) J. Gr. indicate that the gyrogonites are composed in this genus of a magnesium rich calcite accompanied by a variable amount of silica attributable to the presence of diatoms (Soulié-Märsche, 1979). In some living material from Laguna La Salada, several diatoms can be observed partially 'incorporated' in the gyrogonite (Fig. 3, A).

The fossil gyrogonites of *L. haesseliae* and *L. succinctum* identified in Argentinian Quaternary localities, Laguna del Siasgo and Salina del Bebedero, show a great intra-population variability (in addition to inter-specific variability), expressed in the degree of calcification. The more 'immature' and weakly calcified forms show a thick intercellular ridge, concave spiral cells and a more conspicuous basal zone, while the 'mature' and heavily calcified gyrogonites have convex cells and a more rounded basal zone. This variability found in the degree of calcification could be the result of environmental changes, principally salinity. The interpretation of this morphological variability is important because it provides elements for an accurate separation of species and paleoecological reconstruction, as the gyrogonite is the basis of the paleontological systematics of this group.

Ecology and Paleoecology

Knowledge of the environmental requirements of living species permits more certain paleoenvironmental reconstructions. Key characteristics of Laguna La Salada, where living *L. haesseliae* was collected include a salinity of 23 g l^{-1}, an ionic composition dominated by sodium chloride, and the presence of *Lamprothamnium* at depths of 0–2 m.

In the present and in the Quaternary, *Lamprothamnium* is recorded in Argentina from waterbodies directly connected to the sea, such as Laguna Mar Chiquita, in water bodies nearby or influenced by the sea, i.e. Laguna del Siasgo and Laguna Salada Grande, and in inland saline environments such as Laguna La Salada, Laguna La Amarga and Salina del Bebedero. Corillion (1957) described the same kind of biotopes for *L. papulosum* from Europe. Burne *et al.* (1980), recorded living specimens of this last species in Australian waterbodies with salinities ranging from 1–2 to 70 g l^{-1}.

In 1989, an experiment with Argentinian material was carried out. Sediments with 15 extant *L. haesseliae* gyrogonites were put in culture using tap water. Four oospores germinated 15–20 days later and in a few more days the plants were fertile. Only one plant produced gyrogonites, two weakly calcified ones. The other three plants did not survive at this low salinity, and the remaining oospores did not germinate. So, *Lamprothamnium* has a high range of salinity tolerance but did not survive at low salinities and prefers brackish or hypo-mesosaline waters.

It is known that all *Lamprothamnium* species are littoral forms and usually found no deeper than 2 m, probably indicating a need for high light intensities (Corillion, 1975), like the Argentinian *L. haesseliae*.

Similar ecological conditions were recognized for *L. priscum* Castel & Grambast from the Eocene of Corbières, France, one of the earliest fossil record of this genus. Castel & Grambast (1969), based on the association with ostracods and sediment characteristics, concluded that the depositional environment was brackish.

Acknowledgements

I wish to thank gratefully Dr M. Griffin (University of La Plata) for reading, examining the manuscript and correcting the English. I am particularly indebted to the colleagues who permitted me the use of their materials for the micropaleontological analyses and to Dr S. Hurlbert (San Diego State University) for final corrections.

References

Burne, R. V., J. Bauld & P. De Deckker, 1980. Saline charophytes and their geological significance. J. sedim. Petrol. 50: 281–293.

Castel, M. & L. Grambast, 1969. Charophytes de l'Eocène des Corbières. Bull. Soc. geol. Fr. 7: 936–943.

Corillion, R., 1957. Les Charophycées de France et d'Europe Occidentale. Reimp. O. K. Verlag, Koengstein-Taunus, 499 pp. 35 lam., 64 cartes (1972).

Daily, F. K., 1967. *Lamprothamnium* in America. J. Phycol. 3: 201–207.

Dangavs, N. V., 1983. Geología del complejo lagunar Salada Grande de General Lavalle y General Madariaga. Rev. Asoc. geol. arg. 38: 161–174.

Dangavs, N. V., 1988. Geología, sedimentología y limnología del complejo lagunar Salada Grande (Partidos de General Madariaga y General Lavalle, Buenos Aires). Ministerio de Economía de la Provincia de Buenos Aires, La Plata, 143 pp.

Dangavs, N. V. & D. Merlo, 1980. Recursos acuáticos superficiales del Partido de General Paz, Provincia de Buenos Aires. Ministerio de Economía de la Provincia de Buenos Aires, La Plata, 92 pp.

Donterberg, C. C. C. de, 1984. *Lamprothamnium haesseliae* Dont. nov. sp., una nueva Characeae para la Argentina. Comunicaciones Museo argentino de Ciencias naturales 'Bernardino Rivadavia', Bot. 2: 93–102.

Fasano, J. L., 1980. Geohidrología de la Laguna Mar Chiquita y alrededores, Provincia de Buenos Aires. In Publicaciones de la C.I.C. Simposio sobre Problemas geológicos del Litoral Atlántico bonaerense. Mar del Plat: 59–71.

García, A., 1987. Estudio del gametangio femenino de Charophyta actuales de Argentina. Análisis comparado con el registro fósil correspondiente. Tesis. Facultad de Ciencias Naturales y Museo, Universidad Nacional de la Plata, La Plata, 312 pp.

García, A., 1990a. Contribución al conocimiento de las Characeae del Lago Pellegrini, Provincia de Río Negro, Argentina, Candollea 45: 643–651.

García, A., 1990b. Charophyta de las líneas de costa Pleisto-Holocénicas de Salina del Bebedero, Provincia de San Luis, Argentina. Ameghiniana 27: 392. Resumen.

González, M. A., 1981. Evidencias paleoclimáticas en la Salina del Bebedero (San Luis). In VIII Congreso geológico argentino 3: 411–438.

González, M. A., E. A. Musacchio, A. García, R. Pascual & A. Corte, 1980. Sobre la presencia de foraminíferos en sedimentos holocenos de la Salina del Bebedero (San Luis, Argentina). In Publicaciones C.I.C. Simposio sobre Problemas geológicos del Litoral Atlántico bonaerense. Mar del Plata: 253–269.

González, M. A., E. A. Musacchio, A. García, R. Pascual & A. Corte, 1981. Las líneas de costa pleistocenas de la Salina del Bebedero (San Luis, Argentina). Implicancias paleoambientales de sus microfósiles. In VIII Congreso geológico argentino 3: 617–628.

Guerlesquin, M., 1981. Contribution à la connaisance des Characées d'Amérique du Sud (Bolivie, Equateur, Guyane Française). Revue Hydrobiol. trop. 14: 381–404.

Hammer, U. T., 1986. Saline Lake Ecosystems of the World. Dr W. Junk Publishers, Dordrecht, 602 pp.

Musacchio, E. A. & M. A. Moroni, 1983. Charophyta y Ostracoda no marinos eoterciarios de la Formación El Carrizo en la Provincia de Río Negro, Argentina. Ameghiniana 20: 21–33.

Pascual, R., H. E. J. Ortega, D. Ondar & E. P. Tonni, 1965. Las Edades del Cenozoico Mamalífero de la Argentina, con especial atención a aquellas del territorio bonaerense. Anales de la Comisión de Investigaciones Científicas de la Provincia de Buenos Aires VI: 165–193.

Ringuelet, R., 1962. Ecología acuática continental. Eudeba, Buenos Aires, 137 pp.

Ringuelet, R., A. Salibián, E. Calvérie & S. Ilharo, 1967. Limnología química de las lagunas pampásicas (Provincia de Buenos Aires). Physis XXVII (74): 201–221.

Soulié-Märsche, I., 1979. Etude comparée de fructifications de Charophytes actuelles et fossiles et phylogénie des genres actuels. Thése. Universite des Sciences et Techniques du Languedoc, Montpellier, 341 pp. (Edit. Imprim. des Tilleuls, Millau, France, 1989, 237 pp.).

Dunaliella salina from saline environments of the central coast of Peru

Haydee Montoya T. & Alfredo Olivera G.
Faculty of Fishery & Food Engineering, National University of Callao and Faculty of Biological Sciences, Ricardo Palma University, Lima, Peru

Key words: salinity, halophilic *Oscillatoria*, hypersaline life cycle, palmella, aplanospores

Abstract

An evaluation of the algal flora of the Salinas de Huacho on the coast of Peru between 1984 and 1990 was carried out by collection of algal mats and natural waters from a variety of habitats and by microscopic examination. *D. salina* occurred in lagoons and pools with a salinity range of 165‰ to 350‰ and formed planktonic and benthic communities. The benthic palmelloid stage of *D. salina* was found at higher salinities. Aplanospore formation was also observed. Associated halophilic species included *D. viridis*, *Oscillatoria tenuis* and *Pleurocapsa entophysaloides*. *D. salina* was also found at two other salinas on the central Peruvian coast (Chilca and Otuma).

Introduction

Saline ecosystems throughout the world contain halophilic algae such as unicellular Chlorophyta and Cyanobacteria, which often are the major primary producers (Golubic, 1980; Borowitzka, 1981). The growth and persistence of these algal groups under high and variable salinity conditions require physiological tolerance and/or resistant stages in the life cycle. Osmoregulation in algal cells exposed to increased salinity is achieved by accumulation of organic solutes such as glycerol, sucrose, trehalose, glycine betaine, etc. (Mackay *et al.*, 1984; Herbst & Bradley, 1989).

Dunaliella (Chlorophyta) is worldwide in distribution and is found mainly where salinity is greater than 100‰. *D. salina* Teod., the type species, is the most widely reported one. It can grow at salinities ranging from less than sea water (0.1 M NaCl) to NaCl saturation (4.5 M NaCl). It has been used for the commercial production of chemicals such as β-carotene, glycerol and protein (Ben-Amotz & Avron, 1981). Important biological studies of *D. salina* strains include Hamburger (1905), Teodoresco (1905), Lerche (1937), and Borowitzka & Borowitzka (1988).

Cyanobacterial microfloras composed mainly of euryhaline species also colonize and thrive in hypersaline habitats. Benthic cyanobacteria that live in cohesive microbial mats are composed of halophilic filamentous and coccoid species. In habitats undergoing desiccation and air exposure, cyanobacteria are far more successful than any other organism (Erlich & Dor, 1985; Montoya & Golubic, 1991).

In the present study, the *D. salina* strain from the Salinas of Huacho, Lima, is described, with emphasis on its growth and reproductive forms, its life cycle as observed mainly in natural populations, and the cyanobacterial species associated with it. We report the first record of the palmelloid stage at higher salinities (165‰ up to NaCl saturation). We also report *D. salina* from the salinas of Chilca and Otuma.

Study area

Hypersaline environments are common along the desertic plain of the Peruvian Pacific coast. The main study area, the saline plain of Huacho, is located in the central arid coastal region of Peru (11° 10′ S; 77° 30′ W), 110 km north of Lima. It has an areal extent of 160 km², annual precipitation of 12.3 mm, and is located 4.5 km from the Pacific Ocean near the solar evaporation ponds (Ensal Company) used in NaCl production (Petersen, 1977) (Fig. 1). The salinas are formed of several aquatic environments such as small lakes, ponds and pools. The water budget is governed by seawater seepage and evaporation (Maldonado, 1943). As a result of climatic variations, many salinas fluctuate in salinity and volume. In extreme cases they dry out by late summer or fall. These shallow water bodies (0.5–2.5 m) show an extensive salinity gradient between 80 and 280‰ up to NaCl saturation, and pH values of 6.5 to 8.0. Water temperatures range between 22 and 34 °C. The main environment studied was the shallow Laguna Rosada del Sur (Pink South lagoon), which owes its name to its typical color, varies in size with seasons, and had salinities ranging from 165‰ to NaCl saturation. Groundwater brines flow upward, and evaporation leads to precipitation of sodium chloride and, to some extent, magnesium chloride, calcium sulfate and magnesium sulfate (Petersen, 1977). Surrounding areas consist of calcareous, sandy, saline soil covered by a salt crust of undulate profile with numerous fissures and crevices up to 50 cm deep. The crust crumbles during long desiccation periods.

The Salinas of Chilca in the department of Lima and the Salinas of Otuma in the department of Ica (south of Lima) were sampled once (Fig. 1). The athalassic Salina of Chilca, 69 km south of Lima on the Panamerican highway, is located 2 km from the Pacific Ocean and 1 m below sea level. Its water budget is governed by fluvial seepage (Chilca river). Sampling the plankton of an evaporation pond with salinity of 120–200‰ NaCl was done near the main lagoon, Santa Cruz de las Salinas (Chacon, 1980). The Salinas of

Otuma is about 200 m from the Pacific Ocean and its water is seawater seepage. Brine pools with NaCl-saturated waters were sampled.

Materials and methods

Sporadic collections (1–3 times per year) of algae from benthic, planktonic and submerged endolithic habitats in the salinas of Huacho were made between 1984 and 1990. Portions of benthic microbial mats were removed from crystalline substrates with a scalpel and samples were transported in their natural brine. Salinity was measured using an American Optical T/C salinometer. The pH and temperature were recorded with Neutralit 5.5–9.0 indicator paper and a thermometer, respectively.

For culturing algae we used modified f/2 medium of Guillard (1975) with the following components per liter of sea water: Macronutrients: $NaNO_3$, 75 mg; $NaH_2PO_4 \cdot H_2O$, 5 mg; Micronutrients and trace elements: Na_2 EDTA, 4.36 mg; $FeCl_3 \cdot 6H_2O$, 3.15 mg; $CuSO_4 \cdot 5H_2O$, 0.01 mg; $ZnSO_4 \cdot 7H_2O$, 0.022 mg; $CoCl_2 \cdot 6H_2O$, 0.01 mg; $MnCl_2 \cdot 4H_2O$, 0.18 mg; $Na_2MoO_4 \cdot 2H_2O$, 0.006 mg; Vitamins: biotin 0.5 μg; B_{12}, 0.5 μg. Cultures were maintained at c. 21 °C under 40 white fluorescent lamps mounted horizontally in pairs at 40 cm distance from the culture tubes. Standard inoculation and subculturing techniques were used.

Taxonomic evaluation was carried out using Teodoresco (1905), Hamburger (1905), Lerche (1937), and Geitler (1932).

Results

Morphology and life cycle

The *D. salina* strain of Huacho was comprised of free biflagellate unicells with variable cell shape: ellipsoidal, elongated, ovoid or pyriform. The anterior end was acutely rounded or broadly rounded. The posterior end was rounded (Fig. 2A). The cell diameter was 5.5 to 12.5 μm

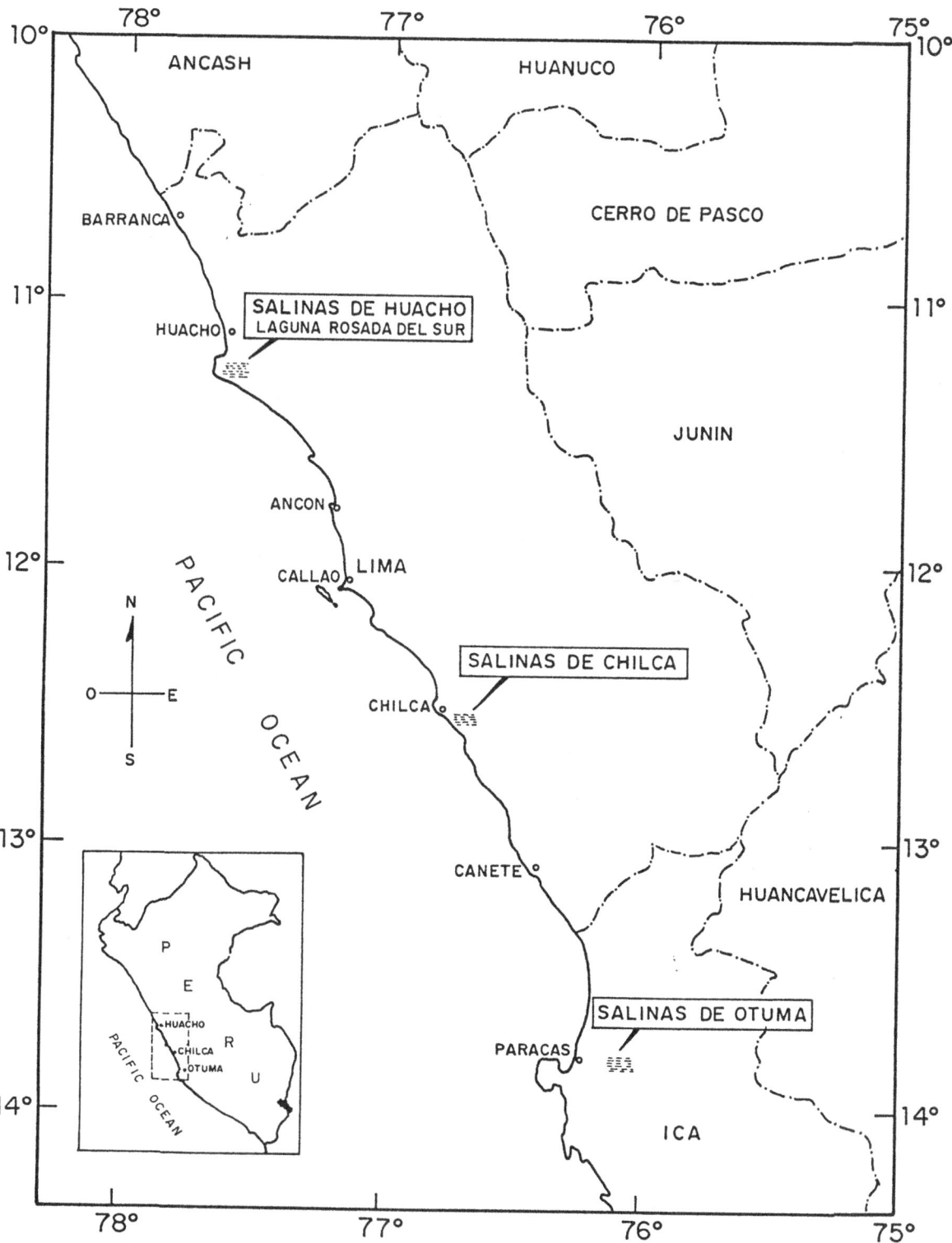

Fig. 1. Location of salinas within the Lima and Ica Departments on the Central Coast of Peru.

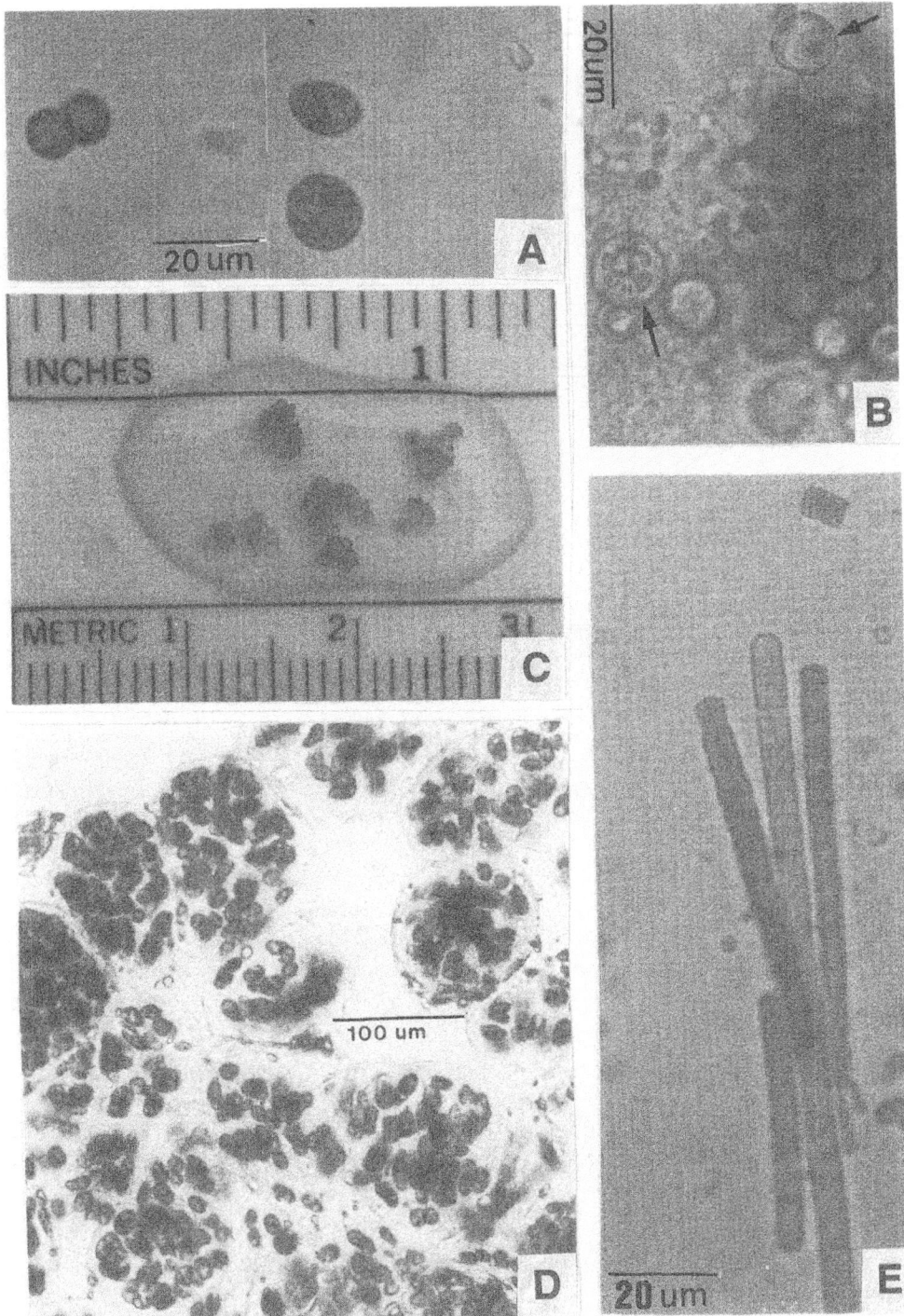

Fig. 2. Algal forms the salinas studied. A. Planktonic reddish vegetative and reproductive (binary fission) cells of *D. salina*. Free unicell of *D. viridis* (upper right); B. Aplanospore groups with empty cell wall (upper arrow) and divided cell contents (lower arrow); C. Irregular and lobulated palmelloid thalli of benthic *Dunaliella salina* attached to the encrusted saline substrate at Laguna Rosada del Sur. D. Palmelloid thalli formed by aggregated colonies with numerous cells embedded within thick mucilage; E. Trichomes of *Oscillatoria tenuis* from submerged endolithic growth in hypersaline lagoon.

and the cell length was 11.3 to 20 μm. The flagella length about equalled cell length (12.5 to 20 μm). One cup-shaped chloroplast occupied the posterior half of the cell, while in some old cells it occupied almost the entire periphery of the cell. The pyrenoid was prominent. The periplast was thin and flexible. Vegetative reproduction was by binary fusion. Cells were coloured red due to the accumulation of large β-carotene droplet globules which, together with halophilic bacteria, imparted a pink coloration to the brine waters. However, the cells turned green 3–4 days after transfer to the laboratory where light intensity was lower than in their natural environments.

Aplanospore (cyst) formation was recognized a few days after plankton samples were collected and brought to the laboratory and maintained in their natural brine. Cells changed to immotile, spherical, reddish aplanospores surrounded by a thick and stratified cell wall (Fig. 2B). Aplanospores were 6.3 to 17.5 μm in diameter, occurred individually or in groups, and changed from red or deep orange to green in the laboratory. They released their contents as free cells or they divided into 4 or 5 cells.

The Huacho *D. salina* also occurred as a palmelloid benthic form on mineral substrates (sodium chloride, magnesium chloride, calcium sulfate) at depths of up to about 1 m, and mainly in winter (August, early September). This macroscopic, irregular and lobulate stage was up to 0.3 cm in diameter and originated in the confluence of cell aggregations embedded in a thick, mucilaginous matrix (Figs 2C & 2D).

The vegetative and reproductive forms of *D. salina* in the Huacho salinas are morphological adaptations to extreme conditions. Under the same light and temperature conditions *D. salina* in lower salinity ponds (about 165‰) appeared red and free swimming. In high salinity lagoons and pools (close to saturation point), it was also red but occurred mainly in the palmelloid form attached to the substrate. When placed in the modified f/2 medium (35‰ salinity) the palmelloid form eventually released free red cells, suggesting that nutrient depletion and high salinity trigger palmelloid formation in this strain.

Various developmental stages with suggested life cycle connections, as observed in natural populations at salinas of Huacho, are shown in Fig. 3.

Habitat and associated species

D. salina was mainly associated with *D. viridis*, including both the planktonic and palmelloid forms of the latter. *D. viridis* remained green in color irrespective of weather conditions and salinity changes, and flourished along with *D. salina*. In general, the colonial palmelloid form of *D. viridis* was bigger (up to 1 cm diameter) than that of *D. salina*. In pools of lower salinity (about 165‰) *D. salina* was usually free, forming a thin surface film with *D. viridis* in the spring (late September). The planktonic *D. salina* from the Otuma salinas were also associated with planktonic *D. viridis* at 350‰. The planktonic *D. salina* observed in 120–200‰ pools at the Salinas de Chilca was also accompanied by *D. viridis*.

In some situations where salinity approached the saturation point for NaCl, the *D. salina* palmelloid form was associated with irregular mucilaginous coatings of mixed cyanobacterial mats formed by the coccoid species *Aphanothece halophytica* Frémy and *Pleurocapsa entophysaloides* Setch. & Gard., and the filamentous species *Oscillatoria tenuis* and *O. amphibia* Agardh.

O. tenuis (Fig. 2E) was the principal benthic filamentous cyanobacterium associated with palmelloid thalli of *D. salina* and *D. viridis* on crystalline substrates. It occurred in leathery, surface microbial mats and also as conspicuous mucilaginous layers within the crystalline substrate. *O. tenuis* had curved or straight trichomes with cylindrical cells 3–7 μm in diameter, and 1–7 μm in length, slightly constricted at the transverse walls with some conspicuous granules and with apical cells rounded or elongated. The trichomes were free or intermingled forming green coatings; they formed spherical masses when they grew within the gelatinous matrix of *Dunaliella* palmelloid forms or the mucilaginous matrix of the colonial coccoid species *A. halophytica* and *P. ento-*

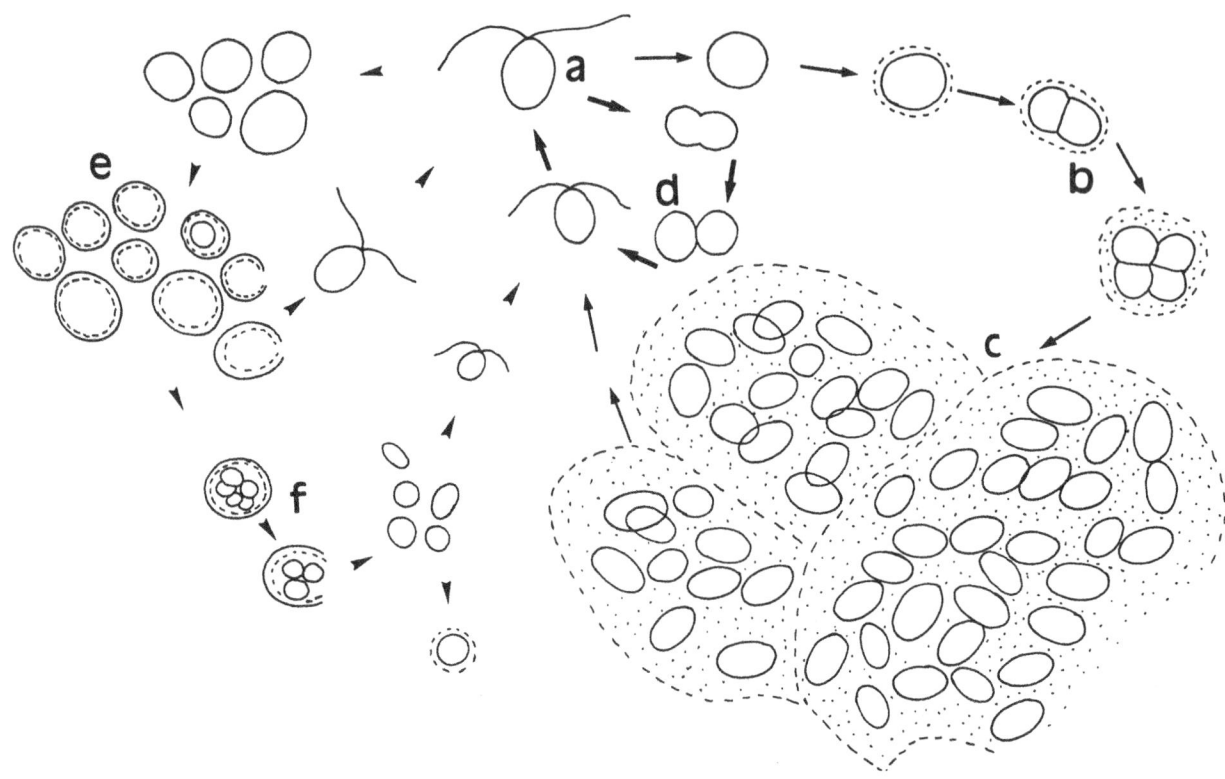

Fig. 3. Apparent life cycle pathways as determined mainly from natural populations of *Dunaliella salina* at the salinas of Huacho. a: biflagellated cells; b: successive cell divisions of immotile rounded cell with gradual increase of colonial mucilage; c: palmelloid thalli formation with multicellular aggregated colonies within a common mucilage, and later cell release; d: binary fission of free cells; e: formation of aplanospores with thick cell walls; these produce motile cells directly or divide before release of small free cells.

physaloides. In some cases the filamentous *O. amphibia* was intermixed, forming a densely interwined mat. These microbial mat assemblages at lower salinities (90–100‰) included *Spirulina subsalsa* (Menegh.) Gom., *Tetraselmis* sp., and diatoms.

Discussion

There is confusion about whether the typical red cells of *Dunaliella* that are able to turn green under specific growth conditions (low light intensity) are *D. salina* or *D. bardawil*. According to Borowitzka *et al.* (1984), the Australian strain of *D. salina* is identical to *D. bardawil*. The latter species is therefore considered to be 'a nomen nudum' (Borowitzka & Borowitzka, 1988). By these criteria the studied Peruvian strains are *D. salina*.

In hypersaline natural waters exposed to high solar irradiation the major factors limiting growth of algal populations appear to be nitrogen and carbon dioxide (Ben Amotz & Avron, 1990). The *D. salina* strain from Huacho maintains populations in brine waters through selection of a growth strategy (formation of reddish palmelloid form) that facilitates survival under poor growth conditions. The previous records for the palmelloid stage of *D. salina* in nature are for ponds with salinities of <10‰ to 143‰ (Lerche, 1937; Sammy, 1983). Nutrient deficiency (phosphate and nitrogen), low temperature (under 25 °C), and low salinity (below 15‰ w/v NaCl) have been noted to produce palmelloid forms in some strains. Aplanospore formation by *D. salina* can be considered as a strategy for remaining alive as a resistant stage in conditions of light intensity reduction (high irradiation in nature versus low light intensity in lab) and temperature decrease. The simultaneous gradual change in aplanospore color from red to green supports this interpretation since β-carotene accumulation only occurs under high light intensity (Lerche, 1937; Ben-Amotz & Avron, 1983). On the other hand, the gradual evaporation and increasing salinity of the medium in which aplanospores developed contrasts with the idea of dilution as a factor causing aplanospore formation. Lerche (1937), for example, reported round immotile green cells of *D. salina* in concentrations below 10‰ NaCl and Loeblich (1969) considered that aplanospores endure low salinities (optimum salinity for cyst production less than 40‰ NaCl). Our findings are in agreement with Margulis *et al.* (1980), who consider that aplanospores in *Dunaliella* are formed under drastic dilution of the medium or drying up of the habitat.

References

Ben-Amotz, A. & M. Avron, 1981. Glycerol and β-carotene metabolism in the halotolerant alga *Dunaliella*: a model system for biosolar energy conversion. Trends Biochem. Sci. 6: 297–299.

Ben-Amotz, A. & M. Avron, 1983. On the factors which determine massive β-carotene accumulation in the halotolerant alga *Dunaliella bardawil*. Plant Physiol. 72: 593–597.

Ben-Amotz, A. & M. Avron, 1990. The biotechnology of cultivating the halotolerant alga *Dunaliella*. Trends Biotechnol. 8: 121–126.

Borowitzka, L. J., 1981. The microflora. Adaptations to life in extremely saline lakes. Hydrobiologia 81: 33–46.

Borowitzka, L. J., M. A. Borowitzka & T. P. Moulton, 1984. The mass culture of *Dunaliella salina* for fine chemicals: from laboratory to pilot plant. Hydrobiologia 116/117: 115–134.

Borowitzka, M. A. & L. J. Borowitzka, 1988. *Dunaliella*. In M. A. Borowitzka & L. J. Borowitzka (eds), Micro-Algal Biotechnology. Cambridge Univ. Press: 27–58.

Chacon, G. R., 1980. *Chlorella peruviana* sp. nov. y su ambiente altamente salino. Bol. Soc. Per. Bot. 8: 83–96. Lima.

Ehrlich, A. & I. Dor, 1985. Photosynthetic microorganisms of the Gavish Sabkha. In G. M. Friedman & W. E. Krumbein (eds), Hypersaline ecosystems. Ecol. Studies. Springer-Verlag, Berlin 53: 296–321.

Geitler, L., 1932. Cyanophyceae. In Rabenhorst's Kryptogamenflora von Deutschland, Osterreich und der Schweiz. Akad. Verlagsges. Leipzig. Reprinted 1971. Johnson, New York. 14: 1–1196.

Golubic, S., 1980. Halophily and halotolerance in Cyanophytes. Origins of Life 10: 169–183.

Guillard, R. L., 1975. Culture of phytoplankton for feeding marine invertebrates. In W. L. Smith & M. H. Chanley (eds), Culture of marine invertebrate animals. Plenum Press, New York: 29–60.

Hamburger, C., 1905. Zur Kenntnis der *Dunaliella salina* und einer Amobe aus salinenwasser von Cagliari. Arch. Protistenk. 6: 111–130.

Herbst, D. & T. Bradley, 1989. Salinity and nutrient limitations on growth of benthic algae from two alkaline salt lakes of the Western Great Basin (USA). J. Phycol. 25: 673–678.

Lerche, W., 1937. Untersuchungen über Entwicklung und Fortpflanzung in der Gattung *Dunaliella*. Arch. Protistenk. 88: 236–268.

Loeblich, L. A., 1969. Aplanospores of *Dunaliella salina* (Chlorophyta). J. Protozool. Suppl. 16: 22–23.

Maldonado, A., 1943. Las Lagunas de Boza, Chilca y Huacachina y los Gramadales de la Costa del Perú. Act. Trab. 2do. Cong. Per. Quim, Lima, 143 pp.

Margulis, L., E. S. Barghoorn, D. Ashendorf, S. Banerjee, D. Chase, S. Francis, S. Giovanonni & J. Stolz, 1980. The microbial community in the layered sediments at Laguna Figueroa, Baja California, Mexico: Does it have Precambrian analogues? Precambrian Res. 11: 93–123.

Mackay, M., R. Norton & L. Borowitzka, 1984. Organic osmoregulatory solutes in cianobacteria. J. gen. Microbiol. 130: 2177–2191.

Montoya, H. & S. Golubic, 1991. Morphological variability in natural populations of mat-forming cyanobacteria in the salinas of Huacho. Lima, Perú. In B. Hickel, K. Anagnostidis & J. Komarek (eds), Cyanophyta (Cyanobacteria). Morphology, Taxonomy, Ecology. Stuttgart. Arch. Hydrobiol. Suppl. Algological Studies 64: 423–441.

Petersen, G., 1977. Historia maritima de Perú. Geografía y geología general de litoral Peruano. Vol. I. Inst. Estudios Histórico-Maritimos del Perú, Lima. 214 pp.

Sammy, N., 1983. Biological systems in North-Western Australian solar salt fields. In B. C. Schreiber & H. L. Harner (eds), Salt Institute Inc. Virginia, USA. Sixth int. Symp. on salt. 1: 207–211.

Teodoresco, E. C., 1905. Organisation et développment du *Dunaliella* nouveau genre de Volvocacée – Polyblepharidée Beih. Bot. Centralbl. 18: 215–232.

Effects of NaCl and KNO₃ concentrations on the abscisic acid content of *Dunaliella* sp. (Chlorophyta)

Noriko Tominaga [1], Makiko Takahata [1] & Hiroyuki Tominaga [2]*
[1] *Institute of Environmental Science for Human Life, Ochanomizu University, Ohotsuka 2-1-1, Bunkyo-ku, Tokyo 112, Japan;* [2] *Water Research Institute, Nagoya University, Furo-cho, Chikusa-Ku, Nagoya 464-01, Japan;* * *Present address: Department of Biology, Musasigaoka University, Yoshimi, Saitama 355-01, Japan*

Key words: abscisic acid, *Dunaliella*, halotolerant

Abstract

Abscisic acid (ABA) is a hormone which has a number of roles during the life cycle of a plant. We demonstrated the occurrence of ABA in a halotolerant green alga, *Dunaliella* sp. isolated from a salt pond near Adelaide, South Australia, using thin layer chromatography (TLC), high performance liquid chromatography (HPLC) and gas chromatography-mass spectrometry (GC-MS). The variation of cellular ABA and protein content during the growth of an axenic clonal culture of *Dunaliella* sp. was investigated under different concentrations of NaCl and KNO₃.

Experimental results can be summarized as follow: (1) ABA content was changed with the growth stage of culture: A rapid increase in ABA content was observed in the logarithmic phase. After this, the content rapidly decreased to very low values. (2) ABA content was also affected by the NaCl concentration. The content had a minimum value at the NaCl concentration (15%) where growth rate was maximal, and higher values at higher or lower concentrations of NaCl. (3) The ABA content also increased with decreasing nitrogen concentration of the medium.

Introduction

The phytohormone abscisic acid (ABA) is widely distributed in higher plants and its physiological role has been well investigated. The occurrence of ABA in lower plants has been demonstrated in algae (Kingsham & Moore, 1982; Tietz & Kasprik, 1986; Sabbatini *et al.*, 1987; Boyer & Dougherty, 1988; Tietz *et al.*, 1989) and phytopathogenic fungi (Assante *et al.*, 1977; Dahiya *et al.*, 1988). Recently Hirsch *et al.* (1989) demonstrated that ABA is universally distributed within the algal kingdom.

ABA is synthesized from ionylidine derivatives in fungi, but the pathway in higher plants and algae remains to be elucidated (Zeevaart & Creelman, 1988). In higher plants, both the direct and indirect pathways involving precursors derived from farnesyl pyrophosphate and carotenoid, respectively, may operate concurrently (Zeevaart & Creelman, 1988). In either case, mevalonate is the ultimate precursor. The synthesis and metabolism of ABA in algae may not be very different from those in higher plants (Hirsch *et al.*, 1989).

The accumulation of ABA was induced by osmotic stress in *Dunaliella parva* and *Draparnaldia mutabilis* (Tietz *et al.*, 1989; Hirsch *et al.*, 1989) and by pH stress in *Dunaliella acidophila* (Hirsch et al., 1989). There are no reports of changes in endogenous ABA levels during growth in algae.

We isolated the alga *Dunaliella* sp. from a salt pond near Adelaide, South Australia. This algal species was used for the investigation of ABA level in algae since it can tolerate a wide range of salt concentrations and produce high levels of carotenoids which are precursors for ABA synthesis in higher plants.

In this study, we identified ABA in this alga by using TLC, HPLC, and GC-MS and the change of endogenous ABA level during growth under stress was determined.

Materials and methods

Algal material

The algal material used for this experiment was *Dunaliella* sp., which was isolated from an almost saturated salt pond near Adelaide, South Australia. An axenic and clonal culture was obtained by the author by picking and washing single red cells with a micropipette (Guillard, 1973). This *Dunaliella* sp. is probably a strain of *Dunaliella salina* Teodoresco or *D. bardawil* Ben-Amotz and Avron as indicated by the following: (1) It is unicellular and bi-flagellate green alga lacking a cell wall; (2) It has a wide salinity tolerance (3–26% NaCl) and the optimal salinity for growth is 15%; and (3) It can become red through accumulation of large amounts of β-carotene, sometimes having a carotenoid to chlorophyll ratio greater than 6 to 1 under extreme environmental conditions (Massyuk, 1973; Loeblich, 1982).

Growth conditions

The alga was maintained by periodical transfers in a synthetic medium of Johnson et al. (1968) containing 15% (2.57 M) NaCl and grown in incubators (Tayo, IS-3000H) at 20 °C.

Unless otherwise stated the experimental cultures were grown in 20 ml of medium containing specified concentrations of NaCl and KNO$_3$ in 100 ml Erlenmyer flasks. Before the experiment, cultures were repeatedly transferred to specified media for at least three times to acclimate them to the experimental conditions. All flasks used for the experiments were inoculated from each specified preculture in exponential growth phase and each inoculum was adjusted to give an initial cell concentration of approximately 4.8×10^4 cells ml^{-1}. To obtain the various salinities only the NaCl concentration was altered.

For the experiments on the effect of nitrate deficiency, precultured cells were inoculated in 20 ml of medium containing the following KNO$_3$ (mM) and NaCl (%) concentrations, respectively: 10 mM and 15% (optimal); 1 mM and 15%; 0.4 mM and 15%; and 0.4 mM and 23%.

Illumination was provided by fluorescent lamps (National, Type 'Homolux', designed for plant cultivation) that delivered approximately 120 μE m^{-2} s^{-1} at the surface of the culture vessels. The photoperiod was a 12 h light:12 h dark cycle.

Cell number was determined using a microscope (Type BH, Olympus) and a Sedwick-Rafter counting chamber. Protein was determined colorimetrically by the method of Lowry et al. (1951). Chlorophylls and carotenoids were analyzed by the method of Scor/Unesco (1966), after extraction with 90% acetone, using a Shimadzu spectrophotometer (model UV-160A).

Extraction of ABA and determination of endogenous level

Extraction was performed according to the method of Norman et al. (1988) with duplicate flasks and began at the midpoint of the light period. After aliquots (0.5–3 ml) for determination of cell number, protein and pigments were taken from the culture (20 ml), the remaining cells were harvested by centrifugation at 1700 × g for 10 min at 4 °C. The algal pellet was suspended in 15 ml of 80% acetone, sonicated for 5 min and centrifuged at 1700 × g for 5 min at 4 °C. This extraction process was repeated 3 times. The combined acetone extracts were acidified to pH 2.5 with 0.1 N HCl and stirred for 30 min with HCl-washed charcoal (1 g) and celite 545 (0.5 g). The mixture was filtered through a glass fiber filter (GF/C, Whatman) and the filtrate was evaporated to remove acetone under reduced pressure. The resultant aqueous residue was readjusted to pH 2.5 and extracted three times with 15 ml ethyl acetate. After the combined ethyl acetate extracts were dried *in vacuo*, the residue was dissolved in 10 ml of water, the pH was adjusted to 2.5, and this solution was passed through a C_{18} Sep-Pak cartridge (Waters). The sample was eluted with 3 ml of methanol.

The endogenous level of ABA was determined by HPLC with 10 μl of methanol eluate. HPLC was carried out on an instrument (Shimadzu, LC6A) using a Zorbax ODS column (DuPont, 4.6 mm × 25 cm) with a 35-min linear gradient from methanol-2 mM H_3PO_4 (50:50) to 100% methanol. Detection of ABA was carried out by measurement of optical density at 257 nm (Shimadzu, SPD-6AV).

Identification of ABA

The extraction procedure was performed on a large scale for the identification of ABA, and the methanol eluate of C_{18} Sep-Pak was further purified by two-dimensional TLC. The solvent systems used were toluene-ethyl acetate-acetic acid (50:30:4, v/v) in the first direction and chloroform-methanol-water (75:22:3, v/v) in the second direction (Tietz *et al.*, 1979). all TLC plates (Silicagel 60, Merck) were washed with methanol before use. The ABA-containing silica gel identified by marker ABA was scraped off and eluted with ethyl acetate. The eluate was methylated with diazomethane.

Identification of ABA was carried out using GC-MS with FS-WCOT OV-1 column (50 m × 0.25 mm i.d.). A JOEL JMS-DX 300 mass spectrometer was connected to a Hewlett-Packard 5790A gas chromatograph and splitless injection was used. The carrier gas was helium. The GC column conditions were: temperature programming from 80 °C to 250 °C at a rate of 4 °C min^{-1}, injector temperature was 220 °C, and electron impact ionization was 70 eV.

Results and discussion

The occurrence of ABA in extracts of 3-day old cultures of *Dunaliella* sp. was demonstrated by HPLC and TLC. Quantitative determination of ABA was carried out using HPLC and the recovery of synthetic ABA added to sample was about 93% after extraction. The GC-MS data derived from *Dunaliella* sp. cell extract corresponded to that of synthetic methylated ABA (data not shown)(Tietz & Kasprik, 1986).

Figure 1 shows growth curves of *Dunaliella* sp. at different salinities. This alga is halotolerant and grew most rapidly in the presence of 15% (w/v, 2.57 M) NaCl and 10 mM KNO_3. The initial growth rates and final cell concentrations for the cultures at 5% and 23% NaCl were lower than those for cultures at 15% (t-tests, $P < 0.05$).

ABA levels were strongly influenced by NaCl concentration (Fig 2). ABA content per cell increased rapidly during the early logarithmic phase, then decreased rapidly and remained low thereafter. Cells grown at 15% NaCl contained significantly lower (t-tests, $P < 0.05$) levels of ABA than did those grown at 5% and 23%. The only previous report of change in ABA level during

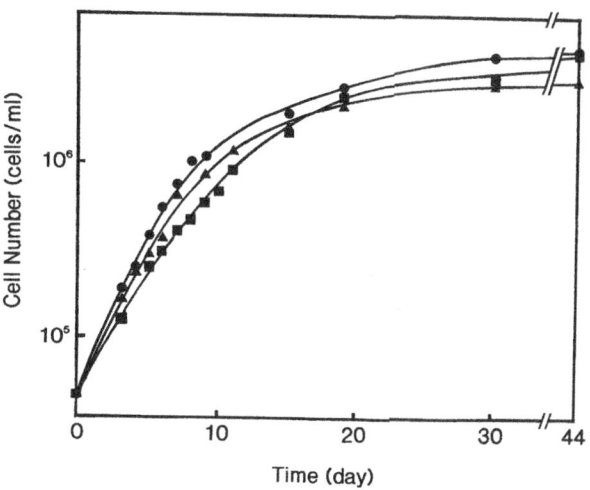

Fig. 1. Growth of *Dunaliella* sp. under different NaCl concentrations. NaCl concentrations were: (▲), 5% (w/v = 0.86 M); (●), 15% (2.57 M); (■), 23% (3.94 M). The concentration of nitrate was 10 mM. Each point represents the average of triplicate experiments with two flasks each (6 measurements).

166

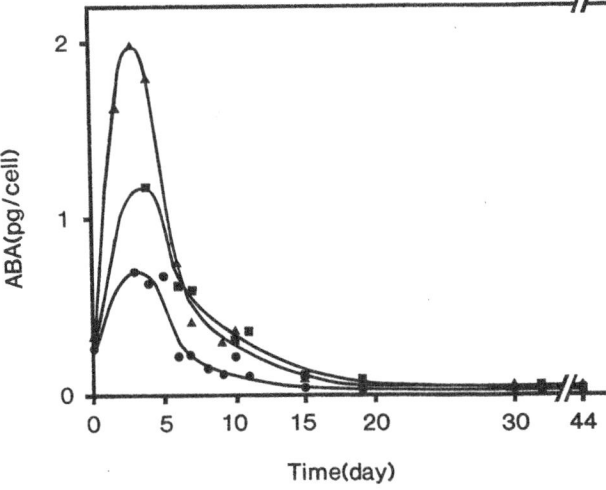

Fig. 2. Endogenous levels of ABA during growth of *Dunaliella* sp. at different NaCl concentration of the medium. Conditions are as specified in the legend to Fig. 1.

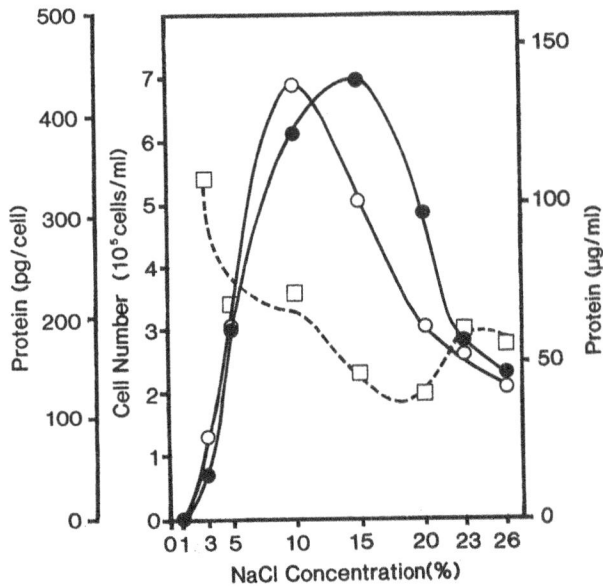

Fig. 3. Cell number (●), protein content per ml of culture (○) and protein content per cell (□) of *Dunaliella* sp. cells as a function of the salinity of the growth per medium. The concentrations of NaCl were 1, 3, 5, 10, 15, 20, 23, and 26% and nitrate was 10 mM. Measurements were made 6 days after inoculation of the culture. Each point represents the average of 4 determinations.

algal growth was made by Tietz & Kasprik (1986), who suggested that young cultures of *Stigeoclonium* had high ABA content. The ABA contents of *D. parva* based on fresh weight, were 40 ng g^{-1} of 60-day cultured cells grown in 1.5 M NaCl (Tietz *et al.*, 1989) and 0.29–1 μg g^{-1} of cells grown in 1.5 M NaCl (Hirsch *et al.*, 1989). *D. bardawil*, which is similar to the species used in the present study, contained 26–370 ng g^{-1} (Hirsch *et al.*, 1989). The levels for *Dunaliella* sp. grown in 2.57 M NaCl were 400 ng g^{-1} for 3-day cells and 60 ng g^{-1} for 44-day cells. Only trace amounts of ABA were found in the external medium of 9-day culture.

The effect of NaCl on the cell number and protein content of 6-day algal cultures was strong (Fig. 3). This alga could grow under a wide range of NaCl concentrations (from 3% to 26%) but could not tolerate 1% NaCl. Cell number was maximal at 15% NaCl, while protein content per ml of culture was maximal at 10% NaCl. Protein content per cell was maximal at 3% NaCl.

The ABA content of cells harvested 6 days after inoculation was lowest at 15% NaCl, which is the optimal concentration for growth (Fig. 4). Lower and higher concentrations of NaCl caused increases in the ABA content. This result is similar to that for *D. parva* (Tietz *et al.*, 1989). Hy-

potonic environments seem to be more stressful for *Dunaliella* sp. than hypertonic ones.

Dunaliella sp. accumulated large amounts of carotenoid under certain conditions, as does *D. bardawil* (Ben-Amotz *et al.*, 1982) and *D. salina* (Borowitzka *et al.*, 1984). Nitrate deficiency, as well as high concentration of NaCl, was a factor associated with the accumulation of carotenoid. The effects of nitrate deficiency on the growth and ABA level of *Dunaliella* sp. cells are shown in Fig. 5 and Fig. 6, respectively. The concentration of nitrate was decreased from 10 mM (optimal) to 1 mM or 0.4 mM while NaCl level was maintained at 15%; additionally, the effects of a stressful condition (0.4 mM KNO$_3$ and 23% NaCl) was examined. At the 15% NaCl concentration, nitrate deficiency caused the ABA content to increase significantly and the final cell yield to be lower, although the growth rates were unaffected. At 0.4 mM nitrate, ABA content was higher and growth rate was lower at 23% NaCl

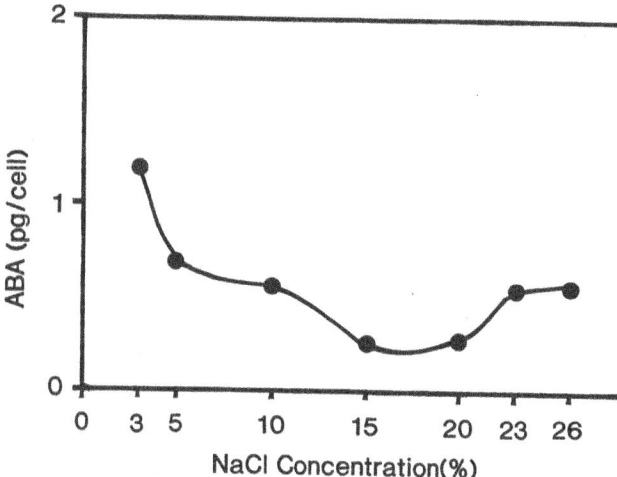

Fig. 4. Effects of NaCl concentration of medium on ABA contents. Measurements were made 6 days after inoculation of the culture. Each point represents the average of 4 determinations.

(saltstress) than 15% NaCl. The maximum ratio (21 to 1) of carotenoids to chlorophyll was found in the cells grown at 0.4 mM nitrate and 23% NaCl.

Although *Dunaliella* sp. accumulated ABA on exposure to osmotic shock (unpublished data), as

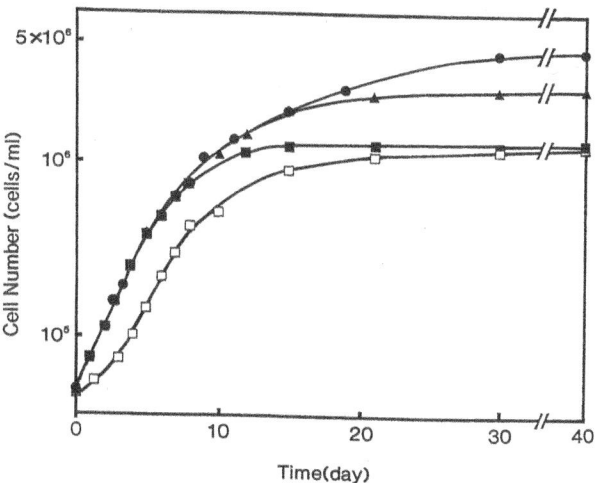

Fig. 5. Growth in the media containing different concentrations of NaCl and nitrate. Concentrations of KNO_3 and NaCl, respectively, were (●), 10 mM and 15%; (▲), 1 mM and 15%; (■), 0.4 mM and 15%; (□), 0.4 mM and 23%. Each point represents the average of triplicate experiments with two flasks each.

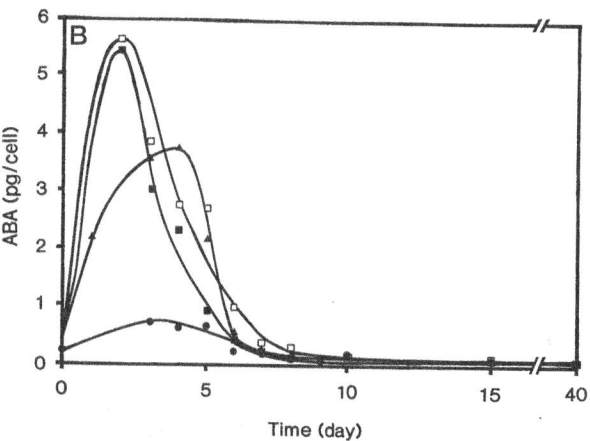

Fig. 6. The effect of nitrate deficiency and NaCl concentration on endogenous levels of ABA. Conditions are as specified in the legend to Fig. 5.

did *D. parva* (Tietz *et al.*, 1989; Hirsch *et al.*, 1989), the change in the endogenous level of ABA in this alga may suggest another role of ABA. Since the inoculated cells had been precultured through at least three transfers in the same medium that was used for the experiments, cells were not exposed to osmotic shock or solute shock. Algal cells nevertheless detected unfavourable conditions such as hypertonic or hypotonic environment or nutrient deficiency with the resultant accumulation of ABA. It is of interest that the accumulation of ABA was observed in the early stage of growth. ABA may regulate growth by inhibition in unknown way.

There are many reports on the effects of ABA on growth in plants. In the elongation zones of single maize roots (*Zea mays* L.), a negative correlation between ABA content and growth rate was demonstrated (Rivier & Saugy, 1986; Reymond *et al.*, 1987). However, the role of endogenous ABA in root growth is obscure because growth rate and ABA content varied independently under application of fluoridine or change of temperature (Reymond & Pilet, 1987). Furthermore, the effects of exogenous ABA on the growth of root and shoot are contradictory.

Singh *et al.* (1987) reported that the maximum ABA level was attained in the lag or early growth phase in NaCl-adapted or -unadapted tobacco cell culture, respectively, then decreased rapidly.

They observed that this increased intracellular accumulation of ABA during cell growth was associated with synthesis of a 26-kD protein and suggested that this protein plays a role in salt adaptation. The change in endogenous level of ABA in tobacco cells is similar to that in *Dunaliella* sp. cells. For studying the role of endogenous ABA on biochemical processes, this alga as well as tobacco cell culture may be better model systems than differentiated and complicated tissues of higher plants.

The accumulation of ABA under stress suggests a similar hormonal function of ABA to that in higher plants (Orr *et al.*, 1986; Robertson *et al.*, 1987; LaRosa *et al.*, 1987; Singh *et al.*, 1987). This study suggests that the accumulation of ABA in *Dunaliella* sp. may inhibit growth, but more detailed investigations are needed.

References

Assante, G., L. Merlini & G. Nasini, 1977. (+)-Abscisic acid, a metabolite of the fungus *Cercospora rosicola*. Experientia 33: 1556–1557.

Ben-Amotz, A., A. Katz & M. Avron, 1982. Accumulation of β-carotene in halotolerant algae: purification and characterization of β-carotene-rich globules from *Dunaliella bardawil* (Chlorophycae). J. Phycol. 18: 529–537.

Borowitzka, L. J., M. A. Borowitzka & T. P. Moulton, 1984. The mass culture of *Dunaliella salina* for fine chemicals: from laboratory to pilot plant. Hydrobiologia 116/117: 115–121.

Boyer, G. L. & S. S. Dougherty, 1988. Identification of abscisic acid in the seaweed *Ascophyllum nodosum*. Phytochem. 27: 1521–1522.

Dahiya, J. S., J. P. Tewari & D. L. Woods, 1988. Abscisic acid from *Alternaria brassicae*. Phytochem. 27: 2983–2984.

Guillard, R. R. L., 1973. Methods for microflagellates and nannoplankton. In J. R. Stein (ed.), Handbook of Phycological Methods – Culture Methods and Growth Measurements. Cambridge Univ. Press: 69–85.

Hirsch, R., W. Hartung & H. Gimmler, 1989. Abscisic acid content of algae under stress. Bot. Acta 102: 326–334.

Johnson, M. K., E. J. Johnson, R. D. MacElroy, H. L. Speer & B. S. Bruff, 1968. Effects of salts on the halophilic alga *Dunaliella viridis*. J. Bact. 95: 1461–1468.

Kingsham, A. R. & J. Moore, 1982. Isolation, purification and quantitation of several growth regulating substances in *Ascophyllum nodosum (Phaeophyta)*. Bot. mar. 25: 149–154.

LaRosa, P. C., P. M. Hasegawa, D. Rhodes, J. M. Clithero, A.-E. A. Watad & R. A. Bressan, 1987. Abscisic acid stimulated osmotic adjustment and its involvement in adapta-

tion of tobacco cells to NaCl. Plant Physiol. 85: 174–181.

Loeblich, L. A., 1982. Photosynthesis and pigments influenced by light intensity and salinity in the halophile *Dunaliella salina* (Chlorophyta). J. mar. Biol. Ass., UK. 62: 493–508.

Lowry, O. H., N. J. Rosebrough, A. L. Farr & R. J. Randall, 1951. Protein measurement with the folin phenol reagent. J. Biol. Chem. 193: 265–275.

Massyuk, N. P., 1973. Morphology, systematics, ecology and geographic distribution of the genus *Dunaliella* Teod. and perspectives of its practical use. Russ. Kiev., 244 pp.

Norman, S. M., S. M. Poling & V. P. Maier, 1988. An indirect enzyme-linked immunosorbent assay for (+)-abscisic acid in *Citrus, Ricinus* and *Xanthium* leaves. J. agric. Food Chem. 36: 225–231.

Orr, W., W. A. Keller & J. Singh, 1986. Induction of freezing tolerance in an embryogenic cell suspension culture of *Brassica napus* by abscisic acid at room temperature. J. Plant Physiol. 126: 23–32.

Reymond, P. & P. E. Pilet, 1987. On the importance of ABA in maize root growth. In Abstracts, General Lectures, Symposia Papers and Posters, p. 135. 14th Int. Bot. Congr., Berlin.

Reymond, P., M. Saugy & P. E. Pilet, 1987. Quantification of abscisic acid in a single maize root. Plant Physiol. 85: 8–9.

Rivier, L. & M. Saugy, 1986. Chemical ionization mass spectrometry of indol-3yl-acetic acid and *cis*-abscisic acid: evaluation of negative ion detection and quantification of *cis*-abscisic acid in growing maize roots. J. Plant Growth Regul. 5: 1–16.

Robertson, A. J., L. V. Gusta, M. J. T. Reaney & M. Ishikawa, 1987. Protein synthesis in bromegrass (*Bromus inermis* Leyss) cultured cells during the induction of frost tolerance by abscisic acid or low temperature. Plant Physiol. 84: 1331–1336.

Sabbatini, M. R., J. A. Arguelo, O. A. Fernandez & R. A. Bottini, 1987. Dormancy and growth inhibitor levels in oospores of *Chara contraria* A. Braun ex. Kuetz (Charophyta). Aquat. Bot. 28: 189–194.

Scor/Unesco, 1966. Determination of photosynthetic pigments in sea water. In: Monographs on oceanographic methodology 1. Unesco Publication center, N. Y., 69 pp.

Singh, N. K., P. C. LaRosa, A. K. Handa, P. M. Hasegawa & R. A. Bressan, 1987. Hormonal regulation of protein synthesis associated with salt tolerance in plant cells. Proc. nat. Acad. Sci. USA 84: 739–743.

Tietz, D., K. Dörffling, D. Wöhrle, I. Erxleben & F. Liemann, 1979. Identification by combined gas chromatography-mass spectrometry of phaseic acid and dihydrophaseic acid and characterization of further abscisic acid metabolites in pea seedlings. Planta 147: 168–173.

Tietz, A. & W. Kasprik, 1986. Identification of abscisic acid in a green alga. Biochem. Physiol. Pflanzen, 181: 269–274.

Tietz, A., U. Ruttkowski, R. Köhler & W. Kasprik, 1989. Further investigations on the occurrence and the effects of abscisic acid in algae. Biochem. Physiol. Pflanzen 184: 259–266.

Zeevaart, J. A. D. & R. A. Creelman, 1988. Metabolism and physiology of abscisic acid. Ann. Rev. Pl. Physiol. Plant Mol. Biol. 39: 439–473.

Spatial heterogeneity of macrophytes in Lake Gallocanta (Aragón, NE Spain)

F. A. Comín, X. Rodó & M. Menéndez
Department of Ecology, University of Barcelona, Diagonal 645, 08028 Barcelona, Spain

Key words: macrophytes, biomass, distribution, ecophysiology, saline lake

Abstract

The spatial distribution of macrophytes (*Ruppia drepanensis* Tineo and *Lamprothamnium papulosum* (Wallr.) J. Gr. was studied along transects perpendicular to the shoreline in Lake Gallocanta (Aragón, NE Spain) in 1988 and 1990. In the shallow zone, a gradient from the shoreline to offshore waters was clear: Small *R. drepanensis* plants were the only colonizers of nearshore waters affected by wave action and desiccation. *R. drepanensis* and *L. papulosum* coexisted at intermediate depths in the shallow zone. *L. papulosum* reached higher biomasses than *R. drepanensis* in the deepest parts of the shallow zone. In the deepest zone of the lake, stands of the two species did not overlap. Individual plants of *R. drepanensis* occured patchily within a sparse prairie of *L. papulosum*.

This spatial pattern was observed at different lake levels, suggesting that macrophytes are adapted to fluctuations of environmental conditions. In very shallow water the macrophytes decay as indicated by negative net production and low chlorophyll a/chlorophyll b and plant K^+ concentration/water K^+ concentration ratios.

Introduction

Populations living in saline lakes experience strong environmental stress because water level fluctuations are very common (Comín *et al.*, 1991). Submerged macrophytes have morphological, physiological and life-cycle adaptations for coping with the large and frequent changes related to water level fluctuations (Brock, 1986). *Ruppia* spp., an angiosperm, and *Lamprothamnium papulosum* (Wallr.) J. Gr., a charophyte, are two of the most common and quantitatively important macrophyte genera living in saline lakes all over the world (Brock, 1981). They occur in similar environments and have been reported to have numerous adaptations that suit them for saline waters that fluctuate in level (Brock, 1986).

The spatial distribution of submerged macro-phytes in saline lakes has not been studied in detail. Several authors have reported the co-existence of *Ruppia* spp. and *L. papulosum* in the same stands in saline lakes, particularly in shallow waters, and also the spatial segregation of species in deeper waters (Brock & Lane, 1983; Comín *et al.*, 1983). However, detailed quantitative studies of the spatial distribution of these species and their metabolic activity are not available in the literature in spite of their outstanding importance for the overall community ecology (Hammer, 1986).

The aims of this study were to asses the spatial distribution of macrophytes in relation to water level changes, and to use the physiological characteristics of the plants to attempt to explain the patterns observed. This paper presents two types of data on submerged macrophytes in Lake

170

Gallocanta. The first documents differences in the spatial distribution of *R. drepanensis* Tineo and *L. papulosum* along a transect from the shore to deep waters. The second data set refers to ecophysiological characteristics of the two species studied in the field.

Material and methods

Three samples for biomass estimations of submerged macrophytes were collected in June 1988 and monthly during the growing season in 1990 with a plastic tube (20 cm diameter) from each of up to eight stations (labelled from 1 to 8) located along the first 50 m of a transect (2 m wide, 300 m long) perpendicular to the shoreline in Lake Gallocanta (Figs 1 and 3). The transect site was selected as representative of the gentle slopes which

constitute most of the shoreline of Lake Gallocanta (Comín *et al.*, 1990b). Two additional sampling stations (labelled A and B) were selected as representative of the deepest part of the lake. These two stations, established at the same depth, were located 300 m from the shore at the end of the transect, and were about 50 m apart so as to permit collection of both *Ruppia* and *Lamprothamnium*, which, in this part of the lake, do not occur in mixed stands.

Sampling stations were selected on the first sampling date at roughly equal depth intervals of 5–10 cm. These same sampling stations were used on all dates except that a fall in water level in 1990 gradually put some of the stations on dry ground. Samples were collected in a manner that avoided bare patches created by previous samplings.

Biomasses were estimated after drying samples

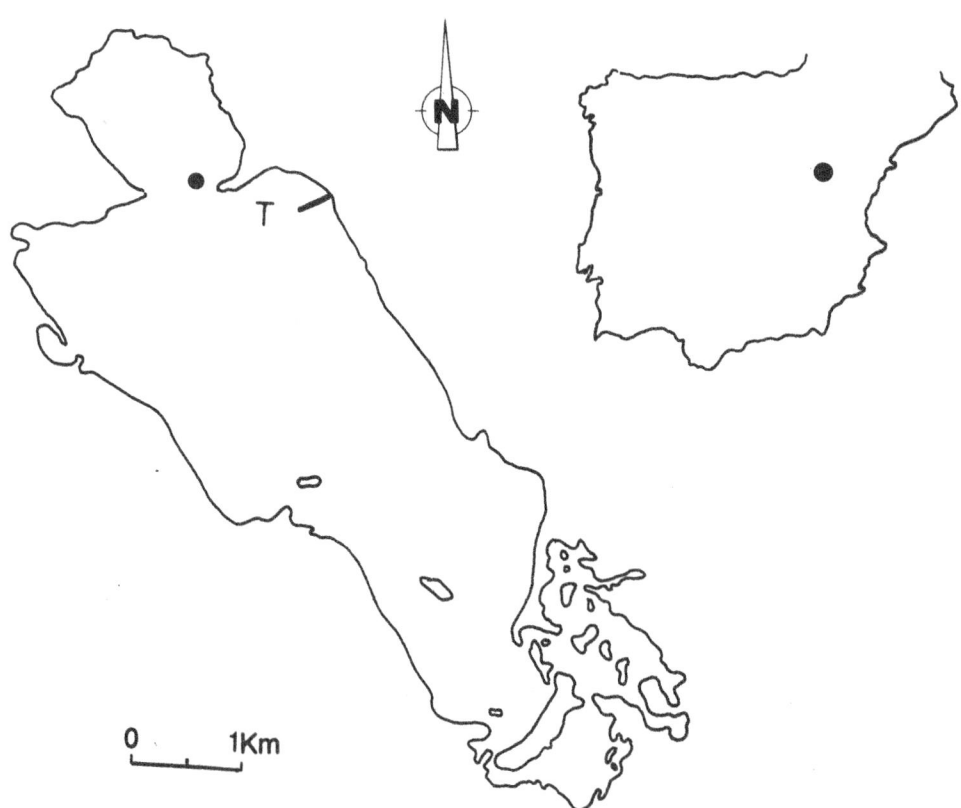

Fig. 1. Location of the study transect in Lake Gallocanta (T) and location of the point where maximum depth of the Lake is recorded (●).

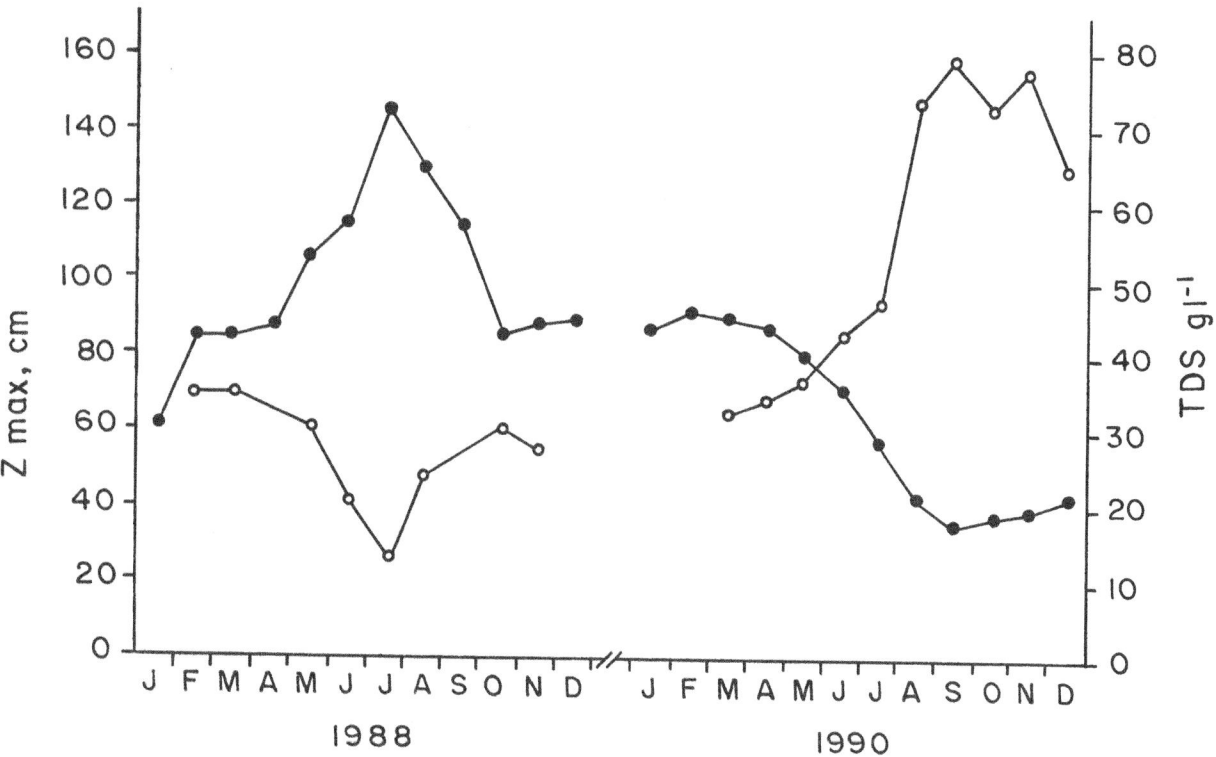

Fig. 2. Changes in maximum depth (●) and total dissolved solids (○) in Lake Gallocanta during 1988 and 1990.

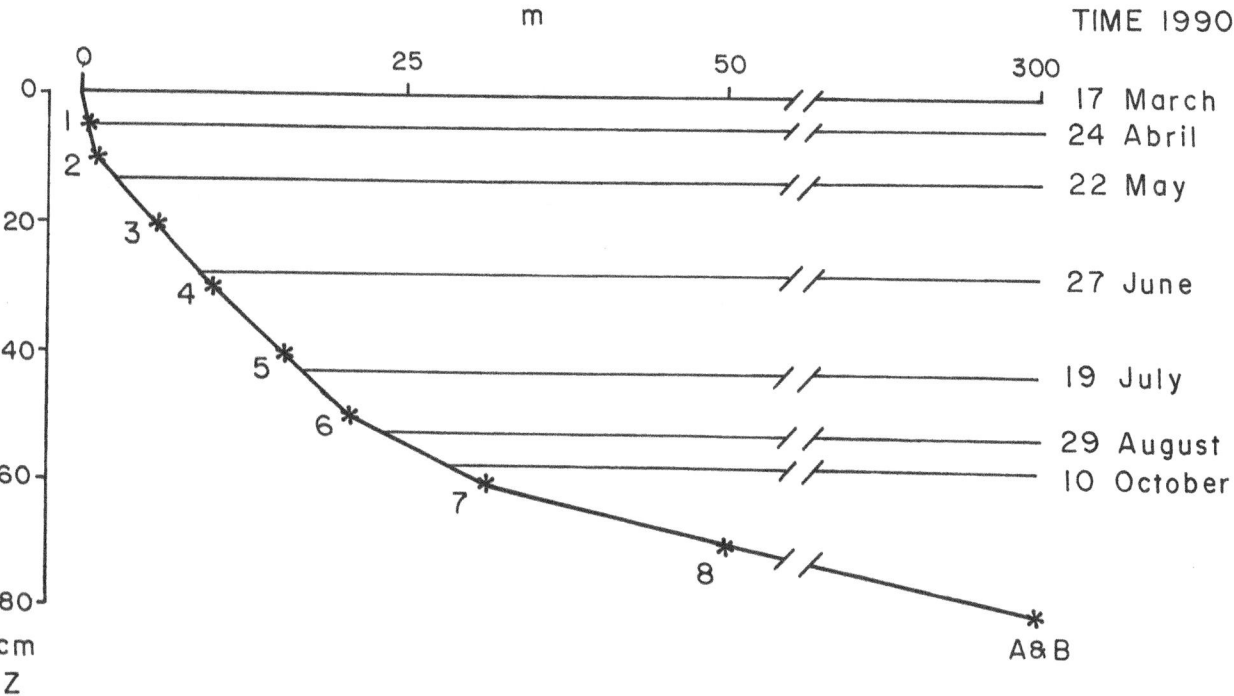

Fig. 3. Changes of the water level during the growing season of 1990 along the transect perpendicular to the shoreline (see Fig. 1).

172

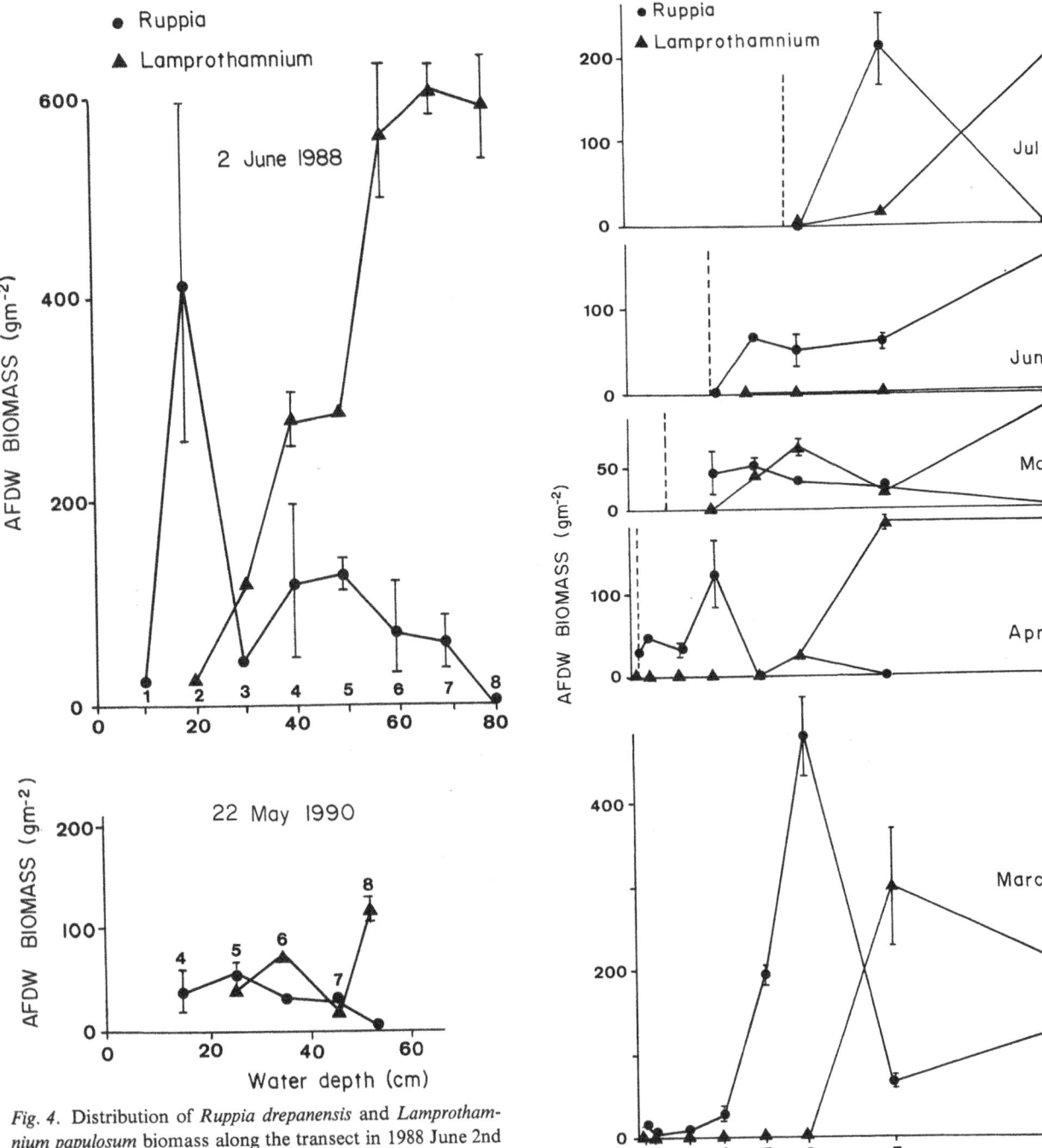

Fig. 4. Distribution of *Ruppia drepanensis* and *Lamprothamnium papulosum* biomass along the transect in 1988 June 2nd and in 1990 (May 22th). The bars indicate standard deviations.

Fig. 5. Changes in the spatial distribution of macrophyte biomasses along the shallow zone (stations 1 to 8 represented in the horizontal axis) of the transect during the growing season 1990. The discontinuous line indicates the position of the shoreline. In March it coincides with the vertical axis. Bars indicate standard deviations.

with paper towels (fresh weight-FW), after drying at 60 °C during 48 hours (dry weight-DW) and after discounting mineral components (ash free dry weight-AFDW) estimated as the residue after combustion of plant material at 450 °C. All samples included above- and below-ground parts of the plants.

Samples for pigment analysis were frozen (−10 °C) in the field immediately after collection, and analyses after extraction with 90% acetone were carried out in the laboratory one week later. Three samples were analyzed for each species on each date, each sample consisting of a set of young, apparently healthy leaves of *R. drepanensis* or thalli of *L. papulosum* collected at stations A and B. Pigment concentrations were calculated following Jeffrey & Humphrey (1975).

Primary production measurements (oxygen method, Vollenweider, 1969) were made at station 7 (see Fig. 3), on the same dates that biomass was sampled. Production measurements were made *in situ*, enclosing the plants in glass bottles (two transparent and two dark bottles of 250 ml capacity each). Water from the lake filtered through a 50 μm mesh net was used to fill the bottles. Simultaneous estimates of oxygen change in bottles filled only with filtered water were used to correct the estimates for plankton activity. Rate of change in standing crop was also estimated from the monthly carbon contents (estimated in a Carlo Erba NA 1500 elemental analyzer) and biomasses, as follows: g C m^{-2} d^{-1} = $[(C_t.B_t) − (C_{t+1}.B_{t+1})]/n$, where C is the carbon content (g mg^{-1} of plant material), B is the plant biomass (mg m^{-2}), t and $t+1$ are successive sampling times, and n is number of days between t and $t+1$.

Total dissolved solids were estimated for a single sample of lake water on each date after evaporation of 20 ml of water filtered through 0.45 μm pore size filters. Potassium concentration in water was determined by atomic absorption spectrophotometry (Philips SP 1900). Potassium concentration in plant tissues was determined for a single sample of each species, collected on each date at station 7, by the same method after acid digestion (Grasshoff *et al.*, 1983). It serves, when compared to potassium concentration in the water, as an indicator of the plant sensitiveness to changes in external potassium concentration (Bisson & Kirst, 1990).

Results

Water level fluctuations

The water level in Lake Gallocanta increased during the growing season in 1988, while it decreased during 1990 (Figs 2 and 3). Small water level fluctuations caused large changes in the area flooded because the lake bottom is very flat. So, during the 1990 growing season most of the sampling stations were progressively exposed. In contrast, during the 1988 growing season the increasing water level progressively flooded a wide littoral zone. In association with these water level changes, salinity decreased from 40 to 10 g l^{-1} in 1988, while it increased from 32 to 72 g l^{-1} in 1990 (Fig. 2).

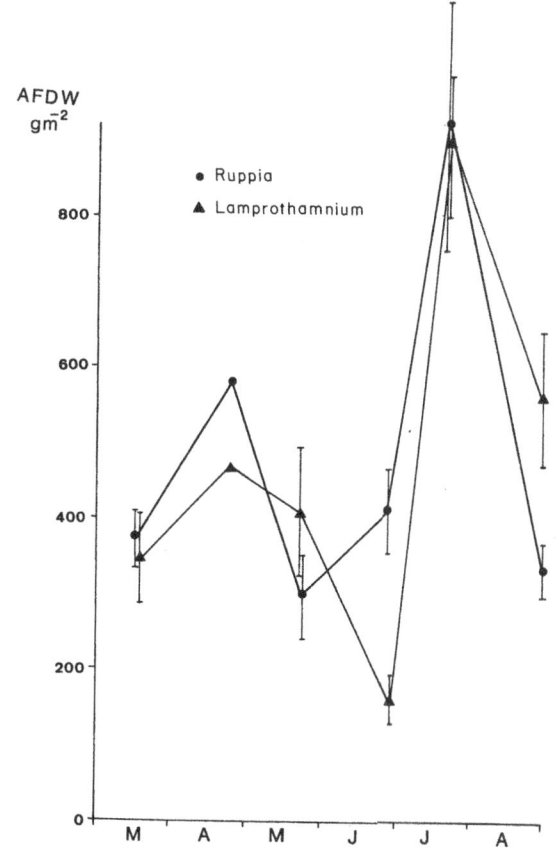

Fig. 6. Changes of macrophyte biomasses in stations A (*Ruppia drepanensis*) and B (*Lamprothamnium papulosum*) located in the deep zone of the lake during 1990.

Spatial distribution of macrophytes

Higher biomasses of *R. drepanensis* and *L. papulosum* were observed in 1988 than in 1990 for most of the stations along the transect (Fig. 4). The spatial distribution of the two species relative to each other followed the same pattern in both years.

In the shallow zone (stations 1 to 8), *R. drepanensis* was the only species present very close to the shoreline. In both years, *L. papulosum* and *R. drepanensis* overlapped along part of the

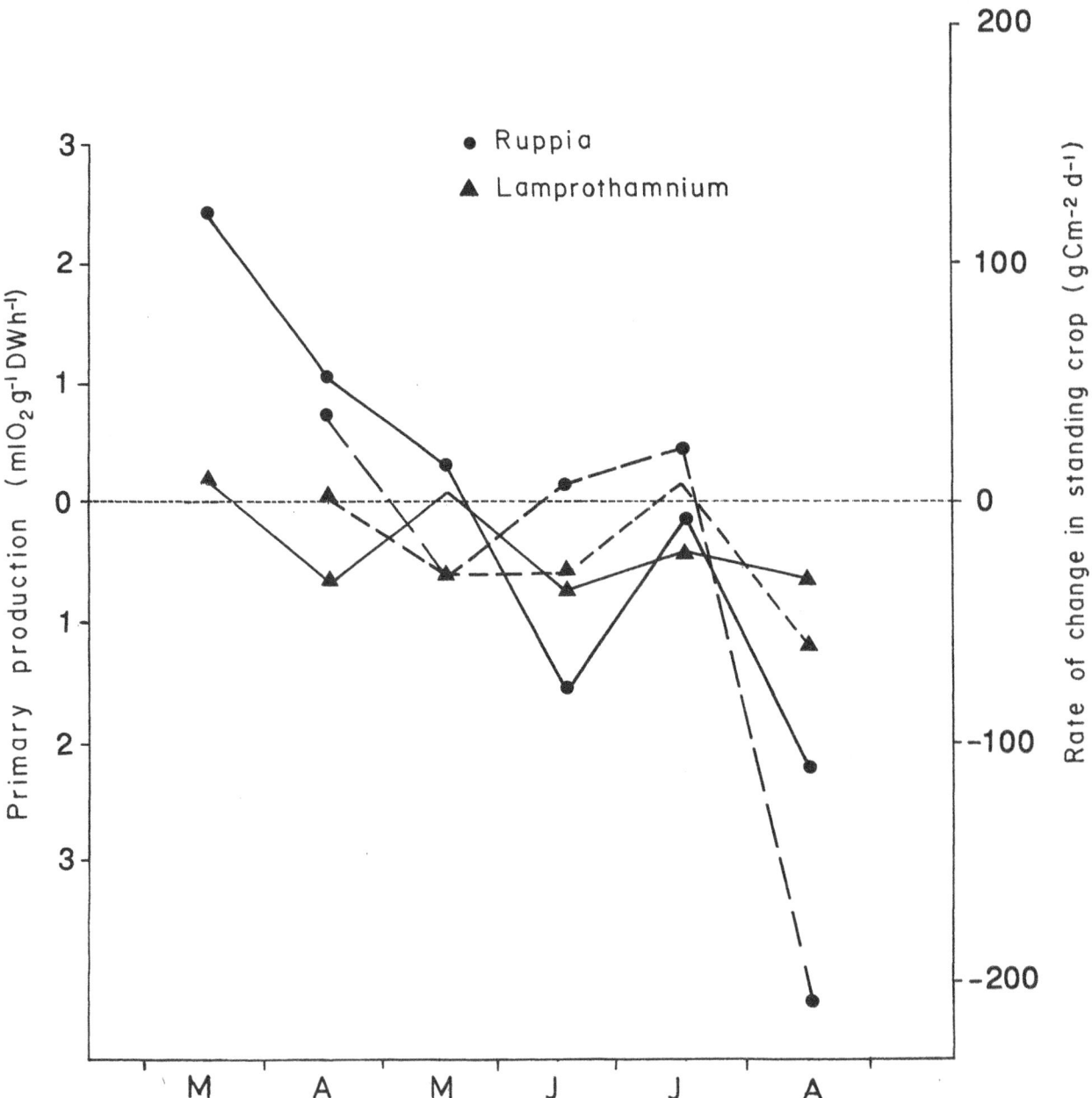

Fig. 7. Primary production and rate of biomass change of *R. drepanensis* and *L. papulosum* estimated in station 7 during 1990 by the oxygen method (continuous line) and by the change of net carbon content (discontinuous line), respectively.

transect. *L. papulosum* biomass was higher than *R. drepanensis* biomass in stations located at greater depths within the shallow zone. This spatial pattern was maintained during the growing season as water level decreased (Figs 3 and 5). *R. drepanensis* had its highest biomass in the shallow zone in March, early in the growing season. *L. papulosum* had its highest biomass at the deepest parts within the shallow zone of the transect, also early in the growing season. As water level decreased from April to July, *R. drepanensis* stands close to the shoreline decreased in density and those in deeper zones, e.g. station 7, increased in density, while *L. papulosum* density decreased.

In general, where *Ruppia* and *Lamprothamnium* occurred together either they were both scarce or only one of them was abundant.

In the deep zone (stations A and B), the two species were distributed separately, *R. drepanensis* forming isolated stands surrounded by a sparse prairie of *L. papulosum*. The biomass of each species was higher than its respective biomass in the littoral zone and, in general, they both species

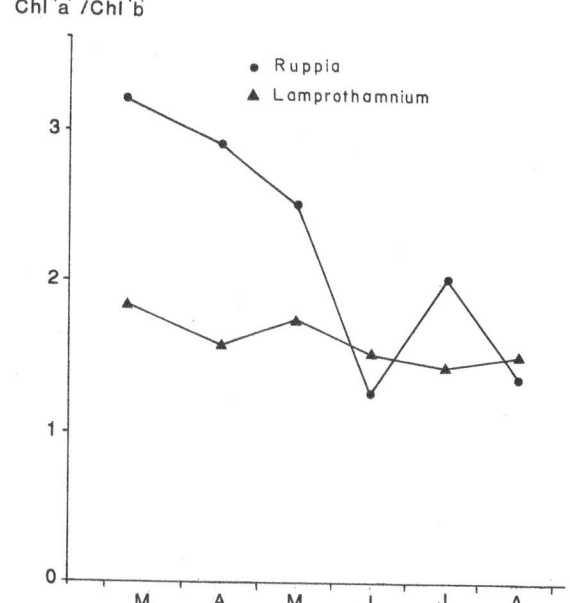

Fig. 8. Chlorophyll '*a*'/Chlorophyll '*b*' ratio of *R. drepanensis* and *L. papulosum* in Lake Gallocanta during 1990.

exhibited the same trend during the growing season (Fig. 6).

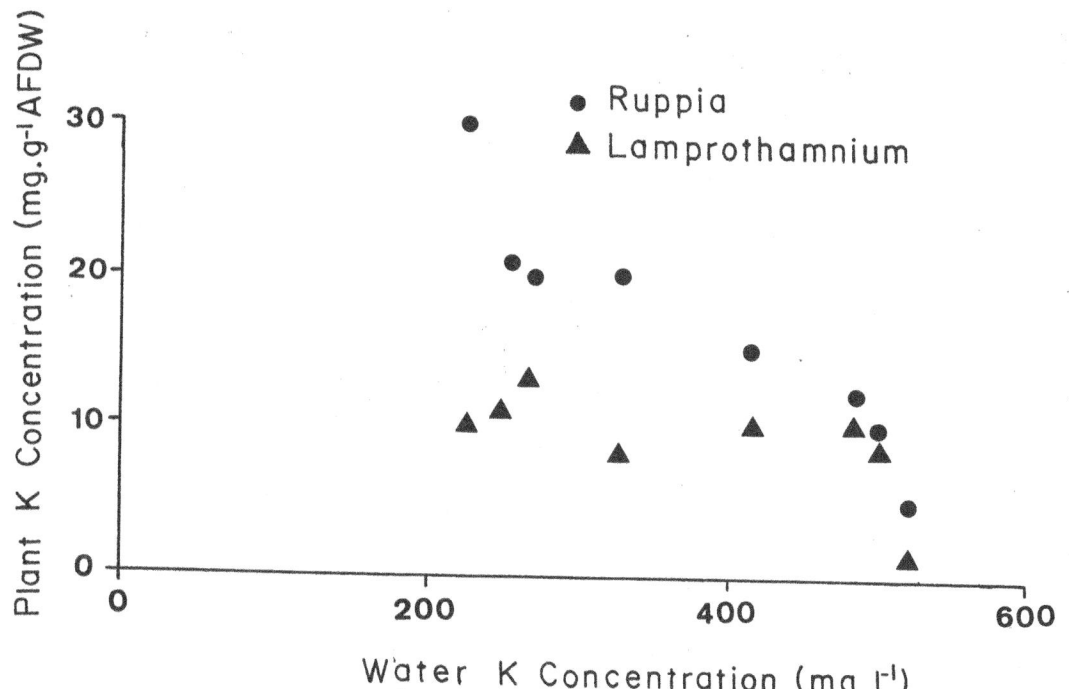

Fig. 9. Potassium content of the macrophytes vs. potassium concentration of the water in Lake Gallocanta in 1990.

Ecophysiological characteristics

All the variables studied as indicators of the physiological activity of the plants indicated a progressively declining state of *R. drepanensis* from March to August 1990: primary production and standing crop decreased (Fig. 7), as did the chlorophyll *a*/chlorophyll *b* (Fig. 8), and plant K^+ concentration/water K^+ concentration (Fig. 9) ratios. For *L. papulosum*, the decreasing trend was not so clear, and primary production values (Fig. 7) were mostly negative, indicating that this species may have been experiencing greater physiological stress than was *R. drepanensis*.

Neither *R. drepanensis* nor *L. papulosum* could withstand the hard environmental conditions created by the decreased water level in August and began to decay. The physiological decline of these plants was evidenced by the decline in tissue K^+ concentrations that was observed as the external K^+ concentration exceeded 470 g l^{-1} (corresponding to a TDS of ± 70 g l^{-1}) (Fig. 9). During years of higher water level, we have noted that dense meadows of macrophytes persisted in Lake Gallocanta through the autumn and winter.

Discussion

The two species of submerged macrophytes living in Lake Gallocanta had different distributions along a transect perpendicular to the shoreline during 1990. Two zones could be differentiated: a shallow zone, close to the shore, where the slope of the bottom was relatively high and a deeper zone where the bottom was very flat.

In the nearshore waters within the shallow zone only small plants of *R. drepanensis* grew. These may be more resistant than *L. papulosum* to water level fluctuations and wave action. Starting a few m from the water's edge, *R. drepanensis* and *L. papulosum* distributions overlapped in an area of variable width, that is, the two species were found together in the same core samples. *L. papulosum* was present alone only on certain dates at some stations located in the deepest waters of the shallow zone.

In the deepest zone, both species were distributed in separate stands: Isolated *R. drepanensis* patches of 10–20 m diameter occurred among sparse stands of *L. papulosum*.

These spatial patterns occurred in Lake Gallocanta within a few hundred m of the shoreline. Salinity did not vary significantly along the transect, either in the water column or in the interstitial water (Comín & Rodó, unpublished data). Moreover, the range of variation of salinity in Lake Gallocanta during the study period is within the tolerance range cited by several authors for both species (Hammer, 1986). Therefore, other factors must be involved in determining the differences observed in the spatial distribution.

In the nearshore zone, the effects of wave action and desiccation caused by water level changes are tolerated by the rooted macrophyte *R. drepanensis* but not by the charophyte *L. papulosum*, which lacks a root system. The species coexist where both find adequate conditions for growth. Interspecific competition will be determined by the ecophysiological response of each species to the environmental conditions.

Light, by itself and combined with other variables, is a key factor. Garcia *et al.* (1991) showed experimentally in the laboratory that *R. drepanensis* from Lake Fuente de Piedra (a saline lake that dries up every year, in southern Spain) is photoinhibited at light intensities over 695 μE m^{-2} s^{-1}. The incident photosynthetically active radiation at these latitudes is usually 2000 μE m^{-2} s^{-1} at the surface of the water column at midday in July and over 1000 μE m^{-2} s^{-1} in the upper 60 cm of the water column in a shallow coastal lagoon in eastern Spain at the same latitude as Gallocanta, where *R. cirrhosa* (Petagna) Grande is the dominant macrophyte (Comín *et al.*, 1990b). Therefore, high light intensity could have an inhibitory effect on *R. drepanensis* photosynthesis. Similar data for *L. papulosum* are not available in the literature. The comparison of these types of data for populations of both species coexisting in the same lake is essential in order to interpret the result of interspecific competition in a changing environment such as the littoral zone of saline lakes.

The data presented here indicate the photosynthetic activity of the two species was very low for most of the determinations made *in situ* during 1990. As positive biomass increases occurred during the study period, however, several possibilities can be suggested. First, the data may not be representative of the periods they represent (e.g. if environmental conditions were, by chance, especially poor or stressful on days measurements were made). We consider this possibility unlikely: no sharp changes in the climatic conditions or in lake characteristics were observed at these times. Second, the measurements were carried out at midday and the plants may have been photoinhibited at this time of day. This possibility has been commented on above, for *R. drepanensis*. The relatively high chlorophyll concentrations and chl *a*/chl *b* ratios of *R. drepanensis* and *L. papulosum* compared to taxonomically close species (Wetzel, 1964; Menendez & Comín, 1989) support this idea.

The patchy spatial distribution of *R. drepanensis* and *L. papulosum* in the deepest zone of Lake Gallocanta must be attributed to unexamined factors. No detectable differences between the salinities of zones occupied by *R. drepanensis* and *L. papulosum* were observed. Different responses of the two species to other unstudied variables such as organic matter content of the sediment and derivative consequences (inhibitory effects of H_2S on germination, colonization and growth) may account for the pattern observed.

An interpretation should take into account the spatial distribution over time of populations living in fluctuating environments (Comín *et al.*, 1991). We believe that during periods of changing water level, the shallow zone is one where *R. drepanensis*, *L. papulosum*, and probably other species, are subjected to the stress of changing environmental conditions (incident radiation, wave action, desiccation). The effects are clear both on the early growing stages and even more so on the germination of propagules, as well as on established plants. In the deepest zone of the lake, the fluctuations in these factors are reduced.

However, high water level fluctuations can occur in saline lakes on different time scales (Comín *et al.*, 1992). Thus the spatial distribution of macrophytes should be considered to be the result of dynamic processes operating over this same range of time scales, as suggested by Valiela (1984) for salt marsh ecosystems. We believe that, as water level increases, the spatial interspersion of macrophytes species, such as that observed in the shallow zone of Gallocanta, tends to give way to spatially segregated species distributions. In deeper water, life cycle characteristics, biotic interactions and sediment-macrophyte relationships take on more importance than does tolerance of extreme physical and chemical conditions in the control of the spatial distribution of macrophytes. The reverse occurs during periods of decreasing water level, during which the relatively high self organization reached by the biological community would revert to initial stages controlled by physical and chemical factors.

Acknowledgements

This work was funded by Diputación General de Arag6n, Servicio de Conservación del Medio Natural. X.R. was in receipt of a fellowship (F.P.I.) from the Ministry of Education and Science.

References

Bisson, M. A. & G. O. Kirst, 1990. *Lamprothamnium*, an euryhaline charophyte. I. Osmotic relations and membrane potential at steady state. J. exp. Bot. 31: 1223–1235.

Brock, M. A., 1981. The ecology of halophytes in the Southeast of South Australia. Hydrobiologia 81: 23–32.

Brock, M. A., 1986. Adaptation to fluctuations rather than to extremes of environmental parameters. In P. De Deckker & W. D. Williams (eds), Limnology in Australia. Dr W. Junk Publ., Dordrecht: 131–140.

Brock, M. A. & J. A. K. Lane, 1983. The aquatic flora of saline wetlands in Western Australia in relation to salinity and permanence. Hydrobiologia 105: 63–76.

Comín, F. A., M. Alonso, P. Lopez & M. Comelles, 1983. Limnology of Gallocanta Lake, Aragón, NE Spain. Hydrobiologia 105: 207–221.

Comín, F. A., M. Menendez & J. R. Lucena, 1990a. Proposals for macrophyte restoration in eutrophic coastal lagoons. Hydrobiologia 200/201: 427–436.

178

Comín, F. A., R. Julia, M. P. Comín & F. Plana, 1990b. Hydrogeochemistry of Lake Gallocanta (Aragón, NE. Spain). Hydrobiologia 197: 51–66.

Comín, F. A., R. Juliá & P. Comín, 1991. Fluctuations, the key aspect for the ecological interpretation of saline lake ecosystems. Oecologia Aquatica 10: 127–135.

Comín, F. A., X. Rodó & M. P. Comín, 1992. Lake Gallocanta, a paradigm of fluctuations at different time scales. Limnetica 8: 79–86.

García, C. M., J. L. Perez-Llorens, F. X. Niell & J. Lucena, 1991. Pigment estimations and photosynthesis of *Ruppia drepanensis* Tin. ex Guss. in an hypersaline environment. Hydrobiologia 220: 147–153.

Grasshoff, K., M. Ehnhardt & K. Kremling, 1983. Methods of seawater analysis. Verlag Chimie, 419 pp.

Hammer, U. T., 1986. Saline lake ecosystems of the world. Dr W. Junk. Publ., Dordrecht, 616 pp.

Jeffrey, S. W. & G. F. Humphrey, 1975. New spectrophotometric equations for determining chlorophylls a, b, c_1 and c_2 in higher plants, algae and natural phytoplankton. Biochem. Physiol. Pflanzen 167: 191–194.

Menendez, M. & F. A. Comín, 1989. Seasonal pattern of biomass variation of *Ruppia cirrhosa* (Petagna) Grande and *Potamogeton pectinatus* L. in a coastal lagoon. Scientia marina 53: 633–638.

Valiela, I., 1984. Marine ecological processes. Springer Verlag, New York, 541 pp.

Vollenweider, R. A., 1969. A manual on methods for measuring primary production in aquatic environments. Blackwell Scientific Publ., IBP Handbook 12, Oxford, 213 pp.

Wetzel, R. G., 1964. A comparative study of the primary productivity of higher aquatic plants, periphyton and phytoplankton in a large, shallow lake. Int. Revue. ges. Hydrobiol. 49: 1–61.

The importance of nitrogen in Pyramid Lake (Nevada, USA), a saline, desert lake[1]

John E. Reuter, Cathryn L. Rhodes, Martin E. Lebo, Mandy Kotzman & Charles R. Goldman
Institute of Ecology and Division of Environmental Studies, University of California-Davis, Davis, California 95616, USA

Key words: saline lakes, desert lakes, nitrogen limitation, nitrogen fixation, nutrient bioassay, Pyramid Lake

Abstract

The increase in human development in the downstream portion of the Pyramid Lake drainage basin has resulted in increased nutrient loading to the lake. Since this is a deep, terminal lake, concern over nutrient build up and change in trophic status exists. On the basis of lake chemistry which shows consistently high concentrations of total reactive-P (mean = 55 μg P l^{-1}) relative to dissolved inorganic-N (DIN) (mean = 15 μg N l^{-1}), it has been hypothesized that Pyramid is N-limited. However, no systematic study of nutrient limitation had been undertaken. Nutrient enrichment bioassays conducted throughout an entire year clearly showed that additions of DIN resulted in a 350–600% stimulation of chlorophyll production. Phosphate, when added singly or in combination with DIN, had no effect. This positive response to N-addition was significant at all times of the year except, (1) immediately after complete lake mixing in February when a large pool of hypolimnetic nitrate was injected into the euphotic zone, and (2) during a fall bloom of the nitrogen fixing species *Nodularia spumigena*. The positive response to N-addition in the bioassay experiments was strong between March and November. However, the seston exhibited only a gradual depletion of nitrogen relative to carbon over this same period. PN:PC ratios suggested no N-deficiency in phytoplankton biomass in February, March and April, moderate N-deficiency in May, June and July and, severe N-deficiency from August until winter turnover. The appearance of nitrogen fixing blue-green algae in September supports the hypothesis of N-limitation in the summer-autumn. In evaluating the nutrient status of a lake, the concepts of nutrient stimulation *versus* nutrient deficiency *versus* nutrient limitation must clearly be defined.

Introduction

The relationship between nutrient availability and algal growth represents a cornerstone in limnological research effort over the past 50 years (e.g. Goldman, 1960; Edmondson, 1972; Vollenweider, 1976; Schindler, 1988). In particular,

considerable emphasis has been placed on the importance of phosphorus in lake eutrophication. This attention to phosphorus has largely been based on the following, (1) the popular empirical models of Vollenweider (1968, 1976) and Rast & Lee (1978) which provide strong statistical relationships between lake phosphorus concentration and chlorophyll in a variety of lakes, (2) experimental evidence showing that changes in phosphorus loading and concentrations can affect

---

[1] This paper is dedicated to G. Evelyn Hutchinson who first visited Pyramid Lake in 1933.

phytoplankton productivity and biomass (Schindler, 1974; Edmondson & Lehman, 1981), (3) the fact that removal of phosphorus from wastewater is much easier to achieve than nitrogen removal (Wetzel, 1983) and, (4) the hypothesis that lakes can respond to nitrogen reduction by increasing nitrogen fixation (Schindler, 1977). While there is little doubt that phosphorus is an important limiting nutrient in freshwaters, the role of nitrogen in lake eutrophication has also been documented (Goldman, 1981) and may be important on a regional basis. Recently, Elser *et al.* (1990) reviewed the existing bioassay literature on phosphorus and nitrogen in lakes of North America and concluded that, in freshwaters, nitrogen as a limiting factor may have a more important role than previously recognized.

Pyramid Lake, Nevada is one of the largest, saline lakes in North America and is located in the semi-arid environment of the Great Basin. It occupies a closed basin (*i.e.* without a channelized outflow) and receives 84% of its annual water input from a single major tributary, the Truckee River (Van Denburgh *et al.*, 1973). Increased development and urbanization in the immediate upstream drainage basin is influencing water quality of the Truckee River and consequently, Pyramid Lake (e.g. nutrient loading from regional wastewater treatment facilities as well as non-point source discharge from local agricultural land) (U.S. EPA 1980; unpub. data, Nevada Division of Environmental Protection). Galat (1990), notes that this is becoming particularly critical to lakes and reservoirs in the arid western United States, where population growth is accelerating and demand for water is intense. Management of nutrient inputs to terminal lakes like Pyramid is especially important since these systems rely primarily on sedimentation and permanent burial for nutrient removal.

In desert environments, nitrogen is the macronutrient which classically limits terrestrial plant production (West & Klemmedson, 1978). Since transport of allochthonous organic matter to desert lakes is low, the amount of terrestrial nitrogen loading to these waters will also be reduced (Galat & Verdin, 1988). Furthermore, higher rates of erosion from sparsely vegetated soils in these regions carry increased sediment and associated phosphorus. In mountainous areas, such as the basin occupied by Pyramid Lake, this condition will be exaggerated. These conditions result in the characteristically low nitrogen and high phosphorus content in Pyramid Lake and in arid and semi-arid lakes in general (Goldman & Home, 1983). Given this, and the observation that nitrogen fixation by the blue-green alga *Nodularia spumigena* can account for over 80% of the annual combined nitrogen input, Pyramid Lake should be nitrogen deficient (Horne & Galat, 1985; Galat & Verdin, 1988).

In this paper we report the results of nutrient limitation studies performed in Pyramid Lake between October 1989 and October 1990. We have, (1) examined water column ratios of dissolved and total nitrogen and phosphorus, (2) compared seston ratios of carbon and nitrogen, (3) conducted monthly nutrient enrichment bioassays using natural phytoplankton populations, and (4) measured seasonal rates of nitrogen fixation. We examine the data in an attempt to distinguish between nutrient stimulation, deficiency and limitation. We also briefly compare our findings with other studies in an attempt to identify regional similarities in lake nutrient status.

Description of study site

Pyramid Lake is a terminal lake located in northwest Nevada entirely within the boundaries of the Pyramid Lake Paiute Indian Reservation. It is a remnant of the ancient Lake Lahontan which existed in the western Great Basin of the United States. It has a salinity of 4.0–4.2‰ and a total dissolved solids (TDS) of approximately 5000–5300 mg l^{-1}. Sodium, chloride, and bicarbonate-carbonate are the principal ions contributing to its salinity (Galat & Verdin, 1988). Compared with other saline lakes in the world, Galat *et al.* (1981) classified Pyramid Lake as slightly saline.

The lake has a surface area of 446 km^2 and a volume of 26.4 km^3, with a mean depth of 59 m and a maximum depth of 103 m. It is 40 km long and its width varies from 16 km at the north to 6.5 km at the south. It lies approximately 1160 m above sea level. Pyramid Lake is bounded by sparsely vegetated mountain ranges along its main axis, with summits exceeding 2490 m (Galat *et al.*, 1981).

The headwaters of the Truckee River, the lake's only permanent water source, arise in the Sierra Nevada mountains of eastern California and western Nevada. The river's main stream begins at the outlet of oligotrophic Lake Tahoe, travels 192 km northeast, while descending 715 m, and forms a delta at the south end of Pyramid Lake. The Truckee River passes through the City of Reno enroute to Pyramid Lake, and effluent from the local wastewater treatment facility is discharged into the river. Currently, this effluent receives tertiary treatment for nitrogen and phosphorus. Non-point sources also contribute to the total nutrient load in the Truckee River and are currently being investigated. Between 1973 and 1987, total-P (TP) and total-N (TN) loading to the lake from the Truckee River have ranged from 28–882 mg P m^{-2} yr^{-1} and 126–3775 mg N m^{-2} yr^{-1}. The TN:TP ratio in the river discharge is typically 4–6:1 by weight (Galat, 1990).

Pyramid Lake is classified as a warm, monomictic lake. In 1989–1990 it exhibited strong, stable thermal stratification between June and November with a thermocline at about 20 m.

Complete mixing occurred in February. Epilimnetic waters were well oxygenated, with oxygen concentrations ranging from 8.0–9.5 mg O_2 l^{-1}. Deeper, hypolimnetic waters (>60 m) were progressively depleted in oxygen until winter turnover. The average rate of oxygen depletion at 100 m in 1988, 1989 and 1990 was 0.61 mg O_2 l^{-1} $month^{-1}$ (Lebo *et al.*, 1991). In 1989–1990, dissolved oxygen in the bottom waters remained greater than 4 mg O_2 l^{-1}. The extinction coefficient for light varied from 0.18–0.33 m^{-1}. During the thermally stratified period, the depth of the euphotic zone (1% surface intensity) was similar to the depth of the thermocline.

Based on levels of primary productivity, chlorophyll and light transparency, Pyramid is considered oligotrophic-mesotrophic (Wetzel, 1983).

However, massive surface blooms of the nitrogen fixing *Nodularia spumigena* often occur in midsummer or early autumn (Galat *et al.*, 1981; Horne & Galat, 1985; Galat & Verdin, 1988; Galat *et al.*, 1990).

Material and methods

Samples were collected in the open water of the main lake basin at the location of maximum depth. Water samples for nutrient chemistry, chlorophyll biomass and seston N and C were taken at monthly intervals from October, 1989 through October, 1990. Values presented in this paper represent the mean of discrete samples taken in the euphotic zone at 0, 5, 10 and 20 m. Nutrient concentrations were determined spectrophotometrically using standard colorimetric methods as follows: ammonium – modified indophenol reaction (Solorzano, 1969; Liddicoat *et al.*, 1975), 2 μg N l^{-1} limit of detection (LOD); nitrate + nitrite – modified cadmium reduction (Jones, 1984), 1–2 μg N l^{-1} LOD; total Kjeldahl nitrogen (TKN) – acid digestion of raw sample followed by ammonium measurement (APHA, 1985), 30–40 μg N l^{-1} LOD; particulate nitrogen and carbon were determined using a Perkin Elmer model 2400 CHN analyzer, 1 μg N and 5 μg C LOD; total reactive phosphorus (TRP) – measured on raw water using the acid molybdate technique (Strickland & Parsons, 1972), 1 μg l^{-1} LOD; total phosphorus (TP) – measured as TRP following acid-persulfate digestion of unfiltered sample, 1 μg l^{-1} LOD [note: TRP and TP methods were modified to remove arsenic interference (Johnson, 1971)]; particulate phosphorus – high temperature combustion followed by acid hydrolysis (Solorzano & Sharp, 1980), <0.05 μg P LOD. Total nitrogen (TN) was calculated as the sum of TKN plus nitrate and nitrite. Chlorophyll-*a* (chl-*a*) was measured fluorometrically after homogenization and a 24 hour extraction into cold 90% acetone in the dark (APHA, 1985), 0.2 μg chl-*a* l^{-1} LOD. Concentrations were corrected for phaeopigments (Strickland & Parsons, 1972).

Nutrient bioassays using natural phytoplankton communities (Goldman, 1978) were done monthly, except in October and December 1989, and January and May 1990. For each assay, sample water was collected at 5 m and returned to the laboratory for processing. Experiments were incubated in an environmental growth chamber at ambient lake temperature. Treatment flasks were swirled and randomly repositioned every two days. Lighting was supplied by 'cool white' fluorescent bulbs at 175 μE m^{-2} sec^{-1} on a 12 hour light:dark cycle. Treatment flasks ($n = 3$ for each treatment) were inoculated with 450 ml of lake water, pre-screened through a 84 μm Nitex net to remove zooplankton. If a large fraction of phytoplankton were colonial or filamentous, raw water was used since pre-screening would remove these cells. This was particularly true in September 1990 when the filamentous blue-green alga *Nodularia spumigena* was present. Nutrients were added to five sets of treatment flasks with one set serving as a control. Nutrients were added as 5 μg N l^{-1}, 20 μg N l^{-1}, 50 μg N l^{-1}, 20 μg P l^{-1}, and 20 μg N + 10 μg P l^{-1} above background. Algal response was measured as the percent increase in chl-*a* relative to the controls following a six day incubation. Three replicate samples of the initial chl-*a* concentration were taken for each experiment. Final chl-*a* concentration in the controls was never significantly ($P < 0.05$) lower than the initials. For clarity of presentation, the data reported herein were from the + 50 μg N l^{-1} treatments. For those months when phytoplankton showed a positive response to N-addition, the mean ratio of chl-*a* in the + 50 versus the + 20 μg N l^{-1} treatments was 2.6:1. The Student's *t*-test was used to determine statistical differences between the N-enriched and control treatments. Chemical analysis of DIN concentration at the beginning and termination of a bioassay conducted in late autumn suggested that substrate depletion over the course of the incubation occurred only in the + 5 μg N l^{-1} treatment.

Primary productivity was estimated by measuring the incorporation of ^{14}C-HCO$_3^-$ into particulate matter during *in situ* incubations at depths of 0, 2.5, 5, 7.5, 10, 12.5, 15, 20, 25 and 30.5 m

(Goldman, 1963). Filters were acidified for 12 hours with 1 ml of 0.5 N HCl after incubation to drive off residual bicarbonate (Jellison & Melack, 1988). Studies with time-zero aliquots and filtered lake water showed that this procedure was also satisfactory for Pyramid Lake waters. Nitrogen fixation was estimated using the acetylene reduction technique of Stewart *et al.* (1967) as modified by Flett *et al.* (1976). Measured rates were converted to nitrogen using a theoretical 3:1 ratio of acetylene to nitrogen (as N$_2$) reduction (Hardy *et al.*, 1968).

Results

The existence of nitrogen limitation was not suggested by the TN:TP ratio. Potential N-limitation is often evaluated by comparing the concentrations of N and P in the lake to the Redfield ratio for phytoplankton biomass (40.6C:7.2N:1P, by weight; Redfield, 1958). In Pyramid Lake, mean TN:TP was 7.4 with an annual range of 6–9 (Table 1). This ratio indicated balanced growth with respect to these two nutrients. However,

Table 1. Summary of 1989–1990 Pyramid Lake euphotic zone nutrient chemistry. Included are ratios of nitrogen to phosphorus abundance commonly used to assess nutrient deficiency. Means were calculated from the individual depths of 0, 5, 10 and 20 m. All parameters were measured directly except for DON = TKN-NH$_4$-PN and TN = TKN + NO$_3$ + NO$_2$. Values in μg l^{-1} except for ratios.

Parameter	Mean	Range
Nitrate-N (NO$_3^-$ + NO$_2^-$)	15	5–60
Ammonium-N (NH$_4^+$)	5	1–15
Dissolved Inorganic-N (DIN)	20	5–65
Dissolved Organic-N (DON)	610	440–650
Particulate-N (PN)	40	10–70
Total Kjeldahl-N (TKN)	655	460–940
Total-N (TN)	670	535–950
Total Reactive-P (TRP)	55	40–70
Total-P (TP)	90	70–105
Particulate-P (PP)	6	3–9
Silicate (SiO$_2$)	350	150–525
Particulate-C (PC)	465	125–1400
DIN/TRP	0.4	0.2–1.6
TN/TP	7.4	6.9

DIN:TRP was very low with a mean of only 0.4 and an annual range of 0.2–1.6 (Table 1). The discrepancy between the ratios calculated on the basis of TN and DIN was due to the large contribution of dissolved organic-N (DON) to TN. Even if the mean ratio of inorganic N:P is modified to include all sources which likely contribute to the biologically available-N pool (DIN + 40% of PN + 15% of DON, Harrison, 1978), it would still only be 2–3. Given the low bioavailability of PN and DON for algal growth, use of the inorganic N:P ratio is more logical, and in Pyramid Lake this ratio strongly suggested the potential for N-limitation.

Natural phytoplankton assemblages from Pyramid Lake showed a dramatic response to the addition of nitrogen. When 50 μg N l^{-1} of DIN was added to lake water, there was a statistically significant (P < 0.01) increase in algal chl-a, relative to controls during seven of the nine months that these experiments were performed (Fig. 1A). The average stimulation of the phytoplankton for these seven months was striking, at 475% over controls. Chlorophyll production in the controls was positive with a mean increase of 25% for all experiments. Only in November was the final chl-a in the control significantly less (P < 0.05) than the initial. However, at that time, growth in the 50 μg N l^{-1} treatment was 2.6 times the initial concentration. Additions of 20 μg l^{-1} phosphorus to lake water between November and June, when TRP concentrations were 40–70 μg P l^{-1}, had no statistically significant effect on chl-a increase (P > 0.18). Furthermore, the addition of nitrogen and phosphorus in combination (20 μg N l^{-1} + 10 μg P l^{-1}) did not enhance the stimulatory effect of adding 20 μg N l^{-1} alone (unpub. data).

N-addition did not lead to a stimulation of chlorophyll biomass during February, the month when complete vertical mixing occurred. Throughout most of the year, nitrate concentrations, the most abundant form of DIN, in the epilimnion were 5–15 μg N l^{-1} (Table 1; Fig. 1C). The February period of holomixis resulted in a large increase of nitrate concentration in surface waters due to the mixing of nitrate-rich waters

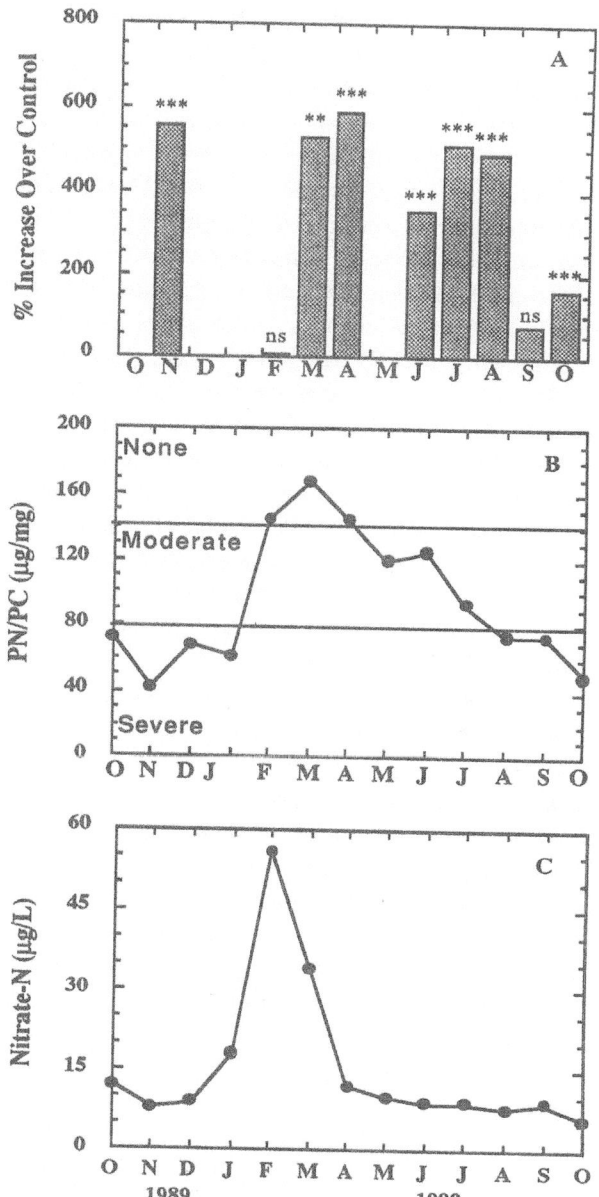

Fig. 1. A. Response of Pyramid Lake phytoplankton to additions of 50 μg N l^{-1} DIN. Reported as %-increase in treatment flasks ($n = 3$) relative to control. ns – $p > 0.05$, ** – P < 0.01, *** – P < 0.001. B. Change in seston PN/PC ratio. Values are as monthly mean of the euphotic depths. None, moderate and severe refers to degree of N-deficiency (Healy & Hendzel, 1980). C. Seasonal cycle of nitrate concentration in euphotic waters, October, 1989 – October, 1990.

from the deep hypolimnion (125–150 μg NO$_3$-N l^{-1}; Lebo et $al.$, 1991). At this time nitrate concentrations were 55–60 μg N l^{-1} throughout the

184

water column. When this source of DIN was provided, phytoplankton did not respond to experimental N-additions (P > 0.20; Fig. 1A). From late February through April a sharp and steady decline in nitrate concentration occurred at a rate of 0.7 μg N d^{-1} in the euphotic zone. During this period when nitrate was being depleted in surface waters, rates of areal primary productivity and chlorophyll were increasing in the lake (Fig. 2), and N-addition became stimulatory to phytoplankton (+531% over control) in March.

The only other time when N-addition did not significantly stimulate chl-*a* accumulation was in September, when heterocystous blue-green algae comprised a large fraction of the phytoplankton assemblage. In September, *Nodularia spumigena* was present in the lake, fixing atmospheric nitrogen. Appreciable rates of nitrogen fixation were measured on 29 August (1028 μg N m^{-2} d^{-1}), with maximum rates (81 200 μg N m^{-2} d^{-1}) occurring between 4–14 September (Fig. 3). While the September bioassay showed a 71% increase in chl-*a* over the controls (Fig. 1A), this response was not statistically significant (P > 0.20). The lack of statistical significance in the September bioassay was in part due to a higher degree of variation between the replicate flasks in this experiment. The coefficients of variation for control and N-addition treatments were, 58% and 34%, respectively. A nonhomogeneous distribution of

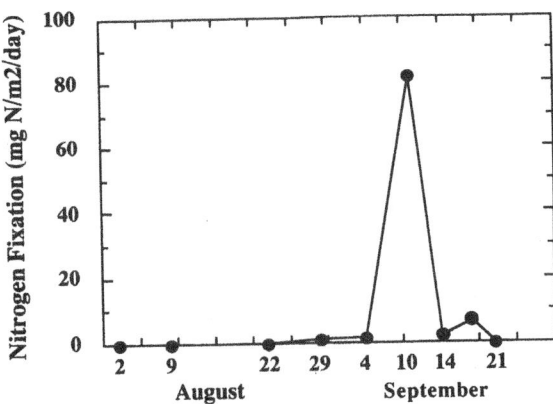

Fig. 3. Mean areal rates of nitrogen fixation in Pyramid Lake. Activity was only measured in August and September, 1990 when *Nodularia spumigena* was present. Assays were done with acetylene reduction and converted to nitrogen using the theoretical 3:1 conversion. Values represent average, lake-wide rates measured during synoptic surveys.

Nodularia filaments between flasks may have contributed to this variability. Vigorous shaking disrupts these fragile filaments and can effect the physiology of nitrogen fixation. As a result of the patchy distribution of *Nodularia* in the lake, the rates of areal nitrogen fixation do not represent a lake-wide average, rather they are site specific and are intended to show the occurrence and timing of this phenomenon.

A condition of nitrogen deficiency in the phytoplankton is also supported by data on the elemental composition of seston. Figure 1B shows mean seston PN:PC (particulate-N to particulate-C ratio) for euphotic zone waters during the study period. Based on both laboratory culture and field studies with natural communities, Healey (1975) and Healey & Hendzel (1980) suggested using PN:PC (μg N: mg C) as an indicator of N-deficiency. A ratio <80 indicates severe N-deficiency, while intermediate (80–140) and large (>140) ratios indicate moderate and no N-deficiency respectively. The presence of detrital and non-algal material can interfere with these interpretations; however, Hamilton-Galat & Galat (1983) reported that epilimnetic PC in Pyramid Lake was primarily living phytoplankton.

From October to January, PN:PC was <80,

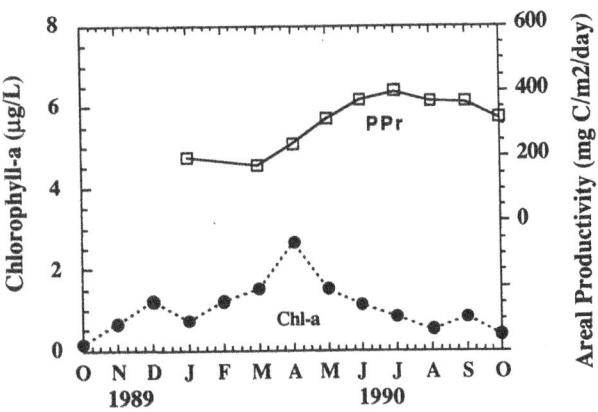

Fig. 2. Monthly chlorophyll *a* and areal primary productivity values for euphotic waters in Pyramid Lake, October, 1989 – October, 1990. Productivity measurements initiated in January.

suggesting that phytoplankton biomass was severely N-deficient. This corresponds to the period of very low nitrate concentration in the euphotic zone and followed the summer growing season (Fig. 1C). PN:PC reached a minimum value of 52 in December and at the same time, minimal annual nitrate concentrations were measured.

With the onset of complete mixing and the resulting increase in the nitrate concentration of surface waters, there was a dramatic increase in PN:PC from 71 in January to 145 in February (Fig. 1B). Similar to the bioassay results, this ratio indicated phytoplankton were not N-deficient at that time. The elemental composition of seston revealed a progressive change following holomixis as evidenced by the uniform decline in PN:PC. Between March and September this ratio decreased at a rate of approximately 10% per month. PN:PC indicated no N-deficiency in lake seston in February, March and April, a moderate N-deficiency in May, June and July, followed by severe N-deficiency in August, September, October, and presumably until winter turnover. Nitrogen fixation in 1990 began in September, when PN:PC was minimal at 53. Unfortunately there is no data on nitrogen fixation for 1989, therefore, a comparison between nitrogen fixation and PN:PC is not possible for that year. However, *Nodularia* was not noticed during casual visual observations of the lake at that time, and inspection (using microscopy) of the phytoplankton samples collected in October 1989 showed no remnants of a previous bloom.

Discussion

The first measurements of nutrient concentration in Pyramid Lake were made in July 1933. In this initial survey, ammonia was below detection (reported to be at least $20 \mu g \, N \, l^{-1}$) in the euphotic zone while soluble phosphorus was much higher at $85 \mu g \, P \, l^{-1}$ (Hutchinson, 1937). No values for nitrate were presented. Hutchinson also observed that there were no obvious terrestrial sources of combined nitrogen within the basin, which lead him to conclude that nitrogen should be the most

important chemical factor restricting the growth of phytoplankton in this lake. Our present studies confirm this hypothesis by showing that additions of dissolved inorganic-N (DIN) nearly always stimulated algal chl-a, and that seston PN:PC ratios were at sufficiently low levels to indicate N-deficient growth.

The best water chemistry indicator of N-deficiency was the ratio of the inorganic pools, DIN:TRP which ranged from <1.5–2.5. Using the lake TN:TP ratio as an indicator of nutrient deficiency was clearly misleading. The TN:TP ratio in Pyramid Lake surface waters suggested balanced growth throughout the year despite contrary evidence from the bioassay experiments and the chemical composition of seston. Dissolved organic-N, which is largely refractory, comprised >90% of TN pool, while organic-P accounted for <40% of the TP (Lebo et al.,1991). Thus, for Pyramid Lake the TN:TP ratio poorly represents the biologically available nutrients and nor was it a valid measure of seston composition.

Water column nutrient ratios should be used with caution and in conjunction with other indicators of nutrient deficiency. For example, in February, the DIN:TRP ratio was <3, suggesting strong N-limitation of growth, but the bioassay showed minimal stimulation by N-addition. At this time, DIN increased to $>50 \mu g \, N \, l^{-1}$ due to vertical mixing, and provided nitrogen for late-winter and spring phytoplankton growth. Annual primary productivity reached its seasonal minimum at this time, largely due to low light and low temperature conditions and therefore algal N-demand of *in situ* algae was probably reduced. Even though the DIN:TRP was less than the Redfield ratio, the new input of nitrate from the hypolimnion during winter mixing was sufficient to meet physiological demands. Consequently, experimental treatment of adding $50 \mu g \, N \, l^{-1}$ had no effect. Apparently, the absolute abundance of biologically available N and P as well as their ratio is important.

It appears that the level of N-stress on phytoplankton growth in the euphotic waters of Pyramid lake can be divided into four distinct periods. The first period immediately followed turnover in

February, when levels of DIN in the euphotic zone increased from 10–15 μg N l^{-1} to 55–60 μg N l^{-1}. This was the only time during this study that such a large increase in available-N was observed. DIN was mainly input to surface waters as nitrate, since ammonium buildup in the deep hypolimnion was minimal (Lebo *et al.*, 1991). Concomitant with this injection of DIN into surface waters, the seston PN:PC ratio changed from a condition of severe N-deficiency to no N-deficiency. Clearly, nitrogen starved cells were utilizing this newly available resource.

The second period was from March through August during which time N-stress became increasingly acute. The response to N-additions was very large during this entire period (358–591% over controls) and highly significant ($P < 0.01$). However, the seston showed a gradual decline in PN:PC indicating a uniform progression from no N-deficiency in March to severe N-deficiency in August. In March and April, PN:PC indicated that the seston was not N-deficient. Furthermore, the seston C:N ratios for these months were 6.0 and 6.9, and very close to Redfield proportion of 5.6C:1N (by weight). PN:PC continued to decline after its maximum in March until August when this ratio indicated severe N-deficiency in the seston. Presumably, between March and August when inputs of 'new'-N are minimal, phytoplankton depend on 'regenerated'-N to sustain production, as shown by Dugdale & Goering (1967) and Axler *et al.* (1982) for low-N environments.

The third period occurred in September when a bloom of N-fixing *Nodularia* appeared. These annual blooms occur frequently in Pyramid Lake between early July and early October (Galat *et al.*, 1990), but in some years, the blooms may not occur as in 1989. While the growth of N-fixing organisms is not limited by nitrogen, low-N environments offer a competitive advantage to these species. In the September bioassay, chl-*a* in the control flasks increased by 2.0 μg l^{-1} in comparison to <0.5 μg l^{-1} during the other months when bioassays were conducted. Clearly, N-fixation can provide a temporary relief from N-stressed growth. The incorporation of 'new' atmo-

spheric-N into phytoplankton was not reflected in the PN:PC of seston. This is most likely due to the fact that the *Nodularia* bloom was low relative to other years (Horne & Galat, 1985) and the distribution of biomass was sporadic and patchy. Consequently, the seston samples taken in September were not dominated by *Nodularia*. Indeed, the concentration of PC was only 20% higher than the annual average (Lebo *et al.*, 1990). Hamilton-Galat & Galat (1983) provide evidence that when large *Nodularia* blooms occur in Pyramid Lake, the relative nitrogen content of seston increases. The 1979 *Nodularia* bloom was one of the largest on record (Galat & Verdin, 1988) and between August and September, when the bloom occurred, the PC:PN dropped sharply from 8 to 5.

The fourth period begins after the *Nodularia* bloom disappears in the fall and continues until winter turnover. During this time, seston composition indicated N-deficiency and N-additions stimulated chl-*a* production. Years when blooms are high may be characterized by different N-dynamics than we observed in either 1989 or 1990 when *Nodularia* was either absent or low. Again, the October and November, 1979 PC:PN data of Hamilton-Galat & Hamilton showed that this ratio remained low at 5 to 7 despite the fact that green algae and diatoms reappeared in the phytoplankton community. The importance of *Nodularia* as a source of 'new'-N will depend on the magnitude of the bloom and rates of N-mineralization and extracellular excretion.

These results highlight the need to distinguish between nutrient stimulation, nutrient deficiency and nutrient limitation when assessing the status of lakes. We refer to nutrient stimulation as the potential capacity to increase algal biomass as a consequence of external nutrient additions and is measured in enrichment bioassay experiments, *i.e.* level II experiments as defined by Hecky & Kilham (1988). Since these are typically done under controlled conditions using small experimental vessels (1–5 liters) and relatively short incubation times (time scale of days), many of the natural nutrient fluxes and loss terms which would occur *in situ* are eliminated (Hecky & Kilham,

1988). These experiments provide evidence for potential biomass accumulation. Nutrient deficiency can pertain to either the nutrient content of the lake water itself or nutrient composition of phytoplankton cells. As indicated by the data from Pyramid Lake, even though the accumulation of phytoplankton biomass can be stimulated by nutrient addition in level II experiments, the growth of the resident algae in the natural system can be nutritionally balanced with respect to N and P. The steady decline in PN:PC during the late-spring to autumn suggests that growth was less than optimal and that DIN was in increasingly low supply. Nutrient limitation refers to the ability of nutrient supply to regulate biomass accumulation in nature and is a concept which should be applied to whole-lake production rather than the growth of individual algal populations. *In situ* biomass accumulation is also regulated by other factors such as, light, temperature, grazing, sedimentation, etc. True determination of nutrient limitation therefore requires extensive study of each of these processes with confirmation by mesocosm or system-wide experimentation. Our results indicate that during the one year study period, phytoplankton growth in Pyramid Lake was stimulated by nitrogen. However, as suggested by the concomitant increase in primary productivity and decline of chlorophyll biomass between April and June (Fig. 2), under current levels of nitrogen loading, phytoplankton biomass accumulation in the lake is also affected by loss factors (e.g. grazing, natural death, sedimentation, etc.).

The nutrient status of Pyramid Lake is similar to other lakes and reservoirs in the desert environment of northern Nevada. The DIN:TRP ratio in most of these systems is typically <1 and bloom populations of N-fixing blue-green algae have been observed in each of those aquatic systems (NES 1977; Nevada Division of Environmental Protection unpub. data). Nutrient enrichment bioassays which have been conducted at five lakes in northern Nevada (including Pyramid Lake), show that phytoplankton production is most frequently stimulated by N-addition (Cooper *et al.*, 1983; Koch *et al.*, 1984; Reuter *et al.*, 1991). Only one lake for which nutrient enrichment bioassays have been conducted showed P-stimulation. In this case, the phytoplankton assemblage during the experiment was dominated by a heterocystous species of *Anabaena*. This suggests that P-deficiency may have been a temporary phenomenon in an otherwise N-deficient system. Pyramid Lake differs from the other lakes and reservoirs in the region in that it is saline, with a conductivity 1–2 orders of magnitude higher than that of freshwater. As a result, the predominant blue-green alga found in the other systems was *Aphanizomenon flos-aquae*, rather than the more salt tolerant *Nodularia spumigena* found in Pyramid Lake.

This pattern of nitrogen deficiency is less evident in a set of 56 lakes and reservoirs located in the semi-arid southwest United States that were sampled as part of the National Eutrophication Survey in the early-mid 1970's (Thornton & Rast, 1989). Despite the observation that the median TN:TP ratio was 2.9, approximately one-half the lakes were considered N-limited, while the other one-half were considered P-limited. Additional data reported for semi-arid lakes and reservoirs in South Africa by Thornton & Rast (1989) and Australia by Cullen & Small (1981) indicated less of a dependence on N than either lakes in northern Nevada or the southwest United States. Our observations of Pyramid Lake and other systems in northern Nevada suggest that strong regional differences exist for semi-arid environments. When lakes from these regions are grouped together for comparison with temperate lakes, regional differences within each group should not be ignored.

Acknowledgements

This research was funded by the U.S. Environmental Protection Agency under Sections 305 and 106 of the Clean Water Act and was done in cooperation with the Pyramid Lake Paiute Indian Tribe-Pyramid Lake Fisheries. Special thanks are given to Paul Wagner, Lee Carlson, Nancy Vucinich, and Dan Mosley of the Pyramid Lake Fisheries. The dedication of Wendell Smith (US

188

EPA Region IX) is greatly appreciated. David Galat and James Cooper provided us with valuable background information and discussions on Pyramid Lake. Editorial comments by Liza Sater were appreciated. S. H. Hurlbert and two anonymous reviews provided comments and suggestions on this manuscript.

References

APHA, 1985. Standard methods for the examination of water and wastewater. 17th edn. American Public Health Association, N.Y.

Axler, R. P., R. M. Gersberg & C. R. Goldman, 1982. Inorganic nitrogen assimilation in a subalpine oligotrophic lake. Limnol. Oceanogr. 27: 53–65.

Cooper, J. S., Vigg, R. W. Bryce & R. L. Jacobson, 1983. Limnology of Lahontan Reservoir, 1980–1981. Desert Research Institute, Univ. Nevada, Reno. Publication #50021.

Cullen, P. & I. Smalls, 1981. Eutrophication in semi-arid area – the Australian experience. Water Qual. Bull. 6: 79–83, 90–91.

Dugdale, R. C. & J. J. Goering, 1967. Uptake of new and regenerated forms of nitrogen in primary productivity. Limnol. Oceanogr. 12: 196–206.

Edmondson, W. T., 1972. Nutrients and phytoplankton in Lake Washington. Limnol. Oceanogr. Special Symp. Vol. I, pp. 172–193.

Edmondson, W. T. & J. L. Lehman, 1981. The effect of changes in the nutrient income on the conditions of Lake Washington. Limnol. Oceanogr. 26: 1–29.

Elser, J. J., E R. Marzolf & C. R. Goldman, 1990. Phosphorus and nitrogen limitation of phytoplankton growth in the freshwaters of North America: a review and critique of experimental enrichments. Can. J. Fish. aquat. Sci. 47: 1468–1477.

Flett, R. J., R. D. Hamilton & N. E. R. Campbell, 1976. Aquatic acetylene-reduction techniques: solutions to several problems. Can. J. Microbiol. 22: 43–51.

Galat, D. L., 1990. Seasonal and long-term trends in Truckee River nutrient concentrations and loads to Pyramid Lake, Nevada: A terminal saline lake. Wat. Res. 24: 1031–1040.

Galat, D. L., E. L. Lider, S. Vigg & S. R. Robertson, 1981. Limnology of a large, deep, North American terminal lake, Pyramid Lake, Nevada, USA. Hydrobiol. 82: 281–317.

Galat, D. L. & J. P. Verdin, 1988. Magnitude of blue-green algal blooms in a saline desert lake evaluated by remote sensing: evidence for nitrogen control. Can. J. Fish. aquat. Sci. 45: 1959–1967.

Galat, D. L., J. P. Verdin & L. L. Sims, 1990. Large-scale patterns of Nodularia spumigena blooms in Pyramid Lake, Nevada, determined from Landsat imagery: 1972–1986. Hydrobiol. 197: 147–164.

Goldman, C. R., 1960. Primary productivity and limiting factors in three lakes of the Alaska Peninsula. Ecol. Monogr. 30: 207–270.

Goldman, C. R., 1963. The measurement of primary productivity and limiting factors in freshwater with Carbon-14. In M. S. Doty (ed.), Conference on Primary Productivity. U.S. Atomic Energy Commission. TID-7633.

Goldman, C. R., 1978. The use of natural phytoplankton assemblages in bioassay. Mitt. int. Ver. Limnol. 21: 364–371.

Goldman, C. R., 1981. Lake Tahoe: two decades of change in a nitrogen-deficient oligotrophic lake. Verh. int. Ver. Limnol. 21: 45–70.

Goldman, C. R. & A. J. Horne, 1983. Limnology. McGraw Hill Book Publishers, 464 pp.

Hamilton-Galat, K. & D. L. Galat, 1983. Seasonal variation of nutrients, organic carbon, ATP, and microbial standing crops in a vertical profile of Pyramid Lake, Nevada. Hydrobiol. 105: 27–43.

Hardy, R. W., R. F. D. Holsten, E. K. Jackson & R. C. Burns, 1968. The acetylene-ethylene assay for nitrogen fixation: laboratory and field evaluations. Plant Physiol. 43: 1185–1207.

Harrison, W. G., 1978. Experimental measurements of nitrogen remineralization in coastal waters. Limnol. Oceanogr. 23: 684–694.

Healey, F. P., 1975. Physiological indicators of nutrient deficiency in algae. Fish. Mar. Serv. Tech. Rep. 585: 30 p.

Healey, F. P. & L. L. Hendzel, 1980. Physiological indicators of nutrient deficiency in lake phytoplankton. Can. J. Fish. aquat. Sci. 37: 442–453.

Hecky, R. E. & P. Kilham, 1988. Nutrient limitation of phytoplankton in freshwater and marine environments: A review of recent evidence on the effects of enrichment. Limnol. Oceanogr. 33: 796–822.

Horne, A. J. & D. L. Galat, 1985. Nitrogen fixation in an oligotrophic, saline desert lake: Pyramid Lake, Nevada. Limnol. Oceanogr. 30: 1229–1239.

Hutchinson, G. E., 1937. A contribution to the limnology of arid regions. Trans. Connecticut Acad. Sci. 33: 47–132.

Jellison, R. & J. M. Melack, 1988. Photosynthetic activity of phytoplankton and its relation to environmental factors in hypersaline Mono Lake, California. Hydrobiol. 158: 69–88.

Johnson, D. L., 1971. Simultaneous determination of arsenate and phosphate in natural waters. Envir. Sci. Technol. 5: 411–414.

Jones, M. N., 1984. Nitrate reduction by shaking with cadmium: an alternative to cadmium columns. Wat. Res. 18: 643–646.

Koch, D. L., S. Howell-Cooper, J. J. Cooper, S. Vigg, E. L. Lider & S. R. Robertson, 1984. Limnology of Topaz Lake, Nevada/California. Desert Research Institute, Univ Nevada, Reno. Publication #5031.

Lebo, M. L., J. E. Reuter, C. L. Rhodes & C. R. Goldman,

1991. Limnological and nutrient cycling studies at Pyramid Lake, Nevada: October 1989–October 1990. Project Rept. Division of Environ. Studies, Univ. California, Davis, CA., 60 pp.

Liddicoat, M. I., S. Tibbits & E. I. Butler, 1975. The determination of ammonia in seawater. Limnol. Oceanogr. 20: 131–132.

National Eutrophication Survey (NES), 1977. Working Paper Series. U.S. Environmental Protection Agency, Corvallis, OR.

Rast, W. & G. F. Lee, 1978. Summary analysis of the North American (U.S. portion) OECD Eutrophication Project; Nutrient loading-lake response relationships and trophic state indices. U.S. Environ. Protection Agency Rept. EPA 600/3-78-008. 455 pp.

Redfield, A. C., 1958. The biological control of chemical factors in the environment. Am. Sci. 46: 205–221.

Reuter, J. E., H. Boriss, C. R. Goldman & J. J. Cooper, 1991. Analysis of water quality data for selected lakes and reservoirs in northern Nevada. State of Nevada, Division of Environmental Protection, Carson City, NV, USA.

Schindler, D. W., 1974. Eutrophication and recovery in experimental lakes: Implications for lake management. Science. 184: 897–899.

Schindler, D. W., 1977. Evolution of phosphorus limitation in lakes. Science. 195: 260–262.

Schindler, D. W., 1988. Experimental studies of chemical stressors on whole lake ecosystems. Verh. int. Ver. Limnol. 23: 11–41.

Smith, V. H., 1982. The nitrogen and phosphorus dependence of algal biomass in lakes: an empirical and theoretical analysis. Limnol. Oceanogr. 27: 1101–1112.

Solorzano, L., 1969. Determination of ammonia in natural waters by the phenolhypochlorite method. Limnol. Oceanogr. 14: 754–801.

Solorzano, L. & J. H. Sharp, 1980. Determination of total dissolved phosphorus and particulate phosphorus in natural waters. Limnol. Oceanogr. 25: 754–758.

Stewart, W. D. P., G. P. Fitzgerald & R. H. Burris, 1967. *In situ* studies on nitrogen fixation using the acetylene reduction technique. Proc. Nat. Acad. Sci. USA 58: 2071–2078.

Strickland, J. D. H. & T. R. Parsons, 1972. A practical handbook of seawater analysis. Bulletin 167. Fish. Res Board Can., Ottawa, Ontario, Canada.

Thornton, J. A. & W. Rast, 1989. Preliminary observations on nutrient enrichment of semiarid, manmade lakes in the northern and southern hemispheres. Lake and Reservoir Manag. 5: 59–66.

U.S. EPA, 1980. Draft Environmental Impact Statement: Reno/Sparks Joint Water Pollution Control Plant Master project. (EPA-9-CA-C-32-0114), U.S. EPA, Region 9, San Francisco, California.

Van Denurgh, A. S., R. D. Lamke & J. L. Hughes, 1973. A brief water resources appraisal of the Truckee River Basin, western Nevada. Wat. Resour.-Recon. Ser. Rep. 57. State of NV, Dept. of Cons. and Nat. Resour., Div. of Wat. Resour., Carson City, Nevada.

Vollenweider, R. A., 1968. Scientific fundamentals of the eutrophication of lakes and flowing waters, with particular reference to nitrogen and phosphorus as factors in eutrophication. Tech. Rept. No. DAS/CSJ/68.27. Organization for Economic Cooperation and Development, Paris, 159 pp.

Vollenweider, R. A., 1976. Advances in defining critical loading levels for phosphorus in lake eutrophication. Mem. Ist. ital. Idrobiol. 33: 53–83.

West, N. E. & J. O. Klemmedson, 1978. Structural distributions of nitrogen in desert ecosystems. In N. E. West and J. Skujins (eds), Nitrogen in desert ecosystems. Dowden, Hutchinson and Ross, Stroudsburg, PA.: 1–16.

Wetzel, R. G., 1983. Limnology (2nd edn). Saunders College Publishing, Philadelphia, PA., 767 pp.

A population model for the alkali fly at Mono Lake: depth distribution and changing habitat availability

David B. Herbst & Timothy J. Bradley
Sierra Nevada Aquatic Research Laboratory, University of California, Star Route 1, Box 198, Mammoth Lakes, CA 93546, USA; Department of Ecology and Evolutionary Biology, University of California, Irvine, CA 92717, USA

Key words: habitat conservation, *Ephydra hians*, Mono Lake, population model, saline lakes, tufa

Abstract

The densities of alkali fly larvae and pupae were measured in relation to depth and substrate type at six locations around Mono Lake. Samples representing a mixture of different bottom features were taken to a depth of 10 m (33 ft) using SCUBA. This is at or near the depth limit of fly larvae and pupae. The biomass of larvae and pupae on hard substrate were maximum and approximately equal at depths of 0.5 m and 1 m, substantially lower at intermediate depths of 3 m and 5 m, and over an order of magnitude further reduced at 10 m. Densities of flies on hard or rocky substrates (mainly calcareous tufa deposits), were significantly greater than those found on soft substrates such as mud or sand, at all but the greatest depth surveyed.

Bathymetric maps of the areas of hard and soft substrate occurring at different lake depths were used to estimate the fly population size over the whole lake, based on the density distribution of larvae and pupae with depth on different substrates. The mapped areas of soft and hard substrates were also calculated for different lake levels, and applying the same procedure, a population model comparing the abundance of flies at different lake levels was developed. This habitat-based population model predicts that the abundance of the alkali fly is maximized at 6380 ft (1945 m) lake surface elevation. Most of the tufa substrate submerged at this lake level will become exposed and unavailable as habitat as the lake declines to 6370 ft (1942 m). In late 1991, the lake level was just over 6374 ft (1943 + m).

Introduction

In the eastern Sierra Nevada of California, streams have been diverted from the Mono Lake Basin for 50 years. While these streams provide water and power to the city of Los Angeles, the deficit in freshwater supply to saline Mono Lake is lowering the level and increasing the salt concentration of this productive wildlife habitat. One of the ecological impacts originating from dropping lake level is exposure of littoral habitat that supports production of one of the major food organisms of the lake, the alkali fly *Ephydra hians* Say (National Academy of Sciences, 1987; Botkin *et al.*, 1988). This insect is a dietary staple for the many migratory shorebirds that use Mono Lake as a feeding and breeding site.

The objective of this research was to provide a predictive population model of the extent to which declining lake levels will alter the abundance of the alkali fly as its habitat becomes exposed. The model is based on surveys of the distribution of the aquatic larval and pupal stages of the fly on different bottom surface features over varied depths. It uses this information to provide a prediction of how the abundance of the alkali fly would change over a range of projected lake levels. The range of lake levels providing maximum habitat for flies could thus be identified.

Benthic aquatic organisms are often found associated with specific substrate types. Rocky and soft bottom environments have distinctive faunal assemblages, and these physical substrates circumscribe the habitats to which the epifauna or infauna are usually restricted. Since the substrate inhabited often serves as the template on which development and production occur, habitat area may be used to define distribution and abundance. Examples of methods that employ habitat-based estimates of population size or standing stock among aquatic invertebrates include habitat-stratified sampling (Elliot, 1977; Wrona et al., 1986) and instream flow models (Gore, 1978; Gore & Judy, 1981). These methods have been used to predict population density based on usable physical habitat available. The present study follows this approach in modeling the lakewide population size of a benthic insect based on samples stratified by depth and substrate type.

Larvae and pupae of the alkali fly (Ephydra hians Say) are aquatic. Larvae hatch from eggs and develop through three instars before pupating. When mature, larvae attach to submerged, stable substrate, and form pupae. Adults emerge from the pupal case and float to the water surface in an air bubble. Adult flies (mainly females) re-enter the water by crawling down partly submerged rocks to feed on algal films and deposit eggs. Algae forms most of the diet for both larvae and adults.

Previous studies (Herbst, 1990; Little et al., 1989, unpublished) established that tufa rock is the primary substrate on which larvae and pupae are found. This reef-like limestone has a complex surface, providing protected microhabitats and attachment sites for larvae and pupae. Since these earlier studies were restricted to shallow water sampling (0.5 m or less), it was not known to what depth larvae and pupae could occur, or

whether tufa was still the predominant substrate used in deep water.

Other research on the alkali fly has established that larvae osmoregulate (Herbst et al., 1988), using an unusual modification of the Malpighian tubule to regulate carbonate (Herbst & Bradley, 1989). Though osmotic and ionic regulation permit salt tolerance, these physiological adaptations are not without cost. Salinities above 10 g l^{-1} slow growth and development and reduce body size at maturity (Herbst, 1992), and those above 150 g l^{-1} are lethal to early instars.

Methods

Sampling locations and procedures

Six locations around the lake, representing a mixture of different bottom features, were sampled to a depth of 10 meters (about 33 feet) using SCUBA (Fig. 1). This depth was determined to be at or very near the limit of fly larvae and pupae – below this, the thermocline is typically established and the water quickly becomes prohibitively cold and dark. Some 20 dives and a total of 40–50 hours of submerged sampling and observation time were logged. The diving began in early August and was completed in mid-October. There were 358 total samples taken: 210 on mixed hard substrates, including tufa, pumice and mineral-encrusted rock and wood; and 148 on soft substrates, including mixed mud, sand and detritus. Sites varied in the amount and type of substrate present and sampled (Table 1). Based on previous, more widespread surveys of the lake (Herbst, 1990), the six stations selected for this study were located in areas including the greatest range of alkali fly standing crop densities and substrate-habitat types.

Sampling was conducted along five depth contours at each location sampled – 10 m, 5 m, 3 m, 1 m, and 0.5 m (15 m was also examined initially but no larvae or pupae could be found and as this depth was usually below the thermocline, sampling was discontinued here). During the sampling period, depth of the thermocline varied be-

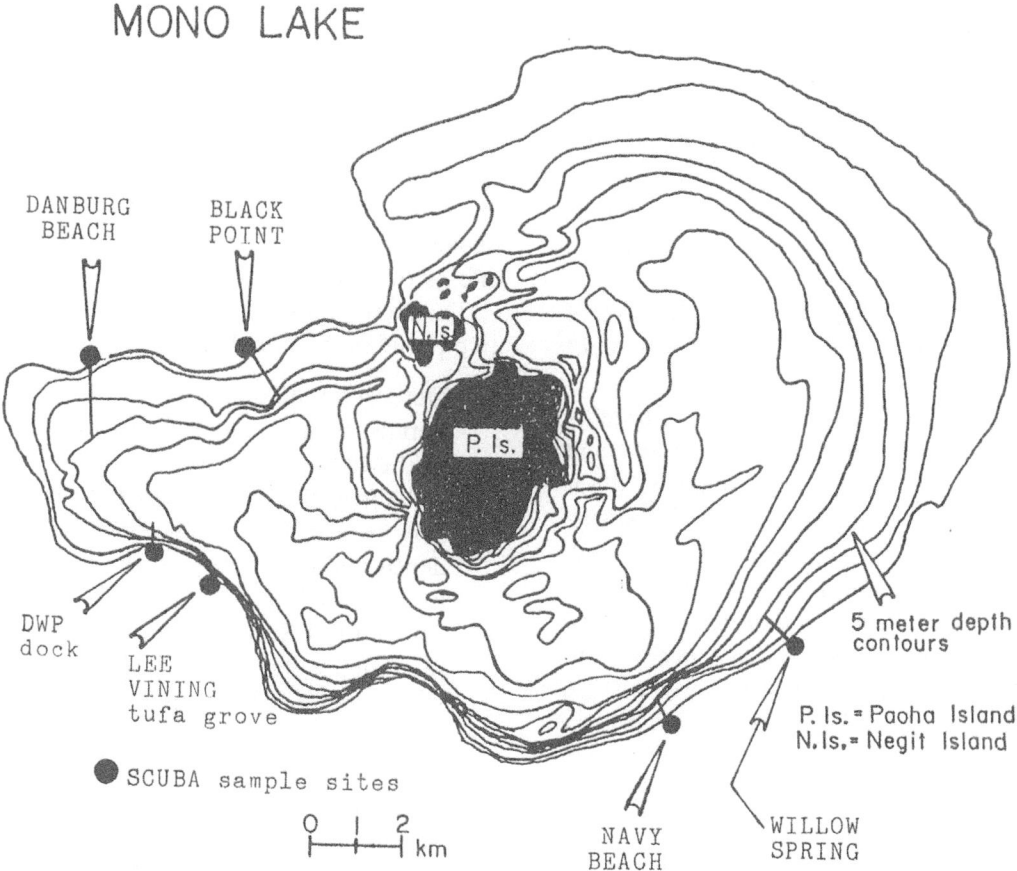

MONO LAKE

DANBURG BEACH

BLACK POINT

N.Is.

P. Is.

5 meter depth contours

DWP dock

LEE VINING tufa grove

P. Is.= Paoha Island
N.Is.= Negit Island

● SCUBA sample sites

0 1 2 km

NAVY BEACH

WILLOW SPRING

Fig. 1. Map of Mono Lake bathymetric contours and SCUBA sample sites and depth transects. Willow Springs and Navy Beach sites are predominantly sand and mud, Lee Vining tufa grove and Black Point tufa shoals both have high tufa cover over sand and mud, DWP dock and Danburg Beach possess mixed substrates with substantial encrusting tufa and gaylussite.

tween 12 and 20 m, and water temperature varied from 20 to 13.5 °C. The sampling procedure consisted of descending to the sample depth and searching for substrates of appropriate nature (conforming to size and consistency criteria), which were removed as encountered along a compass-guided course parallel to shore while swimming at a constant depth. Depth was maintained by the tension of a surface float line tied around the divers wrist. Hard substrates were

Table 1. Distribution of samples by depth and station site (hard substrate/soft substrate). Each sample represents an individual rock or core removal.

Depth (m)	DWP dock	Lee Vining tufa grove	Danburg Beach	Navy Beach	Willow Spring	Black Point	No. sites (N_{sd})	No. samples $(\Sigma_j N_{jsd})$
0.5	15/5	10/0	10/0	0/8	0/8	8/8	4/4	43/29
1.0	20/0	10/0	10/0	0/8	0/8	8/8	4/3	48/24
3.0	19/0	10/0	6/10	0/8	0/5	10/8	4/4	45/31
5.0	14/0	20/0	3/8	0/8	5/9	4/8	5/4	46/35
10.0	15/0	10/0	0/8	0/7	0/8	0/8	2/4	25/31

any loose rock of 5–25 cm diameter, sampled by wrapping the substrate with fine mesh netting and enclosing the sample in a sealable plastic bag. For soft substrates, a coring tube (8 cm diameter) was pushed into undisturbed sediment, capped and withdrawn and the intact core put into a plastic storage bag. For sediments that were not cohesive enough to permit this, the substrate within the emplaced coring tube was removed with a large suction pipet into fine mesh bags. Successive cores were taken at least 5 meters apart to avoid diver-disturbed areas.

Samples were processed by placing sediments or rock substrates in buckets of saturated salt solution, separating larvae and pupae by flotation. The low density organisms float to the surface where they can be skimmed and collected, while the substrate material sinks to the bottom. Following flotation, rock substrates were also submerged in hot tap water to drive any remaining larvae from crevices. The surface area of hard substrates was determined in two ways: (1) by outlining the projected upper surface onto a grid (2 dimensional cover area) and (2) by wrapping the entire exposed upper surface with aluminum foil and measuring the area of the foil used (3 dimensional surface area). Processed samples were preserved in 80% ethanol with 5% glycerol and counted into life stage age classes (three larval instars and pupae).

Stations were selected not to represent a particular region of the lake, but to represent the most varied physical habitat types and alkali fly densities found around the lake (Herbst, 1990). Since sampling of both hard and soft substrate types within each station was conducted over a wide and uneven geographic area, at different times within the season of production, and included unequal mixtures of different types of rock, mud and sand substrates, variability is probably exaggerated.

Modeling procedures

The goal of this project was to produce a model of alkali fly abundance in relation to changes in benthic substrate area at different lake elevations. These changes in habitat availability provide one approach to predicting the abundance of insect food supply to birds at varied lake levels. The model derived is probably a conservative assessment of the potential impact of dropping lake level on alkali fly abundance at Mono Lake because it does not incorporate the growth-limiting effects of increased salinity.

The steps followed in constructing this predictive population model were as follows:

(1) Determine the density of alkali fly larvae and pupae on hard and soft substrates at depths to the limit of distribution.

(2) Calculate the total area of both hard and soft substrates within depth zones centered on the depth contours sampled in (1) for lake levels from 6360 ft to 6390 ft (1939 m to 1948 m). The data of Stine (1988), who determined by planimetry the area of hard and soft substrates from bathymetric and aerial surveys, were used to calculate the amount of substrate habitat available within the zone between the surface and 15 m depth for each lake level.

(3) Population abundance at a given lake level (P_L) was estimated as the product of density and the area occupied by each substrate type, summed over all depth intervals for a given level:

$$P_L = \sum_d \sum_s D_{sd} A_{sd},$$

where

D_{sd} = density (no. or g/m^2) of larvae and pupae on substrate s at depth d

and

A_{sd} = area (m^2) of substrate type s within depth interval d;

d = depth interval (5 intervals);

s = substrate type (2 types).

Depth intervals were delimited by the mid-point between sample depths: 0–0.75 (represents 0.5 m), 0.75–2 (= 1 m), 2–4 (= 3 m), 4–7.5 (= 5 m), and 7.5–15 (= 10 m). No estimates are available for the areas that would be occupied by different substrate types at elevations above 6390 ft (1948 m), so estimates of P_L were not possible above this lake elevation. Elevation is reported in feet (though metric equivalents will also be given) because of the widespread convention of reference to these units in water management planning (e.g. in the environmental impact report in preparation for the California State Water Resources Control Board).

Densities on hard or soft substrates are expressed both as the numerical density (total number of individuals m^2) and as biomass density (grams dry weight of all life stages m^2). Station means showed positively skewed distributions, with some zero values, and so were log-transformed to normalize distributions and equalize variances. The unweighted geometric mean density for a given substrate (s) at a given depth interval (d) was calculated as:

$$D_{sd} = \text{antilog}\left[\frac{\sum_j \log(M_{jsd} + k)}{N_{sd}} + (1.15)(V_M)\right] - k$$

where

N_{sd} = no. stations with substrate s present at depth d,

M_{jsd} = arithmetic mean density per m^2 on substrate s at depth d, for station j,
   = $\sum_i X_{ijsd}/N$

V_M = variance of the log $(M_{jsd} + k)$ values; 1.15 V_M is a correction factor to make D_{sd} closer to the arithmetic mean (Elliot, 1977, p. 33),

X_{ijsd} = density (per m^2) in ith sample from substrate s and depth d at station j,

N_{jsd} = no. samples taken at station j for substrate s at depth d,

and

k = minimum non-zero value of M_{jsd} possible. For biomass density (mg dry wt/m^2), $k = 0.13$ on hard, 0.4 on soft; for numerical density (no./m^2), $k = 7$ on hard, 20 on soft (assuming 10 cores of 50 cm^2 each, or 20 rocks of 75 cm^2 each, per station). Bio-

mass conversions (mg dry wt ind^{-1}) were as follows: instar 1- 0.02, instar 2- 0.16, instar 3- 2.5, and pupae- 2.0.

Results

Depth distribution of larvae and pupae

On hard substrates, biomass and numerical densities of larvae and pupae were greatest at depths of 0.5 and 1 m (about 100 g m^2), lower at 3 and 5 m (about 30–40 g m^2), and more than an order of magnitude lower at 10 m (less than 1 g m^2) (Fig. 2). Soft substrates harbored far lower densities of larvae and pupae than hard substrates at any depth except 10 m (Fig. 2). Biomass on soft substrates was somewhat higher at the most shallow depth (5 g m^2 at 0.5 m), and more uniform in deeper water (1.5 to 2.5 g m^2 from 1 to 10 m depth). The distribution of sampling by station, depth and substrate type are given in Table 1.

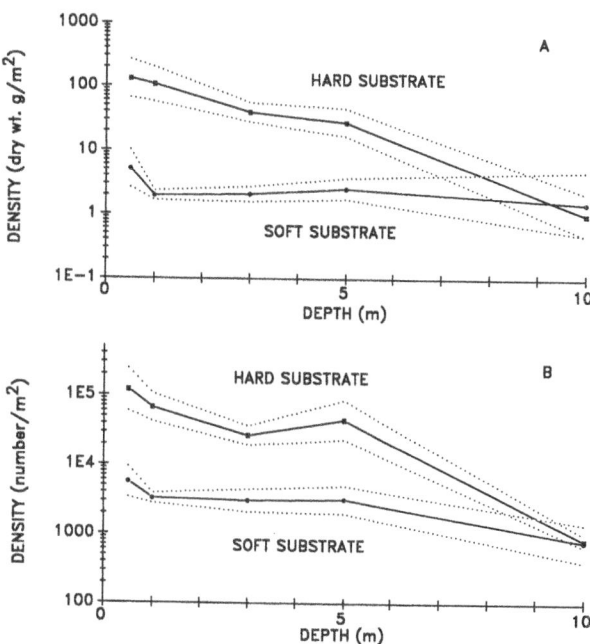

Fig. 2. Depth distribution of biomass density (A) and numerical density (B) for *Ephydra hians* larvae and pupae on hard and soft substrates. Dashed lines connect values representing $D_{ds} \pm 1$ SE. Sample size equal to the number of stations sampled for that depth and substrate type (N_{sd}, Table 1).

196

Eggs were also collected in samples, but not quantified. A few eggs were found even as deep as 10 m, but they were far more common on shallow substrates. Though eggs appear to be deposited by adult females crawling underwater, eggs can also be found in the water column, presumably dislodged by wave action or perhaps laid by adults on the water surface. In any case, eggs may drift down into deeper water, as may larvae, but subsequent survival is probably poor.

The condition of pupae also deteriorates with increased depth. Fully formed pupae found at depths of 5 m or more often contained unemerged adults in a decomposed condition. Few healthy pupae or eclosed pupa cases were found, suggesting pupae forming in deep water often fail to complete development and emerge.

Substrate area at elevations between 6360 ft and 6390 ft

The total lake bottom area of Mono Lake is reduced by about 33% for a drop in lake level from 6390 ft to 6360 ft elevation (Fig. 3). However, 83% of the hard substrate area is exposed by the same drop in lake level, and nearly half of the area

of this habitat is lost when the level drops from 6380 ft to 6370 ft. This restriction of most hard substrate within a narrow range of elevation contours is a prominent feature of the physical environment of Mono Lake. The association of fly larvae and pupae with tufa and other hard substrates portends the importance of this elevation interval as habitat for the production of *E. hians*.

Population model: density × substrate area

A model of alkali fly abundance was derived by combining the results of the density distribution with depth and the area of substrate available at different lake levels (Fig. 4). The model varies with physical habitat availability and predicts that the abundance of the alkali fly is maximized at 6380 ft lake surface elevation. This population maximum at 6380 ft coincides with the elevation where there is the greatest area of hard substrate in shallow water, where densities were found to be highest and most variable. Above and below this level, abundance is projected to decrease in association with the limits of hard substrate habitat availability. Below an elevation of 6372 ft the mean abundance is projected to drop to less than half that predicted for maximum abundance at 6380 ft.

Discovery of the mineral gaylussite

During early dives into deep water sampling locations, large crystals of an unknown mineral were discovered encrusting the surface of a variety of substrates. Subsequently, it was determined from mineralogical tests that these crystals were the evaporite mineral gaylussite [$Na_2Ca(CO_3)_2$ $5H_2O$]. After further diving and observation, it became clear that the crystal size and extent of gaylussite deposited varied with depth. With increasing depth, crystals were both larger and more extensive in coverage on rock surfaces such as pumice. In addition, encrustation on pumice in shallow water (0.5 and 1 m) consists mainly of tufa rather than gaylussite. This, coupled with the observation of large aragonite tufa crystals in deep

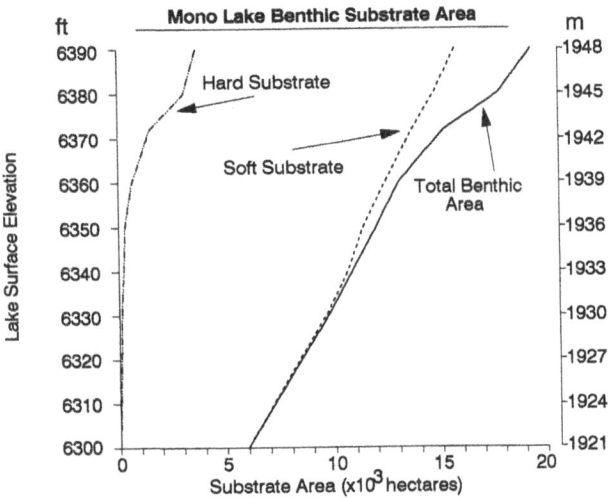

Fig. 3. Estimated areas of benthic hard and soft substrates for Mono Lake between elevations 6300 ft to 6390 ft. Hard substrate assumed to be absent when lake elevation at 6300 ft. Data from Stine (1988, Table 2).

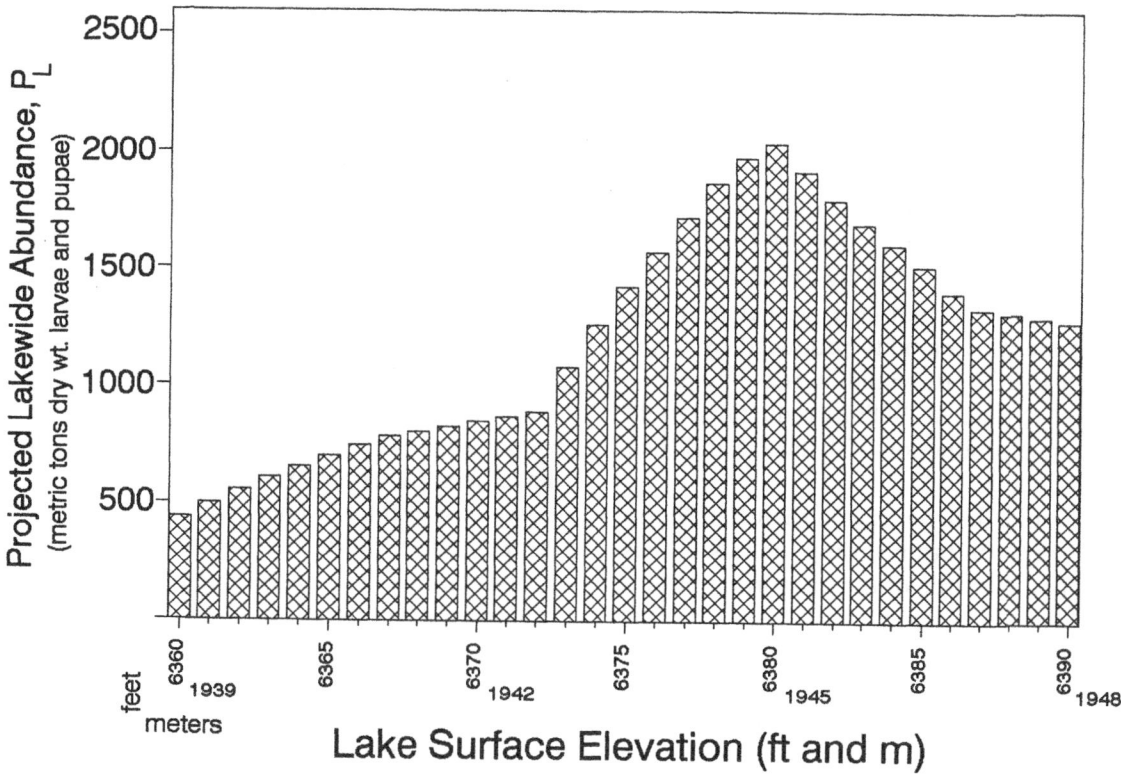

Fig. 4. Predicted lakewide standing stock of the alkali fly at Mono Lake for different lake elevations.

water, occurring inside gaylussite crystals of identical shape, suggests that much of the encrusting tufa formed in Mono Lake originates through the process of pseudomorphism from gaylussite (Bischoff *et al.*, 1991).

Discussion

As depth increases, habitat conditions were expected to become more unfavorable, so it is not surprising that we observed a decrease in the density of larvae and pupae with depth. Among the factors limiting this distribution are decreased light and thus decreased algal production, decreased temperature and oxygen availability, and reduced area of tufa available as habitat. Using vertical temperature profiles, Herbst (1990) constructed a degree-day model for alkali fly development that predicted sufficient heating for one or more generations per year at depths down to

5 m, but cumulative temperature usually did not permit production of even one generation at 10 m depth.

Hard substrates clearly harbor the highest densities, usually 10–20 times those found on soft substrates, and are particularly important for the attachment of the sessile pupa stage. Even considering the greater area of soft substrates found in the lake, the biomass present on soft substrate comprises less than 10% of the total at all but the lowest lake level projected (6360 ft), where it approaches only 20% of the total.

An evaluation of the model developed here to project alkali fly abundance over varied lake levels would be possible if past census information on flies, or diets of their avian predators were available. Unfortunately, earlier observations of the numbers of flies or birds at Mono Lake are largely anecdotal. Entomologists and naturalists visiting the lake in the 19th and early 20th century reported wide bands of adult flies and vast wind-

198

rows of pupae along the shore (Brewer, 1930; Aldrich, 1912). These observations provide no quantitative basis on which comparisons can be made however. Herbst (1988) observed an increase in the abundance of *Ephydra hians* larvae and pupae from 1983 to 1984 coinciding with a period of rising lake levels and reduced salinity. Lake level rose from 6377 ft to 6380 ft over this period and salinity declined by about 5 to 10 g l^{-1}. The model developed here predicts that the increased habitat availability produced by a rise from 6377 ft to 6380 ft, would increase fly population size by 15 percent. Salinity reduction would further enhance production of fly larvae and their algal food sources.

The California Gull, nesting on islands in Mono Lake, feeds both on brine shrimp and alkali flies (all life stages). Records of the proportion of flies in the gull diet (D. Shuford, pers. comm.) show that from 1976 until 1982, while lake level and lake bottom habitat area had been declining, gull diet was never more than about 5% flies. With a dramatic rise in lake level beginning in 1982 and persisting through 1988, the proportion of flies in the diet increased to between 20–50% of the total, and was about 20% in 1989 when the lake level began dropping again (still well above the low in 1982). Bird diets may provide a useful sampling tool for evaluating the availability and quality of prey. Both my own observations between 1983–1984 and the records of gull diet changes over a longer time period are consistent with the predictions of the population model developed here.

The model makes some simplifying assumptions about the relation between physical substrate availability and population abundance. In order to define limitations and identify refinements to the model, an explicit examination of assumptions is necessary.

(1) Substrate is limiting to population density, and as tufa availability changes with fluctuating lake level, population size will respond proportionately. The question is, can crowding occur on the limited hard substrates as lake level declines, without harming survival or growth rate? The model assumes that the present density is at some constant equilibrium carrying capacity. An alter-

native assumption, that population size remains constant, would require that as lake level rises above or falls below elevation 6380 ft, densities increase as larvae and pupae crowd onto limited areas of shallow hard substrate. Between the reference elevation of 6375 ft and 6380 ft, density would actually have to decrease for a lake level rise over this range if population size were to remain constant.

(2) No salinity effect has been incorporated over the elevation range examined by the model. Salinity is 73.8 g l^{-1} at 6390 ft and 127.2 g l^{-1} at 6360 ft (Vorster, 1985). This is a wide salinity range, over which impaired growth and survival have been observed (Herbst, 1992), so this is not a valid assumption. The effect of salinity over this range could be added as a refinement to the population model by considering relative larval survival and other developmental effects. Since densities measured at 6375 ft (the elevation during this study) were taken as the reference point of the model, a model incorporating the effects of salinity would predict total population size to be lower at elevations < 6375 ft, and higher at elevations > 6375 ft relative to the predictions in Fig. 4. The further the lake level is from the reference elevation, the greater is the disparity between the two models. The net effect of such a refinement would be to elevate the lake level at which the predicted population size would be maximized.

(3) There is no new hard substrate forming (or disappearing) as lake level either rises or falls. Tufa formation from springs may occur only around the shallow margins of the lake according to a recent model suggesting that dense interstitial saline water in lake sediments prevents fresh water from entering via sublacustrine springs in all but the shallows where the gradient is more favorable (S. Dreiss & D. Rogers, pers. comm.). Springs would migrate with the changing shoreline and deposit new tufa at lower or higher lake levels.

During these studies, primary formation of the mineral gaylussite was discovered in Mono Lake. This may have important implications for both the origin of tufa habitat and brine evolution in alkaline lakes (Bischoff *et al.*, 1991). Through a process of mineral transformation known as pseudomorphism, the sodium and water are lost from the gaylussite parent mineral and the calcium and carbonate redistributed without loss of the original structure, to form aragonite tufa crystals. Because tufa has been shown to be the most important physical habitat feature for the alkali fly, this process of formation from gaylussite may play an important role in determining where and how tufa substrate becomes available in Mono Lake. Furthermore, precipitation of gaylussite requires the removal of dissolved ions from solution into the solid phase crystal form. This could affect the proportions of different dissolved salts present as the salt concentration of Mono Lake changes.

The rate at which gaylussite transformation and tufa encrustation occurs is unknown. It is uncertain that enough new habitat would ever be created to be make a significant addition to that already mapped. Furthermore, gaylussite precipitation requires that some hard substrate already be present to provide a nucleation surface, and would only add layers to formations already present.

Submerged vegetation can also serve as 'hard' substrate in that it provides attachment sites for pupae (Herbst, 1990). At high lake levels where tufa becomes less abundant in shallow water areas, substantial areas of vegetation could become submerged (e.g. *Distichlis spicata*), contributing to habitat enhancement. From the middle of the last century until about 1920, the level of Mono Lake was rising (Herbst, 1988), inundating large areas of dense vegetation, and providing extensive alternative habitat for the attachment of pupae.

(4) The area of hard and soft substrate above 6390 ft is unknown. This could be determined from aerial photographs (as Stine did for 6375 ft to 6390 ft), and would permit the population model developed here to be extended to elevations above 6390 ft. Such expanded mapping could also incorporate the zones of vegetation that would become submerged.

(5) Standing stock abundance (the population index of the model) is representative of produc-tivity. Productivity, the annual sum of biomass produced by a population, is the most useful measure against which changes in a population should be compared. Standing stock may not represent how productivity is changing. A productivity model algorithm would be based on factors affecting rates (e.g. salinity, food and temperature affects on growth) rather than density. Physical habitat availability is appropriately modeled here as the driving variable for density, yielding standing stock as a relative indicator of population size. A productivity model should be developed for forecasting the effects of factors affecting population growth rates. Density could remain constant while the productivity varied, if turnover rates change with salinity.

(6) Lake surface area is an adequate representation of actual lake bottom area. The planimetered contours of lake surface area at different elevations is the basis for lake bottom area estimates (this is a 2-dimensional projection onto an inclined surface – the lake bottom). The steeper the lake bottom, the closer together are contour lines, and the greater the underestimate of the actual surface area of the lake bottom. This is actually a simple trigonometric problem that could be corrected (the actual area is the hypotenuse rather than the surface leg of the right triangle). However, the underestimate may not affect absolute values much (1.5% for the steepest slopes of 10 degrees), and are otherwise negligible because slopes are similar over the 6360 ft to 6390 ft elevation range.

(7) Maps indicating an outlined area is covered entirely by one substrate type may not be accurate. A map showing a certain area as 'tufa' is unlikely to be 100% tufa, and mud/sand areas are also unlikely to consist entirely of that substrate class. Finer resolution of actual area covered by different substrates could be provided by transect surveys of these areas.

The standing stock densities found in this study are similar to previous estimates (Herbst, 1990), and place *Ephydra hians* density at Mono Lake among the highest of any saline aquatic ecosystem known (Herbst, 1988). An important ecosystem management question that remains however, is how do these densities and overall abundance compare with those that are limiting to bird feeding? This issue, and refinement of the population model presented here, will be addressed in an environmental impact assessment being prepared for the California State Water Resources Control Board.

A balancing of resource values, including both the ecological and economic values of water, will underly the development of a management plan for Mono Lake. The results of the present study may be used to determine the water needed to sustain the ecological value of Mono Lake as a wildlife habitat. Though other refinements to the lake bottom habitat model developed here should be incorporated (such as the effect of salinity), the lake level predicted in the absence of salinity effects to provide habitat conditions maximizing fly abundance, and therefore food abundance for birds, is 6380 ft. Outside the elevation range 6373 ft to 6385 ft little population change is predicted. These levels provide one guideline for balancing management policy.

Over the short-term, the exposure of lake bottom habitat by dropping lake levels is likely to be a more significant impact than the more slowly developing long-term effects of increased salinity. Five feet above or below 6375 ft for example, would increase or decrease the availability of hard substrate habitat by 40 percent. Over the same range, salinity would decrease or increase by only about 10 percent. For this reason, the results of the depth distribution population model may provide useful short-term management objectives for defining the optimum range of lake levels that will sustain a productive wildlife habitat.

A study commissioned by the California Legislature concluded that an elevation of 6382 ft would protect ecosystem values at Mono Lake (Botkin *et al.*, 1988). The US Forest Service, in its management plan for the Mono Basin Scenic Area (1989), recommends a lake level range of 6377 ft to 6390 ft be maintained. The results of the present study are consistent with, and provide independent support of these conclusions and recommendations. This is the first model to simulate population changes at different levels of Mono Lake and establishes a conceptual basis for further defining the conditions that will preserve ecological values.

Acknowledgements

This research was supported by grants from the Policy Research Program of the California Policy Seminar (University of California), and the Mono Lake Foundation.

Without aid of the following dive partners and helpers, research into this unknown realm could not have been completed: Joel Axelrad, Dave Carle, Larry Ford, Larry Miller, Curtis Milliron, Stephen Osgood, and Richard Perloff. For use and handling of boats we thank Dave Carle and the State Tufa Reserve, Chuck Culbertson, Gayle Dana, Lee Dyer, and Curtis Milliron and the California Department of Fish and Game. Helpful assistance in several shallow water dives was provided by Melanie Findling, Cindy Findling, and Dan Tolson. Stuart Hurlbert contributed valuable advice on data analysis.

References

Aldrich, J. M., 1912. The biology of some Western species of the dipterous genus *Ephydra*. J. N.Y. ent. Soc. 20: 77–99.

Bischoff, J. L., D. B. Herbst & R. J. Rosenbauer, 1991. Gaylussite formation at Mono Lake, California. Geohim. Cosmochim. Acta 55: 1743–1747.

Botkin, D. B., W. S. Broecker, L. G. Everett, J. Shapiro & J. A. Wiens, 1988. The Future of Mono Lake. Water Resources Center, University of California, Riverside. Report no. 68.

Brewer, W. H., 1930. Up and Down California in 1860–1864. Berkeley: University of California Press. Repr. 1966.

Elliot, J. M., 1977. Some methods for the statistical analysis of samples of benthic invertebrates. 2nd edition. Freshwater Biol. Assoc. publication no. 25. Ambleside, England. 160 pp.

Gore, J. A., 1978. A technique for predicting in-stream flow requirements of benthic macroinvertebrates. Freshwat. Biol. 8: 141–151.

Gore, J. A. & R. D. Judy, 1981. Predictive models of benthic macroinvertebrate density for use in instream flow studies and regulated flow management. Can. J. Fish. aquat. Sci. 38: 1363–1370.

Herbst, D. B., 1988. Comparative population ecology of *Ephydra hians* Say (Diptera: Ephydridae) at Mono Lake (California) and Abert Lake (Oregon). Hydrobiologia 158: 145–166.

Herbst, D. B., 1990. Distribution and abundance of the alkali fly (*Ephydra hians* Say) at Mono Lake, California (USA) in relation to physical habitat. Hydrobiologia 197: 193–205.

Herbst, D. B, 1992. Changing lake level and salinity at Mono Lake: Habitat Conservation Problems for the Benthic Alkali Fly. In: C. Hall (ed.) History of Water in the Eastern Sierra. Volume 4 of the White Moutain Research Station Symposia. University of California, Los Angeles 4: 198–210.

Herbst, D. B., F. P. Conte & V. J. Brookes, 1988. Osmoregulation in an alkaline salt lake insect, *Ephydra (Hydropyrus) hians* Say (Diptera: Ephydridae) in relation to water chemistry. J. Insect Physiol. 34: 903–909.

Herbst, D. B. & T. J. Bradley, 1989. A Malpighian tubule lime gland in an insect inhabiting alkaline salt lakes. J. exp. Biol. 145: 63–78.

National Academy of Sciences, 1987. The Mono Basin Ecosystem: Effects of Changing Lake Levels. National Academy Press, Washington D.C. 272 pp.

Stine, S., 1988. Geomorphic and geohydrographic aspects of the Mono Lake controversy. Appendix report to The Future of Mono Lake. D. Botkin *et al.* (eds). Report No. 68, Water Resources Center, University of California.

US Forest Service, 1989. Mono Basin National Forest Scenic Area – comprehensive management plan. Inyo National Forest. 58 pp.

Vorster, P. 1985. A water balance forecast model for Mono Lake, California. US Forest Service Region 5. Monograph No. 10, 350 pp.

Wrona, F. J., P. Calow, I. Ford, D. J. Baird & L. Maltby, 1986. Estimating the abundance of stone-dwelling organisms: a new method. Can. J. Fish. aquat. Sci. 43: 2025–2035.

Salinity tolerance of the copepod *Apocyclops dengizicus* (Lepeschkin, 1900), a key food chain organism in the Salton Sea, California

Deborah M. Dexter
Department of Biology, San Diego State University, San Diego, California 92182, USA

Key words: salinity tolerance, Salton Sea, copepod, *Apocyclops dengizicus*

Abstract

The copepod *Apocyclops dengizicus* is a key item in the food chain of the Salton Sea where the salinity is currently 45 g l^{-1}. The salinity of the Salton Sea may reach 90 g l^{-1} within the next 20 years. This study examined the salinity tolerance of this copepod.

Large copepodite and adult *A. dengizicus* were introduced into various salinities with and without acclimation. The 96 h LC_{50} without acclimation was 101 g l^{-1}. Mortality (at 96 h) without acclimation was low at salinities of 90 g l^{-1} or less.

Copepod cultures were maintained, with successful reproduction of at least one new generation, at salinities of from 0.5 to 68 g l^{-1} for at least 120 days. Copepods maintained at higher salinities, up to 79 g l^{-1}, remained alive up to 90 days, but a new generation was not produced. In laboratory studies of larval production and survivorship, few nauplii were released at salinities of 68 g l^{-1} or higher, and none survived to the copepodite stage.

Introduction

The Salton Sea (33° 25′ N, 115° 50′ W), the largest lake in California, was created between 1905–1907 by the accidental diversion of flood water from the Colorado River. The salinity of the Salton Sea has gradually increased from 3.6 in 1907 (Walker, 1961) to 45 g l^{-1} in 1991. Currently almost all freshwater input is wastewater from agricultural uses. Due to U.S. governmental mandated water conservation measures, the lake salinity may reach 90 g l^{-1} by the year 2010 (Black, 1983). This will create a major change in the lake ecosystem.

The current ecosystem is a simple one consisting of several species of phytoplankton, an unknown number of protozoans, a few species of metazoan zooplankters, at least 3 benthic invertebrates of marine origin, a few species of insects, and 7 species of fish. The Salton Sea also supports the greatest diversity of birds of any U.S. National Wildlife Refuge. Present predictions suggest that a key benthic invertebrate, the polychaete *Neanthes succinea* (Frey & Leuchart) and the 3 main sport fishes – orangemouth corvina, (*Cynoscion xanthulus* Jordan & Gilbert), sargo (*Anisotremus davidsoni* (Steindacher)), and croaker (*Bairdiella icistius* (Jordan & Gilbert)) – will become extinct by the time the salinity reaches 50 g l^{-1} (Black, 1983).

What organisms will survive these predicted salinity changes? Will an adequate food chain exist so that the remaining fish species can support large populations of birds?

The rotifer *Brachionus plicatilis* (Müller) and the copepod *Apocyclops dengizicus* (Kiefer) are permanent members of the zooplankton. *Brachionus plicatilis* is widely distributed occurring on 6 continents in salinities from brackish to 250 g l^{-1} (Hammer, 1986). The major components of the merozooplankton are fish larvae, larvae of the barnacle *Balanus amphitritite saltonensis* (Darwin) and larvae of the polychaete *N. succinea*. Studies by Kuhl and Oglesby (1979) suggest that reproduction in *N. succinea* will not be successful at salinities exceeding 50 g l^{-1}. *B. amphitrite* has been collected in salinities up to 75 g l^{-1} (Simmons, 1957), but its tolerance to higher salinities has not been examined.

The other major zooplanktonic species is *Apocyclops dengizicus*. This species was described by Keifer (1931) as *Cyclops dimorphus*; but subsequent taxonomic revisions placed the species in the genus *Apocyclops* (Lindberg, 1940). Kiefer (1967) synonymized *A. dimorphus* with *A. dengizicus* (Lepeschkin, 1900). *A. dengizicus* is also known from inland saline lakes in Australia, Egypt, Haiti, India, Iran, Iraq, and Kazakhstan (Kiefer, 1967). While most species of *Apocyclops* occur at salinities less than 30‰, an undescribed Australian species occurs in salinities as high as 152 g l^{-1} (Hammer, 1986; Timms, 1993). *A. dengizicus* lives in Australian lakes with salinities ranging from 4 to 69 g l^{-1} (Timms, 1993). Carpelan (1961) suggested that development from egg to sexual maturity in *A. dengizicus* took between 10–14 days, and that 10–15 generations of copepods were produced each year at the Salton Sea when it had a salinity of 34 g l^{-1}.

This study focuses on several aspects of salinity relationships in *Apocyclops dengizicus*. These include determination of (1) short term salinity tolerance, (2) long term salinity tolerance, (3) the relationship between the size of gravid females and the number of eggs carried, and (4) the effect of salinity on larval survivorship and metamorphosis.

Methods

Water collected at the Salton Sea was transported to San Diego State University and evaporated outdoors until maximum experimental salinities were obtained. The water was filtered through a 120 μm mesh net, then through a 35 μm mesh net, diluted with de-ionized water to obtain the desired salinities, placed in 4 liter plastic containers, and kept aerated until the start of the experiments. In short term salinity tolerance studies pure de-ionized water was used for the lowest salinity (0 g l^{-1}). In long term salinity cultures water of 0.5 g l^{-1} salinity was obtained by mixing pond water filtered through a 35 μm mesh net as a source for nutrients and phytoplankton with de-ionized water. Phytoplankton were present at all salinities and growth was encouraged with the addition of fish food pellets (Pet Co., Koi's Choice) as a nutrient source.

Salinity was determined with a Reichert-Jung refractometer (0–160‰). This refractometer is calibrated for pure NaCl solutions. Using additions of appropriate salts to, and dilutions, of Salton Sea water and gravimetric determinations, C. Hart & M. Gonzalez (pers.comm.) obtained correction factors for converting to true salinity over the range 10–95 g l^{-1} and these have been used in this study.

Short term salinity tolerance

Salton Sea water was evaporated to obtain a maximum salinity of 102 g l^{-1} in Experiment 1 and 136 g l^{-1} in Experiment 2. Three experimental units (glass dishes) were set up at each salinity in each experiment. Salinities in Experiment 1 were: 0, 11, 23, 34, 45, 57, 68, 79, 90 and 102 g l^{-1}; salinities in Experiment 2 were: 45, 79, 90, 102, 113, 124, and 136 g l^{-1}. Glass dishes 10 cm in diameter filled with 200 ml of water were randomly placed (5 rows of 6 dishes in Experiment 1; 3 rows of 7 dishes in Experiment 2) under a light bank providing 12 hour day/12 hour night cycle with a range of 60–70 microeinsteins m^{-2} s^{-1}. Neither food nor aeration were provided during the experiment. Water temperature during the experiment varied between 23 and 25 °C.

Copepods were collected from Bombay Beach, Salton Sea on Oct. 2, 1990 and maintained in water from that site (45 g l^{-1}) for 2 days (Experiment 1) or 10 days (Experiment 2). Adult and late stage copepodites attracted by light to the top of the container were collected for experimentation.

A 2 ml sample of these concentrated copepods was introduced into each experimental unit in random order; the total number of copepods introduced was determined at the end of the experiment, but averaged 65 (Experiment 1) and 113 (Experiment 2) per container.

At 6, 12, 24, 48, 72, and 96 h experimental units were checked under a dissecting microscope for mortality. Organic debris and molts were removed at each observation, and dead copepods were counted and removed. Upon termination of the experiment the samples were fixed, and the surviving copepods counted.

The results from the three highest salinities in both experiments were used to determine the 96 h LC_{50} using probit analysis (Litchfield, 1949), *i.e.* the estimated salinity at which exactly 50% of the original individuals would remain alive after 96 h.

Long term salinity tolerance

Zooplankton tows were taken with a 110 μm net along the Salton Sea shoreline, usually at Red Hill Marina. Within 24 hours of collection, 100 ml of concentrated zooplankton, mostly *A. dengizicus*, were introduced into aerated plastic containers with 1.6 liters of Salton Sea water adjusted to various salinities (methods of obtaining salinities described above). Each container received 300 ml of mixed phytoplankton culture previously maintained at that salinity for at least 2 weeks. In initial cultures copepods were introduced without acclimation and without replication into salinities of 0.5, 1, 6, 11, 17, 23, 28, 34, 40, 45, 51, 62, 68, 73, 79, 85, 90, 96, 102, and 107 g l^{-1}. In subsequent cultures, three replicates were set up at 0.5, 1, 11, 28, 45, 51, 68, and 79 g l^{-1}, and single cultures at 90, 102, and 107 g l^{-1}. All copepods to be maintained in cultures at salinities greater than 57 g l^{-1} were first introduced into a salinity of 57 g l^{-1}. Salinity was increased by approximately 11 g l^{-1} every 3–7 days by salt addition until final salinities were attained.

Salinities were monitored every 15 days and adjusted by addition of deionized water. Water temperature varied between 20 and 25 °C. Phytoplankton and protozoans were present at all salinities and growth was encouraged with the addition of approximately 0.2 grams biweekly of fish food pellets as a nutrient source. At 30, 60, 90, and 120 days, each culture was gently filtered through a 35 μm mesh net, examined under a dissecting microscope for abundance and presence of life history stages, and returned to its respective container.

Gravid females maintained in the long term 2 liter batch cultures at salinities of 1, 11, 28, 45, 57, and 68 g l^{-1} for at least 60 days were collected for examination. Twenty gravid females selected haphazardly from each salinity were isolated into depression slides, and methyl cellulose was added to slow movement. Right and left egg clutches were separated under dissecting microscope at 60× and counted at 120×. Egg size was not measured. The female was placed under a compound microscope at 100× and the length of the cephalothorax measured with an ocular micrometer to the nearest 12 μm.

Larval survivorship and metamorphosis

A total of 124 gravid females freshly collected from Red Hill Marina, Salton Sea were isolated within 24 h of collection and directly immersed into small dishes containing 6 ml of Salton Sea water adjusted to salinities of 1, 11, 45, 57, 68 and 79 g l^{-1}. Containers were examined daily to determine the presence of nauplii. Females were removed from the containers as soon as nauplii were observed, as predation upon nauplii by adults is well known. Nauplii were followed until metamorphosis into the copepodite stage or death (0–40 days).

Results

Short term salinity tolerance

After 96 h, mean cumulative mortality was 10% or less for all salinities < 102 g l^{-1} (Fig. 1). It was

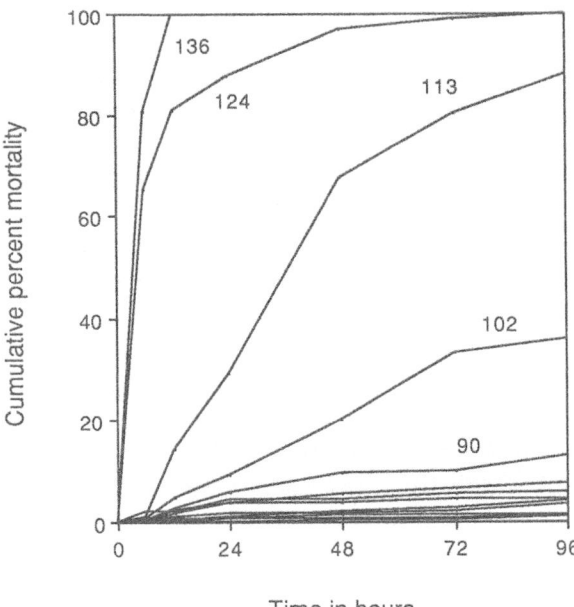

Fig. 1. Cumulative mortality of *Apocyclops dengizicus* through time at various salinities. The lowermost curves are for 79, 0.5, 68, 57, 45, 11, 23, and 34 g l⁻¹, respectively, and are not detectably different from each other.

38.5% at 102 g l⁻¹, 92.4% at 113 g l⁻¹, and 100% at both 124 and 136 g l⁻¹. The 96 h LC_{50} was estimated to be 101 g l⁻¹ with a 95% confidence interval of 89–112 g l⁻¹. The mean number of copepods initially present per experimental unit,

determined at the end of the experiment, was 63.5 (range 31–105) in Experiment 1 and 112.8 (range 41–256) in Experiment 2.

Long term salinity tolerance

Copepod cultures at salinities from 0.5 to 45 g l⁻¹ contained individuals of all life history stages (nauplii, copepodites, males, females, and gravid females) throughout the duration of the 120 day experiment (Table 1). Density of copepods was low in cultures at 0.5 g l⁻¹ in which chironomid larvae, introduced as a contaminant with the pond water, were present. When the rotifer *Brachionus plicatilis* was dense in cultures at various salinities, the density of *A. dengizicus* was often low. Otherwise, throughout the 120 days, densities >100 copepods l⁻¹ were maintained at salinities from 1 to 45 g l⁻¹.

Mating was observed at salinities from 1 to 68 g l⁻¹, but gravid females were extremely rare at salinities of 68 g l⁻¹. Cultures survived for 120 days at this salinity, but at very low densities. When adults maintained at 68 g l⁻¹ for more than 60 days were returned to 45 g l⁻¹, however, successful reproduction and larval development occurred and within 30 days, all life history stages were present, and overall density was high. At

Table 1. Long term salinity tolerance of *Apocyclops dengizicus*.

Life history stage	Abundance[1] at				Salinity (g l⁻¹)
	30 days	60 days	90 days	120 days	
Gravid females	+ +	+ +	+ +	+ +	0.5, 1, 6, 11, 17, 23, 28, 34, 40, 45, 51
	+	+	+	+	57, 62, 68
	0	0	0	0	73, 79, 85, 90, 96, 102, 107
Nauplii larvae	+ + +	+ + +	+ + +	+ + +	0.5, 1, 6, 11, 23, 34, 40, 45, 51
	+ +	+	+	+	57, 62, 68
	0	0	0	0	73, 79, 85, 90, 96, 102, 107
Small copepodites	+ + +	+ + +	+ + +	+ + +	0.5, 1, 6, 11, 17, 23, 28, 34, 40, 45, 51
	+ +	+	+	+	57, 62, 68
	+	0	0	0	73, 79, 85, 90, 96, 102, 107
Large copepodites and adults	+ + +	+ + +	+ + +	+ + +	0.5, 1, 6, 11, 17, 23, 28, 34, 40, 45, 51
	+ + +	+ + +	+ +	+	57, 62, 68, 73, 79
	+ +	+	0	0	85, 90, 96, 102, 107

[1] Abundance categories are denoted as follows: 0 not seen, + 1–5/container, + + 6–20/container, + + + >20/container.

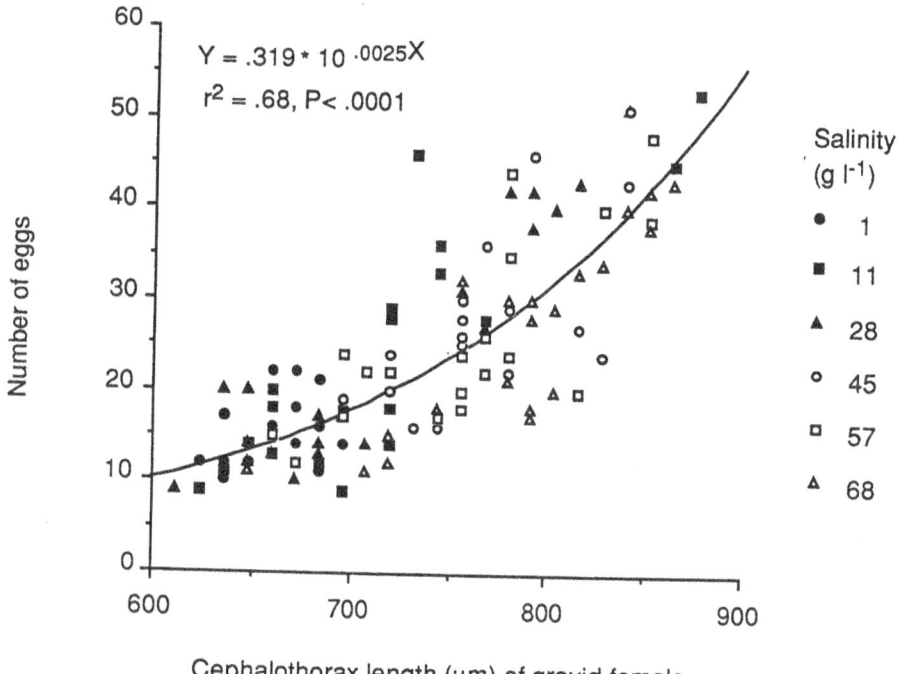

Fig. 2. The relationship between size of gravid *Apocyclops dengizicus* females, number of eggs, and salinity.

salinities ≥ 79 g l^{-1} only large copepodites and adults were present after 30 days, and neither gravid females nor nauplii were seen. Copepods survived salinities up to 79 g l^{-1} for 90 days, and up to 107 g l^{-1} for 60 days, but density rapidly declined with time.

Size of gravid females, number of eggs carried, and relationship to salinity

The number of eggs carried by a female increased with female size at all salinities examined (Fig. 2). Egg number ranged from 9 to 53 and gravid fe-

Table 2. Larval release, survivorship, and metamorphosis, in relation to salinity.

Salinity g l^{-1}	Number of gravid females	Number of nauplii released	Mean days until metamorphosis (range)	Number of copepodites	% of nauplii surviving to copepodite stage
1	27	110	15.2 (5–31)	5	4.5
11	15	104	10.1 (4–18)	63	60.5
45	15	88	7.9 (6–15)	11	16.4
57	18	84	17.8 (6–25)	23	27.4
68	30	68	–	0	0
79	19	1	–	0	0

males ranged from 612–876 μm in cephalothorax length. Salinity did not seem to effect the relationship between female size and number of eggs produced.

Larval survivorship and metamorphosis

There was variability in survivorship of gravid females, nauplii, and copepodites at all salinities (Table 2). Release of nauplii was low at 68 g l^{-1} and only 1 nauplius was released at 79 g l^{-1}. Duration of the nauplius stage ranged from 5–31 days before successful metamorphosis to the copepodite stage. There was no metamorphosis into the copepodite stage at salinities ≥ 68 g l^{-1}.

Discussion

Some of the life history features seen in this study differ from those suggested in earlier studies. Carpelan (1961) reported that *A. dengizicus* was found in the offshore zooplankton in the Salton Sea only during the warmer months of the year (June through December/early January). My studies indicate that it is present all year in the shallow waters along the shoreline, although its density is reduced during in the colder months. Secondly, Carpelan (1961) suggested that generation time was between 10–15 days. In the laboratory experiments *A. dengizicus* took as little as 2 weeks and as long as 2 months to complete the progression from nauplius to sexual maturation and production of a new generation. Thirdly, Johnson (1953) reported a range of 12–16 eggs/ sac while during my studies as many as 27 eggs/ sac were observed. These differences, the slower developmental rate, higher fecundity, and year round presence, suggest a more flexible life history than originally reported.

Robertson *et al.* (1974) concluded that clutch size in *Cyclops vernalis* varied with age of the female, and that food quality influenced fecundity. I made no attempt to determine clutch order in the experiments on *A. dengizicus*, so perhaps some of the variation in the relationship between size of

female and number of eggs is caused by differences in clutch order. Differences in fecundity at the various salinities have not been statistically analyzed, because food quality and quantity were not held constant across salinity levels.

Such variables can confound salinity effects on survival, generation time, and other life history features. Food originated from the Salton Sea at all salinities except at 0.5 and 1 g l^{-1}. Presence and abundance of the rotifer *B. plicatilis* in the cultures was unpredictable and was not correlated with salinity levels. *B. plicatilis* is suitable food for *A. dengizicus* but in cultures where it was very dense, density of *A. dengizicus* was reduced. Perhaps some competition for phytoplankton occurs between these species or perhaps it was the decreased predation by *A. dengizicus* that allow *B. plicatilis* to increase.

Although *A. dengizicus* is able to complete its life cycle at low salinities (0.5 and 1 g l^{-1}) the results suggest poorer performance at these salinities (higher short term mortality – see Fig. 1; lower densities in long term cultures, fewer nauplii surviving to copepodite stage – see Table 2, and small gravid females – see Fig. 2). Greater physiological stress at low salinities is a possible explanation. It is most probable that the experimental conditions were not as good at the lowest salinities. Sources of water (de-ionized and pond) and food (pond phytoplankton) differed from all other cultures where Salton Sea water and phytoplankton were mixed with de-ionized water.

A. dengizicus is quite tolerant to a wide range of salinities. Successful reproduction, larval development, sexual maturation, and production of additional generations occurs easily at salinities from 0.5 to 45 g l^{-1}, and less frequently at salinities up to 68 g l^{-1}. This species has survived short term introduction under laboratory conditions into salinities 2.5 times the salinity in which they were collected. With acclimation *A. dengizicus* adults can live in salinities as high as 107 g l^{-1} for at least 60 days. Completion of the entire life history at salinities above 68 g l^{-1} was not observed in the laboratory studies.

If *A. dengizicus* was introduced into the Salton Sea shortly after its creation, then it survived a

10-fold increase in salinity over a period of 85 years. The predicted increase in salinity at the Salton Sea with subsequent loss of at least 3 of the 7 fish species now present (Black, 1983) will cause a decline in fish predation on *Apocyclops*. Given its short length of life, the occurrence of many generations yearly, and reduced predation it is likely that *A. dengizicus* will remain in the plankton of the Salton Sea for at least a few years.

These initial salinity laboratory experiments on *Apocyclops* show successful acclimation to increased salinity. Whether there is sufficient genetic variability to provide genetic adaptation of this species to even higher salinities is unknown. Natural selection for tolerance to increasing salinity is likely, considering the recent history of the species in the Salton Sea and its short generation time. What is the maximum limit? My data indicate that acclimation allows survival of this species up to salinities of at least 68 g l^{-1}. However, laboratory conditions provide stable physical conditions, constant food, absence of predation, etc. and thus are more optimal for survival than the abiotic and biotic conditions experienced in the natural habitat. Furthermore, as the salinity of the Salton Sea increases, it will likely be colonized by more salt tolerant invertebrates which may prey upon or outcompete *A. dengizicus*. Additional studies are required for accurate predictions of the effect of increasing salinities on this and other Salton Sea species.

Acknowledgements

This research was supported in part by University of California Water Resources Center Grant No. S22625(07) to S. H. Hurlbert. I thank J. Green, U. T. Hammer, D. B. Herbst, S. Twombly, and B. Timms for their positive critical review of this manuscript. The numerous corrections required by Stuart Hurlbert, some accepted by the author only under extreme duress, extended manuscript preparation by at least 6 months, disrupted the serene disposition of the author, and otherwise provided a challenge to maintenance of a long standing friendship.

References

Black, G. F., 1983. Prognosis for water conservation and the development of energy resources at the Salton Sea: Destruction or preservation of this unique ecosystem? In Aquatic Resources Management of the Colorado River Ecosystem, V. D. Adamas & V. A. Lamarra, Ann Arbor Science, Ann Arbor, Michigan, pp. 363–382.

Carpelan, L. H., 1961. The ecology of the Salton Sea, Calif. in relation to the sportfishery. Zooplankton. Calif. Fish. Game Bull. 113: 49–62.

Hammer, U. T., 1986. Saline lake ecosystems of the world. Dr W. Junk Publishers, Boston, 616 pp.

Johnson, M. W., 1953. The copepod *Cyclops dimorphus* Kiefer from the Salton Sea. Am. Midl. Nat. 49: 188–192.

Kiefer, F., 1931. Zur Kenntnis der freilebenden Süsswassercopepoden, insbesondere der Cyclopiden Nordamerikas. Zool. Jahrb. Abt. Syst. Band 61, Heft 5/6: 269–271.

Kiefer, F., 1967. Cyclopiden aus salzhaltigen binnengewässern Australiens (Copepoda). Crustaceana 12: 292–302.

Kuhl, D. I. & L. C. Oglesby, 1979. Reproduction and survival of the pileworm *Neanthes succinea* in higher Salton Sea salinities. Biol. Bull. 157: 153–165.

Lindberg, K., 1940. Cyclopides (Crust. Cop.) de l'Inde. IV. Une révision des représentants indiens et iraniens du sousgenre *Metacyclops* Kiefer du genre *Cyclops* Müller. Rec. Ind. Mus. 42: 567–588.

Litchfield, J. T., 1949. A method for rapid graphic solution of time-per cent effect curves. J. Pharmacol. Exp. Thera. 96: 399–408.

Robertson, A., C. W. Gehrs, B. D. Hardin & G. W. Hunt, 1974. Culturing and ecology of *Diaptomus clavipes* and *Cyclops vernalis*. U. S. Environmental Protection Agency. Research Reporting Series EPA-660/3-74-006, 226 pp.

Simmons, E. G., 1957. An ecological survey of the upper Laguna Madre. Publ. Inst. Mar. Sci. (Port Aransas, Texas) 4: 156–200.

Timms, B. V., 1993. Saline lakes of the Paroo, inland New South Wales, Australia. Hydrobiologia 267: 269–289.

Walker, B. W., 1961. The ecology of the Salton Sea, California in relation to the sportfishery. Calif. Fish Game Bull. 113: 1–204.

Seasonal change in a saline temporary lake (Fuente de Piedra, southern Spain)

Carlos M. García[1] & F. X. Niell
Departamento de Ecología, Facultad de Ciencias, Campus Universitario de Teatinos, 29071 Malaga, Spain
[1] *Present address: Dpto. Biología, Facultad de Ciencias del Mar, 11510 Puerto Real, Cádiz, Spain*

Key words: saline lakes, biomass, seasonality, shallow lakes, phytoplankton, zooplankton

Abstract

Fuente de Piedra saline lake is located in an endorheic basin in the south of Spain. This lake is very shallow (0.5 m max. depth during 1987–88) and relatively large ($\pm$ 1350 ha). It is a temporary playa lake, showing irregular cycles, with frequent seasonal drought and a high degree of unpredictability. The lake was sampled monthly during a relatively rainy year (1987–88, 10.5 months permanence). The result of combined analyses for environmental variables (salinity, temperature and soluble inorganic forms of nitrogen and phosphorus), variables related to biological activity (chlorophyll 'a', sediment organic matter and redox potential) and the direct analysis of the planktonic community, shows the existence of two periods of dominance by autotrophs. The first occurs during winter, exhibits a progressively higher surface to volume ratio for phytoplankton and is followed in the spring by high zooplankton densities (*Moina salina, Fabrea salina*) and very low phytoplankton densities, suggesting the existence of a period with a detritus-based food web. The summer period coincides with a community better adapted to high salinities that is dominated by *Dunaliella salina, D. viridis*, diatoms and the ciliate *Fabrea salina*, and associated with high ammonium concentrations. A new period of organic matter accumulation could be facilitated, in the last moments before the lake dries, by a progressive decrease in zooplankton abundance.

Introduction

Fuente de Piedra lake is, in spite of its shallowness, an important nature reserve in the South of Spain because of its population of nesting flamingoes (*Phoenicopterus ruber*) that reached 12 500 pairs in 1988 (Rendón *et al.*, 1991). Although its vertebrate fauna has been extensively studied (Vargas *et al.*, 1983), there are few data on its aquatic community (Herberg, 1986; Comín & Alonso, 1988; Alonso, 1990). The objective of the present study was to document the large seasonal changes that take place in this system.

Seasonality has been extensively studied in permanent, deeper saline lakes (Hammer, 1986), but few studies have been done on shallow, temporary systems. On the other hand, the phytoplankton to zooplankton biomass ratio (P:Z) has been already used to characterize the community structure among different pelagic ecosystems (Sheldon *et al.*, 1977; Sprules & Knoechel, 1984), including deeper saline lakes (Drabkova *et al.*, 1978; Letanskaya, 1980, Hammer, 1986). Fuente de Piedra Lake is a shallow, saline lake that undergoes seasonal flooding and usual complete drying. The present study of events during a relatively rainy year (1987–88) was interesting because the lake exhibited a wider range of states than usual. The changes in the phytoplankton and zooplankton assemblages are analysed in relation to the seasonal forcing by salinity, nutrients, and other factors and in relation to certain variables that help describe biological activity (chlorophyll, redox potential, organic matter percentage).

Material and methods

Characteristics of the lake

Fuente de Piedra is a saline lake located in an endorheic basin ($\pm$ 152 km^2) in southern Spain (37° 6′ N, 4° 44′ W). Sediments from the Superior Triassic with a high saline content are abundant in the surface lithology of the basin, especially its southeastern part (I.G.M.E., 1986). Chloride and sulphate predominate in the lake waters. This lake has a relatively large surface area ($\pm$ 13.5 km^2) but it is very shallow (maximum depth of 0.5 m during 1987–88), clearly polymictic and with an extremely high mixing potential *sensu* Melack (1981). This year (Oct. 1987–Oct. 1988) was relatively rainy (557.5 mm) compared with the mean 1962–1988 precipitation in the region (463 mm), and had an evaporation rate lower than average (1300 mm *vs* 1313 mm) (I.G.M.E., 1988). The lake contained water longer in 1987–88 (10.5 months) than is normal (average duration for 1962–1988 = 9.4 months, Linares, 1990). Rainfall exceeded evaporation from October to February; this filling period was followed by an evaporation phase that ended when the lake became completely dry on July 19, 1988.

Sampling sites and periodicity

Samples for temperature and salinity analysis were taken at five sites (Fig. 1). Three of them were situated in the north (sites 1, 2, 3) separated by 400 m moving gradually offshore; these samples were taken before noon (10^{00}–12^{00}h). The other two sites were situated in the south, separated by 250 m (sites 4, 5) and were sampled around 13^{00}–14^{00}h. Nutrients and biological variables were measured only for samples from sites 2 and 3. The northern series of sites was closer to influences of the main water inputs. Water depths at the sites varied from 1 to 46 cm. Routine sampling was done monthly, but during autumn and late spring additional samples were taken since changes in the system were faster then.

Sampling and analyses of water and sediment

Water was taken in 1 l flasks at the sampling sites, from the bottom to the surface, except in the warmer and shallower months when a 150 ml syringe had to be used to fill the flasks. Water for nutrient analysis was filtered (Whatman GF/C) immediately after sampling and stored in a freezer. Total dissolved solids (TDS) were measured by drying 100 ml aliquots of filtered lake water (Whatman GF/C). Nutrients were measured with a Technicon II autoanalyzer. Soluble reactive phosphate was analysed following the automated method of Fernández *et al.* (1985). Nitrate and nitrite concentrations were determined after Shinn (1941) and Wood *et al.* (1967). Ammonium concentration was determined after Slawyk & McIsaac (1972).

Concentration of total chlorophyll '*a*' was estimated following Talling & Driver (1963) on the same Whatman GF/C filters used to obtain the particle-free samples for nutrient analysis. The filters were extracted immediately in a refrigerator with sodium carbonate neutralized acetone for 24 h. Chlorophyll concentration in the first cm of sediment was determined after extraction by the same procedure.

Percent organic matter in the first cm of sediment was estimated by loss on ignition at 550 °C (Håkanson & Jansson, 1983). This percentage is calculated on a sediment dry weight basis (105 °C, 24 h). Redox potential was measured with CRISON standard electrodes. Eh at the

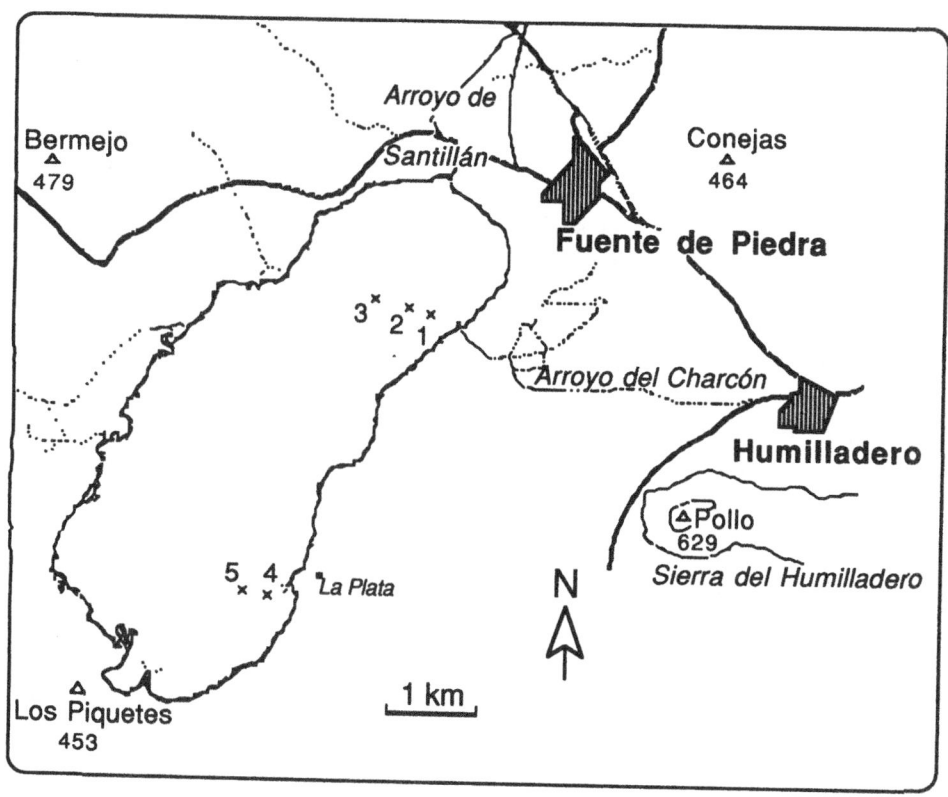

Fig. 1. Fuente de Piedra saline lake and location of the sampling sites. Only data for sites 2 and 3 were used for most analyses reported here.

sediment surface was measured by allowing the poles of the electrode to lie horizontally on the surface of sediment.

Sampling of phytoplankton and zooplankton

Phytoplankton and ciliates samples were collected at sampling site 2 by immersing a 150 ml dark glass bottle 5–20 cm beneath surface, when it was possible. A 150 ml syringe was used to fill the bottles when depth was less. This sample was taken simultaneously with water samples and was preserved *in situ* with a Lugol-acetic acid solution (Parsons *et al.*, 1984).

To sample zooplankton, we drove a 30 cm diameter cylinder into the lake bottom and pumped the water column of known depth through two consecutive meshes with 500 and 100 μm mesh size. This process was repeated on each sampling data at six different points along a 1000 m long transect connecting sites 1 and 3, so each sample was a composite of the six subsamples. Filtered volumes of composite samples varied, volumes being about 100 l from February to April, about 60 l in December, January and May, about 30 l in June and 3 l in July. Samples were concentrated *in situ* and preserved in 5% formol. The 100 μm mesh size fraction of zooplankton samples was refiltered to remove detrital and inorganic particles after cleansing it using an ultrasonic bath (Ultramet II Sonic cleaner) to disperse the particles (Marcus, 1984).

Analysis of plankton samples

Counting and measuring of the planktonic organisms was carried out simultaneously using an A.M.S™ VIDS-IV semi-automatic image analysis system. The image analysis program receives data from a digitizing tablet which is connected to a monitor. The screen shows the plankton sample and the measurements are carried out by the operator using a 'mouse'.

Phytoplankton ($>2\,\mu m^3$) and ciliates were analysed after sedimentation of 2.5–15 ml sub-samples following the Utermöhl (1958) method on an inverted microscope. Crossed diameters were carried out at $1000\times$, $100\times$ and $40\times$ following the recommendations of Lund *et al.* (1958).

The $100\,\mu m$ fraction of zooplankton was analysed in the same way as phytoplankton but taking a subsample volume enough to count 400 to 800 individuals of the more abundant species in an inverted microscope transect at $40\times$.

All of the $500\,\mu m$ mesh size fraction of zooplankton was enumerated in a Bogorov tray (Parsons *et al.*, 1984). Contours of specimens were traced using a drawing tube and biovolumes estimated with the image analysis program (see below). Although all *Moina salina* Daday individuals in the sample for April and May were counted, only 30 of them were measured, using their mean to estimate total biovolume in the sample.

Biovolume was used to express biomass of phytoplankton, ciliates and larger zooplankton since densities approximate $1\,g\,cm^{-3}$. Semi-automatic methods of image analysis can avoid some of the problems that arise using automatic devices (Rolke & Lenz, 1984) and those that arise with a Coulter Counter (Rassoulzadegan, 1979; Rodríguez *et al.*, 1987). Fixation, however, may reduce cell volumes, leading to an underestimate of the real value, at least in ciliates (Choi & Stoecker, 1989).

The image analyser measures directly lengths and projection areas of the organisms in the microscope field. Biovolume was estimated using algorithms that assumed more or less regular shapes of the cells. The operating mode that was habitually used consisted of measuring the long and short axis of the organisms with the 'mouse', and a rotation volume was then calculated. This mode was used for the more extended type of

ellipsoidal and cylindrical shapes. For long filaments a single linear measurement was made and a mean width used to obtain biovolume.

In the same way as biovolume was calculated, the surface area of each cell was obtained from the long and short axis measurements. This allowed estimation of the surface to volume (S:V) ratio for each cell as well as total surface area per unit volume of water sample.

Results

The lake began filling in September and maximum depth (0.49 m) was reached on February 17. The lake became completely dry on July 19. Average depth for the sampling sites (Fig. 2) was always less than 40 cm. Water level varied inversely with salinity and water temperature (Fig. 2). Salinity ranged from $30\,g\,l^{-1}$ in February to $220\,g\,l^{-1}$ in June. TDS could not be measured directly in July because of interference from an oily substance, but was estimated from conductivity measurements as $270\,g\,l^{-1}$. High standard deviations in the warm season indicated a higher spatial heterogeneity, whereas the spatial differences were low in winter, especially in March.

Soluble reactive phosphorus (SRP) concentration was high during winter (December, January),

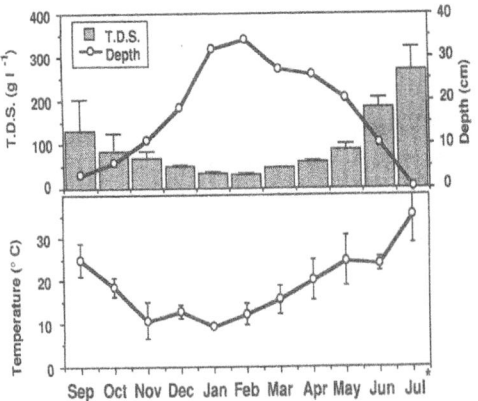

Fig. 2. Seasonal changes in mean depth for sampling sites, total dissolved solids (T.D.S.), and water temperature during 1987–88. Vertical bars show sample standard deviations ($n = 5$).

lower during spring and summer, and high again under the extreme conditions of July, just before the lake dried up (Fig. 3a).

Nitrate concentration was usually low, peaked in autumn and was undetectable in winter (Fig. 3b). Nitrite concentrations were usually

lower than nitrate levels and showed maxima in July and autumn (Fig. 3c). Ammonium was the main form of inorganic nitrogen in the water, with high concentrations in the warmer months (Fig. 3d). Dissolved inorganic nitrogen values followed an almost identical pattern as ammonium, except in early November, when nitrate values exceeded those for ammonium.

The ratio of dissolved inorganic nitrogen to soluble reactive phosphorus (DIN:SRP) was usually very high in the lake (Fig. 3e). The lowest values occurred in January and December, coinciding with a higher runoff and lower Eh values. DIN:SRP was higher during the warm season (16 May) because of the high ammonium concentration. DIN:SRP for July, however, was not measured.

Chlorophyll 'a' concentration in the water was high during winter; a secondary maximum occurred in June which was relatively much lower when expressed per square metre (Fig. 4a). Chlorophyll 'a' in the water reached its lowest concentrations in the spring (28 April). High standard deviations reflected considerable spatial heterogeneity of chlorophyll concentration in the water and sediment (Fig. 4), perhaps attributable to the variable distance of sites, 1, 2 and 3 from the main stream input (Fig. 1).

Sediment organic matter content was high and spatially heterogeneous (high standard deviations) (Fig. 4b). It decreased with the first rains (September to October), then increased and remained high from October to April and declined to low levels for the summer.

Redox potential at the sediment surface (Fig. 4c) increased from negative to positive during the fall, became strongly negative for the duration of winter, then increased to positive values in spring and then decreased to near zero as the lake dried.

Values for sediment chlorophyll concentration fluctuated greatly, especially in autumn, were maximal in March and declined towards the end of the hydrologic year (Fig. 4d). Chlorophyll concentrations in the sediment did not show the sharp minimum in April shown by chlorophyll concentration in the water.

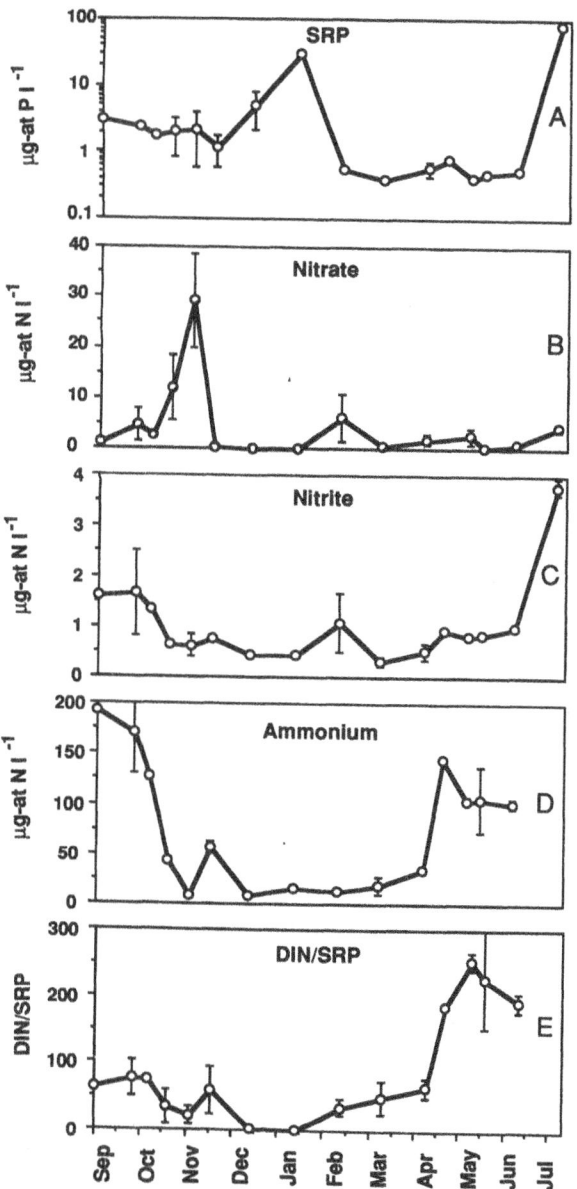

Fig. 3. Seasonal variations in nutrient concentrations and the atomic ratio of dissolved inorganic nitrogen (DIN) to soluble reactive phosphorus (SRP) during 1987–88. Vertical bars represent sample standard deviation (n = 2).

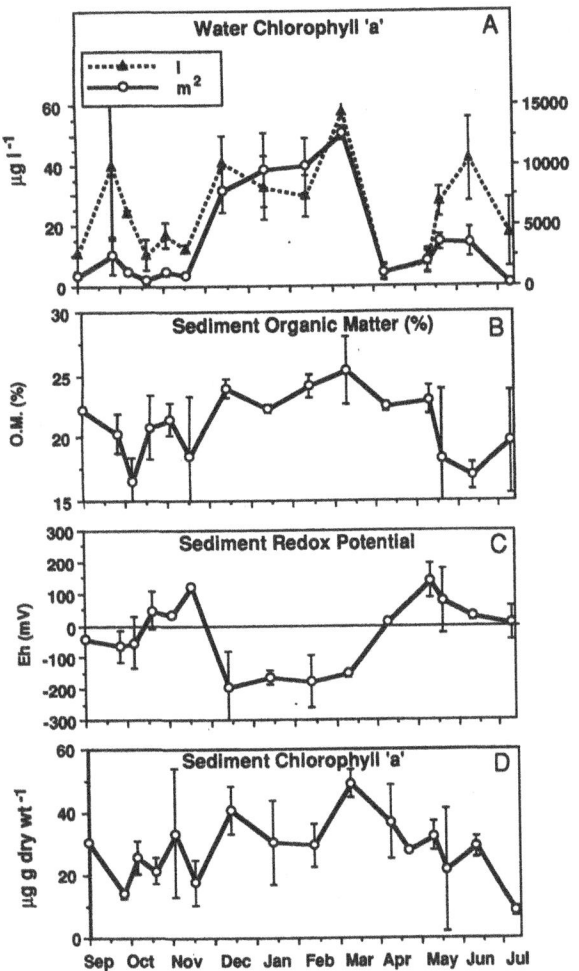

Fig. 4. (a) Chlorophyll '*a*' concentration in the water (b) Percent organic matter in the first centimetre of sediment. (c) Redox potential at the sediment surface. (d) Chlorophyll '*a*' concentration in the first centimetre of sediment. Vertical bars are sample standard deviations (*n* = 2).

Total plankton biovolume per m^2 averaged about 5 cm^3 m^{-2} (Fig. 5a). It was greatest in February when the water column was deeper, and also high in May, coinciding with an increase in the cladoceran *Moina salina* Daday. When water depth decreased, biovolume per m^2 decreased too, reaching a very low value in July (0.05 cm^3 m^{-2}). Biovolume per litre (Fig. 5a) was, however, greatest in June, reflecting a high concentration of plankton in the thin layer of water.

Although total plankton biovolume did not exhibit a distinct trend over most of the year, the proportions represented by the zooplankton and phytoplankton fractions (P:Z ratios) varied considerably (Figs 5b, 5c and 5f). During winter, when the lake was filling, phytoplankton predominated and represented from 50% to 80% total biovolume. From April to June, during the drying period, consumer biovolume was much higher than phytoplankton biovolume. This imbalance was especially remarkable in April and May, when *Moina salina* abounded, but was no longer evident by July, when zooplankton had nearly disappeared.

The ratio of total phytoplankton surface area to total phytoplankton biovolume (S:V; Fig. 5g) usually was < 1 mm^2 mm^{-3} throughout the year, but increased during winter to a maximum near 2 mm^2 m^{-3} in March, when the extremely slender diatom *Nitzschia closterium* (Ehr.) W. Smith predominated.

The biovolume and relative abundance of the main taxa has been shown for both the phytoplankton and zooplankton in Fig. 5b–e. Absolute numerical and biovolume densities of the main taxa are also shown in Fig. 6.

At the beginning of the hydrological cycle (October), *Dunaliella salina* Teod. and *D. viridis* Teod. were abundant (Figs 5d, 6). Their populations decreased as the water level rose, and then increased again as salinity increased the following spring (Figs 2, 6).

Gymnodinium cf. *excavatum* Nyg. was abundant from November to January and again in April, being the dominant phytoplankter at those times (Figs 5d, 6). Filamentous cyanobacteria (*Oscillatoria* spp.) were also detected from October to December and from May to drought (Fig. 5d).

Tetraselmis apiculata (*Platymonas apiculata*) Butcher, was first detected in January and became the dominant phytoplankter in February, when salinity was low.

Diatoms constituted a large fraction of the phytoplankton all through the year. Many species of diatoms were found: *Hantzschia amphioxys* (Ehr.) W. Smith was especially abundant in the autumn, *Amphora coffeaiformis* (Ag.) Kütz. in the summer; *Stauroneis amphioxys* Greg., *Cocconeis placentula*

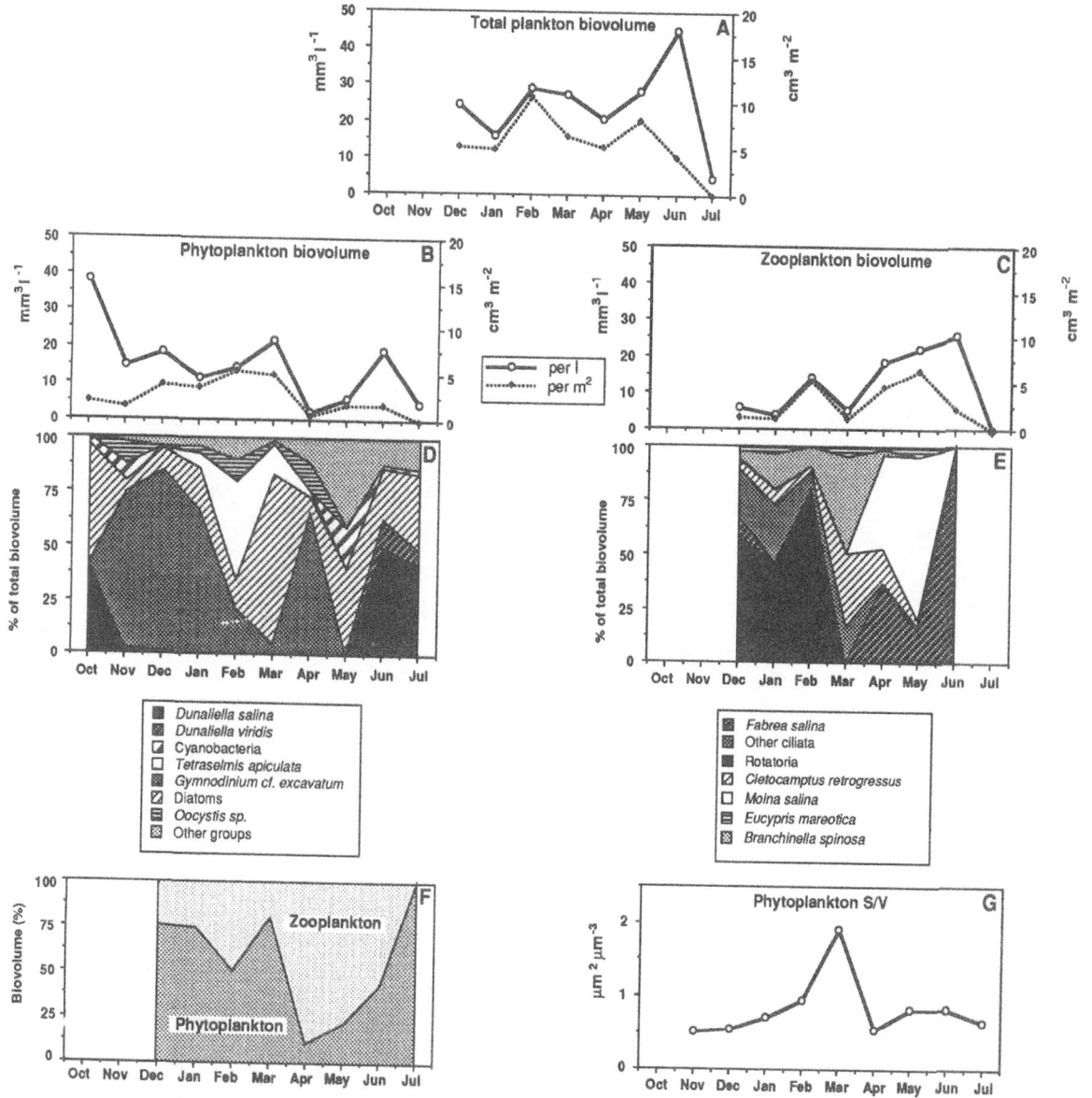

Fig. 5. Seasonal variation in the plankton assemblage. (a) Total plankton biovolume per litre and square metre. (b) Biovolume of the phytoplankton. (c) Biovolume of the zooplankton. (d) Taxonomic composition of the phytoplankton. (e) Taxonomic composition of the zooplankton. (f) Relative abundances of phytoplankton and zooplankton. (g) Ratio of total phytoplankton surface area to total phytoplankton biovolume.

Ehr. *Entomoneis* sp. and several species of *Navicula* and *Nitzschia* were also found. *Nitzschia closterium* (Ehr.) W. Smith achieved a higher population density than did any other phytoplankter, being represented by 209×10^6 ind. l^{-1} in March.

Ciliates were also abundant throughout the year. *Fabrea salina* Henneguy was the most abundant at times of high salinity (Fig. 6). Other ciliates were occasionally abundant (*Strombidium* sp., cf. *Cyclidium* sp., *Euplotes* sp., *Trachelocerca* sp., *Uroleptus* sp., cf. *Trithigmostoma* sp., cf. *Holophrya* sp., *Podophrya* sp.). Small amoebae (10–20 μm) were also found, especially in October.

218

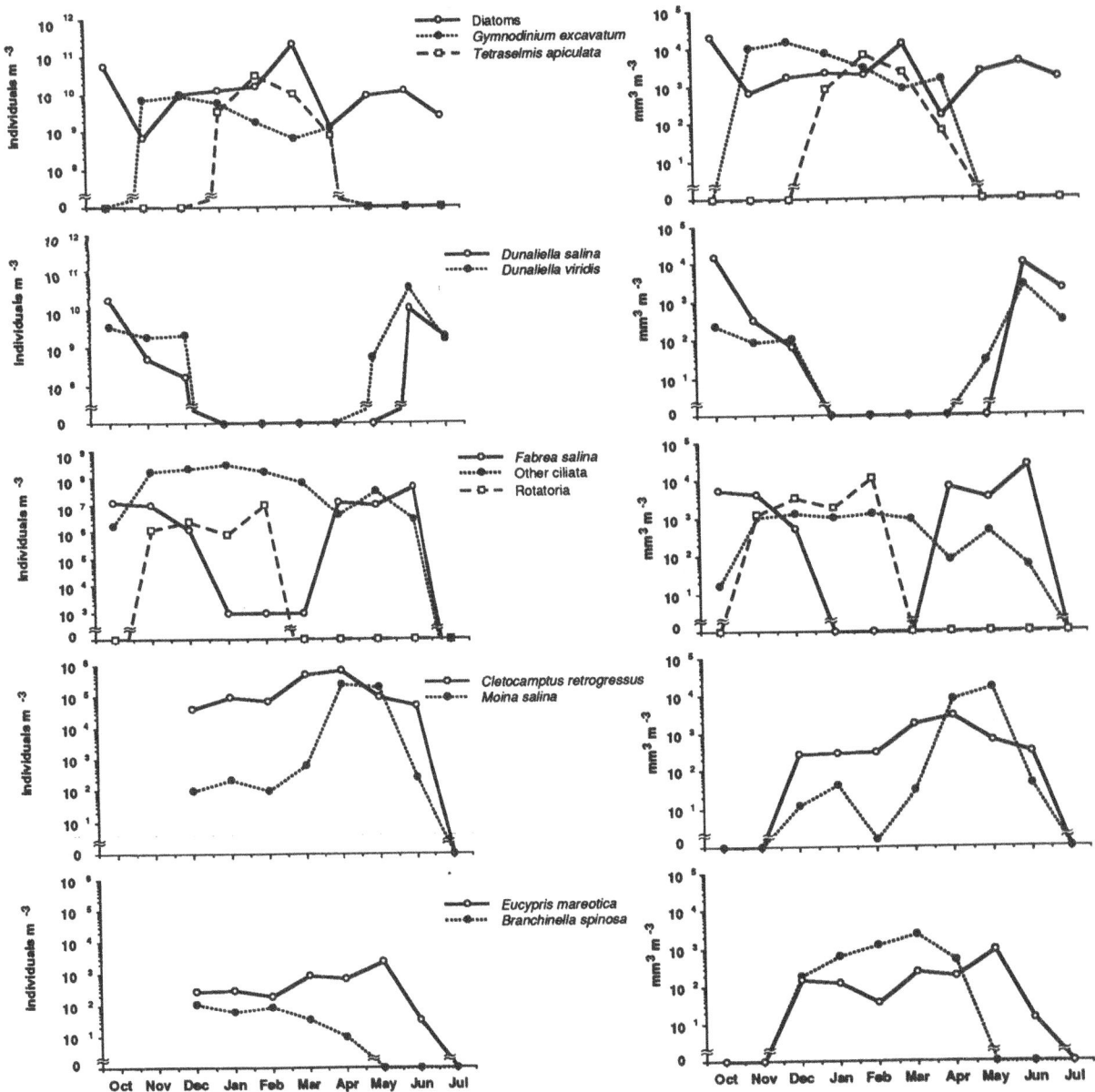

Fig. 6. Seasonal changes in absolute numerical densities (left) and biovolume (right) for the more important taxa.

Rotifers contributed substantially to zooplankton biomass in winter (Fig. 5e). *Hexarthra fennica* Levander predominated from November to January while *Synchaeta* sp. (cf. *tremula*) was the only rotifer species detected in February.

Among the crustacea (Figs 5e, 6), *Cletocamptus retrogressus* Schmankevitch was present all throughout the year and was most abundant in early spring. *Branchinella spinosa* Milne-Edwards

appeared during the winter. *Moina salina* Daday achieved the highest biomass among the crustaceans, reaching 16.9 mm^3 l^{-1} in May, but decreasing drastically by June when the ciliate *Fabrea salina* became the most important consumer and reached the highest observed biovolume (25.2 mm^3 l^{-1}) (Fig. 6). The ostracod *Eucypris mareotica* (Fischer) Gauthier was also observed throughout the year.

Discussion

Temporary lakes such as Fuente de Piedra are usually very shallow, and strongly controlled by seasonal and shorter term environmental changes to which their high surface to volume ratio makes them especially sensitive. The seasonality of climatic forcing is also enhanced by the large changes in salinity during the filling and drying cycle. Shallow lakes usually also exhibit marked interannual variability (Melack, 1981; Vareschi, 1987; Hammer, 1986) reflecting high interannual variation in their hydrological budgets. It is difficult, therefore, to extrapolate from the results of a single study: the present results would have been very different if another year had been studied.

Observed successional patterns

The results show that there was a phytoplanktonic predominance during the winter filling phase. During December, chlorophyll in the water column and organic matter content of the surface sediments increased, and Eh values became more negative than in the fall. This was the start of a productive period with autotrophic predominance which was maintained from December to March. This was followed by a spring predominance of zooplankton (high P:Z ratio, Fig. 5f). Finally, in late spring and in summer, another phytoplankton assemblage better adapted to high salinity became abundant. The decrease in zooplankton during the summer indicated a progressively important role of physical control as the lake dried up.

Nutrients in the water also showed clear seasonal trends. In the autumn nitrate increased and ammonium decreased. This may have been due, in part, to degradation of previously produced organic matter, since aerobic production of nitrate will occur with a lag following release of ammonium from decomposing organic matter and excretions (Margalef, 1974). Zooplanktonic excretion probably was responsible for the high ammonium concentrations during spring, with

a possibly important contribution via excretion by flamingoes that are more abundant at this time.

DIN:SRP ratio and the hypothesis of control by nutrient limitation.

The productive winter period was characterized by a low DIN:SRP ratio (0.24) which suggested possible limitation by phosphorus. During the rest of the year the ratio was always high (>20), reaching values >150 in late spring and during summer. Although these values could indicate a limitation of algal production by phosphorus in the warmer months and by nitrogen in December and January (Forsberg et al., 1978; Tominaga et al., 1987), these values are not the only determining factor (Hecky & Kilham, 1988). Thus, absolute phosphate and nitrogen concentrations seemed to be moderately high for most of the year and mean concentrations were a little lower than those typical of other saline lakes (Hammer, 1986). Grazing may have had an important controlling role, especially during the spring, and strong abiotic control was also likely during summer. On the other hand, phosphorus has usually a higher turnover rate than does nitrogen (Hopkinson, 1987; Holtan et al., 1988) and this can result in N-limitation under higher DIN:SRP in the water than would be expected on the basis of particulate matter N:P ratios such as that of Redfield et al. (1963). Studying relationships between salinity and nutrient limitation, Tominaga et al. (1987) found for several Australian salt lakes that the more saline the lake, the lower the DIN:SRP ratio. An inverse relationship was apparent, however, in Fuente de Piedra lake, where the highest DIN:SRP values occurred during times of highest salinity (Figs 2, 3). This was due to a phosphate increase in winter caused by increased stream inflow. However, self-eutrophication may have also occurred because of the highly negative redox potentials in winter that would have facilitated release of phosphorus from the sediments. This process could be similar to the feedback mechanism suggested for shallow lakes by Oláh

220

(1975): organic matter sedimentation causes a more negative redox potential and a release of phosphate that increases organic matter production. The effect of the high summer DIN:SRP ratios on phytoplankton production could be partially compensated also by a higher sediment–water interchange rate at high salinities (Clavero et al., 1990).

Succession and size of phytoplankters

Algal nutrient assimilation rate and potential internal storage must be also taken into account in discussing nutrient limitation. Both assimilation rate and reserves depend on cell size and taxon (Capblancq, 1990). Large short term fluctuations in nutrient concentrations in a lake can favor larger species with high internal storage capabilities and a consequently lower phytoplankton surface to volume ratio (S:V). This may occur especially at the beginning and end of the flooding-drying cycle when small environmental changes can influence more easily the water column. The high S:V ratio associated with the small diatom *Nitzschia closterium* in March could have represented an advantage at the end of winter, when external nutrient concentrations were low, the water level was still high, allochtonous inputs decreased, and the system was relatively stable. Affinity for nutrients is related to phytoplankton cell size and, therefore, the S:V ratio (Eppley et al., 1969). Determination of the S:V ratio at more frequent intervals from February to April would have been interesting. After March the increasingly abundant cladocerans had probably caused the dramatic decrease of phytoplankton densities and, through size-selection, the decrease in the phytoplankton S:V ratio as well.

The low springtime P:Z ratios

During winter, phytoplankton biovolume represented 50–80% of total plankton biovolume, but in April the P:Z ratio and chlorophyll concentration in water decreased sharply (Figs 4, 5f). In-creased redox potential at the sediment surface at this time may have been due to decreased sedimentation of organic matter from the water column and the maintenance of bacterivorous activity, which together would have produced a decreasing rate of heterotrophic activity by bacteria in the sediments. A switch to predominantly benthic photosynthetic activity possibly occurred in the spring as chlorophyll in the water column declined much more than did chlorophyll in sediments. Sediment organic matter percentage declined in late spring, perhaps reflecting reduced primary production and acceleration of decomposition by high temperatures. Zooplankton biovolume exceeded that of phytoplankton by a factor of 8 in April and of 3.7 in May. Similar imbalances were observed by Drabkova et al., 1978 and Letanskaya, 1980 (Hammer, 1986) in saline lake Bolshoy Shantropy. Three explanations of how high zooplanktonic biomass could occur in the presence of low phytoplankton biomass can be suggested: utilization by zooplankton of the phytobenthos, a very high turnover of phytoplankton cells, or consumption by zooplankton of detritus derived from earlier primary production.

Phytoplankton species were basically the same in spring as in winter but with lower densities, so it seems unlikely that they had higher spring production rates. Increasing salinity stress would also have made this unlikely. However, high nutrient regeneration rates by zooplankton may have enhanced phytoplankton production and thus contributed to high zooplankton biomass. A phytobenthos-based maintenance of zooplankton was possible, though variations in sediment chlorophyll seem not clearly correlated with zooplankton densities.

The decline in sediment organic matter towards June, which probably reflected a summer enhancement of respiration parallel to temperature and salinity increase, may also suggest a pressure on algae, detritus and attached bacteria. This was also related to the gradual rise of Eh values in the same period. These effects could have been facilitated also by a possible increase in water circulation and resuspension processes that, in turn, stimulated production of the 'microbial loop' (Wainright, 1987). *Moina salina* and ciliates were important in April and May, and *Fabrea salina* reached its highest biomass in June. The wide trophic adaptability of cladocerans (Lehman, 1976; Poulet, 1982; Echevarría, 1987) and *Fabrea salina* may allow a high grazing rate on phytoplankton and the maintenance of high zooplankton biomass. Zooplankton omnivory has already been pointed out by Por (1980) as a feature of hypersaline environments. Drabkova *et al.* (1978) and Letanskaya (1980) attributed relative low biomass of phytoplankton in Bolshoy Shantropy lake to intensive zooplankton grazing. In fact, if one accepts the filtering rates found by Anderson (1958) for *Moina hutchinsoni*, the high densities of *Moina* in April could have filtered 30% of the total water of the lake each day. This intensive grazing could have been supplemented by feeding on benthic detritus or bacteria. Drabkova *et al.* (1978) and Letanskaya (1980) reported variations in the contribution of bacteria to the diet of zooplankton during the year, and *Fabrea salina* is known to feed on *Dunaliella* (Post *et al.*, 1983) and bacteria (Gervais, 1969).

Although annual average of P:Z ratio was near 1, during the spring this ratio was lower than the lower limit proposed by Sheldon *et al.* (1977). This indicated a temporal uncoupling that is characteristic of stressed systems. The lake had separate autotroph-dominant and heterotroph-dominant phases. During the dry months of summer and early autumn, remineralization of earlier production will occur, followed by a phase of phytoplanktonic production during filling, a heterotrophic phase of grazing of benthic algae, bacteria and detritus and, finally, a phase of production by halotolerants as *Dunaliella*. All these processes correlate with the filling and drying cycle and hence with salinity.

Salinity, succession stages and trophic chains

The results suggest an important role of salinity forcing in the determination of the successional patterns in Fuente de Piedra lake. Por (1980) proposed a system for classification of hypersaline waters that was based on communities characteristic of different salinities. This model can also serve, however, to describe seasonal stages in temporary saline lakes. The amplitude and number of these stages will depend mainly on precipitation and on the annual hydrological regimen. The evolution of planktonic communities could be related to this classification of hypersaline waters. Under the classification of Por (1980), Fuente de Piedra lake has characteristics ranging from 'limnogenic α-hypersaline' to 'δ-hypersaline'. Temporary saline lakes may exhibit this spectrum of change throughout a single year. Community complexity usually increases gradually during the filling period, with frequent perturbations related to allochtonous inputs. Simplification of the community occurs during the drying period. Maximal rates of secondary production could be reached at this time if an omnivorous-detritivorous zooplankton assemblage appears capable of exploiting the previously produced organic matter. A characteristic of the summer period is the gradual narrowing of the planktonic size spectrum as salinity rises. Crustaceans tend to disappear and a simple community with *Fabrea salina*, diatoms and *Dunaliella* spp. is established. This could be considered as the rise of a 'summer community' similar to that of γ-type waters proposed by Por (1980). In the middle of July, when the lake is about to dry, detrital organic matter accumulates again. This accumulation is related to a gradual decline of zooplankton. This organic matter will be degraded to ammonium and possibly to nitrate in the early stages of next hydrological year. During the dry phase, degradation activity by essentially terrestrial invertebrates can be important (García & Niell, 1991).

Further study of detritus-based food webs and benthos-plankton interactions would contribute importantly to an understanding of the trophic behaviour of shallow, temporary lakes such as Fuente de Piedra.

222

Acknowledgements

We acknowledge the Patronato de la Laguna de Fuente de Piedra who allowed the sampling in the nature reserve and especially Dr Lucena; and the Andalusian Agencia de Medio Ambiente, especially Mr Rendón. The final manuscript was improved with helpful suggestions by S. H. Hurlbert and an anonymous referee. This work was supported by a grant from the Junta de Andalucía programme 'F.P.I.'/85 and by the CICYT project NAT90-0355.

References

Alonso, M., 1990. Anostraca, Cladocera and Copepoda of Spanish saline lakes. Hydrobiologia 197: 221–231.

Anderson, G. C., 1958. Seasonal characteristics of two saline lakes in Washington. Limnol. Oceanogr. 3: 51–68.

Capblancq, J., 1990. Nutrient dynamics and pelagic food web interactions in oligotrophic and eutrophic environments: an overview. Hydrobiologia 207: 1–14.

Choi, J. W. & D. Stoecker, 1989. Effects of fixation on cell volume of marine planktonic protozoa. Appl. Envir. Microbiol. 55: 1761–1765.

Clavero, V., J. A. Fernández & F. X. Niell, 1990. Influence of salinity on the concentration and rate of interchange of dissolved phosphate between water and sediment in Fuente de Piedra lagoon (S. Spain). Hydrobiologia 197: 91–97.

Comín, F. & M. Alonso, 1988. Spanish salt lakes: their chemistry and biota. Hydrobiologia 158: 237–245.

Echevarría, F., 1987. Análisis de la alimentacíon fitófaga 'in situ' de Ceriodaphnia spp (Cladocera): Variaciones a corto plazo. Tésis de Licenciatura. Universidad de Málaga, 134 pp.

Eppley, R. W., J. N. Rogers & J. J. McCathy, 1969. Half-saturation constants for uptake of nitrate and ammonium by marine phytoplankton. Limnol. Oceanogr. 14: 912–920.

Fernández, J. A., F. X. Niell & J. Lucena, 1985. A rapid and sensitive automated determination of phosphate in natural waters. Limnol. Oceanogr. 30: 227–230.

Forsberg, C., S. O. Ryding, A. Claesson & A. Forsberg, 1978. Water chemical analyses and/or algal assay? Sewage effluent and polluted lake water studies. Mitt. Int. Ver. Theor. Angew. Limnol. 21: 352–363.

García, C. M. & F. X. Niell, 1991. Burrowing beetles of the genus Bledius (Staphylinidae) as agents of bioturbation in the emergent areas and shores of an athalassic inland lake (Fuente de Piedra, southern of Spain). Hydrobiologia 215: 163–173.

Gervais, C., 1969. Influence de la concentration saline du milieu sur l'éclosion des kystes de Fabrea salina Henneguy (cilié hétérotriche). Protistologica V, fasc. 1: 109–114.

Hammer, U. T., 1986. Saline lake ecosystems of the world. Dr W. Junk Publishers, Dordrecht, 616 pp.

Hecky, R. E. & P. Kilham, 1988. Nutrient limitation of phytoplankton in freshwater and marine environments: A review of recent evidence on the effects of enrichment. Limnol. Oceanogr., 33: 796–822.

Herberg, O., 1986. Valoración del impacto provocado por el arroyo Santillán en la Laguna de Fuente de Piedra (Málaga). Tésis de Licenciatura. Universidad de Málaga, 122 pp.

Holtan, H., L. Kamp-Nielsen & A. O. Stuanes, 1988. Phosphorus in soil, water and sediment: an overview. Hydrobiologia 170: 19–34.

Hopkinson Jr., C. S., 1987. Nutrient regeneration in shallow-water sediments of the estuarine plume region of the nearshore Georgia Bight, USA. Mar. Biol. 94: 127–142.

Håkanson, L. & M. Jansson, 1983. Principles of lake sedimentology. Springer Verlag, Berlin, 316 pp.

I.G.M.E., 1986. Mapa geológico de España E. 1: 50.000. Hoja 1023. Ministerio de Industria y Energía, Madrid.

I.G.M.E., 1988. Observaciones climatológicas e hidrogeológicas en la cuenca de Fuente de Piedra durante el año 1987–88. Nota técnica 336. Ministerio de Industria y Energía, Madrid, 39 pp.

Lehman, J. T., 1976. The filter-feeder as an optimal forager, and the predicted shapes of feeding curves. Limnol. Oceanogr. 21: 501–516.

Linares, L., 1990. Hidrogeología de la laguna de Fuente de Piedra (Málaga). Tesis Doctoral, Dpto. Geodinámica, Universidad de Granada, 343 pp.

Lund, J. W. G., C. Kipling & E. D. Le Cren, 1958. The inverted microscope method of estimating algal numbers and the statistical basis of estimations by counting. Hydrobiologia 11: 143–170.

Marcus, N. H., 1984. Recruitment of copepod nauplii into the plankton: importance of diapause eggs and benthic processes. Mar. Ecol. Prog. Ser. 15: 47–54.

Margalef, R., 1974. Ecología. Omega, Barcelona, 951 pp.

Melack, J. M., 1981. Photosynthetic activity of phytoplankton in tropical African soda lakes. Hydrobiologia 81: 71–85.

Oláh, J., 1975. Metalimnion function in shallow lakes. In: Limnology of shallow waters. J. Salánki & J. E. Ponyi (eds), Symp. Biol. Hung. 15: 149–155.

Parsons, T. R., Y. Maita & C. M. Lalli, 1984. A manual of chemical and biological methods for seawater analysis. Pergamon, Oxford, 173 pp.

Por, F. D., 1980. A classification of hypersaline waters, based on trophic criteria. Mar. Ecol. 1: 121–131.

Post, F. J., L. J. Borowitzka, M. A. Borowitzka, B. Mackay & T. Moulton, 1983. The protozoa of a Western Australian hypersaline lagoon. Hydrobiologia 105: 95–113.

Poulet, S. A., 1982. Nutrition du zooplankton marin: interface particules-copepodes. Thèse de doctorat d'etat. Université Pierre et Marie Curie. Paris VI, 158 pp.

Rassoulzadegan, F., 1979. Cycles annuels de la distribution

de differentes catégories de particules du seston et éssai d'identification des principales poussés phytoplanctoniques dans les eaux néritiques de Villefranche. J. exp. mar. Biol. Ecol. 38: 41–56.

Redfield, A. C., B. Ketchum & F. Richards, 1963. The influence of organisms on the composition of seawater. In M. Hill (ed.), The sea 2: 6–77. Interscience, New York.

Rendón, M., J. M. Vargas y J. M. Ramírez, 1991. Dinámica temporal y reproducción del flamenco común (*Phoenicopterus ruber roseus*) en la laguna de Fuente de Piedra (sur de España). In: Reunión técnica sobre la situación y problemática del Flamenco rosa (*Phoenicopterus ruber roseus*) en el Mediterráneo occidental y Africa noroccidental: 135–153.

Rodríguez, J., F. Jiménez, B. Bautista & V. Rodríguez, 1987. Planktonic biomass spectra dynamics during a winter production pulse in Mediterranean coastal waters. J. Plankton Res 9: 1183–1194.

Rolke, M. & J. Lenz, 1984. Size structure analysis of zooplankton samples by means of an automated image analyzing system. J. Plankton Res. 6: 637–645.

Sheldon, R. W., W. H. Sutcliffe Jr. & M. J. Paranjape, 1977. Structure of pelagic food chain and relationship between plankton and fish production. J. Fish. Res. Bd Can., 34: 2344–2353.

Shinn, J. A., 1941. Ind. Eng. Chem. (Annual edition), 13: 33. In J.D.H. Strickland and T.R. Parson. A practical handbook of seawater analysis. Fish. Res. Bd Can. Bull. 167.

Slawyk, G. & J. J. McIsaac, 1972. Comparison of two automated ammonium methods in a region of coastal upwelling. Deep-Sea Res. 19: 521–524.

Sprules, G. W. & R. Knoechel, 1984. Lake ecosystem dynamics based on functional representations of trophic components. In D.G. Meyers & R. Strickler (eds), Trophic interactions within aquatic ecosystems. A.A.A.S. Selected Symp. 85. Westview-Boulder, Colorado: 383–403.

Talling, J. F. & D. Driver, 1963. Some problems in the estimation of chlorophyll 'a' in phytoplankton. Proc. Conf. prim. Prod. Meas. Mar. Freshwat. Hawaii. 1961: 142–146.

Tominaga, H., N. Tominaga & W. D. Williams, 1987. Concentration of some inorganic plant nutrients in saline lakes on the Yorke peninsula, South Australia. aust. J. mar. Freshwat. Res. 38: 301–305.

Utermöhl, H., 1958. Zur Vervollkomnung der quantitativen Phytoplankton-Methodik. Mitt. int. Ver. Limnol. 9: 1–38.

Vareschi, E., 1987. Saline lake ecosystems. In: E. D. Schulze & H. Zwölfer (eds), Potentials and limitations of ecosystem analysis. Ecol. Stud. 61 Springer-Verlag, Berlin, 347–364.

Vargas, J. M., M. Blasco & A. Antúnez, 1983. Los vertebrados de la Laguna de Fuente de Piedra (Málaga). ICONA. Monografias n°28.

Wainright, S. C., 1987. Stimulation of heterotrophic microplankton production by resuspended marine sediments. Science 238: 1710–1712.

Wood, E. D., F. A. J. Armstrong & F. Richards, 1967. Determination of nitrate in seawater by cadmium-copper reduction to nitrite. J. mar. biol. Ass. U.K. 47: 23–31.

The fauna of athalassic saline waters in Australia and the Altiplano of South America: comparisons and historical perspectives

I. A. E. Bayly
Department of Ecology and Evolutionary Biology, Monash University, Clayton, Vic. 3168, Australia

Key words: athalassic, salt lakes, Australia, South America, Altiplano, *Boeckella* (Copepoda), *Daphniopsis*

Abstract

Similarities and differences between the fauna of inland saline waters in Australia and on the Altiplano are explored and explanations sought.

Elements common to both continents include the calanoid copepod genus *Boeckella* (*B. triarticulata* in Australia, *B. poopoensis* and *B. meteoris* in South America) and the cladoceran genus *Daphniopsis*. Salinity data for Altiplano lakes are given for six species of *Boeckella* and for *Daphniopsis*.

Ostracods have adapted to the open water of saline lakes in Australia but not in South America, a difference that may reflect past differences in the degree of predation by birds. In South America, diatoms are grazed by the flamingos *Phoenicoparrus andinus* and *P. jamesi*, while in Australia the main diatom grazer is probably the aquatic oniscoid isopod *Haloniscus searlei*. However, at least four species of flamingos were present in Australia during the late Cenozoic and one or more of these may well have grazed diatoms. The extinction of diatom-grazing or carnivorous flamingos, or both, in Australia may have been factors in the unique colonization of inland saline waters by *H. searlei*.

Introduction

Although the fragmentation of Gondwana commenced in the Late Jurassic (*ca* 150 Ma B.P.), and Africa, Madagascar and India were separated from Antarctica by the Early Cretaceous (120 Ma B.P.) (Owen, 1983), direct connection between Australia and South America via Antarctica was maintained until the Late Cretaceous (70–66 Ma B.P.) (BMR Palaeogeographic Group 1990). As a result of this long-enduring connection, a closer similarity might be expected between the inland water faunas of Australia and South America than that between either of these countries and Africa. Close zoogeographic relationships between Australia and South America are known to exist within several freshwater taxa: Crustacea [Anaspidacea and Parastacidae] (Williams, 1981a), Plecoptera [Antarctoperlaria] (Zwick, 1981) and Amphibia [Hylidae and Leptodactylidae] (Tyler, 1979; Tyler *et al.*, 1981). The present paper seeks to show that, also with respect to athalassic saline waters, some close faunal similarities exist between Australia and South America. Differences between the fauna of inland saline waters on the latter two continents are also explored.

Review and discussion of some major faunal groups

Copepoda

Calanoid copepods belonging to the family Centropagidae and the genus *Boeckella* occur in athalassic saline waters of Australia and the Altiplano of South America (Table 1). *Boeckella* is considered (Bayly, 1992b) to have invaded southern fresh waters from marine-estuarine ancestors after Africa, Madagascar and India separated from Antarctica in the Early Cretaceous (*ca* 135–120 Ma B.P.) (Owen, 1983), but before Australia separated from Antarctica in the Late Cretaceous (70–66 Ma B.P.) (BMR Palaeogeographic Group 1990). Subsequently, *Boeckella* invaded inland saline waters.

There are 19 species of *Boeckella* in the Australasian region (Bayly, 1979) and all but one of these are restricted to fresh waters, or nearly so. The exception is *B. triarticulata* (Thomson), which is mainly a freshwater species but which also tolerates saline waters with a salinity of up to about $22 \, g \, l^{-1}$ (Bayly, 1969). Similarly, there are 14 species of *Boeckella* in South America (Bayly, 1992a), nearly all of which are confined to fresh waters. The most significant exception is the halobiont species, *B. poopoensis* Marsh, which tolerates salinities of up to about $80 \, g \, l^{-1}$ (Tables 1 and 2) and constitutes a significant prey for the Chilean flamingo, *Phoenicopterus chilensis* Molina, and Wilson's Phalarope, *Phalaropus tricolor* (Viellot) (Hurlbert *et al.*, 1984). *B. meteoris* Kiefer is a freshwater form showing a slight degree of salt tolerance on the Altiplano (Tables 1 and 2). *B. occidentalis* Marsh and *B. titicacea* Harding have a salinity tolerance that extends from fresh waters into waters that are marginally saline (Table 2).

The most important calanoids in Australian athalassic saline waters are *Calamoecia clitellata* Bayly and *C. salina* (Nicholls), both of which are halobiont species. Unlike *Boeckella*, the genus *Calamoecia* is not shared with South America and almost certainly evolved in Australia after this country separated from Antarctica in the Late Cretaceous (70–66 Ma B.P.) (BMR Palaeogeographic Group 1990). The evolution of halobiont species of *Calamoecia* from freshwater ancestors (12 of the 14 species of *Calamoecia* are confined to fresh water) probably occurred at a much later date, more recently than mid-Miocene (16 Ma B.P.), after which time saline waters became more

Table 1. Some faunal similarities and differences in athalassic saline waters of Australia and the Altiplano of South America. Taxa marked with an asterisk are fossils.

Taxon	Australia[a]	South America (altiplano region)
COPEPODA		
Centropagidae		
Boeckella[b]	*B. triarticulata* (fresh-*ca* $22 \, g \, l^{-1}$)	*B. meteoris* (fresh-$6 \, g \, l^{-1}$)
		B. poopoensis (5–$80 \, g \, l^{-1}$)
Calamoecia	*C. clitellata* (6–$132 \, g \, l^{-1}$)	–
	C. salina (7–$195 \, g \, l^{-1}$)	–
CLADOCERA		
Daphniopsis	*Daphniopsis*[c] (6–$68 \, g \, l^{-1}$)	*Daphniopsis*[d] (4–$11 \, g \, l^{-1}$)
OSTRACODA		
(Limnetic taxa)	*Australocypris*	–
	Diacypris	–
	Platycypris	–
	Trigonocypris	–
AVES		
Phoenicopteridae		
Xenorhynchopsis*	*X. tibialis**	–
	X. minor*	–
Ocyplanus*	*O. proeses**	–
Phoenicopterus	*P. ruber**	*P. chilensis*
Phoenicoparrus	–	*P. andinus* (diatom feeder)
	–	*P. jamesi* (diatom feeder)
ISOPODA		
Haloniscus	*H. searlei* (4–$161 \, g \, l^{-1}$, diatom feeder)	–
GASTROPODA		
Coxiella	*Coxiella* (6–$124 \, g \, l^{-1}$, ? diatom feeder)	–

[a] Salinity ranges in parentheses (except *Boeckella*) are taken from De Deckker & Geddes (1980), Williams (1981b, 1983) and Williams & Mellor (1991).

[b] Salinity ranges for South American species from author's unpublished data arising from the examination of collections from the Altiplano made by S. H. Hurlbert.

[c] There are now three Australian species of *Daphniopsis*, *D. pusilla* Serventy, *D. australis* Sergeev and Williams and *D. queenslandensis* Sergeev, the first two of which have probably been confused in previous ecological studies. Consequently data are here restricted to the generic level.

[d] Recorded from three salt lakes on the Altiplano (see Table 2).

Table 2. Occurrence of *Boeckella* species and *Daphniopsis* in Altiplano lakes in relation to salinity (Plankton collections and TDS determinations by S. H. Hurlbert, identifications of taxa by I.A.E. Bayly).

Locality name (Laguna)	Position (S. lat., W. long)	Date	TDS (g l^{-1})	Bp	Bt	Bm	Bo	Bpa	Bc	D
Suches	16° 56′, 70° 24′	20.vi.76	0.23	–	–	–	x	–	–	–
		21.xi.76	0.30	–	–	–	x	–	–	–
Colorada II	15° 22′, 70° 21′	25.vi.76	0.27	–	x	–	x	–	–	–
Loripongo	16° 50′, 70° 05′	21.vi.76	0.46	–	x	–	–	x	–	–
		20.xi.76	0.62	–	x	–	–	x	–	–
Cotacotani	18° 14′, 69° 13′	1.vii.76	0.60	–	x	–	x	–	–	–
Totoral	22° 32′, 67° 17′	2.ii.79	0.65	–	x	–	–	–	–	–
Conchostraca	22° 18′, 67° 14′	1.xii.77	0.69	–	x	–	–	x	x	–
Viscacha	16° 53′, 70° 14′	21.vi.76	0.90	–	x	–	–			–
Pampamarca	14° 08′, 71° 29′	23.vi.76	0.80	–	–	–	x			–
		17.xi.76	1.0	–	–	–	x			–
Saracocha	15° 47′, 70° 38′	25.vi.76	0.92	–	–	–	–			–
		19.xi.76	1.1	–	–	–	x			–
Chungara	18° 15′, 69° 09′	1.vii.76	1.2	–	–	–	x			–
Campo Grande	22° 33′, 67° 12′	2.xii.77	3.4	–	x	–	–			–
Huancaroma	17° 40′, 67° 30′	6.vii.76	3.5	–	x	–	x			–
Penitas Blancas	22° 25′, 67° 15′	29.xi.77	3.7	–	x	–				–
Pelada	22° 45′, 67° 10′	2.xii.77	4.1	–		x				x
Parinacochas	15° 17′, 73° 42′	28.vi.76	5.6	x		–				–
Pozuelos	22° 20′, 66° 00′	27.v.77	6.2	–		x				–
Santa Rosa	27° 05′, 69° 10′	21.xi.75	7.5	x						–
Catalcito	23° 31′, 67° 15′	2.xii.77	8.1	x						x
Puripica Chico	22° 31′, 67° 30′	18.ii.79	8.2	x						–
Khara	21° 54′, 67° 52′	26.xi.77	8.7	x						–
Calientes II	23° 31′, 67° 34′	5.xii.74	10.0	x						–
Loriscota	16° 52′, 70° 02′	21.vi.76	10.4	x						–
		21.xi.76	11.6	x						–
Soledad	17° 44′, 67° 22′	7.vii.76	11.0	x						–
Chojllas	22° 22′, 67° 06′	30.xi.77	11.1	x						
		18.ii.79		x						
Polques	22° 32′, 67° 37′	18.vii.76	13.0	x						
		11.xii.86	10.0	x						
		6.ii.79		x						
Hombre Muerto	25° 30′, 66° 51′	29.v.77	21.0	x						
Calientes III	25° 00′, 68° 38′	29.xi.75	24.2	x						
Ramaditas	21° 38′, 68° 05′	17.xii.75	35.1	–						
		12.xii.76	25.7	x						
		23.xii.79	–	x						
Guacha	22° 33′, 67° 31′	2.i.79	36.0	x						
Collpacocha	15° 15′, 70° 03′	24.vi.76	38.6	x						
Calientes I	23° 08′, 67° 24′	15.i.79	47.0	x						
Verde II	22° 48′, 67° 48′	13.xii.75	57.1	x						
		11.xii.76	58.1	x						
		3.xii.77	–	x						
Chulluncani	21° 32′, 67° 52′	12.xii.76	64.5	x						
Este	22° 31′, 67° 29′	25.xii.78	86.0	x						

[a] Bp = *Boeckella poopoensis* Marsh, Bt = *B. titicacae* Harding, Bm = *B. meteoris* Kiefer, Bo = *B. occidentalis* Marsh, Bpa = *B. palustris* (Harding), Bc = *B. calcaris* (Harding); D = *Daphniopsis* sp.

abundant in Australia. Kershaw (1981) and Bowler (1982) describe a general, but rather irregular, increase in aridity from the Middle Miocene, with a maximum aridity event occurring in the Late Pleistocene around 18 000 yrs B.P. In line with evidence obtained from the well-documented oceanic record for the Pleistocene, other arid maxima must have occurred previously in Australia.

Cladocera

Bayly & Morton (1978) suggested that the occurrence of *Daphniopsis* in Antarctica, Australia and Tibet is explicable on the basis that all three of these regions were part of a Gondwana that was larger than is commonly recognized (Crawford, 1974). At that time *Daphniopsis* had not been recorded from South America. However, Hann (1986) and Valdivia & Burger (1989) recorded the genus from fresh water in Chile and Peru, respectively, and Bayly (unpublished) observed it (in 1983) in collections made by S. H. Hurlbert from three salt lakes on the Altiplano (Table 2). The latter three localities have an altitude of *ca* 4500 m and all are within 80 km of the freshwater lake from which Hann (1986) recorded the genus. These new records from South America reinforced the view that the genus evolved after the separation of India and Africa from the remainder of Gondwana through which it became widely distributed before the further fragmentation of this supercontinent. The occurrence of *D. ephemeralis* (Schwartz & Hebert, 1985) in North America does not conform with this interpretation, and, if this species is correctly assigned, one can only invoke subsequent dispersal of the genus to this region. However, a world revision of the genus (or subgenus) *Daphniopsis* is needed and may reveal that the genus *sensu stricto* does not, in fact, occur in North America.

Ostracoda

The distinction between plankton and benthos in very shallow saline lakes may be rather artificial (cf. Hurlbert *et al.*, 1984), but deep saline lakes, in which the usual distinction may be reasonably maintained, do occur on both continents. One of the most striking biotic differences between salt lakes in Australia and those on the Altiplano concerns the degree of occupancy of open waters by ostracods (Table 1). In Australia, the predominantly halobiont genera *Australocypris*, *Diacypris*, *Platycypris* and *Trigonocypris* include several species that are sufficiently good swimmers to lead a planktonic existence in the limnetic region of saline lakes (De Deckker, 1983). In the case of *Australocypris* and *Trigonocypris*, this is true despite the fact these genera contain 'giant' ostracods (>3 mm in length) (De Deckker, 1983). There has been no corresponding invasion of the open water in South American saline lakes. This suggests that planktonic crustaceans in Altiplano lakes may have been subject to a greater degree of predation than those in Australia. In the more saline lakes at least, it is birds rather than fish that are likely to have been the more significant predators. Leaving aside waters that are only slightly saline, there are presently only three Australian birds that prey on salt lake invertebrates to a significant degree: the Banded Stilt (*Cladorhychus leucocephalus* (Vieillot)), the Red-necked Avocet (*Recurvirostra novaehollandiae* Vieillot) and the Red-capped Plover (*Charadrius ruficapillus* Temminck). On the Altiplano the corresponding list is: the Chilean Flamingo (*Phoenicopterus chilensis*), Wilson's Phalarope (*Phalaropus tricolor*), and the Andean Avocet (*Recurvirosta andina* Philipi and Landbeck). Could it be that the extinction of carnivorous flamingos in Australia (see below), and the absence in that country of any bird equivalent to Wilson's Phalarope, were factors that enabled some ostracods to colonize the limnetic region of salt lakes?

Dodson & Egger (1980) showed that the Red Phalarope (*Phalaropus fulicarius* (L.)) is a size-selective predator on freshwater zooplankton and is especially effective in consuming adult *Daphnia* with a length of *ca* 3 mm. Phalaropes feed on the zooplankton only in the top few centimetres of the water column, and their predation has its greatest effect where the water depth is less than 20 cm (Dodson & Egger, 1980). In the long term (say over a span of several hundred years) nearly all athalassic saline lakes (even those presently regarded as deep and stable) may be regarded as astatic and periodically shallow enough for predation by phalaropes, in these parts of the world in which they occur, to have exerted a major impact on zooplankton communities.

Cypridid ostracods are comparatively slow swimmers and lack the rapid escape reflexes that are characteristic of copepods. Large free-swimming ostracods belonging to the Australian tribe Mytilocypridini (e.g. *Australocypris* and *Trigonocypris*) would seem particularly susceptible to phalarope predation if that were possible.

The only two ostracod genera known to occur in athalassic saline waters on both continents are *Cyprideis* and *Limnocythere* (P. De Deckker, pers. comm.) These two genera are much more widely distributed than the four genera mentioned above, which are endemic to Australia except possibly for one species of *Diacypris* recorded from New Zealand.

Diatom-grazers: birds and isopods (? and gastropods)

Hurlbert & Keith (1979) documented the occurrence of three species of flamingos [*Phoenicopterus chilensis*, *Phoenicoparrus andinus* (Philippi) and *P. jamesi* (Sclater)] on the Andean altiplano of South America, mainly in association with shallow lakes. All of these species depend mainly on athalassic saline waters for their food. *P. andinus* and *P. jamesi* feed mainly on benthic diatoms, whereas *Phoenicopterus chilensis* feeds on invertebrates such as brine shrimp, calanoid copepods, chironomid larvae, amphipods and corixids (Hurlbert, 1982; Hurlbert *et al.*, 1984). Hurlbert & Chang (1983) used exclusion experiments to demonstrate the large impact that *Phoenicoparrus andinus* has on the density and biomass of microorganisms such as large diatoms, amoebae, ciliates and nematodes. The diatom genus *Surirella* is a dominant component of the microbenthos in many Altiplano lakes.

In Australia, there are no birds that rival *Phoe-*

nicoparrus in their capacity to graze diatoms in salt lakes. There is some evidence (M. Brock pers. comm.) that the Black Swan (*Cygnus atratus* (Latham)) may be capable of nourishing themselves in part from diatoms epiphytic on *Ruppia*. At higher salinities, where macrophytes are absent, diatom-grazing birds are absent (but may have become extinct only recently). However, the unique Australian isopod, *Haloniscus searlei* Chilton, is a very effective grazer of epilithic diatoms (Blinn *et al.*, 1989). *H. searlei* is one of the most abundant macro-invertebrates in the littoral region of salt lakes in southern Australia. The mean length of these isopods is *ca* 1 cm, and densities in the range 40–500 individuals m^{-2} have been recorded (Blinn *et al.*, 1989). *H. searlei* may be preyed on by birds such as the plover *Charadrius ruficapillus*, but the diatom-to-bird linkage is not as direct as on the Altiplano of South America.

It is likely that another invertebrate endemic to Australia, the gastropod *Coxiella*, is capable of grazing diatoms, but presently the evidence is only indirect. *Coxiella* is widely distributed in southern Australian ephemeral and permanent salt lakes and often occurs in great abundance (Williams & Mellor, 1991). The radula of *Coxiella* has a rasping structure typical of a gastropod (W. D. Williams, pers. comm.).

Although flamingos are absent from the modern avifauna of Australia, their extinction on this continent is quite a recent event, probably occurring during the Late Pleistocene (Rich *et al.*, 1987). A reasonably diverse flamingo fauna was present in Australia during the late Cenozoic; Rich *et al.* (1987) recognise four species (including the elsewhere extant *Phoenocopterus ruber* L.) and three genera from Quaternary deposits (Table 1). These four species were apparently restricted to the Lake Eyre sub-Basin, and one or more of them may well have grazed diatoms. Rich *et al.* (1987) suggest that three species may have occurred contemporaneously at permanent saline lakes in the Lake Kanunka region during the Late Pliocene. Extant *Phoenocopterus ruber* is widely distributed with two subspecies: the Caribbean Flamingo (the Caribbean region and the Atlantic side of northern South America) and the Greater

Flamingo (southern Europe, central Asia, India and the eastern side of Africa). *P. ruber* feeds on a wide variety of invertebrates including *Artemia*, brine-fly and chironomid larvae, and crustaceans (Ogilvie, 1986).

The disappearance of saline lakes from the inland during one or more of the arid phases associated with the extreme climatic oscillations that occurred during the latter half of the Quaternary is very probably implicated in the extinction of Australian flamingos. De Deckker (1986) pointed out that Australia experienced an intense period of aridity around 18 000 years ago, during which time much of the aquatic biota probably survived in a narrow coastal strip where rainfall was more abundant. De Deckker stated that 'there is no evidence that any important element of the fauna of Australian lowland lakes became extinct at that time' (p. 493), but in making this comment he was considering truly aquatic animals, not terrestrial animals with a trophic dependence on inland saline waters. Flamingos may well represent an exception.

De Deckker (pers. comm.) points out that flamingos were abundant in Australia in the Miocene when the climate was wetter, and there were extensive strongly alkaline, carbonate-saline lakes in central Australia. Presently, however, the deposits associated with saline lakes in this region are mainly gypsum and halite, and the waters are only slightly alkaline. De Deckker (pers. comm.) suggests that flamingos became extinct in Australia not so much because of the disappearance of inland waters, but because of a lowering of pH as the lakes changed from being highly alkaline to only slightly alkaline. However, of the extant flamingos only the Lesser Flamingo, *Phoeniconaias minor* (Geoffrey), appears to have a requirement for high alkalinity (pH commonly in excess of 9.0) via its planktonic cyanobacterial food (*Spirulina platensis* (Nordst.) Gomont). The latter is cited as a true natronophilic species by Bayly & Williams (1973). There is no evidence that the invertebrate-feeders, *Phoenicopterus ruber* and *P. chilensis*, require abnormally high alkalinity. Furthermore, the great majority of Altiplano lakes where *Phoenicoparrus andinus* and *P. jamesi*

feed on benthic diatoms are not highly alkaline (pH < 8.6), and have chloride (or sulphate), not bicarbonate, as the dominant anion (Ballivian & Risacher, 1981; S. H. Hurlbert pers.comm.).

The unique colonization of inland saline waters by *Haloniscus* may well have been a fairly recent event in which a significant factor was possibly a super-abundance of diatoms following the extinction of diatom-grazing flamingos in Australia. The release of predation pressure on salt-lake macroinvertebrates with the demise in Australia of *P. ruber* may provide an alternative explanation for, or represent an additional factor in, the advent of *Haloniscus*.

Acknowledgements

I wish to thank Dr Patrick De Deckker and Dr Patricia Rich for their constructive criticism of an early draft of the manuscript. Discussions with Prof. J. Green and Dr J. R. Jehl following the oral presentation of my paper were also helpful. I am indebted to Dr S. H. Hurlbert for allowing me access to his zooplankton collections from the Altiplano and providing me with laboratory facilities in San Diego in October 1983.

References

Ballivian, O. & F. Risacher, 1981. Los salares del Altiplano boliviano. Metodos de estudio y estimacion economica. ORSTOM, Paris: 246 pp.

Bayly, I. A. E., 1969. The occurrence of calanoid copepods in athalassic saline waters in relation to salinity and anionic proportions. Verh. int. Ver. Limnol. 17: 449–455.

Bayly, I. A. E., 1979. Further contributions to a knowledge of the centropagid general *Boeckella*, *Hemiboeckella*, and *Calamoecia* (athalassic calanoid copepods). Aust. J. mar. Freshwat. Res. 30: 103–127.

Bayly, I. A. E., 1992a. The non-marine Centropagidae (Copepoda:Calanoida) of the world. In: H. J. Dumont (ed.) Guides to the Identification of the Microinvertebrates of the Continental Waters of the World. SPB Academic Publishers, The Hague.

Bayly, I. A. E., 1992b. Fusion of the genera *Boeckella* and *Pseudoboeckella* (Copepoda) and revision of their species from South America and sub-Antarctic islands. Rev. Chilena Hist. Nat. 65: 17–63

Bayly, I. A. E. & D. W. Morton, 1978. Aspects of the zoogeography of Australian microcrustaceans. Verh. int. Ver. Limnol. 20: 2537–2540.

Bayly, I. A. E. & W. D. Williams, 1973. Inland Waters and Their Ecology. Longman, Melbourne.

Blinn, D. W., S. L. Blinn & I. A. E. Bayly, 1989. Feeding ecology of *Haloniscus searlei* Chilton, an oniscoid isopod living in athalassic saline waters. Aust. J. mar. Freshwat. Res. 40: 295–301.

BMR Palaeogeographic Group, 1990. Australia: Evolution of a Continent. Bureau of Mineral Resources, Australia, 96 pp.

Bowler, J. M., 1982. Aridity in the late Tertiary and Quaternary of Australia. In W. R. Barker & P. J. M. Greenslade (eds), Evolution of the Flora and Fauna of Arid Australia. Peacock Publication, Adelaide: 35–45.

Crawford, A. R., 1974. A greater Gondwanaland. Science 184: 1179–1181.

De Deckker, P., 1983. Notes on the ecology and distribution of non-marine ostracods in Australia. Hydrobiologia 106: 223–234.

De Deckker, P., 1986. What happened to the Australian aquatic biota 18 000 years ago? In De Deckker, P. & W. D. Williams (eds), Limnology in Australia. CSIRO, Melbourne: 487–496.

De Deckker, P. & M. C. Geddes, 1980. Seasonal fauna of ephemeral saline lakes near the Coorong Lagoon, South Australia. Aust. J. mar. Freshwat. Res. 31: 677–699.

Dodson, S. J. & D. L. Egger, 1980. Selective feeding of Red Phalaropes on zooplankton of Arctic ponds. Ecology 61: 755–763.

Hann, B. J., 1986. Revision of the genus *Daphniopsis* Sars, 1903 (Cladocera: Daphniidae) and a description of *Daphniopsis chilensis*, new species from South America. J. Crustacean Biol. 6: 246–263.

Hurlbert, S. H., 1982. Limnological studies of flamingo diets and distributions. Nat. Geogr. Soc. Res. Rep. 14: 351–356.

Hurlbert, S. H. & C. Y. Chang, 1983. Ornitholimnology: effects of grazing by the Andean flamingo (*Phoenicoparrus andinus*). Proc. Natl. acad. Sci. USA 80: 4766–4769.

Hurlbert, S. H. & J. O. Keith, 1979. Distribution and spatial patterning of flamingos in the Andean altiplano. Auk 96: 328–342.

Hurlbert, S. H., M. Lopez & J. O. Keith, 1984. Wilson's Phalarope in the Central Andes and its interaction with the Chilean Flamingo. Rev. Chilena Hist. Nat. 57: 47–57.

Kershaw, P., 1981. Climate and Australian flora. Aust. Nat. Hist. 20: 231–234.

Ogilvie, M. & C., 1986. Flamingos. Alan Sutton, Gloucester, 121 pp.

Owen, H. G., 1983. Atlas of continental displacement, 200 million years to the present. C.U.P., Cambridge.

Rich, P. V., G. F. van Tets, T. H. V. Rich & A. R. McEvey, 1987. The Pliocene and Quaternary flamingos of Australia. Mem. Qd Mus. 25: 207–225.

Schwartz, S. S. & P. D. N. Hebert, 1985. *Daphniopsis ephemeralis* sp. n. (Cladocera: Daphniidae): a new genus for North America. Can. J. Zool. 63: 2689–2693.

Tyler, M. J., 1979. Herpetofaunal relationships of South America with Australia. In W. E. Duellman (ed.), The South American Herpetofauna: its Origin, Evolution, and Dispersal. University of Kansas, Lawrence: 73–106.

Tyler, M. J., G. F. Watson & A. A. Martin, 1981. The Amphibia: diversity and distribution. In A. Keast (ed.), Ecological Biogeography of Australia. Dr W. Junk Publishers, The Hague: 1277–1301.

Valdivia, R. & L. Burger, 1989. Descripción de *Daphniopsis marcahuasensis* sp. nov. (Cladocera: Daphniidae) del Perú, con la inclusión de una clave de identificación de las especies del Género. Amazoniana 10: 439–452.

Williams, W. D., 1981a. The Crustacea of Australian inland waters. In A. Keast (ed.), Ecological Biogeography of Australia. Junk, The Hague: 1103–1133.

Williams, W. D., 1981b. The limnology of saline lakes in western Victoria: A review of some recent studies. Hydrobiologia 82: 233–259.

Williams, W. D., 1983. On the ecology of *Haloniscus searlei* (Isopoda, Oniscoidea), an inhabitant of Australian salt lakes. Hydrobiologia 105: 137–142.

Williams, W. D. & M. W. Mellor, 1991. Ecology of *Coxiella* (Mollusca, Gastropoda, Prosobranchia), a snail endemic to Australian salt lakes. Palaeogeogr. Palaeoclim. Palaeoecol. 84: 339–355.

Zwick, P., 1981. Plecoptera. In A. Keast (ed.), Ecological Biogeography of Australia. Dr W. Junk Publishers, The Hague: 1171–1181.

The penetration of cladocerans into saline waters

David G. Frey
Department of Biology, Indiana University, Bloomington, IN 47405, USA

Key words: Australia, South Africa, Saskatchewan, Iran, Germany, thalassic saline waters, athalassic saline waters, chydorids, salinity

Editor's note: with sadness I report that Dr Frey died on April 1, 1992. He was very active to the last and working on, among other things, a revision and expansion of this article. In particular he had undertaken to base his quantitative analysis on conductivity intervals defined as equal increments on a logarithmic scale of conductivity. He only partially finished this work. Using his well-ordered notes and data sheets, kindly sent me by his wife Elizabeth, I have taken it a bit further, especially in revising his Figs 1–3, and adding brief comment on patterns evident in these. Though the conductivity intervals employed in the several figures and tables are not exactly the same, enough resolution is provided in all of these that problems of interpretation should not exist. I thank Cheryl Hart for assistance in preparing the figures. Requests for reprints of this article may be addressed to myself.

STUART H. HURLBERT

Abstract

Cladocerans are essentially freshwater organisms, many of which have been able to penetrate slightly saline waters (up to 5‰ salinity), both thalassic and athalassic, some of which occur at higher salinities, and a few of which, mostly non-chydorids, penetrate still higher salinities (15–30‰ and even higher) and may be confined to these salinities. Three previous studies from Saskatchewan, Iran, and Germany (the latter including thalassic waters) have been analyzed, and records for the athalassic saline waters of the World have been summarized; all results show a decline, at some point, in species number against increasing concentration of salinity.

Examination of samples for 67 waterbodies in southern Australia and 167 in South Africa, covering the full salinity range over which cladocerans occur, reveals much the same relationships. Graphical analysis was carried out using salinity intervals defined along a logarithmically-scaled conductivity axis.

The mean number of chydorid taxa per site was considerably greater in Australia than in South Africa, and this was true for both freshwater ($< 5 \, \mathrm{mScm}^{-1}$) and saline ($> 5 \, \mathrm{mScm}^{-1}$) sites. In both countries, the number of chydorid taxa per site showed little variation with conductivity over the freshwater range but declined rather abruptly at conductivities $> 5 \, \mathrm{mScm}^{-1}$. For South Africa, there was also some indication of reduced numbers of chydorid taxa in the most dilute ($< 0.2 \, \mathrm{mScm}^{-1}$) waters.

Non-chydorid taxa, which were analysed only on a generic basis, averaged much more numerous in Australian saline sites than in either South Africa saline sites or Australian freshwater ones. Mean number of non-chydorid taxa per site was about the same for Australian and South African freshwater sites.

Plots of total number of taxa observed per conductivity interval had maxima in the 0.2–0.7 mScm^{-1} conductivity range and decreased at higher and lower conductivities. This trend to some extent only reflected the effect of the variation in number of sites per interval; a negative impact on chydorids of very low conductivities ($< 0.2 \, \mathrm{mScm}^{-1}$) is nevertheless suggested.

Introduction

The four orders of cladocerans contain primarily freshwater taxa. There are a few genera – *Podon*, *Pseudoevadne*, *Evadne*, *Penilia*, and *Pleopis* – that are strictly marine and a number of other taxa in the Podonidae and Cercopagidae that occur primarily in the Caspian Sea, but nearly everything else is freshwater in occurrence. A number of taxa are able to push into weakly saline thalassic or athalassic waters, but only a few can push into higher salt concentrations. Because the cladocerans have no real ability to maintain an acceptable internal ionic concentration against higher external concentrations, each taxon eventually reaches an external concentration to which it cannot adjust physiologically and hence is eliminated.

There are two different kinds of saline waters into which cladocerans and other freshwater organisms can penetrate. The first is essentially diluted seawater, such as occurs in estuaries, where the freshwater input at the head maintains a gradient of increasing salinity toward the mouth. Freshwater organisms are able to enter the upper part of such thalassic systems and to push into progressively higher salinities according to their abilities. At most salinities in such estuarine systems, however, the relative ionic composition of the water is much the same, only the concentration is different.

In athalassic saline waters, which are those that develop out of contact with the oceans, the salts in them derive from surface inflow, ground water, and the atmosphere, and then become concentrated by evaporation. The system is enormously complex. The major catonic species (Na, K, Mg, and Ca) and the anionic ones (Cl, SO_4, and HCO_3/CO_3) can vary greatly from one waterbody to another, both as to relative abundance and as to total concentration. If the concentration increases over time, as by evaporation, various salts are precipitated or crystallized out as evaporites, thus leading to still further changes in relative concentrations of the major ions. Freshwater organisms can enter such saline waters as well as thalassic waters, but there is a good possibility they are influenced as much by the particular ionic composition and pH of the water as by the total concentration of dissolved substances. Thus two saline water bodies are rarely chemically equivalent.

Athalassic saline waterbodies differ further from thalassic waters in that many of them are seasonal or intermittent, sometimes not containing any water for years. When they fill with water from exceptional precipitation and runoff, the salinity tends to be low at first, then increases from evaporation. Even permanent saline waterbodies can experience large changes in concentration seasonally or over some years (cf. Bayly & Williams, 1966), thereby providing further stress for organisms of freshwater origin that have penetrated into the systems.

The chief question is, how do freshwater cladocerans react overall to the changing salinity of thalassic systems and to the changing salinity and highly variable ionic composition of athalassic systems? Literature reports on the cladocerans of Saskatchewan (Moore, 1952), Iran (Löffler, 1961), and Western Europe (Flössner, 1972), and my own detailed observations in Australia and South Africa provide partial answers.

Previous studies

1. Moore (1952) sampled 48 waterbodies in southern Saskatchewan with a salinity range of 0.17 to 107‰. Samples of entomostracans were collected with a 20-liter trap and with nets used for littoral tows, surface tows, and vertical tows. Twenty-four lakes were sampled once each in 1938, of which 16 plus an additional 23 were sampled in 1939. Eight of these lakes, covering the full salinity range, plus one new lake were sampled up to three times each in 1940, and one of these at 13.1% salinity was sampled at roughly weekly intervals in 1941. A total of 31 cladoceran taxa was collected, of which the littoral taxa admittedly were sampled inadequately.

A partial summary of the results is presented in Table 1. The lower salinity ranges selected are meant to coincide with the various limits proposed to separate freshwater from saline water –

Table 1. Number of occurrences of non-chydorid and chydorid taxa by salinity interval in 48 lakes from southern Saskatchewan. Data are from Table IV in Moore (1952).

Taxon	Salinity ‰ (number of lakes)							Total occurrences
	<0.5 (10)	0.5–1 (10)	1–3 (17)	3–10 (3)	10–15 (4)	15–25 (2)	>25 (2)	
Non-chydorid taxa								
Daphnia longispina O. F. Müller	8	9	16	1	3	2	2	41
Moina hutchinsoni Brehm						1		1
Diaphanosoma leuchtenbergianum Fischer	7	6	12	2	1			28
Daphnia pulex Leydig	4	4	5	2	2			17
Ceriodaphnia quadrangula (O. F. Müller)	1	2	8	2	2			15
Bosmina obtusirostris Sars	10	9	16	3	2			40
Scapholeberis mucronata (O. F. Müller)	1	1			1			3
Simocephalus vetulus Schoedler	1	1		1				3
Daphnia magna Straus			1	1				2
Ceriodaphnia lacustris Birge	3	1	8					12
Leptodora kindtii (Focke)	3	2	5					10
Macrothrix laticornis (Jurine)			1					1
Sida crystallina (O. F. Müller)	3	1						4
Polyphemus pediculus (Linnaeus)	1	1						2
Bosmina longispina Leydig	1							1
Ceriodaphnia reticulata (Jurine)	1							1
Ilocryptus acutifrons Sars	1							1
Mean no. non-chydorid taxa per lake	4.5	3.7	3.7	4.0	3.0	1.5	1.0	(3.8)
Chydorid taxa								
Chydorus sphaericus (O. F. Müller)	10	9	14	1	2			36
Pleuroxus denticulatus Birge		2	4		1			7
Alona affinis (Leydig)			2					2
Pleuroxus aduncus (Jurine)			3					3
Alona rectangula Sars	2	6	5					13
Alona costata Sars	3	1	2					6
Alona guttata Sars	1	1	1					3
Leydigia quadrangularis (Leydig)			2					2
Eurycercus lamellatus (O. F. Müller)	1	1	1					3
Camptocercus rectirostris Schoedler		1	2					3
Graptoleberis testudinaria (Fischer)	1		1					2
Chydorus gibbus Lilljeborg	1	1						2
Acroperus harpae Baird	2							2
Alonella nana (Baird)	1							1
Mean no. chydorid taxa per lake	2.2	2.2	2.2	0.3	0.8			(1.8)
Mean total no. cladoceran taxa per lake	6.7	5.9	5.8	4.3	3.8	1.5	1.0	(5.6)

0.5‰ (S.I.L., 1959), 1.0‰ (Löffler, 1961), and 3‰ (Williams 1981, and earlier). The last mentioned is now generally accepted as a conventional boundary between fresh and saline water, in part because so many freshwater organisms seemingly can function well at salinities up to about 3‰.

There is no means of identifying all the lakes that were sampled more than once, and likewise there are no separate data for cladocerans present on each sampling date and the salinity on those dates. Moore states, however, that the salinity of one lake had increased appreciably since 1920 and that the salinity in all or most of the lakes sampled had apparently increased perceptibly between 1938 and 1941. Salinity in any lake was reduced during the summer. Seasonal variations were small in lakes with low salinities but up to 5–8‰ in more saline waters. All that can be done in this analysis is to consider the taxa reported for a lake in relation to the one salinity datum reported for that lake.

Table 1 shows the 17 non-chydorids and 14 chydorids recovered from the lakes sampled and the number of lakes in each salinity range that contained each taxon. Both sets of cladocerans are arranged in order of decreasing maximum salinity at which the taxa were collected. The lowest three ranges of salinity were selected to coincide with the values of 0.5, 1, and 3‰ proposed in the past to separate freshwater from saline water. The scientific names are those used by Moore, even though quite a few changes have been made since his study; these changes do not invalidate the type of analysis undertaken here.

Of the 48 lakes studied, 37 would be considered freshwater (salinity <3‰). Nine of the 17 non-chydorids and 2 of the 14 chydorids had managed to push into saline water, with one species of *Daphnia* (called *D. longispina* in Moore's paper) and *Moina hutchinsoini* occurring in strongly saline water. Three additional species of non-chydorids and 10 chydorids extended into the 1–3‰ interval, which is saline according to the criterion Löffler (1961) used but freshwater according to the presently accepted limit (3‰). The rest of the taxa had their upper limits of occurrence somewhere in the freshwater range. Both sets of cladocerans showed decline in number of species per lake with increasing salinity.

Chydorid taxa are underrepresented, probably because they were inadequately sampled. If one calculates the mean number of chydorids and of non-chydorids in each salinity interval, the number of non-chydorids is always greater than the number of chydorids, which is the reverse of what one usually finds if the littoral zone is adequately sampled. The mean number of chydorids or non-chydorids per lake is fairly uniform for salinities up to 3 to 10‰.

All lakes in prairie (7) and parkland (22) tended to have salinities greater than 1‰. Lakes in the forest region (19) tended to be fresh. The prairie and parkland lakes were dominated by Na, Mg, and SO_4 ions, those of the forest by Ca, Mg, and HCO_3. Thus, these lakes are appreciably different from the Na and Cl dominated lakes in many parts of the world. Moore claimed that total salinity is the controlling factor in the distribution of these entomostracans, not the dominance of particular ions. He suggested that the maximum salinity that can be tolerated by most freshwater animals lies between 4 and 15‰.

2. Löffler (1961) spent six months in Iran, studying the organisms in 16 waterbodies or groups of waterbodies out of a national total of no more than 70, half of which are hypo- to mesosaline (approx. 5–30‰). The collections yielded a total of 39 species of cladocerans, 7 of which occurred only in the Caspian Sea. Fourteen additional species had been reported by previous investigators, yielding a total of 53 (or 46 without the Caspian forms) known from the country. Of the 39 species Löffler identified, 12 are chydorids, and four more chydorids were added by earlier investigators. Of the chydorids, *Alona rectangula* was the most frequent, occurring in six of the waterbodies. All the other species occurred only in one or two of the 16 waterbodies. *Monospilus dispar* Sars, *Dunhevedia crassa* King, and *Chydorus* (= *Ephemeroporus*) *barroisi* (Richard) occurred only in the samples from the Lake Hamun region. *Alona affinis* (Leydig) and *Alonella excisa* (Fischer) occurred only in Lake Zeribar. Chydorid species per sample ranged from 0 to 6, with a mean of 1.3, whereas non-chydorid taxa ranged from 0 to 7 per sample, with a mean of 2.8.

Arranging these taxa against salinity is difficult. For some samples complete major ion analyses were given, for other samples only chloride, and for a couple samples nothing. Yet it is apparent that lakes Hamun and Zeribar, which had the freshest water, had 17 and 8 species of cladocerans, respectively, and that Lulunar-Göl with a salinity of roughly 31‰ had only *Moina salinarum* Gurney. Lake Famur with 2.76‰ salinity and Tatawi-Tschai with 170 mg Cl l^{-1} had 8 and 6 species, respectively. All the other waterbodies had just 1 or 2 species, except for Binah, which had 4, all non-chydorids. These are very low numbers of cladoceran taxa, even for the salinities reported.

What Löffler did in addition was to determine the maximum salinity at which each entomostracan taxon had been recorded from papers on saline environments down to 1‰, which he took as the dividing line between freshwater and saline water. Only four such maxima were at lesser salinities, down to 0.4‰. The 183 species in the papers reviewed, including 51 species of cladocerans of which 14 are chydorids, were tabulated against salinity, chlorinity, and $CO_3 + HCO_3$. Salinity was considered to be roughly twice chlorinity, so that when exact data for the former were lacking they could be approximated for purposes of tabulation. Fourteen salinity records of cladocerans were obtained by doubling the chlorinity.

The number of entomostracan salinity maxima per salinity category when plotted against salinity yielded a strongly descending curve, with the largest number of maxima at low salinities. No evident break occurred in the curve at any salinity.

Löffler listed 54 cladoceran species, but for 4 non-chydorids no salinity estimates were available. Thirty non-chydorids (out of a total of 37) had their maximum recorded salinities in saline water (>3‰, mostly <13‰), but with *Daphnia atkinsoni* Baird at 19.7‰, *Moina macrocopa* Straus at 22.2‰, and four other species of *Moina* plus *Macrothrix hirsuticornis* Brady & Norman at salinities from 30 to 39‰. Four species of chydorids likewise had maxima at salinities >3‰ – *Leydigia acanthocercoides* (Fischer) at 3.5‰, *Chydorus sphaericus* at 4‰, *Alona tenuicaudis* Sars at 6‰, and *Alona rectangula* at 12.6‰. Using Löffler's criterion of 1‰, all species in his list would be classified as occurring in saline water. Löffler stated that some of the non-chydorids, such as *Moina salinarum*, seemed confined to saline water. Löffler also pointed out that many cladocerans

are excluded from waters of very low salinity, e.g. <0.09‰.

3. Flössner (1972). The two previous works treated cladocerans found in athalassic waterbodies. In his book on the branchiopods of Germany, which includes the cladocerans, Flössner has a total of 107 species, 5 of which (in the genera *Podon* and *Evadne*) occur only in marine waters. For most of the other 102 species, Flössner presents information from the literature concerning the highest salinity at which the species has been collected. This includes thalassic records from the Baltic Sea, Barents Sea, Black Sea, Zuidersee, Bremerhaven and Pellworm Island in the North Sea, plus the estuaries of the Elbe and Rhine rivers. The Baltic Sea is the region cited most frequently. For the athalassic waters that are fresh, no localities are given, but the saline athalassic waters are variously listed as occurring in 'steppe regions', Saskatchewan, Hungary, Greenland, and lakes marginal to the Sea of Azov.

Table 2 summarizes the distribution of the 102 German species by maximum salinity reported and by the nature of the waterbodies. For both thalassic and athalassic waters, few species had maxima at salinities >10‰. The few chydorids reported from higher salinities were *Alona quadrangularis* (O. F. Müller), *A. rectangula* Sars, *Chydorus sphaericus* (O. F. Müller), and *Dunhevedia crassa* King. The ten non-chydorids from these higher salinities are *Daphnia atkinsoni* Baird, *D. galeata galeata* Sars, *D. cristata* Sars, *D. similis* Claus, *D. schoedleri* Sars, *Scapholeberis mucronata* (O. F. Müller), *Moina macrocopa* (Straus), *M. brachiata* (Jurine), *Bosmina coregoni* Baird, and *B. longispina* Leydig. Waterbodies in the 1–3‰ salinity range are certainly underrepresented, and hence the number of species surviving into this range but no higher is probably also underrepresented.

These data are of limited value in the present context, because there usually was given only one salinity value per species.

4. Hammer (1986) attempted to summarize everything about athalassic saline lakes from the world literature and from his great personal

Table 2. Numbers of chydorid (C) and non-chydorid (N) taxa from Germany with maximum recorded salinities in the ranges listed and the sources of these salinities. Adapted from Flössner, 1972.

		Salinity ‰					
		<1	1–3	3–10	10–15	15–25	>25
Barents Sea	C			0		0	
	N			3		3	
North Sea, including outer Elbe, Pellworm Island, Bremerhaven, Rhine delta, and Spiekeroog	C		0	2			1
	N		1	0			1
Zuidersee	C						1
	N						1
Baltic Sea, including Kattegat	C		2	10			
	N		0	8			
Total thalassic (33 taxa)	C		2	12		0	2
	N		1	11		3	2
Athalassic, including steppic waters, Greenland, Saskatchewan, and Darsser Bodengewasser (69 taxa)	C	18	3	3	1	0	1
	N	26	1	11	1	2	2

experience. He considered 3‰ salinity as the boundary between subsaline (0.5–3‰) and hyposaline (3–20‰) waters and included no organisms that are restricted to the subsaline range. For salinities >3‰ he listed 32 species of cladocerans in his Table 7.18, and mentioned an additional 18 species in the text. Seven cladocerans from Western Australia are difficult to include in this analysis, because only the generic names are given. The 43 named species are distributed by family as follows: sidids – 4, daphniids – 15, moinids – 8, bosminids – 1, macrothricids – 5 (+3 from Western Australia), chydorids – 10 (+4 from Western Australia). The three chydorids in his table and the maximum salinity from which each was recorded are *Oxyurella tenuicaudis* – 6‰, *Alona rectangula* – 12‰, *Chydorus sphaericus* – 59‰. The additional 11 chydorids mentioned in the text, including the generic designations from Western Australia, are reported from salinities between 3 and 18‰. These are: *Chydorus letourneuxi* Richard (Algeria); *Alona costata* (Pyramid Lake); *Euryalona occidentalis* Sars and *Leydigia quadrangularis* (Buenos Aires); *Alona*, *Chydorus*, *Biapertura*, and *Monospilus*

(Western Australia); and *Biapertura rigidicaudis*, *Alona cambouei*, and *Alona davidi* Richard from Australia.

Thus, worldwide there is a fair number of taxa that push into athalassic saline waters of relatively low salinity, including members from all the families of the Anomopoda and from the Sididae of the Ctenopoda. Species occurring at higher salinities are chiefly daphniids, moinids, and macrothricids. A few, such as *Daphniopsis* and *Moina baylyi* Forró in Australia, and *Moina microcephala* and *Moina mongolica* Daday elsewhere, seem to be confined to mesosaline waters (20–50‰ salinity) and to the upper part of the hyposaline range (3–20‰).

In his discussion of fishes in athalassic saline waters, Hammer noted that chloride waters are less stressful than sulfate waters, which in turn are much less stressful than bicarbonate waters, so that salinity tolerance may be controlled more by specific ion effect than salinity per se. He also noted that in Australia there is a host of invertebrate species 'that are well adapted and more tolerant of salinity than anywhere else in the World'.

Field studies in Australia and South America

In 1986–87 I spent 3-1/2 months in southern Australia and in 1990, 2-1/2 months in South Africa making numerous littoral collections in the littoral zone of a great variety of surface waters. My main objective was to study the chydorid anomopods on these two southern continents and to compare them morphologically for discerning any evidences of Gondwana relationships. I was attempting to collect the complete chydorid fauna of each region, and hence I collected lakes, ponds, pools, ditches, swamps, marshes, and flowing waters of various size and discharge. Sites were selected not for their salinity but for their likelihood of containing chydorids. Because I worked only in the littoral zone I undoubtedly missed planktonic non-chydorids occurring offshore in large waterbodies, but I probably have all or nearly all the cladocerans momentarily present in the smaller waterbodies.

Up to this time I had been working mainly in humid regions, where salinity is probably a minor influence on the distribution of anomopods except at extremely low salinities. Australia was an awakening because of the few waterbodies in its southern quarter that are strictly fresh. With borrowed instruments I measured the conductivity (K_{25}) of about a third of the waterbodies visited in Australia. In South Africa, with the obliging field assistance of various limnologists, I obtained measurements of conductivity in nearly all the waterbodies visited. Each waterbody, except for Diep Rivier in the Knysna/George region of Cape Province, was sampled only once. All the sites on both continents were visited in the austral spring–early summer.

Samples at each site were collected from whatever substrate was present with a polemounted, 110 μm -mesh nitex net, the mouth of which was protected by a brass screen having 5 mm meshes. Any aquatic vegetation present was particularly sampled. The one sample collected from each waterbody took about 20 minutes to collect, after which it was preserved immediately with concentrated formalin to which sugar had been added, yielding a final concentration of about 3–5‰

formaldehyde. Collecting methods and intensity were the same on the two continents. In the laboratory an aliquot of each sample, representing about 1/10 to 1/50 of the total volume of the sample and usually containing at least a couple hundred chydorids, was examined at a magnification of $25\times$ with a stereomicroscope. Taxa were listed as encountered, along with estimates of their relative abundance and notes on the occurrence of gamogenetic stages and distinctive features of morphology. No actual counts were made. If one were after the complete faunule present in each waterbody, examination of remains in the sediments would be best, except that this requires aknowledge of what species are present in the region and how they differ from one another morphologically. This was not the purpose of the present collecting, which was to obtain intact animals of all instars and reproductive stages.

Taxa found were identified to species where possible, though all non-chydorids and some chydorids, especially in the genera *Alona*, *Tylopleuroxus*, and *Chydorus*, are listed only to genus.

Table 3 shows the distribution of collecting sites by region, type, and presence of cladocerans. Australian sites were mainly within a couple hundred kilometers of the south and southwest coasts, in the vicinity of Perth, Adelaide, Melbourne, Albury, and Hobart. The Murray River region at Albury is the farthest from the coast any collections were made. The South African sites were distributed over the entire country, although largely from near the bases of operation at Pretoria, Bloemfontein, Cape Town, Knysna, Grahamstown, and Pietermauritzburg.

The standing water sites are the focus of the analyses in this paper, as no stream sites were saline and as Australian stream sites were too few to permit a strong Australia-South Africa comparison. Some of the very softwater ($<0.1\,\mathrm{mScm^{-1}}$) streams in the Cape Town, Knysna/George and Natal Drakensberg regions yielded a number of species of chydorids not found in any of the lentic waterbodies. These included *Monospilus* cf. *dispar* Sars, *Alona* cf. *rustica* Scott, *Alonella* cf. *nana* (Baird), and *Chydorus* cf.

Table 3. Number of standing water sites sampled in Australia and South Africa where conductivity measurements were made. The Cape Province talleys include some sites in the Grahamstown region that were sampled in December 1989.

Country	Number of sites	
State or province	Positive[a]	Blank[b]
Australia (1986–1987)		
Western Australia	22	0
South Australia	25	4
Victoria	11	0
New South Wales	0	0
Tasmania	5	0
Totals	63	4
South Africa (1989–1990)		
Transvaal	9	1
Orange Free State	41	4
Cape Province	70	14
Natal	26	2
Totals	146	21

[a] Sites in which one or more cladoceran species was found.
[b] Sites in which no cladoceran species were found.

piger Sars. *Acroperus* sp was found in standing water sites but occurred with greater frequency in streams.

The frequency distributions of standing water sites over the conductivity spectrum are shown in Fig. 1. Vertical dashed lines are drawn at conductivity values of 0.77, 1.5, and 4.6 mScm^{-1}. As estimated by the approximate relation

$$S = aK_{25},$$

where S = salinity (g l^{-1}), K_{25} = conductivity (mScm^{-1} at 25 °C), and a = 0.65 (Hem, 1970), these three conductivity values correspond to salinities or total dissolved solids (TDS) of 0.5, 1.0, and 3.0 g l^{-1} or ‰, the salinities proposed in the past as dividing lines between fresh and saline waters.

Both frequency distributions are irregular, positively skewed, and somewhat suggestive of the underlying bimodal distributions (S. Hurlbert, in preparation). The best 'dividing line' between modes would lie at about 1–2 g l^{-1} for the Australian waterbodies and at 2–3 g l^{-1} for the

South African ones. The freshwater sites in South Africa included proportionally more very low conductivity (e.g. <0.4 mScm^{-1}) sites than did the Australian freshwater sites.

Cladocerans were found in nearly all sites (Fig. 1). The four 'blank' Australian sites all were saline. 'Blank' South African sites included saline or near-saline waters in the Bloemfontein and Cape Agulhas regions, newly created farm ponds, recently flooded seasonal waterbodies, and sites where possible reasons for the absence of cladocerans were less clear.

From standing water sites, a total of 84 cladoceran taxa was recovered, 41 from Australia (27 chydorids) and 43 from South Africa (31 chydorids). Many of these taxa contained more than one species. The non-chydorid genera of *Daphnia*, *Ceriodaphnia*, *Moina*, *Macrothrix*, and *Echinisca* certainly contain more than one species in the samples from each continent.

Moreover, quite a number of the chydorid taxa listed are multispecific or composite and not yet sorted out, such as *Biapertura* cf. *rigidicaudis* Smirnov, *Leydigia* sp., *Alona* sp., *Tylopleuroxus* sp., and *Chydorus* cf. *sphaericus*. Australia has many endemic genera and species (Smirnov & Timms, 1983; Smirnov 1989a, 1989b; Frey, 1991a, 1991b). South Africa has few. Most of the South African species closely resemble related European species and currently share the European names. Because morphological comparisons that have been made for other cognate pairs involving other continents show that the cognates on different continents are usually different species (see Frey, 1986 for a review), any taxon on either continent that bears a name originating on some other continent is qualified in this paper by a cf. designation.

The frequency of occurrence of each taxon in each conductivity interval is given in Table 4, in which the non-chydorids and chydorids are both arranged in order of decreasing maximum conductivity at which they were found. Eight genera of non-chydorids had taxa extending well into the saline range. Striking patterns included the occurrence of *Eudaphnia* at low conductivities and *Ctenodaphnia* at high conductivities in Australia,

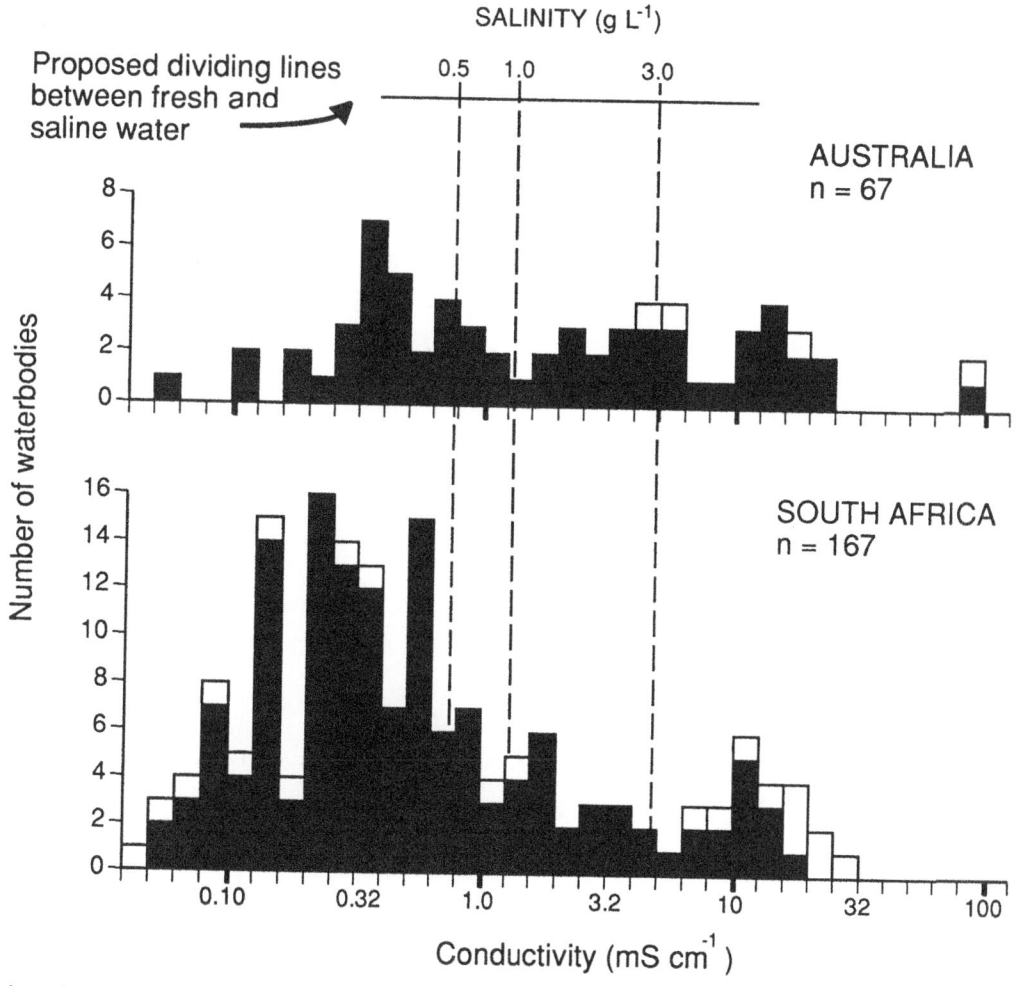

Fig. 1. Number of standing water sites per conductivity interval, for 234 sites in Australia and South Africa. Unshaded portions of bars represent sites in which no cladocerans were found (see Table 3). Boundaries between conductivity intervals correspond to 10^b times 1.00, 1.26, 1.58, 2.00, 2.51, 3.16, 3.98, 5.01, 6.31, 7.94, and 10.0 μScm^{-1}, where b = an integer varying from -2 to $+2$.

and the high frequency of *Echinisca* spp. at high conductivities in Australia. *Moina* in the Grahamstown region of South Africa seems to be mainly a freshwater form, whereas in Australia the genus was found only at two saline sites. *Daphniopsis* was strictly saline, and *Latonopsis* extended into the saline range. The other eight genera, except *Scapholeberis* and *Pseudosida*, were recorded only from waters with conductivities less than 1 mScm^{-1}. However, the frequencies of these taxa were very low, except for *Scapholeberis* and *Ilyocryptus*, and so the records undoubtedly underestimate the true ranges of salinity tolerance.

On the other hand, no taxa, except *Pseudosida*, have isolated occurrences only at conductivities greater than 1 mScm^{-1}. Thus, for the non-chydorids there are a number of genera that occur over almost the whole conductivity range but with strong indications that the species at the upper and lower ends of this range are different species.

For the chydorids the situation is much the same. A fair number of taxa occur at conductivities greater than 4–5 mScm^{-1}, *i.e.* at salinities greater than 3‰. The genera *Leydigia*, *Alona*, and *Tylopleuroxus* are complicated in that each contains a number of species, often at the same site,

Table 4. Number of occurrences of non-chydorid and chydorid taxa, by conductivity interval, in 67 standing water sites in Australia (1) and 167 in South Africa (SA). See text for explanation of sequences and groupings of taxa.

Taxon		Conductivity interval (mS cm^{-1}) (Number of sites, A/SA)														Total occurrences	
		<0.1 (1/15)	0.1–0.2 (5/24)	0.2–0.3 (4/26)	0.3–0.5 (12/24)	0.5–0.7 (4/18)	0.7–1.0 (4/11)	1–2 (5/15)	2–3 (5/5)	3–4 (2/3)	4–6 (6/3)	6–10 (4/6)	10–15 (5/8)	15–20 (5/6)	>20 (5/3)	A	SA
Non-chydorid taxa																	
Daphnia	A	–	–	2	2	–	–	–	–	–	–	–	–	1	1	6	
	SA	5	7	4	9	2	4	2	–	1	–	1	1	1	–		37
Simocephalus	A	–	5	1	3	–	1	2	2	2	4	1	3	1	1	26	
	SA	4	7	12	9	9	6	5	2	2	1	–	–	1	–		58
Macrothrix	A	–	1	–	1	–	–	1	–	1	1	–	1	1	1	8	
	SA	2	7	14	3	5	4	4	1	2	1	1	1	–	–		45
Echinisca	A	–	–	–	1	–	1	1	2	1	2	2	4	4	1	19	
	SA	–	1	–	–	–	–	–	–	–	–	–	–	1	–		2
Ceriodaphnia	A	1	1	1	1	–	1	–	2	1	2	1	3	1	–	15	
	SA	4	10	14	9	8	6	8	3	2	3	2	2	1	–		72
Moina	A	–	–	–	–	–	–	–	–	–	–	–	–	2	–	2	
	SA	2	4	3	6	3	5	2	–	–	1	–	2	2	–		30
Scapholeberis	A	–	–	–	4	1	–	2	–	–	1	–	–	–	–	8	
	SA	–	1	2	2	1	–	3	–	1	–	–	–	–	–		10
Ilyocryptus	A	–	–	–	2	–	–	–	–	–	–	–	–	–	–	2	
	SA	1	–	3	–	3	1	–	–	–	–	–	–	–	–		8
Other non-Chyrdorids[1]	A	–	2	1	3	2	–	–	–	–	1	1	–	–	1	11	
	SA	1	–	2	2	–	–	1	–	–	–	–	–	–	–		6
Chydorid taxa																	
Biapertura cf. rigidicaudis Smirnov	A	1	–	2	4	1	1	1	–	1	4	2	–	2	1	20	
Leydigia spp.	A	–	–	–	–	–	1	–	–	–	–	–	–	–	–	1	
	SA	1	4	4	2	4	2	2	2	–	1	–	1	1	–		24
Alona spp.	A	–	1	1	2	3	–	1	1	–	3	–	–	2	–	14	
	SA	3	1	3	7	6	3	3	2	1	1	–	–	–	–		30
Biapertura setigera (Brehm)	A	1	2	–	5	3	1	3	1	1	2	–	–	1	–	20	
Alona cf. costata Sars	SA	–	–	4	–	–	–	–	–	–	–	–	–	–	–		4
Tylopleuroxus spp.	A	–	–	3	4	–	3	2	4	2	3	–	1	1	–	23	
	SA	1	1	7	7	7	5	6	3	3	1	–	1	1	–		43
Chydorus cf. sphaericus (O. F. Müller)	A	1	3	1	5	3	3	3	3	1	4	–	–	1	–	28	
	SA	8	9	22	12	11	8	7	2	1	1	–	1	–	–		82
Dunhevedia crassa King	A	–	2	1	–	–	1	3	2	1	3	3	1	–	–	17	
Dunhevedia cf. crassa King	SA	1	1	4	1	5	1	4	–	1	1	1	–	–	–		20
Ephemeroporus cf. barroisi (Richard)	A	–	–	–	–	2	1	3	–	2	3	1	1	–	–	13	
	SA	–	–	2	1	3	–	–	–	–	–	–	–	–	–		6
Alonella cf. exigua (Lilljeborg)	SA	–	–	1	2	1	–	1	–	–	–	–	–	1	–		6

Table 4. (Continued)

Taxon		Conductivity interval (mS cm⁻¹) (Number of sites, A/SA)														Total occurrences	
		<0.1 (1/15)	0.1–0.2 (5/24)	0.2–0.3 (4/26)	0.3–0.5 (12/24)	0.5–0.7 (4/18)	0.7–1.0 (4/11)	1–2 (5/15)	2–3 (5/5)	3–4 (2/3)	4–6 (6/3)	6–10 (4/6)	10–15 (5/8)	15–20 (5/6)	>20 (5/3)	A	SA
Alona diaphana King	A	–	–	1	1	–	–	2	2	–	1	1	–	–	–	8	
Alona cf. *diaphana* King	SA	–	–	4	–	–	–	–	–	–	–	–	–	–	–		4
Alona cf. *cambouei* (de Guerne & Richard)	A	–	–	–	–	–	1	2	2	1	1	–	–	–	–	7	
	SA	1	–	–	–	–	–	–	–	–	–	–	–	–	–		1
Alonella cf. *excisa* (Fischer)	A	–	1	–	4	–	–	–	–	–	2	–	–	–	–	8	
	SA	–	–	5	3	2	1	–	–	–	–	–	–	–	–		11
Camptocercus australis Sars	A	–	–	–	4	1	–	1	–	1	–	–	–	–	–	7	
Grapotoleberis testudinaria (Fischer)	A	1	–	–	5	–	–	1	–	1	–	–	–	–	–	8	
	SA	1	–	1	–	1	–	–	–	–	–	–	–	–	–		3
Alona cf. *pulchella* King	SA	–	–	2	1	3	2	1	–	1	–	–	–	–	–		10
Pseudochydorus cf. *globosus* (Baird)	A	–	–	–	1	–	–	2	–	1	–	–	–	–	–	4	
	SA	–	1	1	1	–	1	1	–	–	–	–	–	–	–		5
Tretocephala coletti (Sars)	SA	–	1	4	1	–	–	1	–	–	–	–	–	–	–		7
Euryalona cf. *orientalis* (Daday)	SA	–	–	2	1	2	1	1	–	–	–	–	–	–	–		7
Biapertura kendallensis (Henry)	A	–	1	–	2	–	–	1	–	–	–	–	–	–	–	4	
Biapertura cf. *affinis* (Leydig)	SA	3	2	6	1	–	–	1	–	–	–	–	–	–	–		13
Biapertura cf. *karua* (King)	SA	–	1	2	2	2	–	1	–	–	–	–	–	–	–		8
Alona cf. *guttata* Sars	A	1	1	–	1	1	2	–	–	–	–	–	–	–	–	6	
	SA	3	2	10	1	1	1	–	–	–	–	–	–	–	–		18
Chydorus cf. *pubescens* Sars	SA	1	1	3	1	2	2	–	–	–	–	–	–	–	–		10
Rak spp.	A	1	1	–	3	1	–	–	–	–	–	–	–	–	–	6	
	SA	–	1	1	1	1	–	–	–	–	–	–	–	–	–		4
Biapertura cf. *intermedia* (Sars)	SA	5	–	5	2	1	–	–	–	–	–	–	–	–	–		13
Other chydorids[2]	A	1	2	–	7	–	3	–	1	–	–	–	–	–	–	14	
	SA	–	3	4	–	5	2	1	–	–	–	–	–	–	–		15

[1] Australia: *Neothrix armata* Gurney (1:0.36); *Streblocerus* sp. (2:0.36, 0.38); *Bosmina* sp. (2:0.55, 0.66); Pseudomoina sp. (1:0.17); *Latonopsis* sp. (4:0.17, 0.24, 4.5, 6.6); *Daphniopsis* sp. (1:84.7).
South Africa: *Diaphanosoma* sp. (4:0.22, 0.27, 0.37, 0.38); *Bosmina* sp. (1:0.10); *Pseudosida* sp. (1:1.2).

[2] Australia: *Oxyurella* sp. (1:0.05); *Alona* cf. *rectangula* Sars (1:3.8); *Alona* cf. *inreticulata* Shen *et al.* (1:0.87); *Alona* cf. *quadrangularis* (O. F. Müller) (2:0.48, 0.85); *Biapertura longingua* Smirnov (2:0.17, 0.42); *Biapertura macrocopa* (Sars) (3:0.17, 0.36, 0.91); *Leberis aenigmatosa* Smirnov (1:0.40); *Monope reticulata* (Henry) (2:0.36); *Rhynchochydorus australiensis* Smirnov & Timms (1:0.36).
South Africa: *Eurycercus* (*Eurycercus*) sp. (1:0.13); *Acroperus* sp. (1:0.13); *Kurzia* cf. *longirostris* (Daday) (2:0.62, 0.82); *Alona* cf. *quadrangularis* (O. F. Müller) (1:0.15); *Notoalona* cf. *globulosa* (Daday) (2:0.24, 0.82); *Alonella* cf. *clathratula* Sars (1:0.55); *Alonella* cf. *hamulata* (Birge) (1:0.22); *Picripleuroxus* sp. (2:0.24, 0.62); *Chydorus tilhoi* Rey & Saint-Jean (2:0.24, 0.62); *Oxyurella* sp. (2:0.62, 1.5).

especially in the case of *Alona*. *Biapertura* cf. *rigidicaudis*, *Chydorus* cf. *sphaericus*, and *Ephemeroporus* cf. *barroisi* likewise each include more than one species. *Biapertura setigera* (Brehm) looks very much like *Alona* cf. *costata* except that it has only two median headpores instead of three, thereby providing one of the strong counterarguments against the validity of *Biapertura* as an acceptable genus. The other taxa extending into saline waters – *Dunhevedia crassa*, *Alonella* cf. *exigua*, *Alonella* cf. *excisa*, *Alona diaphana*, *Alona* cf. *cambouei*, and *Alona* cf. *rectangula* – all seem to be individual species on each continent.

At lesser conductivities, five chydorid taxa barely extend into the 3–4 mScm^{-1} interval and hence might marginally be considered saline adapted. The other 25 taxa (or a few more if the cognate pairs are eventually resolved into separate species) are all strictly freshwater. Many of these seem to be habitat specialists, and at least seven from Australia and one (or two) from South Africa are specialized endemics in those geographic regions. If one assumes that all cognate chydorid taxa on the two continents are different species, then the number of taxa in each of four conductivity ranges spanning the fresh-to-saline transition is as follows: <0.5 mScm^{-1}, 54 taxa; 0.5–1 mScm^{-1}, 36 taxa; 1–3 mScm^{-1}, 27 taxa; and >3 mScm^{-1}, 21 taxa. Many taxa, of course, occur in more than one conductivity range. There is a greater number of taxa present at low conductivities, with a considerable number persisting to the upper limit of freshwater, and an impressive number penetrating into saline waters.

To analyse further how species richness varied with conductivity, the mean number of taxa per site was calculated for each conductivity interval (Fig. 2). Patterns exhibited by the resultant curves need to be interpreted with caution. Standard errors are relatively large and particular peaks or depressions may reflect only the vagaries of site selection.

In freshwaters (<4.6 mScm^{-1}), chydorid taxa averaged almost twice as many per site in Australia as they did in South Africa (Table 5). Non-chydorid taxa, however, averaged slightly more numerous per freshwater site in South Africa than in Australia. This difference is not statistically significant (Table 5). But it does at least suggest that the relative paucity of chydorids in South African freshwater sites cannot be attributed to those sites being of 'poorer' quality for cladocerans generally than were the Australian freshwater sites.

Mean number of taxa per site showed no clear trend over the freshwater portion of the salinity spectrum for either country or either group of cladocerans (Fig. 2). A possible exception was the reduced mean number of chydorid taxa that characterized the more dilute (<0.2 mScm^{-1}) South African sites.

At conductivities >4.6 mScm^{-1}, mean number of taxa per site declined rather abruptly, except in the case of Australian non-chydorid taxa (Fig. 2). These latter actually averaged 30 percent more abundant per saline water site than per freshwater site (Table 5, 1.75 *vs.* 1.35). Species of *Simocephalus*, *Ceriodaphnia* and *Echinisca* are those primarily responsible for the high richness of the saline sites.

In Australian saline waterbodies, both chydorid and non-chydorid taxa averaged much more numerous than they did in South African saline waterbodies (Table 5).

When one considers the total number of non-chydorid and chydorid taxa in each conductivity interval, instead of the mean number per site in each interval, a different picture emerges (Fig. 3). It is a less conclusive one, however, in that these values, unlike those in Fig. 2, are strongly influenced by the sample size for each interval, *i.e.* the number of waterbodies in it, and this varies from interval to interval and between the two countries. It is predictable that the number of taxa recorded for an interval will be a positive function of the number of sites examined in that interval.

Thus in the freshwater range, where South African sites per interval were more numerous than Australian sites, the curves for South Africa lie mostly above those for Australia. The decrease in number of taxa per interval at higher salinities surely reflects the smaller percentage of the biota that is adapted to those salinities but also reflects

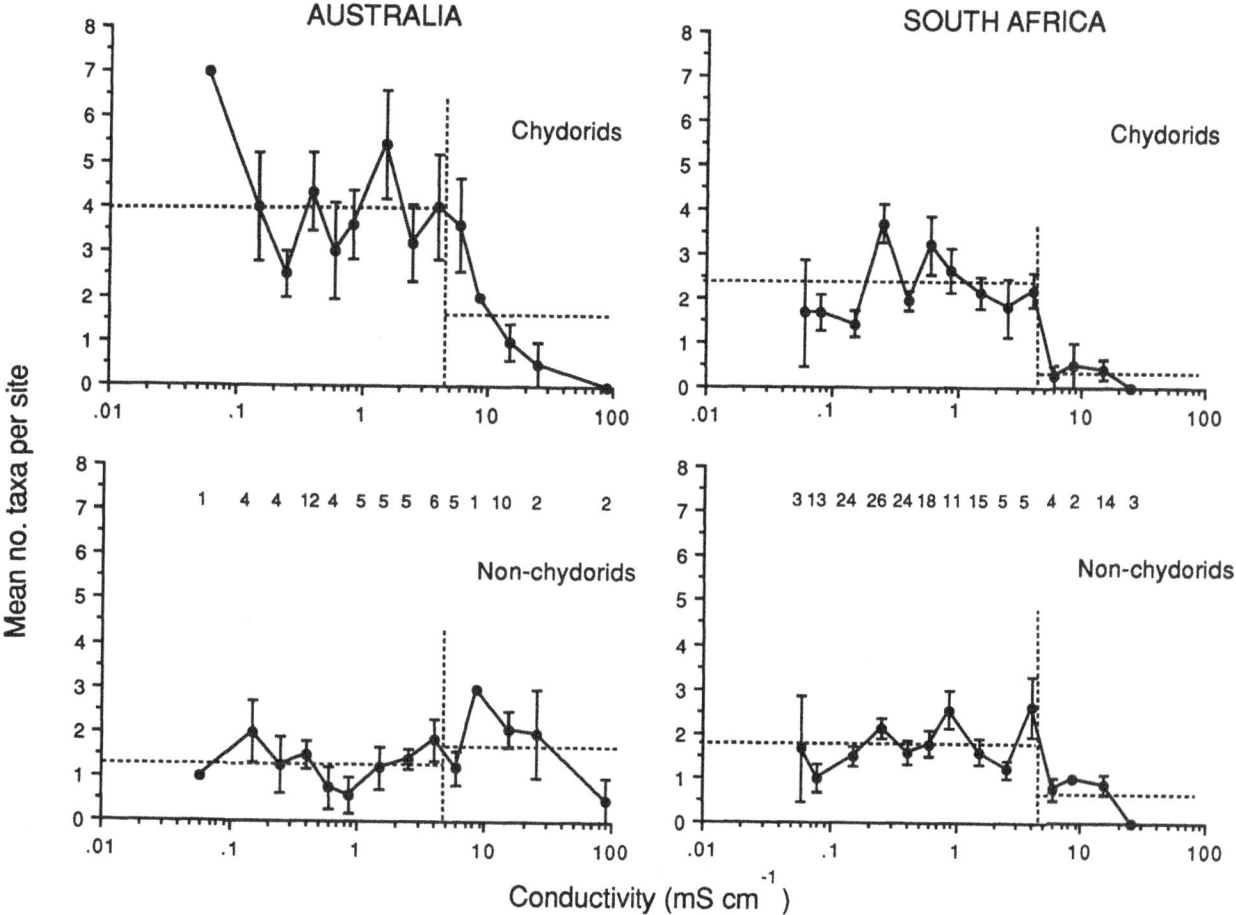

Fig. 2. Mean number of chydorid and non-chydorid taxa per site (± SE), by conductivity interval, for Australia and South Africa. Boundaries between conductivity intervals correspond to 10^b times 1.0, 2.0, 3.0, 5.0, 7.0 and 10.0 mScm^{-1}, where b = an integer varying from −2 to +2. Mean number of taxa has been plotted at values corresponding to the arithmetic mean of the upper and lower boundary for each conductivity interval. Dashed horizontal lines represent the mean number of taxa per site for all sites with conductivities <5 mScm^{-1} and >5 mScm^{-1}, respectively. Dashed vertical lines are drawn at 4.6 mScm^{-1}, which corresponds approximately to a salinity of 3 g l^{-1}. For each country, the number of sites on which the means are based are indicated across the top of the non-chydorid graphs.

the smaller number of waterbodies per interval in the high salinity part of the spectrum. In principle, one could attempt corrections for variation in sample size but this is not done here.

Some biologically interesting patterns are nevertheless discernible. The decline in number of chydorid taxa per interval for intervals below 0.2 mScm^{-1} is especially notable for South Africa (Fig. 3). It cannot be an artifact of low sample sizes because these are rather high, e.g. $n = 24$ for the interval 0.1–0.2 mScm^{-1}. Together with the observed low average number of chydorids per

site for these same intervals (Fig. 2), this suggests that the low ionic strength of these waters may be exerting a negative impact on the chydorids, as Löffler (1961) suggested.

The number of chydorid taxa found in the interval 0.2–0.7 mScm^{-1} was greater in South Africa than in Australia perhaps partly for reasons relating to site distribution. Sites in northern coastal Natal yielded various subtropical chydorid taxa, whereas the Australian taxa were all south temperate in distribution.

The rather flatter curves for the non-chydorid

Table 5. Mean number of taxa per site as a function of conductivity, country, and group.

Salinity category	Chydorid taxa		Non-chydorid taxa	
	Australia	South Africa	Australia	South Africa
Freshwater sites ($K_{25} < 5$ mScm^{-1})				
$\bar{X}$	3.96	2.38	1.35	1.74
s	2.45	1.94	1.06	1.28
n	46	144	46	144
pa	< 0.001		0.55	
Saline sites ($K_{25} > 5$ mScm^{-1})				
$\bar{X}$	1.55	0.35	1.75	0.74
s	1.88	0.71	1.21	0.75
n	20	23	20	23
pa	0.016		0.009	

[a] *P* values signify results of simple t-tests.

taxa (Fig. 3) may reflect the carrying out of this analysis at the generic rather than the specific level, but possibly also reflect real differences between chydorids and non-chydorids. Over the freshwater intervals, the curve for South Africa is mostly above that for Australia. This could be entirely an artifact of the larger sample sizes for South Africa. The curve for Australia shows no tendency to drop off until conductivity exceeds 30 mScm^{-1}.

Discussion

The quantity of substances naturally dissolved in water increases from virtually nothing up to supersaturated brines. The lower concentrations are called fresh water and the higher concentrations saline, with one or another concentration arbitrarily proposed to separate fresh from saline. To operate effectively at higher salinities, organisms need some physiological mechanism to counterbalance the higher ionic concentrations outside the body. Without such mechanisms, the organisms are osmoconformers and eventually are eliminated by external salinities beyond their limited capacity to compensate for.

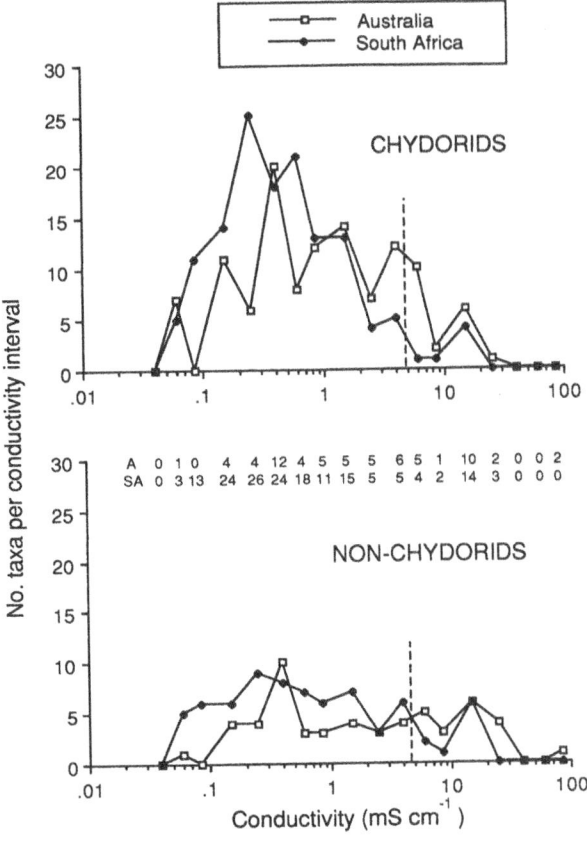

Fig. 3. Number of observed chydorid and non-chydorid taxa per conductivity interval, for Australia and South Africa. Boundaries between conductivity intervals and positioning of plotted values are as in Fig. 2. Dashed vertical lines are drawn at 4.6 mScm^{-1}. Numbers across top of bottom graph give number of sites in each conductivity interval for the two countries. Points are plotted even for intervals where sample size (= no. sites) is zero, so as to draw attention to the sample size dependence of number of taxa found per interval.

Cladocerans are one such group. They occur primarily at low electrolyte concentrations, and many of the species are confined to such concentrations. A large proportion, though, can easily withstand salinities higher than 1‰ or even 3‰. As the salinity concentration increases, the number of species that can remain functional declines, but even at quite high salinities there are a few cladocerans, chiefly non-chydorids, that thrive here and may in fact be confined to these higher salinities. The lowest salinities may also create physiological problems, in that some taxa may not be able to osmoregulate satisfactorily at concentrations less than about 0.05‰ or even somewhat higher. Thus, the cladocerans although freshwater in origin and primarily freshwater in occurrence may tend to be restricted to some intermediate range of salinity, with the extent of penetration into higher salinities varying from one taxon to another. The taxa that can tolerate high salinities are much the same from one part of the world to another.

Toleration of different ionic concentrations may be controlled by the organism's ability to handle particular ions. Potts & Fryer (1979), for example, in studying sodium uptake and retention by *Daphnia magna*, which occurs preferentially in alkaline, high-conductivity waters, and by *Acantholeberis curvirostris* (O. F. Müller), which occurs preferentially in acidic, low-conductivity water, with no overlap in occurrence between the two taxa, found that *Acantholeberis* had a higher uptake and retention of sodium from acidic, low-conductivity waters than did *D. magna*. The uptake in both species declines with pH, more so in *D. magna*, but is not influenced by calcium content. *D. magna* exhibited differences in sodium uptake from one population to another, indicating that there can be adaptation to local conditions. Thus, the physiological cause of an organism being absent from a particular ionic composition and salinity may be quite subtle.

The two categories of saline waters in the world are essentially seawater and diluted seawater, such as occurs in estuaries, and saline waterbodies that occur on land out of contact with the oceans now and back into the relatively distant past. The first group – thalassic waters – has much the same ionic composition, with Na and Cl dominant. The land-based group – athalassic saline waters – are generally in closed basins. Whatever gets into them from the surface or subterranean water sources or from the atmosphere becomes concentrated by evaporation. As salinity increases and depending on the ionic composition, which is largely a function of the water source, various evaporites precipitate out, leading to still further changes in the ionic composition of the remainder. Such changes can occur rapidly over short time intervals (seasons, or a few years). The situation is highly complex. Little work has

been done on this with the cladocerans, but it seems likely that they are influenced not only by salinity but also by the ratio of monovalent to divalent cations and by the dominant anions. Of the latter, Cl is said to be the least stressful, followed by SO_4, and then CO_3/HCO_3. Of course, other changes, such as pH, are also associated with the ionic composition and may exert a strong control over the occurrence of organisms.

The occurrence of freshwater cladocerans in thalassic habitats likewise has not been studied closely. The Baltic Sea, according to Flössner (1972), seems to be the region that has received the most attention thus far, but there are other seas with reduced salinities and large estuaries in the world that likewise would be expected to have many cladoceran taxa in them as immigrants from fresher habitats. Streams must continually transport cladocerans from freshwater into thalassic environments, where various species will be able to survive according to their tolerance of salinity. One can envision an extension and contraction of the ranges of species seasonally and from one year to another as the balance between the receiving saline water and freshwater inputs continually changes.

Some cladocerans from the families Sididae and Podonidae are strictly marine and can at times provide a considerable percentage of the zooplankton biomass. Other specialized predators, mainly from the families Podonidae and Cercopagidae, are important components of the Caspian Sea biota and a few other saline waterbodies in Eurasia. Some freshwater taxa from various non-chydorid families are capable of existing at moderate to high salinities, and a very few chydorids also can exist at quite high salinities. Some of these taxa, both non-chydorids and even a few chydorids (such as the species of *Celsinotum*: Frey, 1991a), seem confined to these saline environments. The great majority of taxa, however, occur in water with lesser salinity, and the number of species present declines with increasing salinity, at least at higher salinities.

Thus the relationship of the occurrence of cladocerans to salinity is not a simple one. This study has not provided definitive explanations for the patterns observed in the Australia and South

Africa data presented here. But it has documented patterns with greater clarity than many prior studies, partly through utilization of a logarithmic conductivity scale. Future studies of these patterns must also give close attention to the need for standardizing sampling effort per site and to the recording of supplementary independent variables that contribute to the 'noise' in the observed relationships between salinity and species richness.

Acknowledgements

I am indebted to many persons for assistance in field work in Australia and South Africa, for providing the use of their conductivity meters, and for many other services essential to the success of the expeditions. In Australia I used W. D. Williams' meter for all measurements on the continent and Peter Tyler's meter for the few measurements in Tasmania. Persons helpful in providing equipment or participating in field work were Jenny Davis at Murdoch University, W. D. Williams and R. J. Shiel at the University of Adelaide, Ian A. E. Bayly at Monash University, Terry J. Hillman and John Hawking at the Murray-Darling Freshwater Research Center, and Peter Tyler at Hobart University. Williams was the primary arranger and coordinator. In South Africa I was variously aided in field work and the measurement of conductivity by F. Mark Chutter at CSIR in Pretoria; Johan V. Grobbelaar, Maitland T. Seaman, and Dawie J. Kok at the University of the Orange Free State; Jackie M. King and Jenny A. Day at Cape Town University; Brian Allanson at Knysna; Jay H. O'Keeffe at Rhodes University and Ferdinand C. de Moor at the Albany Museum; and Rob C. Hart at the University of Natal. Brian Allanson was the general coordinator. In South Africa especially, many of the persons mentioned provided bed and meals as well as expert guidance and companionship in the field. Because I was unable to participate in the symposium at Lake Titicaca, Brian V. Timms generously read the original version of the paper, for which I am indebted. I also am indebted to Stuart H. Hurlbert for his great editorial concern for the manuscript, which resulted in a marked increase in clarity of presentation and interpretation.

References

Bayly, I. A. E. & W. D. Williams, 1966. Chemical and biological studies on some saline lakes of southeast Australia. Austr. J. mar. freshwat. Res. 17: 177–228.

Flössner, D., 1972. Krebstiere, Crustacea. Kiemen- und Blattfüsser, Branchiopoda. Fischläuse, Branchiura. Die Tierwelt Deutschlands, 60 Teil: 1–501.

Frey, D. G., 1986. The non-cosmopolitanism of chydorid Cladocera: implications for biogeography and evolution, pp. 237–256 in *Crustacea Biogeography*, R. H. Gore & K. L. Heck (eds). Balkema: Rotterdam, XI, 292 pp.

Frey, D. G., 1991 a. A new genus of alonine chydorid cladocerans from athalassic saline waters of New South Wales, Australia. Hydrobiologia 224: 11–48.

Frey, D. G., 1991b. The species of *Pleuroxus* and of three related genera (Anomopoda, Chydoridae) in southern Australia and New Zealand. Rec. Austral. Museum 43: 291–372.

Hammer, U. T., 1986. Saline lake ecosystems of the World. Dr W. Junk Publishers, Dordrecht, x, 616 pp.

Hem, J. D., 1970. Study and interpretation of the chemical characteristics of natural water, 2nd edn. U.S. Geol. Surv., Water Supply Paper 1473.

Hutchinson, G. E., G. E. Pickford & J. F. M. Schuurman. 1932. A contribution to the hydrobiology of pans and other inland waters of South Africa. Arch. Hydrobiol. 24: 1–154.

Löffler, H., 1961. Beiträge zur Kenntnis der Iranischen Binnengewässer. II. Regional-limnologische Studie mit besonderer Berücksichtigung der Crustaceen-fauna. Int. Revue ges. Hydrobiol. 46: 309–406.

Moore, J. E., 1952. The Entomostraca of southern Saskatchewan. Can. J. Zool. 30: 410–450.

Potts, W. T. W. & G. Fryer, 1979. The effects of pH and salt content on sodium balance in *Daphnia magna* and *Acantholeberis curvirostris* (Crustaceae: Cladocera). J. comp. Physiol. 129: 289–294.

Smirnov, N. N., 1989a. Tropicheskiye Cladocera. 1. Novyye vidy rodov *Alona* i *Biapertura* (Aloninae, Chydoridae) tropicheskoy Avstralii. Zool. Zhurnal 68: 133–140.

Smirnov, N. N., 1989b. Tropicheskiye Cladocera. 2. Novyye vidy semestv Chydoridae, Macrothricidae i Moinidae tropicheskoy Avstralii. Zool. Zhurnal 68: 5–59.

Smirnov, N. N. & B. V. Timms, 1983. A revision of the Australian Cladocera (Crustacea). Rec. austral. Museum, Suppl. 1: 1–132.

Societas Internationalis Limnologiae, 1959. Symposium on classification of brackish waters. Archiv. Oceanogr. Limnol. Roma 1 (Suppl.): 1–248.

Williams, W. D., 1964. A contribution to lake typology in Victoria, Australia. Verh. int. Ver. Limnol. 15: 158–163.

Williams, W. D., 1981. The limnology of saline lakes in western Victoria. A review of some recent studies. Hydrobiologia 82: 23–259.

Williams, W. D., 1986. Conductivity and salinity of Australian salt lakes. Aust. J. mar. freshwat. Res. 37: 117–182.

Zooplankton associations in East African lakes spanning a wide salinity range

J. Green

Centre for Research in Aquatic Biology, Queen Mary and Westfield College, Mile End Road, London E1 4NS, UK; Present address: 17 King Edwards Grove, Teddington, Middx. TW11 9LY, UK

Key words: Rotifera, Copepoda, Cladocera, species richness, salinity

Abstract

Abstract The zooplankton of 38 East African lakes has been analysed in terms of species richness and dominance. The conductivities of the lakes range from 48 to $72\,500\,\mu$S cm^{-1} 20 °C. The lakes generally contain more species of rotifers than either Copepoda or Cladocera. The number of species of rotifers begins to decline at a conductivity below $1000\,\mu$S cm^{-1}, and falls to 2 or 3 species above $3000\,\mu$S cm^{-1}. Similar reductions occur in the Copepoda and Cladocera.

Many species can be dominant at conductivities below $1000\,\mu$S cm^{-1}, but the range is restricted progressively with increasing salinity. The dominant species of Rotifera, Copepoda and Cladocera change independently along the salinity gradient, but there are indications of interactions and modifications of community structure by predation and competition.

Introduction

East Africa has a wide variety of lakes, with areas ranging from the largest in the Old World (Lake Victoria) to small saline crater lakes only a few hundred metres in diameter. These lakes provide a series in which it is possible to study aspects of community structure in relation to a variety of factors, such as size, altitude and salinity. Pioneering studies on the distribution of animals in the inland waters of East Africa were made by Beadle (1932) and Jenkin (1936). The present paper attempts to integrate the details of the changes found among the planktonic Rotifera, Copepoda and Cladocera along a salinity series. Previous studies have dealt with some of the groups separately. For Instance LaBarbera & Kilham (1974) dealt with the copepods in lakes in Kenya, Tanzania and Uganda, while Green &

Mengestou (1991) gave an account of the rotifers from a variety of salinities in Ethiopia.

Material and methods

Conductivities were measured with a Dionic Water Tester (Evershed and Vignoles).

The samples were collected at various times between 1962 and 1990. Some of the lakes were sampled once only, but others were revisited at varying intervals (cf. Table 2). Long term data were obtained for a few lakes from my own data and information in the literature (cf. Table 1). Nets with a mesh of 55 μm were used for the rotifers, and 250 μm mesh was employed to catch the crustaceans. Wherever possible vertical hauls were made from close to the bottom of the lake at four separate stations. All the samples were

Table 1. Mean momentary and long term numbers of species of planktonic Rotifera in samples from some East African Lakes. Numbers in parentheses give the additional non-planktonic species in the samples.

Lake	Sampling details	Conductivity μS cm^{-1} 20 °C	Momentary No. of spp.	Long-term No. of spp.
Naivasha	1	260–335	13.2	26
Ziway	2	370–430	7.8	24 (+ 18)
Albert	3	700	9.2	35 (+ 27)
Awasa	4	840–1050	8.6	27 (+ 13)
Nakuru	5	11000*	3	3

* Variable

Sampling details

1. Jenkin (1936), Pejler (1974), Nogrady (1983), Own samples (1980).
2. Bryce (1931), Cannici & Almagia (1947), Green & Mengestou (1991).
3–8. Stations sampled at monthly intervals for 1 year (Green, 1967).
4. Samples collected on 19 dates over 4 years (Mengestou, Green & Fernando, 1991).
5. Vareshi & Vareshi (1984) numerous samples over several years; Own samples (1980).

preserved in 5% formaldehyde. For each sample at least 100 rotifers and 100 cladocerans were identified. Copepods posed more of a problem because of the presence of immature stages, so adults were selected and identified. The determination of the dominant species of copepod was thus based only on the adults. This may have led to some small errors in lakes with numerous species, but probably does not affect the main thesis of this paper.

The dominant species in each sample was simply the most abundant species in each of the three groups.

Various indices of diversity have been used in the past. All have certain disadvantages. Here species richness is used rather than a diversity index. This allows useful comparisons when the numbers of individuals examined in each sample are similar.

The basic comparisons are made in terms of the number of species found in a sample, or more generally in a series of four samples taken in a short period of time on one day. If one samples a lake at intervals over a long period of time the list of species accumulates, but it does not follow that all the species can coexist actively at the same time, although all the species could be present in the lake at very low densities or as resting eggs. The relationship between momentary species richness and longterm cumulative species richness is shown in Table 1 for the few East African lakes for which we have longterm data. In the freshwater lakes the longterm number can be two to three-and-a-half times the momentary species number, but in the saline lake there is no change. The momentary samples will tend to underestimate the number of species active in a lake at any one time. This error will be greater in large lakes than in small lakes, and will tend to be greater in fresh lakes than in saline lakes.

Results

Species richness and salinity

Figures 1 and 2 show the momentary number of planktonic species in each group plotted against conductivity. When a lake has a variable conductivity the middle of the range is plotted. It is apparent that rotifers are present in greater richness throughout the range, and continue into higher salinities than either the Copepoda or the Cladocera. It is also clear that major reductions in species richness occur at conductivities above about 1000 μS cm^{-1}. Among the rotifers a reduction in species number appears to start at even lower conductivities, with the highest number of species being found at about 300 μS cm^{-1}.

Table 2. Mean momentary numbers of planktonic species in East Africa lakes arranged in order of increasing area within three ranges of conductivity.

Lake	Area km^2	K_{20} μS cm^{-1}	No. sample dates	Number of species		
				Copepoda	Cladocera	Rotifera
A. Conductivities to 1000 μS cm^{-1}						
1. Mulehe, Uganda	2.8	210	3	2	2.7	6.3
2. Chala, Kenya	4.0	250	1	2	2	6
3. Oloidian, Kenya	5.5	590	1	3	2	14
4. Kundi, Sudan	2–12	110	1	4	4	13
5. Keilack, Sudan	5–30	550	1	4	3	14
6. Ambadi, Sudan	14	48	1	2	2	10
7. Shambe, Sudan	20	250	1	3	3	14
8. No, Sudan	25	200	1	3	5	13
9. Mutanda, Uganda	29	198	3	3	3.7	12.7
10. Bunyonyi, Uganda	60	230	3	2	3	8.3
11. Awasa, Ethiopia	129	945	19	3.6	2	8.6
12. Baringo, Kenya	130	460–530	2	3	4	6
13. Naivasha, Kenya	115–150	260–335	3	3	4	13.2
14. Koka, Ethiopia	200	200	1	5	4	9
15. George, Uganda	250	215	1	2	3	11
16. Fincha, Ethiopia	400	75	1	2	2	7
17. Ziway, Ethiopia	442	370–430	12	4.7	3	7.8
18. Abaya, Ethiopia	1162	698	2	4	5.5	9
19. Edward, Uganda	2325	848	3	5.7	4.7	6.7
20. Kyoga, Uganda	2622	245–365	1	5	5	17
21. Tana, Ethiopia	3156	137	1	3	7	NR
22. Albert, Uganda	5600	700	12	4.8	5	9.2
23. Victoria, Uganda	68800	100	3	5.7	6.3	13.7
Mean no. spp.				3.5	3.7	10.4
Standard deviation				1.23	1.43	3.24
B. Conductivities between 1000 and 3000 μS cm^{-1}						
24. Pawlo, Ethiopia	0.58	1000	1	2	0	5
25. Bishoftu, Ethiopia	0.93	1830	1	3	0	4
26. Biete mengest, Ethiopia	1.03	2340	1	2	0	5
27. Langano, Ethiopia	241	1900	1	3	3	9
28. Chamo, Ethiopia	551	1100	1	2	4	8
29. Turkana, Kenya	7500	2750	1	4	4	3
Mean no. spp.				2.7	1.8	5.7
Standard deviation				0.82	2.04	2.32
C. Conductivities over 3000 μS cm^{-1}						
30. Sonachi, Kenya	0.18	4770	1	0	0	1
31. Aranguadi, Ethiopia	0.54	6000	1	1	0	2
32. Small Dariba, Sudan	0.85	6000	2	1	0	3
33. Kilotes, Ethiopia	0.77	5930	1	2	0	0
34. Chitu, Ethiopia	0.8	28600	1	0	0	3
35. Large Dariba, Sudan	2.2	27000	2	0	0	3
36. Metahara, Ethiopia	3.2	72500	1	0	0	2
37. Nakuru, Kenya	36–49	11000	1	1	0	3
38. Shala, Ethiopia	329	21000	1	1	1	3
Mean no. spp.				0.75	0.13	2.2
Standard deviation				0.71	0.35	1.09

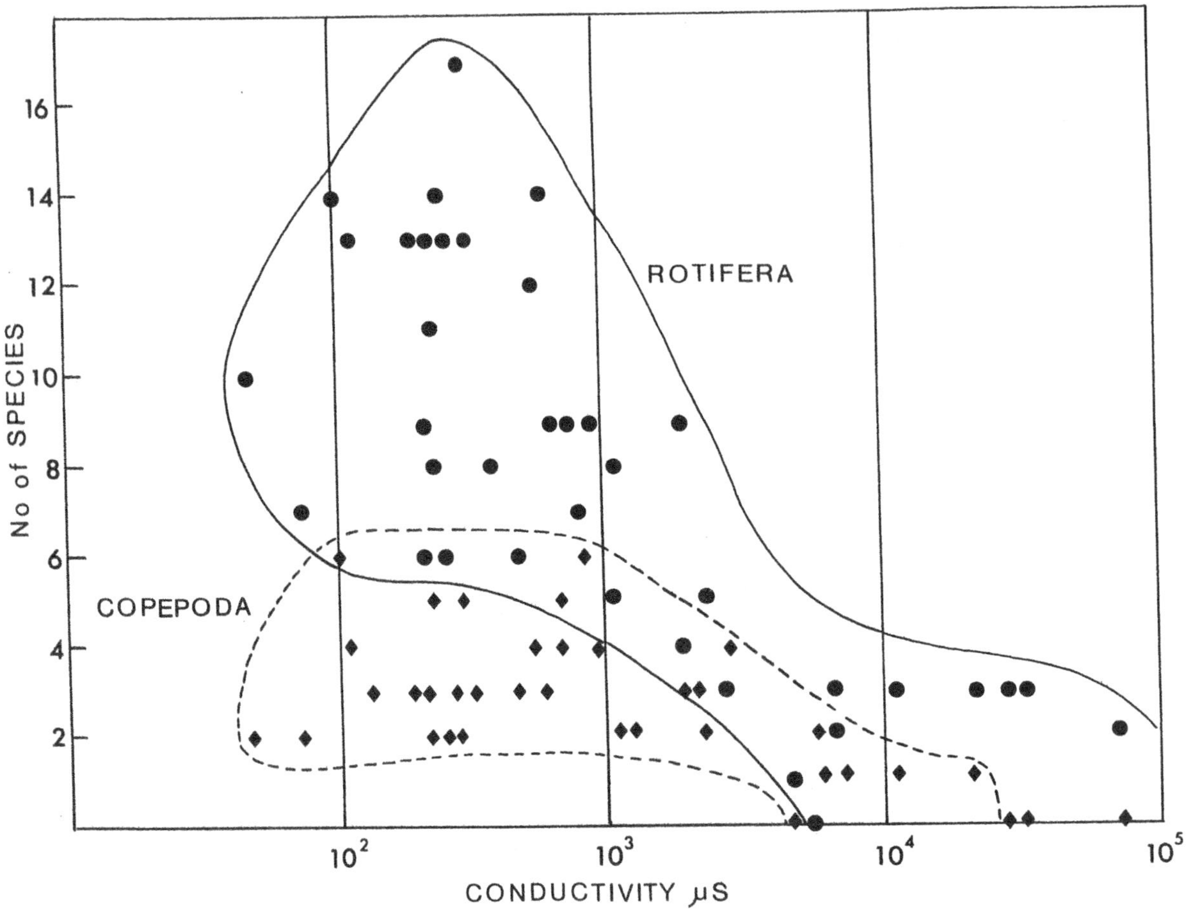

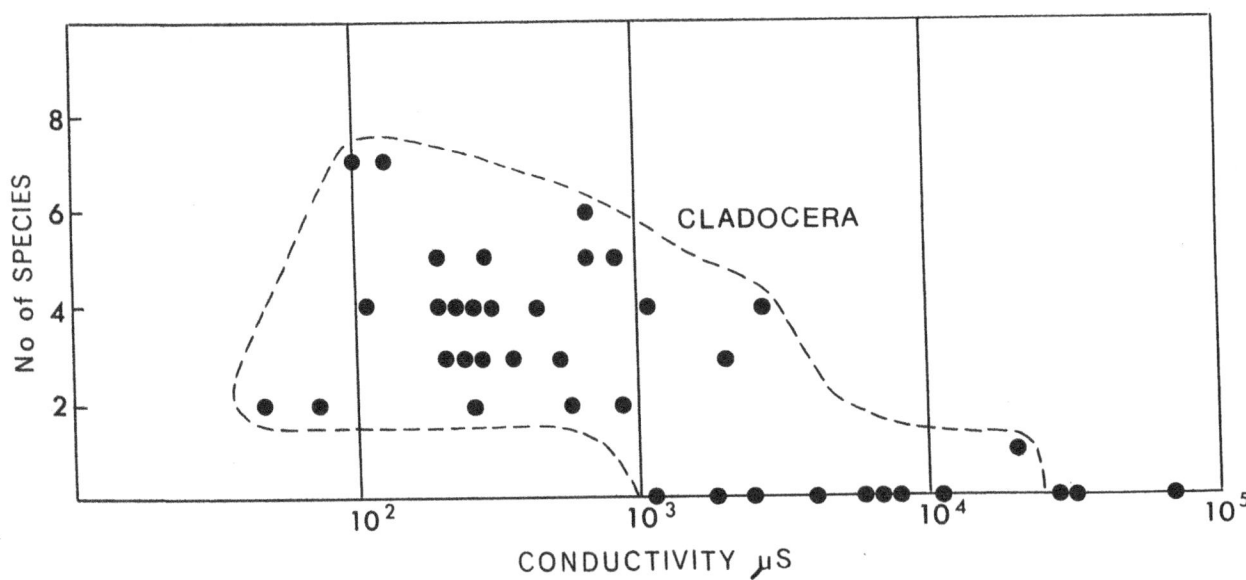

Fig. 1. Numbers of species of planktonic Rotifera (● enclosed in solid line) and Copepods (◆ enclosed in broken line) in relation to conductivity in samples from 38 East African lakes).

Fig. 2. Numbers of species of planktonic Cladocera in relation to conductivity in samples from 38 East African lakes.

The number of species also appears to be influenced to some extent by lake area (Table 2). For lakes with conductivities less than 1000 μS, lakes larger than 1000 km^2 have consistently higher numbers of total planktonic crustacean species.

Of the 14 lakes with conductivities over 1000 μS cm^{-1}, Cladocera were absent from 10, copepods were absent from 3, whilst rotifers were present in all but one. Among lakes with conductivities of 1000–3000 μS cm^{-1}, there is a clear effect of size with respect to Cladocera. The last three lakes in this part of the table, with areas over 240 km^2, have three or four species of Cladocera whereas the first three have none. Lake Turkana has a low number of species of rotifers, but Langano and Chamo have higher numbers than any others in Table 2B. Lakes with conductivities over 3000 μS cm^{-1} average half as many rotifer species and less than half as many copepod and cladoceran species as do lakes with conductivities of 1000–3000 μS cm^{-1}.

Dominant species and salinity

The occurrences and dominance of zooplankters in relation to conductivity are shown in Fig. 3. When a lake was sampled more than once and different species dominated on different dates, each date is represented in the figure.

Up to a conductivity of 1000 μS cm^{-1} any one of 9 species can dominate the rotifers. At higher conductivities 6 species were found as dominants, but the number falls to 3 above 2000 μS cm^{-1}.

Twelve species of copepods were found dominant in various lakes with conductivities below 1000 μS cm^{-1}. At conductivities above 2000 μS cm^{-1} only three species were found as dominants, and of these either *Afrocyclops gibsoni* or *Lovenula africana* dominated in 6 of the 7 lakes with copepods.

The cladocerans in these lakes are less tolerant of raised salinity than the other two groups. They were absent from 9 out of 11 lakes with conductivities over 2000 μS cm^{-1}. *Moina belli* was the most tolerant, occurring only in the anomalous

Lake Shala, which is deep and apparently unproductive.

Discussion

Figures 1 and 2 show that rotifers have more species in the plankton than either the Copepoda or the Cladocera. The solid line in Fig. 1 encloses an area within which the precise position of a lake is only partly determined by salinity. For instance at conductivities between 100 and 1000 μS cm^{-1} there may be between 6 and 17 species of planktonic rotifers. The figure shows that a decline in species of planktonic rotifers begins somewhere below a conductivity of 1000 μS cm^{-1}. This agrees with the data of Green (1986a) from 29 crater lakes with a wide geographical spread, where a decline in species number was found at conductivities over 400 μS cm^{-1}.

The size of a waterbody has little influence on the number of species of rotifers at conductivities under 1000 μS cm^{-1}, but a size effect is found on the numbers of planktonic crustacean species (Table 2). The smaller lakes with conductivities over 1000 μS cm^{-1} lack Cladocera, and those with conductivities below 3000 μS cm^{-1} generally have fewer species than the larger lakes.

Lake Sonachi illustrates another phenomenon. It is the smallest lake examined in this study, and at the time of sampling had a surface conductivity of 4770 μS cm^{-1}. The lake is meromictic (MacIntyre & Melak, 1982), and the conductivity in deeper water is sometimes two or three times that at the surface (Njuguna,1988). One would not have expected the absence of copepods and cladocerans, or the reduction of the planktonic rotifers to a single species. Lowndes (1936) recorded *Paradiaptomus africanus* from this lake, and Beadle (1932) indicated that it was abundant. In my samples there was an abundance of the predatory hemipteran, *Anisops varia* Fieber, which might account for the absence of Cladocera and Copepoda.

Several different factors operate in determining which species becomes dominant in the plankton. Historical biogeography is one factor, exemplified

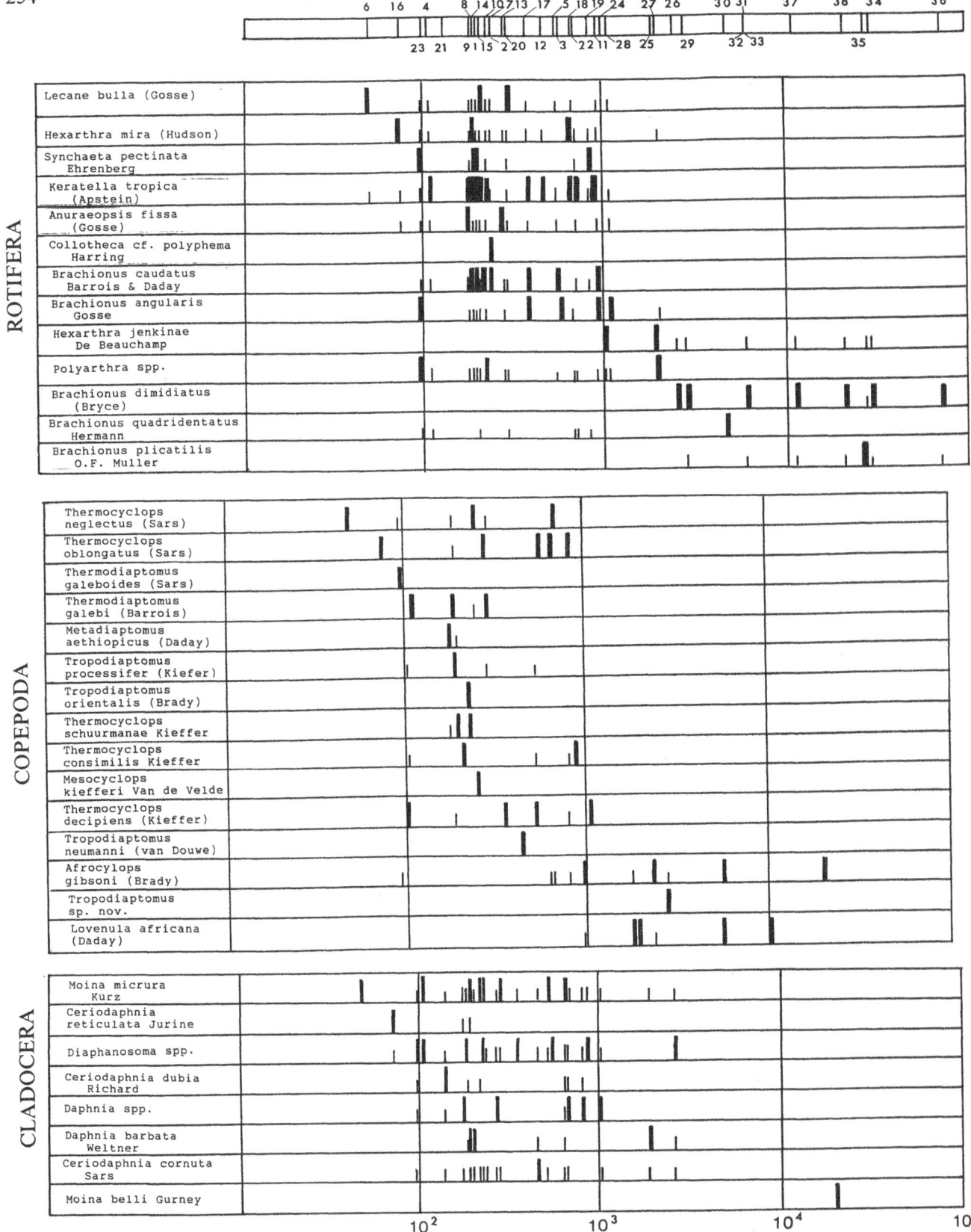

by the dominance of the endemic *Thermodiaptomus galeboides* in Lake Victoria, while *T. galebi* is widespread elsewhere in the Nile system. Another, apparently endemic, species of *Tropodiaptomus*, is found in Lake Turkana. Salinity clearly plays a role in determining the dominant copepod at conductivities over $1000\ \mu S\ cm^{-1}$, where either *Afrocyclops gibsoni* or *Lovenula africana* dominates.

Among the rotifers congeneric species frequently dominate in different parts of the salinity range. Figure 3 shows that although the salinity ranges of occurrence of *Hexarthra mira* and *H. jenkinae* overlap slightly they have not been found together in the same lake. *Hexarthra mira* is only found as a dominant at conductivities below $1000\ \mu S\ cm^{-1}$, while *H. jenkinae* is found as a dominant at higher conductivities, and can extend its range of occurrence at least up to a conductivity of $28\,600\ \mu S\ cm^{-1}$. *Brachionus caudatus* and *B. dimidiatus* do not overlap in occurrence, and dominate at widely different salinities.

The influence of biotic factors has already been mentioned in relation to Lake Sonachi. In more saline lakes there is interaction between the calanoid copepod, *Lovenula africana*, and the rotifers. Vareschi & Vareschi (1984) noted that when the salinity of Lake Nakuru increased above the limit of tolerance for *Lovenula* there was a marked increase in the numbers of rotifers. A parallel to this was found in Ethiopia (Green, 1986b). Lake Pawlo had abundant rotifers and a population density of 10 *Lovenula* per dm^2; Lake Aranguadi had 181–290 *Lovenula* per dm^2 and had fewer rotifers, while Lake Kilotes had 300–400 *Lovenula* per dm^2 and no rotifers. Lake Chitu, with a much higher salinity than the other lakes, lacked *Lovenula* and had three species of planktonic rotifers. Thus there is evidence from two independent sources of interaction between *Lovenula* and rotifers, with high populations of the copepod suppressing rotifer populations.

Acknowledgements

My travels in Africa have been financed by the Royal Society, the Leverhume Trust, the Central Research Fund of London University, the Inter-University Council, the British Council, and the Canadian International Development Agency. My initial forays were in Uganda, where Michael Holden was most helpful in a great variety of ways, Asim Moghraby enabled me to visit many localities in the Sudan, including Jebel Marra. In Ethiopia Arthur Harrison, Seyoum Mengestou, and Getachew Tefera made life much easier than it might otherwise have been. To all these and the native fishermen who lent me their canoes I offer my gratitude.

References

Beadle, L. C. 1932. Scientific results of the Cambridge Expedition to the East African Lakes 1930–31. 4. The waters of some East African Lakes in relation to their fauna and flora. J. linn. Soc., (Zool.) 38: 157–211.

Bryce, D. L., 1931. Report on the Rotifera: Mr Omer-Coopers investigation of the Abyssinian freshwaters (Dr Hugh Scott Expedition). Proc. zool. Soc. Lond. 1931: 865–878.

Cannicci, G. & F. Almagia, 1947. Notizie sulla 'facies' planctonica di alcuni laghi della Fossa Galla. Bolletino Pesc. Pisc. Idrobiol. 2 n.s.: 54–77.

Green, J., 1967. Associations of Rotifera in the zooplankton of the lake sources of the White Nile. J. zool. Lond. 151: 343–378.

Green, J., 1986a. Associations of zooplankton in six crater lakes in Arizona, Mexico and New Mexico. J. zool. Lond. 208: 135–159.

Green, J., 1986b. Zooplankton associations in some Ethiopian crater lakes. Freshwat. Biol. 16: 495–499.

Green, J. & S. Mengestou, 1991. Specific diversity and community structure of Rotifera in a salinity series of Ethiopian inland waters. Hydrobiologia 209: 95–106.

Jenkin, P. M., 1936. Reports on the Percy Sladen Expedition to some Rift Valley lakes in Kenya in 1929. VII. Summary of the ecological results with special reference to the alkaline lakes. Ann. Mag. nat. Hist. 18: 133–181.

LaBarbera, M. C. & P. Kilham, 1974. The chemical ecology

Fig. 3. Occurrence of planktonic species in three major taxa in relation to conductivity, for 38 East African lakes. Heavier marks indicate samples in which the species was the most abundant one in the taxon (Rotifera, Copepoda, or Cladocera). Bar at top of figure shows position of each lake, numbered as in Table 2, along the conductivity gradient.

of copepod distribution in the lakes of East and Central Africa. Limnol. Oceanogr. 19: 459–465.

Lowndes, A. G., 1936. Scientific results of the Cambridge Expedition to the East African Lakes 1930–31. 16. The smaller Crustacea. J. Linn. Soc., Zool. 40: 1–31.

MacIntyre, S. & J. M. Melack, 1982. Meromixis in an equatorial African soda lake. Limnol. Oceanogr. 27: 595–609.

Mengestou, S., J. Green & C. H. Fernando, 1991. Species composition, distribution and seasonal dynamics of Rotifera in a Rift Valley Lake in Ethiopia (Lake Awasa). Hydrobiologia 209: 203–214.

Njuguna, S. G., 1988. Nutrient-phytoplankton relationships in a tropical meromictic soda lake. Hydrobiologia 158: 15–28.

Nogrady, T., 1983. Succession of planktonic rotifer populations in some lakes of the Eastern Rift Valley, Kenya. Hydrobiologia 98: 45–54.

Pejler, B., 1974. On the rotifer plankton of some East African lakes. Hydrobiologia: 389–396.

Vareschi, E. & A. Vareschi, 1984. The ecology of Lake Nakuru (Kenya). IV. Biomass and distribution of consumer organisms. Oecologia (Berlin) 61: 70–82.

Benthic invertebrates of some saline lakes of the Sud Lipez region, Bolivia

Claude Dejoux
ORSTOM, 213 rue La Fayette, F-75480, Paris FAX 52-5-282-5355 (ORSTOM/Mexico City)

Key words: Benthos, macroinvertebrates, faunistical survey, saline lakes, Bolivia, Sud Lipez region

Abstract

The benthic invertebrates fauna of most of the saline lakes of the Sud Lipez region (Bolivia, Altiplano) has been until now quite unstudied. Samples collected during an extensive survey of 12 lakes and two small inflow rivers allow a first list of the main macroinvertebrates living in these biotopes.

The heterogeneous nature of these saline lakes with their freshwater springs and phreatic inflows offers a variety of habitats to macroinvertebrates. The benthic fauna in lakes with salinities $> 10 \, \mathrm{g} \, \mathrm{l}^{-1}$ is not so low in density but includes few species and is dominated by Orthocladiinae and Podonominae larvae. In contrast, the freshwater springs and inflows are colonized by a diverse fauna, with a mixture of both freshwater and saline taxa, but dominated by Elmidae and Amphipoda. The lakes are quite isolated and, apart from some cosmopolitan organisms, their fauna can be quite distinctive.

Extending into the southern part of Bolivia and close to the Chilean border, the Sud Lipez region is a volcanic area of the Bolivian Altiplano located at a mean altitude of 4000 m. Numerous small lakes can be encountered there. They are in general shallow, saline, temporary or permanent, but always subject to wide variations in area.

Since the works of Risacher (1978), Ballivian & Risacher (1981), Hurlbert & Chang (1984, 1988), the physicochemical characteristics of these lakes are quite well known.

Their biota are less well studied and only the algae has been described (Servant-Vildary, 1978, 1983, 1984; Iltis *et al.*, 1984) as well as the aquatic avifauna (Pena, 1961; Hurlbert & Keith, 1979; Hurlbert, 1978, 1981, 1982; Hurlbert & Chang, 1983).

Apart from some nematodes and *Artemia salina* (Iltis *et al.*, 1984; Hurlbert, 1982) no other macroinvertebrates from this desertic region have been recorded. We report here the benthic invertebrates encountered during a survey of these saline lakes carried out in November 1987.

Area studied and methods

Qualitative samples of invertebrates were collected from 10 lakes and 2 small rivers in the region (Fig. 1). Sampling was done using techniques such as brushing and sieving (250 μm screens) of submerged substrates (stones, gravel, plants, sediments) or collecting with fine mesh nets. All samples were preserved in the field with formalin and sorted in the laboratory under a binocular microscope. We tried to sample a variety of biotopes in each lake, including muddy sediments from the most saline areas as well as substrates from the freshwater springs or small brooks running into the lakes. Each type of

258

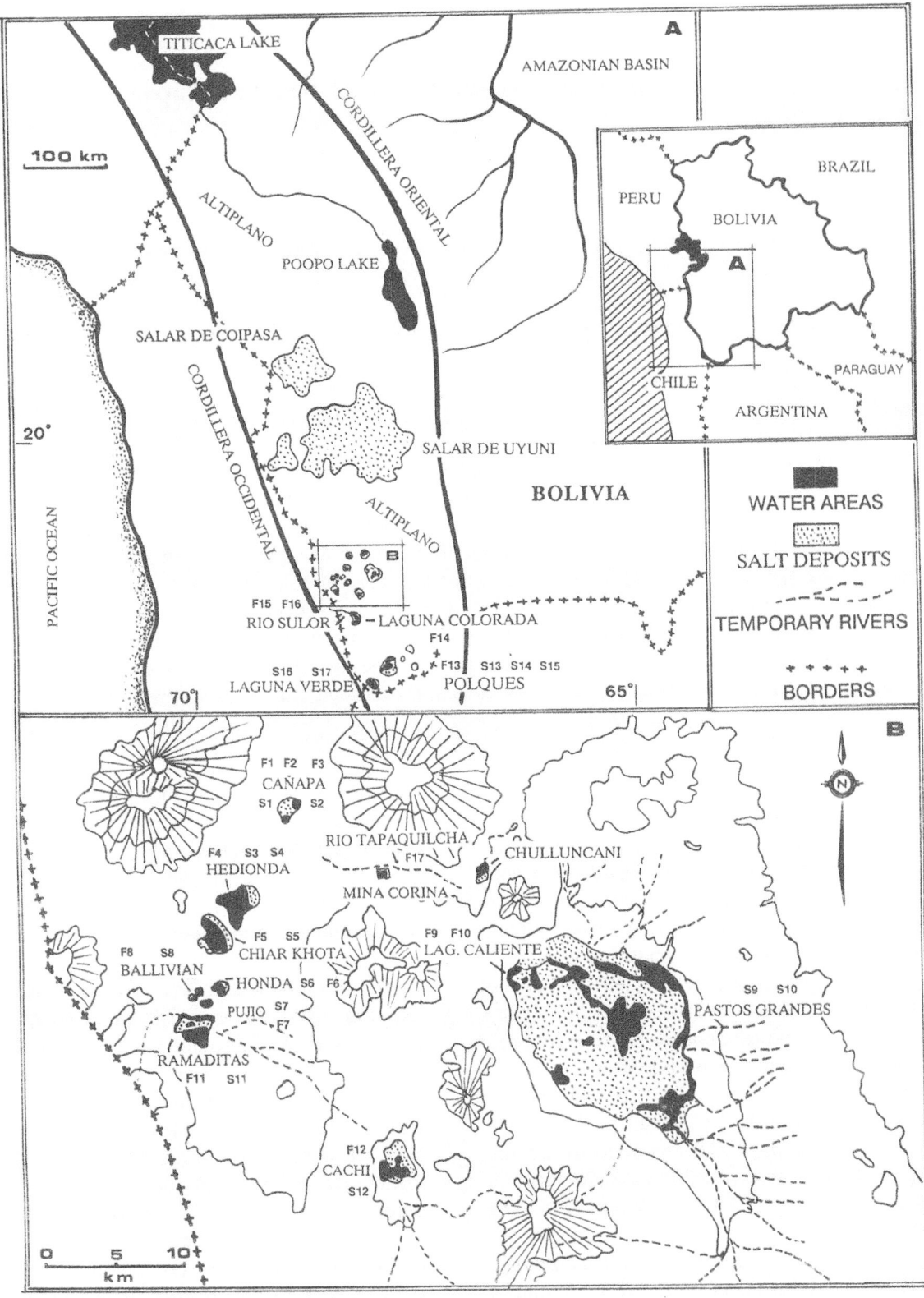

Fig. 1. Map of the studied region and location of studied sites in and around saline lakes of the Sud Lipez.

biotope identified in a lake was considered as a 'site'. Sites are numbered S1 to S17 for the saline habitats (< 3 g l^{-1}). In each site we collected 2 to 6 subsamples, according to the density of organisms present, and these subsamples were combined into a single composite sample. The results (Tables 3 and 4) refer to the composition of these composite samples. Water samples were collected for chemical analysis and some variables (pH, temperature and conductivity) were measured *in situ* at each site. Temperature was measured with a digital K/T BIOBLOCK thermometer, pH with a LED IP/65 BIOBLOCK apparatus and conductivity with a JP BIOBLOCK scientific conductimeter, the value being corrected for a temperature of 20 °C. Total salinity was measured at least at one site for each lake, using a field optical salinometer (Manual refractometer ATAGO, France). For the sites where the total salinity was not measured, an evaluation was made from the conductivily value (C), using the empirical formula:

$$\text{Salinity (mg l}^{-1}) = C\,(\mu\text{S cm}^{-1}) \times 0.76$$

when the conductivity is less than $10\,000\,\mu\text{S cm}^{-1}$. This formula changes for conductivities over

$10\,000\,\mu\text{S cm}^{-1}$ and becomes

$$\text{Salinity (mg l}^{-1}) = C\,(\mu\text{S cm}^{-1}) \times 0.85$$

The greatest difficulty encountered in this study were the taxonomic determinations. Taxa that we could not identify to species were coded with the letter L (= Lipez) and a number. The same code was always applied to specimens nominally identified as the same species and added to the tentative generic or familial name.

Most of the material sent to specialists for identification has not been yet returned and the taxonomy was worked out using the following works: Haas, 1955; Viets, 1955; Roback *et al.*, 1980; Roback & Coffman, 1983.

Oligochaetes were identified by Dr B. Lafont from CEMAGREF Institut (France) and beetles by Dr P. J. Sprangler from the Smithsonian Institution (Washington).

All the material collected remains in the reference collections of the author and is available for taxonomical works.

As our samples are only qualitative, we have only used the following semi-quantitative scale to indicate the relative abundance of each collected taxon: + = only one individual collected,

Table 1. Some physico-chemical parameters from water samples collected in different lakes from Sud Lipez.

Sampling sites *	Temp. (°C)	pH	Cond. in laboratory (μS cm^{-1})	Total salinity (g l^{-1})	Ca (mg l^{-1})	Mg (mg l^{-1})	Na (mg l^{-1})	K (mg l^{-1})	SiO$_2$ (mg l^{-1})
Lag. Colorado	9.0	8.2	96285	81.4	–	1056	82500	6405	30.4
Lag. Verde	17.0	8.6	23820	20.2	164	300	4647	340	65.8
Lag. Hedionda	23.1	8.4	48920	41.6	–	634	17252	1488	47.5
Lag. Cachi (spring)	16.8	8.3	262	0.2	1.8	1.0	128	35	36.8
Lag. Chulluncani	17.1	8.4	44600	38.0	–	–	13250	4074	32.7
Pastos Grandes (spring)	31.1	6.7	1938	1.5	22	26	485	53	82.0
Lag. Pastos Grandes	18.0	8.9	3409	2.6	300	510	896	142	41.9
Lag. Chiar Khota	23.2	8.0	82980	70.5	1184	1690	39932	4069	81.3
Lag. Cañapa	16.2	8.8	61090	52.0	–	–	31332	1242	43.3
Lag. Ballivian	20.0	8.2	36770	31.2	648	352	10304	1068	43.5
Lag. Honda	18.5	8.6	48360	41.1	204	282	15750	1593	32.9
Lag. Ramaditas	15.5	8.2	34770	29.5	1152	405	8531	914	95.7
Lag. Pujio	18.7	8.6	30570	26.0	192	158	8860	693	50.1
Lag. Polques (spring)	40.2	8.2	1435	1.9	54	12	318	31	26.0
Lag. Polques	21.6	8.5	15720	13.2	168	132	5554	334	14.7

* All samples were collected between the 14th and the 19th of November 1987. Analysis were made by the SENAMHI laboratory in La Paz.

++ = between 2 and 10; +++ = between 11 and 100; ++++ = more than 100.

Results

+ Physico-chemical characteristics of the studied environments

As our purpose was not a detailed study of the physico-chemical conditions, analyses of water (Table 1) were carried out only to check conditions at the time of sampling, in order to detect changes that might have occured since the previous studies of Ballivian & Risacher (1981) and Iltis *et al.* (1984).

The lakes are mainly very shallow (0.2–1 m deep). Laguna Verde has a depth of several meters. Bottom sediments are usually mud, covered by a fine layer of silt which can easily be disturbed by the strong winds frequent in that region. Transparency then is very low, especially when the plankton is dense.

Diel temperature variation is great, particularly during the spring (September to December), when the nights are very cold and the days are sunny. A light ice layer can build up during the night, but disappear in the morning as water temperatures can reach 20 °C.

Some of the springs feeding the lakes have outflow temperatures that exceed 40 °C, as at Salar de Chalviri, for example.

Salinity varies notably in time and space, and our data have been compared in Table 2 to those of Iltis *et al.* (1984), showing the interannual variability of this factor. At small spatial scales, salinity variation can be strongly influenced by whether freshwater input is from diffuse seepage or by discrete springs. Such variation can produce corresponding gradients in the distribution of the invertebrate fauna.

The chemical composition of water varies from one lake to another, but three types can be recognized (Iltis *et al.*, 1984):

– the sodium chloride type (Chiar Khota, Ramaditas, Laguna Verde, Pujio, Honda, Ballivian, Colorada);

Table 2. Temporal changes in salinity (in g l^{-1}) for some saline lakes of the Sud Lipez region. All values determined with an optical salinometer.

Lakes studied	November 1987	November 1982	Earlier studies
Lag. Chulluncani	14	7	11.6
Lag. Honda	45	12	21.4
Lag. Cañapa	80	13	11.5
Lag. Pujio	36	18	31.1
Lag. Ramaditas	33	20	27.7
Lag. Ballivian	35	28	45.3
Laguna Verde	20	46.5	66
Lag. Hedionda	60	57	67.2
Lag. Cachi	90	86	43.5–322
Lag. Chiar Khota	124	119	69.3
Lag. Colorada	200	137	120

– the sodium carbonate type (Cachi);

– the sodium sulfate type (Cañapa, Chulluncani).

Lakewater pH varied from 8 to 9. Freshwater springs were mostly basic (pH 7.5–8), but the freshwater hot spring at the Salar de Pastos Grandes had a pH of 6.7.

Benthic invertebrates assemblages

Laguna Cañapa

Located in the northern part of the studied area, this small lake is often divided in two separate basins. Both basins were sampled in different places, with the salinity ranging from less than 1 g l^{-1} in small springs present along the shore (site F1 in the northern basin and site F3 in the southern one) to 80 g l^{-1} in the lake itself (site S2). A shoreline area of the southern basin, fed by a diffuse groundwater was also sampled (site S1), as well as a deep hole receiving a freshwater spring and covered with aquatic macrophytes (site F2).

The greatest diversity of invertebrates at Laguna Cañapa was found in the freshwater habitats (Table 3).

Apart from the amphipod *Hyalella* cf. *dentata inermis* (present at 3 sites) and two species of the midge *Parochlus* (2 sites), each sampling site had

Table 3. Macroinvertebrate assemblages found in some freshwater habitats related to saline lakes in the Sud Lipez region.

Taxa collected	Cañapa			Hedionda	Chiar Khota	Honda	Pujio	Ballivian	Caliente		Ramaditas	Cachi	Polques	Laguna Colorada	Rio Sulor		Rio Tapaquilcha	
	F1	F2	F3	F4	F5	F6	F7	F8	F9	F10	F11	F12	F13	F14	F15	F16	F17	
Salinity (mg l⁻¹)	680	600	720	930	1760	2040	1920	2400	620	870	410	1930	2044	1910	661	820	420	
Hirudinea Glossiphoniidae sp. L38						+++												
Tubificidae Rhyacodrilus sp. L61						+++												
Naididae Nais andina										+++		+++						
Naididae Paranais litoralis	++							+	++		+++							
Naididae Nais simplex	+																	
Euplanaria dorotocephala		++																
Orthocladiinae Cricotopus sp. L17			+++		++	++		++		++		++		+++		+		
Orthocladiinae sp. L3			+++															
Orthocladiinae sp. L4		++													+++			
Orthocladiinae sp. L5															+++			
Orthocladiinae sp. L6															++			
Orthocladiinae sp. L7		++													+		++	
Orthocladiinae sp. L72															+			
Orthocladiinae sp. L80	+++																	
Podonominae Parochlus sp. L1		+++															+++	
Podonominae Parochlus sp. L22		++													++++		+	
Podonominae Parochlus sp. L23			+															
Tanypodinae Procladius sp. L81																	+	
Tanypodinae Ablabesmyia sp. L82																	++	
Simuliidae spp.																	++	
Ceratopogonidae sp. L25		+++			++		+											
Tipulidae sp. L41													+					
Ephydridae sp. L18										++		++						
Ephydridae sp. L19				+			++					++						
Ephydridae sp. L33		++												++			++	
Ephydridae sp. L34						+											+	
Ephydridae sp. L63												+					++	
Corixidae Ectemnostegella quechua Bachman		+++					++					++					++	
Corixidae Ectemnostegella stridulata Hungerford		++										++						
Claudioperla sp. L14																		
Baetidae sp. L12															++			
Leptoceridae sp. L84																	+++	
Hydroptilidae Hydroptila sp. L78						+++											++	
Hydroptilidae Leucotrichia sp. L83						++											++	
Protallagma titicacae Calvert																		
Elmidae Austrelmis consors Hinton	+					++++											+++	
Dytiscidae Lancetes nigriceps Guignot								++	++++	++++	+++	++	++	++		+	+++	
Hydracarina sp. L9																		
Hydracarina sp. L10																++		
Hydracarina sp. L52								++										
Hydracarina sp. L59							+											
Hydrachnella sp. L53																		
Oribatei Hydrozetes sp. L31																		
Oribatei Hydrozetes sp. L62				+														
Oribatei Hydrozetes sp. L64																		
Oribatei Hydrozetes sp. L70		++																
Orchestidae Hyalella cf. dentata inermis Smith	+++					++++		+++	+++	+++								
Planorbiidae Tropicorbis canonicus Cousin	+	+++										++++				++		
Hydrobiidae Littoridina cf. languiensis											++++		++				+++	
Total number of taxa collected	6	10	3	2	3	8	5	5	3	3	3	9	3	8	5	5	16	
Total number of individuals collected	37	116	75	2	54	304	16	39	209	247	231	230	11	532	17	17	199	

a distinct fauna despite their proximity. This situation, commonly encountered is South Lipez aquatic environments, will be discussed later. The saline site had a poor fauna.

Lakes Hedionda and Chiar Khota

Laguna Hedionda has an area of *ca* 5–6 km², and is fed by freshwater springs on its south-eastern and western margins. The spring discharge is high enough to form a shallow channel 10 m long (site F4), ending at the shoreline level in a small branched delta (site S3), where the freshwater rapidly mixes with the saline water of the open lake (site S4). The salinity gradient along this system ranges from $< 1 \, g \, l^{-1}$ to $60 \, g \, l^{-1}$.

The channel, with its strong salinity gradient, had the most diversified biota, especially in the section where the salinity was about $10 \, g \, l^{-1}$ and the bottom covered with algae. In contrast, invertebrates were very scarce in the clean sandy bottom of the spring outflow. A similar situation existed in the near open lake itself, possibly due to its high salinity, and only a few *Artemia salina* were found in open water.

Lying south of and close to Lag. Hedionda, Lag. Chiar Khota was smaller, very shallow, and with a salinity of $120 \, g \, l^{-1}$ (site S5) during our visit there. Only a small area of diffuse outflow occurring near the shoreline was sampled, by brushing small stones half covered by semi-stagnant water (site F5). The fauna there was sparse and only chironomids and ceratopogonids were present in some abundance;

Lakes Honda, Ballivian and Pujio

Very close together, these three lakes are small, have similar salinities (35, 36 and $45 \, g \, l^{-1}$, respectively) and receive freshwater mainly via small shoreline springs or diffuse seeps. Bottoms are muddy with a superficial silt deposit mixed with algae. Small pools supplied by freshwater seeps were occasionally present along the shoreline (sites F6, F8 and F7).

Austrelmis consors and *Cricotopus* sp L17 were the most common invertebrates, the first species occurring mainly in the freshwater habitats and the second one in the saline areas (Tables 3 and 4). It is however interesting to note that *A. consors*, a very common freshwater elmid found throughout the Altiplano, was also present in Laguna Ballivian at a place where the salinity was $31.2 \, g \, l^{-1}$.

Salar de Pastos Grandes

This is the largest saline depression of the Sud Lipez region, but water bodies were only sparsely scattered over it. We sampled the western side of the depression where there was a lake (Laguna Caliente) fed by both cold (14 °C, site F9) and hot (32 °C, site F10) freshwater spring inflows. A low salinity water area located some 10 m outside the hot spring outflow but directly under its influence was also sampled, as well as a site (S10) in the central area of the salar. In contrast to the situation found at the preceding lakes, the fauna at Salar de Pastos Grandes was more diverse in the saline water than in the freshwater outflows (Tables 3 and 4). *Cricotopus* sp. L17 was again dominant in saline water. In freshwater habitats, amphipods, elmids and leeches were dominant and about equally abundant in both cold and hot outflows. In the saline environments, the most diverse fauna was found some 10 m away (site S10: salinity = $3 \, g \, l^{-1}$; 9 species) from the spring outflows, but apart from the abundant *Cricotopus* mentioned above, all species were scarce. At a somewhat higher salinity ($5–10 \, g \, l^{-1}$, site S9) we found only 7 species (5 of them very abundant), fewer substrate types being present here than near the spring outflows (stones, gravel and macrophytes were lacking).

Laguna Ramaditas

The depression of Laguna Ramaditas is about 10 km², but the open water area during our visit was restricted to a narrow area along the north-eastern margin, a small waterbody in the middle of the depression, surrounded by emerged area with muddy sediments covering old ice deposits (Hulbert & Chang, 1988), and a larger open water area along the southern side. The salinity of this last zone (site S11) was more than $30 \, g \, l^{-1}$, and on the water surface we found many dead or dying amphipods. Possibly this resulted

Table 4. Macroinvertebrate assemblages found in some saline lakes from the Sud Lipez region.

Taxa collected

Taxa collected	Cañapa		Hedionda		Chiar Khota	Honda	Pujio	Ballivian	Caliente		Ramaditas	Cachi		Polques		Laguna Verde	
	S1	S2	S3	S4	S5	S6	S7	S8	S9	S10	S11	S12	S13	S14	S15	S16	S17
Salinity (g l^{-1})	8.5	80	9.8	60	120	35	45	36	5–10	38	33	12.4	4.9	12	15	20.2	20.2
Hirudinea Glossiphoniidae sp. L38																	
Naididae *Nais andina*			++													+++	
Naididae *Nais elinguis*			++													++	++
Naididae *Paranais litoralis*			++						+++	++							
Euplanaria dorotocephala Woodworth	+																
Orthocladiinae *Cricotopus* sp. L17							+++	+++	+++				+++				
Orthocladiinae *Paratrichocladius* sp. L15			+++			+++	+++	+++	+++			+++	+++	++	+++	+++	+++
Orthocladiinae sp. L3			+++					+++									
Orthoclodiinae sp. L24									++	++				++			
Orthoclodiinae sp. L40																	
Orthoclodiinae sp. L46		++	+++														
Podonominae *parochlus* sp. L1																	
Podonominae *Parochlus* sp. L22	++		+							++		++	++	++	++		
Podonominae *Parochlus* sp. L23										++		++	++	++			
Podonominae *Parochlus* sp. L27								+									
Podonominae *Parochlus* sp. L35											+						
Ceratopogonidae sp. L25	++											++	++				
Ephydridae sp. L18												+	+				
Ephydridae sp. L19												++	++				
Ephydridae sp. L20	+++								++	++		+	+	++	++		
Ephydridae sp. L33			+									++	++				
Ephydridae sp. L34												++	++				
Ephydridae sp. L56												++	++				
Ephydridae sp. L63				+													
Ephydridae sp. L76			+		++												
Corixidae *Ectemnostegella quechua* Bachman			+						+++								
Corixidae *Ectemnostegella stridulata* Hungerford			+++						+++								
Elmidae *Austrelmis consors* Hinton	+											+					
Dytiscidae *Uvarus* sp. L29								++	+++			+++	++++	+++	++	++	
Hydracarina sp. L28												+++	+++				
Oribatei *Hydrozetes* sp. L55												+	+				
Orchestidae *Hyalella* cf. *dentata inermis* Smith									+++		+++	++++	+++	+++	++	++	
Artemia salina		+++	+++	++++							+++		+++				
Planorbiidae *Tropicorbis canonicus* Cousin													+++			++	
Hydrobiidae *Littoridina* cf. *languiensis*									+								
Total number of taxa collected	5	2	12	2	1	1	2	7	9	3	3	18	7	5	6	2	2
Total number of individuals collected	35	62	199	170	10	26	27	162	191	53	144	452	88	41	68	189	189

from a rapid increase in salinity caused by evaporation.

Apart from *Hyalella* cf. *dentata inermis* which possibly reached at that time its tolerance limit of salinity, we found in that saline area a classical invertebrate assemblage with the presence of Orthocladiinae and Podonominae larvae.

Diffuse groundwater inflows were sampled along the southern shore, as well as a freshwater well fed by a low discharge spring (site F11). In these areas we again found widespread elements such as Elmidae and Oligochaeta, and for the first time in Sud Lipez lakes we found numerous molluscs that we attributed to *Littoridina languiensis*, and also some Dytiscidae larvae.

There were numerous ducks swimming around in the freshwater outflow and feeding on the bottom, perhaps attracted by the abundant benthic fauna (molluscs) or by the aquatic plant *Ruppia* sp., which was present in this lake.

Laguna Cachi

A bit larger than Ramaditas, the saline depression of Cachi Laguna also had a marginal flooded area. We sampled on the western shoreline of this lake, a freshwater spring with a discharge of some litres per second (site F12). The water came out from two well defined outlets (temperature: 17 °C) and then flows over a patchily substratum.

The salinity was about $2 \, g \, l^{-1}$ at the outlet and rapidly increased as the flow ran over evaporites and mixed with the lake water, which had a salinity $> 90 \, g \, l^{-1}$. Invertebrates were present only in the spring pools and in the short outflow channels (site S12). None were found in the lake itself where the black and anoxic sediments appeared to contain abundant decomposing organic matter. The fauna of the low salinity sections (site F12) was dominated by amphipods, elmids and oligochaetes (Tables 3 and 4).

Laguna Polques

This is one of many lakes in Salar de Chalviri. We sampled it on its western shore, alongside the road going from Laguna Colorada to Laguna Verde. Many diffuse inflows were entering the lake there, as well as a freshwater hot spring with

sufficient discharge to create a long outflow channel 2–3 m wide and about 50 cm deep.

At our visit, the water was emerging from the ground at a temperature of 42 °C, with a salinity of about $2 \, g \, l^{-1}$. The temperature rapidly decreased and salinity increased along the outflow channel. Salinity was $12 \, g \, l^{-1}$ at twenty meters away from the spring and $15 \, g \, l^{-1}$ at about thirty meters, where the channelized outflow mixes with the lake water. The benthic fauna was sampled at four sites (F13, S13, S14 and S15) along the salinity gradient (Tables 3 and 4).

No invertebrates were found until about 3 m downstream from the spring outlet. The first species appeared on the sides of the channel where the temperature decreased to about 30 °C, but the most diverse fauna was found only 10 m downstream, where the salinity was $5 \, g \, l^{-1}$ and the temperature not more than 25 °C. The density of organisms was high, with elmids, amphipods and tricladids dominating. Only 5 species were collected at the last and most saline sampling site (S15, $15 \, g \, l^{-1}$), where *Cricotopus* sp. L17 larvae dominated.

The presence of *Euplanaria dorotocephala* in relatively saline, warm water is unusual, the normal habitat of this triclad being mainly fresh and cold water of Altiplano brooks.

Laguna Colorada and Rio Sulor

Supporting a large flamingo population, this is one of the most famous lakes of the Sud Lipez, because of its red waters colored by halobacteria and *Dunaliella*, because of its ancient ice deposits and also because the proximity of an important center for geothermal energy exploitation. This lake was sampled in different places but we never found a single macroinvertebrate, perhaps because of the very high salinity at the time of our visit ($200 \, g \, l^{-1}$). Copepods (*Boeckella* sp.), colored red probably by β-carotene from the *Dunaliella*, were on the other hand so abundant that they formed blood red accumulations several cm thick in many places along the shoreline.

Elmids, ephydrids and *Cricotopus sp.* were

found around a freshwater artificial captage overflowing into a small pool connected with the lake itself (Table 3, site F14). A small river, Rio Sulor, feeds the lake on its western shore and was sampled some 50 m above Laguna Colorada, both in flowing sections (site F15) and in stagnant pools (site F16). This river is permanent and had a discharge of about 2 m³ at the time of our visit. *Austrelmis consors* was the only element common to both biotopes, and apart from the small corixid *Ectemnostegella quechua*, found also in saline environments of the Sud Lipez region, the assemblages found in the river were very different from those normally encountered in small freshwater brooks or spring outflows of the region (Table 3). It was an assemblage more typical of the permanent rivers of the Altiplano (Marin, 1989). The absence of amphipods in the Laguna Colorada aquatic environment was notable and remains without clear explanation.

Laguna Verde

This was the southernmost lake studied. It also has the greatest water volume, with a depth of several m and a length of 3–4 km at the time of our visit. A pebble and sand substratum occupied the shoreline zone, which was regularly washed by waves. Two sites (S16 and S17) were sampled there at depths of 20 and 60 cm. At the deeper site, the sediment was formed of a mixture of sand and compact clay covered by a fine greenish silt layer.

The fauna of soft sediments was dominated by chironomid larvae never found in other lakes (possibly *Paratrichocladius*, Hurlbert pers. comm.). The fauna of pebbles was more diverse, and in addition to common elements such as amphipods and elmids, we also found some living *Littoridina* (cf. *languiensis*) which indicate that this species is able to support saline water conditions. We also found a single uninhabited shell of a snail very close to the form *Littoridina andecola andecola* living in Lake Titicaca. The shell was well preserved, but no living specimens belonging to the same species were found.

Rio Tapaquilcha

Rio Tapaquilcha (site F17) runs close to the Mina Corina encampment (Fig. 1). This river is permanent, unpolluted above the mine encampment, and presents a high variety of habitats such as muddy or sandy bottoms in shoreline pools, gravels in riffle sections, stones in faster currents, and aquatic vegetation along its margin. It can be then regarded as one of the most suitable habitat for freshwater macroinvertebrates in the region.

Of the 16 taxa found in this river (Table 3), only 7 were also collected in other aquatic habitats of the region and the general faunal assemblage was similar to that of central Altiplano and Cordillera rivers (Marin, 1989). Dominant groups were Orthocladiinae, mayflies and beetles (mainly elmids).

Discussion and conclusions

The present study suggests that several ecological factors play a role in the distribution of the benthos in aquatic habitats of the Sud Lipez.

All lakes appear to present two types of biotopes that grade into one another: unstable freshwater-shoreline habitats (springs, phreatic inflows) and saline open water environments. In some cases the limit between these two types of biotopes fluctuates in time and space, according to the action of various abiotic factors. The taxa living in this transition area are mainly the most euryhaline ones. Variation of salinity in lake open waters due to variability in precipitation certainly induces important changes in faunal composition. That was possibly the reason why we found so many amphipods dying at Laguna Ramaditas during our visit and why a dead, but no live, *Littoridina andecola andecola* was found in Laguna Verde. Of course, this single specimen could have been carried into the lake by a bird or other source of transport and died on arrival.

The benthic fauna of biotopes with more than 10 g l^{-1} is in general of low diversity. Orthocladiinae and Podonominae larvae typically are dominant. In some cases ephydrid larvae, oligochaetes and nematodes are also present but are never abundant.

Freshwater springs or diffuse marginal inflows of groundwater origin are usually colonized by a rich and relatively diverse fauna. Exceptions would be the points of outflow of hot springs or of cold springs that churn the sediments. In these cases, the high temperature or the mechanical action of shifting sand around the outlet does not allow the establishment of invertebrates. Low salinity sites ($2–6$ g l^{-1}) with abundant food sources and heterogenous substrates, also often contain a diverse fauna with a mixture of freshwater and euryhaline taxa.

From one lake to another the faunal composition can be very different, apart from some ubiquitous and cosmopolitan taxa. In this respect the lakes appear quite isolated one from another, and cross-colonization may be limited. Insects can be able to colonize a lot of habitats during their aerial adult phase, but after hatching only the most ubiquitous may have chance to survive in diverses biotopes. That is mainly the case for Ephemeroptera, Trichoptera or Plecoptera which are living on the Altiplano and are generally unable to tolerate a salinity of more than 1 g l^{-1}. Such a situation occurs probably in Lake Titicaca, where among those groups, only a few species of Trichoptera, carried in by inflow rivers, can survive in the lake, close to the river entrance (Dejoux, 1991). Strictly aquatic forms like molluscs, amphipods or water mites possess less effective means of colonization; possible, birds commonly living in the Sud Lipez region (flamingos, ducks, phalaropes) transport invertebrates when they move from one lake to another.

It is difficult to compare directly the benthic populations found in the Sud Lipez lakes to those of other saline lakes in the world. Numerous factors are important in determining the presence or absence of a particular taxa. If the salinity level is in many cases one of the most important, others, such as the ionic composition or the climatic environment must also be taken into account.

In the Sud Lipez region the aquatic faunal diversity is enhanced by the presence of relatively permanent sources of freshwater, and gradients of biotopes with different physico-chemical conditions between these sources and the saline lakes are frequent. This allows species with differing ecological requirements to persist in close proximity to one another.

Despite this positive aspect, freshwater habitats created by springs very close to the lake shorelines, are frequently affected by changes in salinity and lake level. In this way such habitats differ from permanent streams like Rio Sulor or Rio Tapachilqua. On the other hand, such larger streams may not contribute very much to the faunal diversity of the small freshwater biotopes found at the margins of the Sud Lipez lakes. It is clear that many aquatic invertebrates would not find in these small habitats sufficiently stable ecological conditions to establish themselves permanently. Such may be the case for example for the Simuliidae, Plecoptera, Trichoptera and Ephemeroptera present in the Rio Tapaquilcha.

The present study gives only a momentary picture of the benthic fauna of these lakes. Longer term studies of population dynamics, diversity and colonization patterns at one or a few lakes in relation to environmental change would be useful.

References

Ballivian, O. & F. Risacher, 1981. Los salares del altiplano boliviano. ORSTOM (off. Rech. Sci. Tech. Outre-Mer), Paris, 246 pp.

Dejoux, C., 1991. VI.4i. Los insectos. In: El lago Titicaca, sintesis del conocimiento limnológico actual. C. Dejoux – A Iltis (eds), ORSTOM/ISBOL co-edition. La Paz, Bolivia: 371–386.

Haas, F., 1955. Mollusca in 'Results of the Percy Sladen Trust Expedition to Lake Titicaca'. Trans. Linn. Soc., 1, 3: 275.

Hurlbert, S. H., 1978. Results of five flamingo censuses conducted between November 1975 and December 1977. Andean Lake and Flamingo Investigations, San Diego State Univ., California, Tech. Rept. No. 1, 16 pp.

Hurlbert, S. H., 1981. Results of three flamingo censuses conducted between December 1978 and July 1980. Andean Lake and Flamingo Investigations, San Diego State Univ., California, Tech. Rept. No. 2, 9 pp.

Hurlbert, S. H. & C. C. Y. Chang, 1983. Ornitholimnology: Effects of grazing by the Andean Flamingo (*Phoenicoparrus andinus*). Proc. nat. Acad. Sci. USA 80: 4766–4769.

Hurlbert, S. H. & C. C. Y. Chang, 1984. Ancient ice islands

in salt lakes of the Central Andes. Science 223: 299–302.

Hurlbert, S. H. & C. C. Y. Chang, 1988. Distribution, structure, composition and thermal environment of ice deposits in Andean salt lakes. Hydrobiologia 158: 271–299.

Hurlbert, S. H. & J. O. Keith, 1979. Distribution and spatial patterning of flamingoes in the Andean Altiplano. The Auk, 96: 328–342.

Iltis, A., F. Risacher & S. Servant-Vildary, 1984. Contribution à l'étude hydrobiologique des lacs salés du sud de l'Altiplano bolivien. Revue Hydrobiolol. trop. 17: 259–273.

Marin, R., 1989. Elementos para una tipología de los ríos de altura de la región de la Paz: Caracterización biológica y potencialidades piscícolas. Tesis UMSA La Paz. 76 pp.

Peña, L. E., 1961. Results of research in the Antofagasta ranges of Chile and Bolivia. I. Birds. Postilla (Yale Peabody Museum) 49: 3–42.

Risacher, F., 1978. Le cadre géochimique des bassins à évaporites des Andes boliviennes. Cah. ORSTOM sér. Géol., 10: 37–48.

Roback, S. S., L. Berner, S. R. Flint Jr., N. Nieser & P. J. Spangler, 1980. Results of the Catherwood Bolivian-Peruvian Altiplano Expedition. Part I. Aquatic insects except Diptera. Proc. Acad. nat. Sci. Philad. 132: 176–217.

Roback, S. S. & W. P. Coffman, 1983. Results of the Catherwood Bolivian-Peruvian Altiplano Expedition. Part II. Aquatic Diptera including montane Diamesinae and Orthocladiinae (Chironomidae) from Venezuela. Proc. Acad. nat. Sci. Philad. 135: 9–79.

Servant-Vildary, S., 1978. Les Diatomées des dépôts lacustres quaternaires de l'altiplano bolivien. Cah. ORSTOM, sér. Géol. 10: 25–35.

Servant-Vildary, S., 1983. Les Diatomées des sédiments superficiels de quelques lacs salés de Bolivie. Bull. Sci. géolog. 3: 74–86.

Servant-Vildary, S., 1984. Les Diatomées des lacs sursalés boliviens. Sous-classe Pennatophycidées. I. Famille des Nitzschiacées. Cah. ORSTOM sér. Géol. 14: 35–53.

Viets, K., 1955. XVI. Hydrachnellae in 'results of the Percy Sladen Trust Expedition to Lake Titicaca'. Trans. Linn. Soc. 3: 249–274.

Saline lakes of the Paroo, inland New South Wales, Australia

B. V. Timms
Sciences Department, Avondale College, Cooranbong, N.S.W., 2265, Australia

Key words: saline lakes, water chemistry, flora, fauna, biogeography

Abstract

Twenty-five lakes from fresh to crystallizing brine in the semi-desert of northwestern New South Wales, Australia, were studied regularly for 27 months. The lakes are small, shallow and ephemeral. Chemically waters are mainly of the NaCl type. Seventy-four species of invertebrate occur in saline waters (> 3 g l^{-1}) with crustaceans such as *Parartemia minuta, Apocyclops dengizicus, Daphniopsis queenslandensis, Diacypris* spp. and *Reticypris* spp. dominant, particularly at higher salinities. The insects *Tanytarsus barbitarsis* and *Berosus munitipennis* are also important in meso- and hypersaline lakes. They are joined in hypo- and mesosaline waters by many others, including more beetles, odonatans, trichopterans, pyralids, notonectids, and corixids. Species richness declines with increasing salinity. There is a prominent inland faunal component mainly of crustaceans, including *P. minuta, D. queenslandensis, R. walbu, Trigonocypris globulosa* and *Moina baylyi*.

Introduction

Salt lakes abound in Australia and after three decades of study much is known (see De Deckker, 1983; Williams, 1981a, 1990, for bibliography). However, lake distribution and knowledge of them is regionally uneven, so that in New South Wales (hereafter NSW) there are not many salt lakes (< 50?) and the only data on them are a few chemical analyses (Johnson, 1980; Williams *et al.*, 1970). Most athalassic saline lakes in NSW lie in the north-west, particularly in a horseshoe-shaped cluster 100–150 km west of Bourke in the Paroo district (Fig. 1). They exhibit a range of salinities from fresh to crystallizing brine, and relative permanency from almost permanent except in major droughts to quite episodic with water present for only a few weeks following unusual rains.

This study aims to provide information on geomorphology, water chemistry and biota for comparative purposes with other Australian saline lakes. In this respect, their isolation is of significance as the Paroo lies *ca* 800 km north of the main cluster of saline lakes in southern Australia, *ca* 900 km southwest of Lake Buchanan in tropical Queensland, and *ca* 800 km east of episodic Lake Eyre. Each of these areas has a characteristic biota (Timms, 1987; Williams 1984, 1990). The central question is to which of these lake systems are the Paroo lakes most closely related?

Methods

The study area (Fig. 1) was visited 12 times regularly over 27 months commencing July 1988. Sometimes lakes were not visited due to logistic impediments or they were dry. Most were sampled by wading, but in those of depth > 0.75 m

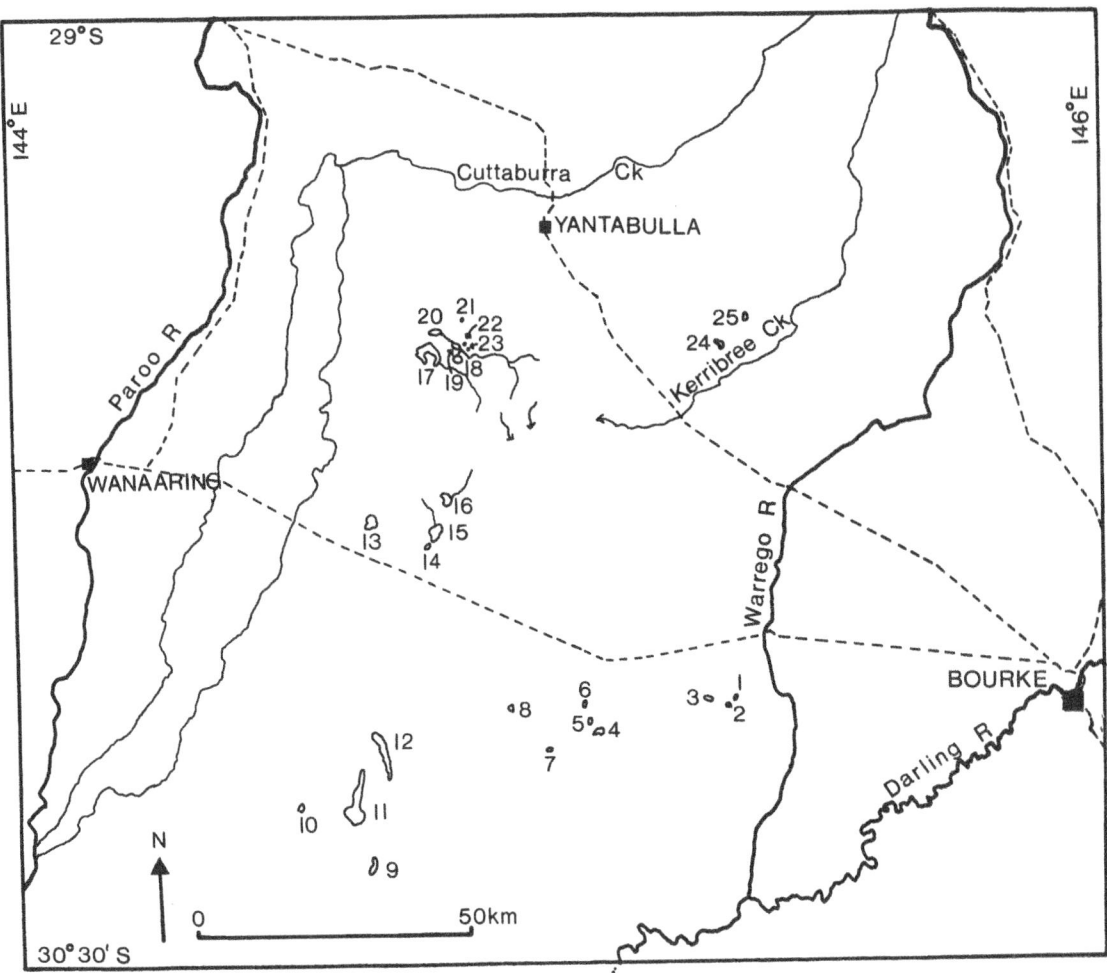

Fig. 1. Map of the Paroo district of northwest N.S.W. The code to lake numbers is in Table 1.

sampling was from a canoe. From stations near mid-lake, water temperature (by a mercury thermometer), light penetration (with a Secchi disc), and pH (by a Hanna Instruments HI 8424 meter) were determined. A water sample was taken for later determination of total dissolved solids (hereafter TDS) by gravimetry. Duplicate samples of 200 ml volume if <5 g l^{-1}, 100 ml if 5.1–20 g l^{-1}, 50 ml if 21–100 g l^{-1} and 20 ml if >101 g l^{-1}, were dried at 105 °C until consecutive weighings were consistent. Time taken varied from *ca* 24 hours for low salinity samples and 72 hours for high salinity samples. Turbidity was measured with a Hach Environmental Laboratory DR/EL1. Ionic analyses were made on samples collected

mainly in July, 1989; Na, K, Ca and Mg were measured by atomic spectrophotometry, Cl by potentiometric tritration against AgNO$_3$, HCO$_3$ by tritration against 0.01N HCl to a pH 4.5, and SO$_4$ by the turbidimetric BaSO$_4$ method. Accuracy of all procedures was $\pm 2.5\%$ or better.

Plankton was collected with a conical net of 30.5 cm diameter and mesh size 159 μm towed at constant speed for one minute (rarely up to 5 minutes if plankton was sparse) from lake edge towards the middle. Generally this caught $>10\,000$ plankters which were preserved in 4% formalin. In the laboratory, each collection was totally examined under a microscope and species identified and listed.

Littoral and epibenthic animals were caught with a rectangular pond net 20 cm by 8 cm and of 1 mm mesh. One man-hour was spent on each occasion using this net. All taxa caught were listed in the field, and representative specimens of each preserved in Carnoys solution for later identification. Generally almost all taxa listed for a lake were caught in the first 10–30 minutes, but in fresher lakes a few additional species were sometimes recovered in the last 30 minutes. Such additional species were rare in the more saline lakes. This method then did not necessarily catch all species present in a lake, but is a compromise as to achieve full representation would take many hours and may not be absolutely achievable. Furthermore it is probable, based on the accrual of additional species in the last 30 minutes of each man-hour, that species lists were most representative in saline lakes and least so in fresher lakes.

Often in the text TDS is loosely referred to as salinity, which is the sum of the seven major ions (Table 3). However salinity is equal to only *ca* 0.8 of the TDS values recorded (Tables 3 and 4), perhaps because of the presence of dissolved organic matter. The lake salinity classification scheme of Hammer (1986) is used because it is universally known and easily described – freshwater (0–0.5 g l^{-1}), subsaline (0.5–3 g l^{-1}), hyposaline (3–20 g l^{-1}), mesosaline (20–50 g l^{-1}) and hypersaline (> 50 g l^{-1}).

Results

Geomorphology

The lakes lie in a sandy plain at *ca* 100 m above sea level, are mostly < 250 ha in area and < 1 m deep (Table 1, Fig. 1). Most arose by deflation by strong southwest and northwest winds, though damming by dunes is often important. Lunette dunes on the eastern shores point to the importance of deflation. These dunes typically consist of a high outer dune of cemented gypsum and one or more lower unconsolidated dunes near the present lake shore. Lake floors are generally of gypsiferous clays.

Some lakes lie on palaeodrainage channels still occasionally but inefficiently used so that ponding and flooding occur. The best example (Johnson, 1980) is the chain consisting of Lake South Nichebulka, Avondale Salt Lake, and Lake Utah on Kerribree Creek. Lake Utah is generally the terminal basin, though water can terminate in the other lakes in drier years (e.g. in 1988), or rarely flow beyond Utah (e.g. in 1974, B. O'Malley, pers. com.). Lake Nichebulka may connect to this system, but did not during 1988–90, and Lake Burkanoko is the terminus on what appears to be a tributary now well isolated by dunes.

Bell Creek northwest of Kerribree Creek provides a smaller example of a old drainage system now dammed to form a terminal lake, here Lower Bell Lake. The two lakes upstream, Gidgee and Middle Bell, are regularly connected (each winter, 1988–90), while water from nearby Horseshoe Lake may reach there by a tortuous route as it did in 1974 (A. Davey, pers. com.). Another example is the unnamed valley leading to the Warrego River in the eastern part of the study area. It carries Lakes Willeroo and Yandaroo, neither of which has overflowed in living memory. Gypsum Lake is probably part of this system but is now isolated by dunes. The final case is Lake Pirillee, which occupies an indistinct waterway leading to upper Kerribree Creek – it connected with this creek in the 1989 and 1990 floods (W. Hann, pers. com.). Lake Strathern seems to lie on a dammed isolated tributary of this system.

Many lakes (Nos. 5–10, 13, 21–23) lie isolated on the sand plain, and Taylors Lake lies just beyond the junction of exposed bed rock and the sand plain. This lake and Nos. 1, 2, and 21 have a high proportion of their drainage from this rock, whereas the remainder gain much water from the sand plain, via channels (e.g. Lake Burkanoko), overland flow (e.g. Ballymere), seepage (e.g. Bells Bore Lake) or in any combination of these.

The lakes varied widely in reliability with respect to water presence (Table 2), both among themselves and secularly (A. Davey, W. Davis, pers. coms.) Only two, Willeroo and Yandaroo, always contained water, though both are known

Table 1. Some physiographic features of lakes of the Paroo.

Lake[1]	Lat & Long.	Elevation[2] (m)	Area (ha)	Max. depth (m)	Geomorphological type[3]
1. 'Gypsum'	30° 07′ S 145° 22′ E	115	35	<0.5	Deflated and dammed
2. Yandaroo	30° 07′ S 145° 21′ E	115	56	<1.0	Dammed on old water course
3. Willeroo	30° 05′ S 145° 17′ E	115	113	<2.5	Dammed on old water course
4. Taylors	30° 07′ S 145° 04′ E	100	62	<2.0	Deflated and dammed
5. 'Ballymere'	30° 06′ S 145° 02′ E	100	32	<0.5	Deflated
6. 'Barakee'	30° 05′ S 145° 02′ E	100	90	<0.2	Deflated
7. Mere	30° 11′ S 144° 59′ E	115	68	<0.5	Deflated
8. 'Kings Bore'	30° 06′ S 144° 55′ E	95	151	<0.5	Deflated
9. 'near L. Pelora'	30° 21′ S 144° 39′ E	90	250	<0.5	Deflated
10. 'Flowing Bore'	30° 15′ S 144° 31′ E	90	150	<0.5	Deflated
11. Utah	30° 15′ S 144° 38′ E	90	1900	<0.2	Deflated; old water course
12. 'Avondale Salt'	30° 09′ S 144° 39′ E	90	520	<0.5	Old water course; deflated
13. 'Rainbar'	29° 48′ S 144° 40′ E	95	190	<0.5	Deflated
14. 'Sth Nichebulka'	29° 51′ S 144° 47′ E	95	54	<1.0	Old water course
15. Nichebulka	29° 49′ S 144° 48′ E	95	322	<0.1	Deflated
16. Burkanoko	29° 46′ S 144° 49′ E	95	280	<1.5	Deflated
17. Horseshoe	29° 32′ S 144° 46′ E	110	746	<0.5	Deflated
18. 'Gidgee'	29° 33′ S 144° 50′ E	110	185	<1.5	Deflated
19. 'Middle Bell'	29° 31′ S 144° 49′ E	110	32	<1.0	Old water course; deflated
20. 'Lower Bell'	29° 30′ S 144° 48′ E	110	160	<1.0	Dammed and deflated
21. 'Freshwater Bw'	29° 29′ S 144° 50′ E	115	19	<1.5	Dammed and deflated
22. 'Bells Bore'	29° 31′ S 144° 50′ E	115	24	<0.1	Deflated
23. 'Woolshed'	29° 32′ S 144° 51′ E	110	4	<0.2	Deflated
24. Pirillee	29° 32′ S 145° 18′ E	115	150	<2.0	Old water course; deflated
25. 'Strathern'	29° 28′ S 145° 19′ E	115	146	<0.5	Deflated; old water course

[1] Names in apostaces are not officially approved by the Geographical Names Board, but are used by local people.
[2] Determined to the nearest 5 m.
[3] The origin of most lakes is complex. The dominant mode is given first, followed, if appropriate by formative process of secondary importance.

to dry in major droughts. Some dry periods must be long, as dead trees stand in Willeroo. Taylors Lake is almost permanent, but the remainder are ephemeral. Generally they fill from autumn rains and water persists till spring (or to summer in some cases). The most reliably filled lakes are Gidgee and Middle Bell while Rainbow, near Pelora, and Woolshed Lakes are the most transient (Table 2). Many lakes have truncated inner lunettes and wave-built beaches 1–2 m above levels observed in 1988–90. These probably relate to a record filling in 1974. At the other extreme in dry years (eg. 1984–86) all lakes in the Bell system remained dry as did probably most others. In these years Taylors and Willeroo became ephemeral (A. Davey & W. Davis, pers. com.).

Water chemistry

In most lakes the dominant ions are Na and Cl (Table 3). The exceptions are the least saline lakes: Lakes Willeroo and Yandaroo are dominated by Na and HCO_3, Lake Gypsum by Na and SO_4, and Lake Strathern by Ca and SO_4. Relative ion importance largely relates to salinity. Potassium reached *ca* 5 equivalent per cent in the fresh lakes, but rarely approached a value of 0.5 in the saline lakes. Bicarbonate was most important in the fresh lakes, though in a few saline lakes its contribution reached 20 per cent, with figures <5 per cent more common. The relative importance of NaCl increased with salinity (Table 3). The ratios Na/Ca and Mg/Ca both increased sig-

Table 2. Seasonal variation in the presence of water and its TDS (g l^{-1}) in some lakes in the Paroo.

Lake \ Date	July 1988	Oct 1988	Dec 1988	Feb 1989	April 1989	June 1989	July 1989	Sept 1989	Nov 1989	Feb 1990	May 1990	Oct 1990
1. 'Gypsum'	?	−	−	−	−	0.79	1.42	4.31	−	−	0.27	1.05
2. Yandaroo	?	0.17	0.14	0.37	0.54	0.21	0.14	0.19	0.24	1.12	0.18	0.27
3. Willeroo	?	0.44	0.12	0.23	0.26	0.15	0.12	0.14	0.17	0.24	0.09	0.17
4. Taylors	?	2.29	1.67	2.77	3.62	2.06	2.55	2.83	4.87	−	0.67	1.61
5. 'Ballymere'	?	?	−	−	−	1.94	3.95	10.12	−	−	0.48	1.42
6. 'Barakee'	?	166.8	−	−	−	42.6	110.5	−	−	−	23.0	127.2
7. Mere	?	4.14	9.6	81.9	−	17.8	71.4	−	−	−	4.7	27.1
8. 'Kings Bore'	?	33.5	186.5	−	−	22.2	39.0	76.0	−	−	22.9	108.2
9. 'near Pelora'	14.1	−	−	−	−	−	−	−	−	−	11.2	−
10. 'Flowing Bore'	29.1	−	−	−	−	16.1	52.4	−	−	−	30.8	−
11. Utah	46.4	−	−	−	−	88.9	263.8	−	−	−	55.5	−
12. 'Avondale Salt'	18.5	−	−	−	−	?	?	?	−	−	9.3	69.0
13. 'Rainbar'	8.4	−	−	−	−	−	−	−	−	−	−	−
14. 'Sth Nichebulka'	6.6	?	−	−	−	6.7	19.3	83.7	−	−	4.0	24.6
15. Nichebulka	68.9	?	−	−	−	95.7	232.1	−	−	−	53.8	361.0
16. Burkanoko	13.4	?	−	−	−	31.5	−	−	−	−	5.6	22.6
17. Horseshoe	30.0	174.0	−	−	−	77.6	−	−	−	−	13.2	31.0
18. 'Gidgee'	?	14.1	51.6	−	−	4.6	9.1	13.4	45.8	−	5.2	18.7
19. 'Mid Bell'	5.3	14.8	151.7	−	−	11.0	26.5	144.4	−	−	5.8	37.5
20. 'Lower Bell'	6.7	15.8	122.9	−	−	32.3	−	−	−	−	8.3	34.5
21. 'Freshwater'	?	6.9	−	−	−	1.29	1.54	2.55	−	−	0.3	0.47
22. 'Bells Bore'	61.7	−	−	−	−	46.6	186.6	−	−	−	28.8	−
23. 'Woolshed'	?	−	−	−	−	19.9	−	−	−	−	16.2	−
24. Pirillee	?	?	?	?	?	?	1.07	1.30	1.93	−	?	1.28
25. 'Strathern'	?	?	?	?	?	?	?	5.2	−	−	?	8.2

Code − = dry; ? = unknown.

nificantly with salinity ($r = 0.65$, and $r = 0.67$ respectively, both significant at $P < 0.001$).

Because of seasonal and secular salinity variations, it is difficult to assign the lakes to salinity classes (Table 4). Assignment based on mean TDS, adjusted to salinity, is probably the most simple and results in a division into 2 freshwater lakes, 5 subsaline, 7 hyposaline, 5 mesosaline and 6 hypersaline lakes. Salinity varied widely in most lakes during 1988–90 (Tables 2 & 5) with coefficients of variation generally between 50 and 125%. This variability was unrelated to salinity ($r = -0.09$, $n = 15$, $P > 0.10$).

All lakes are alkaline, with mean pH between 8.2 and 10.0 (Table 5). Highest individual readings occurred on late sunny afternoons (e.g. a pH of 11.0 in Lake Pirillee) and lowest ones in hypersaline lakes in early mornings (e.g. 7.6 in Lake Barakee).

Some physical variables

Minimum-maximum water temperatures were 8–21 °C in winter, and 25–39 °C in summer, with individual values naturally strongly influenced by time of day when measurements were made.

Most lakes had clear waters of low turbidity (Table 5). The exception was Freshwater Lake which had turbid waters typical of local freshwater pans. The two freshwater lakes and some of the subsaline and mesosaline lakes had slightly opaque waters (Table 5). Given such clear waters, Secchi disc readings could not be obtained for most lakes; only values for opaque to turbid lakes are given (Table 5). These reinforce the differences in turbidities between fresh and saline lakes. Only in the freshwater lakes are there enough data to interpret seasonal variations; these

Table 3. Chemical features of some lakes in the Paroo. [determined on samples collected in July 1989, except for No. 23 (June 1989) and No. 25 (Sept 1989)].

Lake No. and Name	TDS (mg l^{-1})	Salinity[1] (mg l^{-1})	Aspect reported[2]	Na$^+$	K$^+$	Ca^{2+}	Mg^{2+}	Cl$^-$	SO$_4^{2-}$	HCO$_3^-$
3 Willeroo	125	123	A	0.019	0.004	0.008	0.004	0.021	0.004	0.063
			B	49.4	6.6	23.2	20.8	34.9	4.6	60.5
2 Yandaroo	162	158	A	0.013	0.003	0.009	0.004	0.025	0.006	0.081
			B	59.4	4.1	19.6	16.9	33.0	5.5	61.5
21 'Freshwater'	664	420	A	0.074	0.002	0.036	0.016	0.110	0.064	0.118
			B	51.3	3.2	27.7	20.8	48.4	21.1	30.5
24 Pirillee	1068	974	A	0.270	0.006	0.044	0.022	0.398	0.170	0.064
			B	73.9	1.0	13.7	11.4	71.0	22.3	6.7
1 'Gypsum'	1424	1004	A	0.120	0.008	0.100	0.052	0.116	0.496	0.112
			B	35.0	1.6	33.7	29.7	21.2	66.9	11.9
4 Taylors	2250	1600	A	0.37	0.01	0.11	0.05	0.76	0.24	0.06
			B	61.6	0.4	20.5	17.5	78.3	18.1	3.6
5 'Ballymere'	3952	3460	A	1.12	0.01	0.09	0.06	1.76	0.36	0.06
			B	83.7	0.2	7.7	8.4	85.5	12.8	1.7
25 'Strathern'	5180	3630	A	0.38	0.02	0.50	0.17	0.58	1.86	0.12
			B	29.6	0.9	44.6	24.9	28.7	68.0	3.3
19 'Mid Bell'	5340	4180	A	0.96	0.02	0.28	0.16	2.18	0.48	0.10
			B	60.1	0.3	19.8	19.8	84.1	13.7	2.2
14 'Sth Nichebulka'	6620	4910	A	1.26	0.01	0.26	0.14	2.34	0.84	0.06
			B	69.8	0.2	15.6	14.4	78.1	20.7	1.2
20 'Lower Bell'	6760	5420	A	1.19	0.02	0.44	0.18	2.24	1.06	0.14
			B	57.1	0.5	24.9	17.5	73.5	23.9	2.6
13 'Rainbar'	8400	6000	A	1.36	0.02	0.50	0.18	2.82	1.04	0.08
			B	59.3	0.6	25.3	14.8	77.4	21.2	1.4
18 'Gidgee'	9012	7560	A	1.76	0.04	0.44	0.32	4.20	0.56	0.16
			B	61.3	0.3	17.2	21.2	89.4	8.8	1.8
16 Burkanoko	13423	9500	A	2.25	0.05	1.00	0.35	3.95	0.10	1.80
			B	64.5	0.3	16.4	18.8	73.9	1.3	24.8
9 'near Pelora'	14060	11450	A	2.50	0.05	0.90	0.45	5.60	0.15	1.80
			B	56.8	0.3	23.6	19.3	79.8	1.3	18.9
12 Avondale Salt	18540	16270	A	4.85	0.02	0.5	0.55	9.6	0.1	0.65
			B	75.2	0.1	8.9	15.8	94.5	0.7	4.8
23 'Woolshed'	19880	15500	A	4.3	0.1	0.7	0.5	8.0	0.2	1.7
			B	69.8	0.4	13.8	16.0	85.3	1.1	13.6
10 'Flowing Bore'	29090	19300	A	5.2	0.1	0.5	0.9	10.6	0.3	1.7
			B	68.9	0.3	8.2	22.6	88.2	1.5	10.3
17 Horseshoe	30040	25400	A	7.6	0.1	0.6	0.8	14.7	0.1	1.5
			B	77.6	0.2	6.8	15.3	92.4	0.5	7.1
8 'Kings Bore'	39080	33300	A	9.9	0.1	0.5	1.5	18.5	0.2	2.6
			B	74.2	0.2	4.7	20.9	90.0	0.7	9.3
11 Utah	46450	36900	A	12.5	0.1	0.5	0.9	21.6	0.2	1.1
			B	85.0	0.2	3.4	11.4	95.8	0.6	3.6
22 'Bells Bore'	61750	59000	A	19.4	0.2	0.8	2.0	34.8	0.8	1.0
			B	80.0	0.2	4.2	15.6	92.8	1.1	6.1
15 Nichebulka	68920	58000	A	18.2	0.2	0.8	2.2	35.2	0.4	1.0
			B	77.8	0.2	4.3	17.7	97.5	0.6	1.9
7 Mere	71440	69000	A	21.4	0.2	1.6	3.2	35.4	0.6	6.6
			B	80.2	0.3	6.6	12.9	87.1	0.9	12.0
6 'Barakee'	110480	85200	A	24.8	0.4	3.2	2.8	50.8	1.2	2.0
			B	73.1	0.3	11.1	15.5	96.8	0.5	2.7

[1] Determined by summing the values for the seven major ions measured.

[2] A = absolute amount in (g l^{-1}).

   B = relative amount as equivalent percent of total cations or anions.

Table 4. Lake salinity classification and ionic ratios.

Class [1]	Lake No.	Ranking according to salinity on July 89 [2]			Ranking according to mean TDS × 0.8 [3]	
		Salinity ($g\,l^{-1}$)	Na/Ca^{2+}	Mg^{2+}/Ca^{2+}	Lake No.	Salinity ($g\,l^{-1}$)
Fresh ($<0.5\,g\,l^{-1}$)	3	0.12	2.13	0.89	3	0.15
	2	0.16	3.03	0.86	2	0.26
	21	0.42	1.85	0.75		
Subsaline ($0.5–3.0\,g\,l^{-1}$)	24	0.97	5.39	0.83	24	1.12
	1	1.00	1.04	0.75	1	1.26
	4	1.60	3.00	0.85	21	1.76
					4	1.99
					5	2.86
Hyposaline ($3.0–20\,g\,l^{-1}$)	5	3.46	10.87	1.09	25	5.36
	25	3.63	0.66	0.59	13	6.72
	19	4.18	3.04	1.00	9	10.08
	14	4.91	4.47	0.93	23	14.48
	20	5.42	2.29	0.70	16	14.64
	13	6.00	2.43	0.58	18	16.24
	18	7.56	3.56	1.23	14	19.28
	16	9.50	3.91	1.15		
	9	11.45	2.40	0.82		
	23	15.50	5.05	1.16		
	12	16.27	8.45	1.78		
	10	19.30	8.40	2.78		
Mesosaline ($20–50\,g\,l^{-1}$)	17	25.40	11.41	2.27	12	23.12
	8	33.30	15.79	4.54	7	24.72
	11	36.90	25.00	3.33	10	25.68
					20	29.44
					19	39.78
Hypersaline ($>50\,g\,l^{-1}$)	15	58.00	18.09	4.17	8	55.84
	22	59.00	19.05	3.70	17	58.56
	7	69.00	12.16	1.96	22	64.72
	6	85.20	6.59	1.39	6	75.20
					11	90.96
					15	129.84

[1] Classification as per Hammer (1986).
[2] See Table 3.
[3] Mean TDS calculated from Table 2 and multiplied by the average ratio (0.8) of Salinity/TDS in Table 3.

showed maximal values in winter/spring and minimal values after first filling and in summer.

Flora

Almost all lakes had well-vegetated central and marginal areas, except for a month or two after first filling. Only the claypan-like Freshwater Lake never developed vegetation, while most of the hypersaline lakes had sparse vegetation when less saline, but no live plants at high salinities. *Vallisneria?gigantea* Graebner was dominant in the two freshwater lakes; other common species included *Ludwigia peploides* (Kunth.) Raven, *Myriophyllum verrucosum* Lindl., *Chara* sp., and *Nitella* sp. The

Table 5. Some physicochemical features of the lakes.

Lake	Total dissolved solids		Mean pH	Mean turbidities (FTU)	Secchi disc value (cm)	
	Mean* $(g\,l^{-1})$	Coefficient of variation			Range	Mean
3 Willeroo	0.19	52	9.1	20	23–150	80
2 Yandaroo	0.32	90	9.4	36	12–120	53
24 Pirillee	1.40	27 +	10.0	1	b	–
1 'Gypsum'	1.57	101	9.0	17	10–45	24
21 'Freshwater'	2.20	113	9.0	1202	1–12	7
4 'Taylors'	2.49	47	9.6	6	b	–
5 'Ballymere'	3.58	108	9.2	16	b–17–20	18
25 'Strathern'	6.7	32 +	9.0	7	b	–
13 'Rainbar'	8.4	0 +	9.5	5	b	–
9 'near Pelora'	12.6	16 +	9.5	12	b	–
12 'Avondale Salt'	16.8	100 +	9.2	10	b–20	20
23 'Woolshed'	18.1	14 +	8.9	2	b	–
16 Burkanoko	18.3	61 +	9.5	5	b	–
18 Gidgee	20.3	90	9.4	5	b	–
14 'Sth Nichebulka'	24.1	125	9.2	16	b–15–35	25
7 Mere	30.9	105	9.3	10	b	–
10 'Flowing Bore	32.1	47 +	8.8	2	b	–
20 'Lower Bell'	36.8	119	8.9	2	b	–
19 'Mid Bell'	49.6	124	8.8	6	b	–
8 'Kings Bore'	69.8	87	8.9	9	b	–
17 Horseshoe	73.2	100	8.8	9	b	–
22 'Bells Bore'	80.9	89 +	8.9	9	b	–
6 'Barakee'	94.0	64	8.3	9	b	–
11 Utah	113.7	90 +	8.2	5	b	–
15 Nichebulka	162.3	81	8.4	9	b	–

* Calculated from Table 2.

+ Based on less than 5 values, so not used in correlation.

b: Bottom visible from lake surface.

latter three species (including *C. fibrosa* in Lake 5), together with *Lepilaena* sp. (*L. bilocularis* T. Kirk. in Lake 4) dominated in the subsaline lakes. *Lepilaena* sp., with a salinity range of 1–40 g l^{-1} dominated in hyposaline and some mesosaline lakes, and *Ruppia* sp. (*R. megacarpa* Mason in Lakes 11, 15, 16 and 19) dominated in many mesosaline and all hypersaline lakes, with a salinity range of 22–108 g l^{-1}. The charophyte *Lamprothamnion papulosum* (Wallr.) J. Gr. also occurred in the range 5–71 g l^{-1}, but generally was not conspicuous.

Phytoplankton was not studied, but blooms of blue-green algae were observed in the two freshwater lakes in both summers.

Fauna

The lake fauna is listed in Tables 6 (Crustacea), 7 (Insecta) and 8 (miscellaneous groups). Several taxa could not be identified to species level because of incomplete taxonomic knowledge (e.g. Conchostraca), or they were new to science and not yet formally named (e.g. the alonid chydorids), or it was impractical to do so regularly (e.g. *Diacypris*, *Reticypris*). Some details are provided below.

Anostraca

Seven species were encountered (Table 6). Of these, four *Branchinella* spp. were restricted to

Table 6. Crustacea of the Paroo lakes.

Species	Salinity range (g l⁻¹)	Number of records	Lakes from which recorded [1]
Anostraca			
Branchinella australiensis (Richters)	$0.3 \to 11.2$	10	1, 4, 5, 9, 12, 21
Branchinella nichollsi buchananensis Geddes	$1.9 \to 11.2$	10	5, 7, 9, 12, 14, 16, 19
Branchinella spp. [2]	$0.3 \to 2.1$	7	1, 4, 5, 21
Parartemia minuta Geddes	$8.4 \to 255.0$	34	6–8, 10–12, 15–20, 22, 23
Notostraca			
Triops australiensis Spencer and Hall	$0.3 \to 19.3$	13	1, 4, 5, 9, 12, 14, 21
Conchostraca			
Cyzicus sp.	$0.2 \to 5.2$	18	1, 2, 4, 5, 7, 21, 25
Limnadia sp. a	$0.2 \to 11.2$	19	1, 2, 4, 5, 7, 14, 16, 18–21, 25
Limnadia sp. b	$0.2 \to 0.5$	3	1, 2, 4
Lynceus sp.	0.5	1	5
Cladocera			
Daphnia carinata King	$0.11 \to 17.8$	43	1–5, 7, 14, 18–21, 24, 25
Daphniopsis queenslandensis Sergeev	$4.57 \to 71.4$	45	6–10, 12, 14, 16–20, 23
Ceriodaphnia cornuta Sars	$0.14 \to 0.69$	4	3, 5, 21
Ceriodaphnia aff. 'dubia' Richard	$0.18 \to 1.94$	5	2, 3, 5, 24
Ceriodaphnia aff. 'quadrangula' (O. F. Muller)	0.14	1	3
Simocephalus vetulus elizabethae (King)	$0.17 \to 0.19$	2	2
Moina australiensis Sars	$0.79 \to 1.94$	2	1, 5
Moina baylyi Forro	$22.3 \to 51.6$	2	8, 18
Moina micrura Kurz	$0.47 \to 0.54$	2	2, 21
Diaphanosoma unguiculatum Gurney	$0.14 \to 0.33$	4	2, 21
Latonopsis australis Sars	$0.44 \to 0.48$	2	3, 5
New Aloniae 3 spp.	$2.21 \to 18.7$	16	4, 7, 16, 18–20
Other chydorids [3]	$0.11 \to 19.9$	22	1–5, 7, 14, 18–21, 24
Echinisca carinata Smirnov	$0.2 \to 15.9$	26	2–4, 7, 13, 14, 18–21, 24
Macrothrix ?breviseta Smirnov	$0.2 \to 0.4$	5	2, 3
Copepoda			
Boeckella triarticulata Thomson	$0.09 \to 10.1$	46	1–5, 7, 19–21, 24, 25
Calamoecia canberra Bayly	$0.17 \to 0.79$	7	1, 2, 5, 21
Calamoecia lucasi Brady	$0.09 \to 1.29$	9	2–4, 21
Apocyclops dengizicus (Lepeschkin)	$4.0 \to 69.0$	41	4, 7, 9, 13, 14, 16, 18–20, 23
Metacyclops platypus Kiefer	$13.3 \to 77.6$	11	6, 8, 10–12, 15, 17, 18, 22
Microcyclops sp.	$108.0 \to 144.4$	3	8, 19, 20
Microcyclops varicans (Sars)	$0.9 \to 0.67$	14	2–5
Schizopera spp.	$4.57 \to 177.5$	5	7, 12, 18, 19
Ostracoda			
Bennelongia sp.	$0.14 \to 5.3$	2	3, 19
Cypretta sp;	$0.14 \to 9.6$	8	1–5, 7, 14, 17–21, 24, 25
Diacypris spp. [4]	$5.2 \to 263.8$	57	6–8, 10–12, 15–20, 22, 23
Heterocypris n. sp.	$0.3 \to 33.5$	30	4, 13, 14, 16–18
Reticypris spp.	$5.2 \to 122.9$	24	7–10, 12, 14, 16–20, 23
Cyprinotus n. sp.	$0.17 \to 24.6$	32	1, 2, 4, 5, 7, 9, 14, 16, 18, 20, 24, 25
Trigonocypris globulosa De Deckker	$2.2 \to 122.9$	34	4, 5, 8, 10, 12–14, 16–20, 23, 25
Mytilocypridini gen. nov.	$14 \to 122.9$	18	5, 14, 16–20, 23
Decapoda			
Cherax destructor Clark	$0.1 \to 0.3$	5	3

[1] Key to lake numbers given in Table 1.
[2] *B. arborea* Geddes, *B. lyrifera* Linder, *B. occidentalis* (Dakin), and *B. pinnata* Geddes.
[3] Includes at east 9 species, the most salt tolerant being *Biapertura rigidicaudis* Smirnov at 19.9 g/L followed by *Dunhevedia crassa* King at 14.0 g/L.
[4] Includes *D. dictyote* De Deckker, *D. dietzi* (Herbst) and *Diacypris* n. sp.
[5] Includes *R. herbstii* McKenzie and *R. walbu* De Deckker.

Table 7. Insects of the Paroo Lakes.

Species	Salinity range $(g\,l^{-1})$	Number of records	Lakes from which recorded [1]
Ephemeroptera			
Cloeon sp.	0.1 → 4.8	20	2–4, 24
Tasmanocoenis tillyardi (Lestage)	0.1 → 2.2	8	2–4
Odonata			
Diplacodes bipunctata (Brauer)	0.1 → 22.7	20	1–5, 16, 18, 19, 24, 25
Diplacodes haemotodes (Burmeister)	0.1 → 1.9	8	2, 3
Hemianax papuensis (Burmeister)	0.1 → 8.2	6	23, 24 25
Orthetrum caledonicum (Brauer)	0.1 → 24.6	7	2–4, 7, 14, 18
Austrolestes annulosus (Selys)	0.1 → 37.5	37	2–5, 14, 16, 18–21, 24
Xanthagrion erythroneurum Selys	0.1 → 0.5	12	2, 3
Hemiptera – Notonectidae			
Anisops ?calcaratus Hale	1.0	2	3, 24
Anisops gratus Hale	0.1 → 24.6	24	1–5, 7, 14, 21
Anisops stahi Kirkaldy	0.1 → 1.9	4	2–5
Anisops thienemanni Lundbald	0.1 → 24.6	71	1–5, 7, 14, 18–21, 24
Anisops sp.	3.6	1	4
– Corixidae			
Agraptocorixa eurynome Kirkaldy	0.1 → 18.7	36	1–5, 7, 14, 18–21, 24
Agraptocorixa hirtifrons Hale	0.1 → 27.1	19	1–5, 18, 20
Agraptocorixa parvipunctata Hale	0.1 → 6.7	24	1–5, 7, 14, 18, 19, 21, 24
Micronecta spp.	0.1 → 53.8	81	1–7, 9, 10, 12–21, 24, 25
Sigara truncatipala Hale	0.1 → 10.1	5	2, 5
Sigara sp. (females only)	0.1 → 19.3	8	2, 7, 14, 18, 25
– Naucoridae			
Naucoris congrex Stal	0.1 → 0.3	3	2, 3
– Nepidae			
Ranatra dispar Montandon	4.1	1	7
Trichoptera			
Oecetes ?australis Banks	0.1 → 0.2	3	2
Oecetes sp.	0.1 → 4.8	6	3, 4, 21
Notolina sp.	0.1 → 10.1	11	2–5
Triplectides ?australicus Banks	0.1 → 13.3	37	1–4, 7, 18, 21, 24
Lepidoptera			
Pyralidae	0.1 → 37.5	14	2, 4, 7, 14, 16, 18, 19, 24
Diptera – Chironomidae			
Coelopynia pruinosa Freeman	0.4	1	21
Procladius paludicola group	0.2 → 26.5	8	1, 5, 18, 19, 21
?Cricotopus sp.	0.1 → 1.9	9	2–5
Chironomus tepperi Skuse	0.1 → 13.3	47	1, 4, 5, 7, 9, 13, 14, 17–21, 24, 25
Dicrotendipes sp.	0.1 → 1.3	6	2–4
Polypedilum nubifer Skuse	0.1 → 2.0	14	2–5
Tanytarsus barbitarsis Freeman	11.2 → 255.0	22	6–8, 10–12, 16, 17, 19, 22, 23

Table 7. (Continued).

Species	Salinity range (g l)$^{-1}$	Number of records	Lakes from which recorded[1]
– Culicidae			
Anopheles annulipes Walker	0.1 → 11.2	11	1–3, 5, 9, 24
Anopheles amictus Edwards	0.1	1	2
Aedes sp. near *sagax* (Skuse)	11.2	1	9
Culex australicus Dobrotworsky & Drummond	0.2 → 0.5	2	5
– Others			
Ceratopogonidae	0.1 → 33.5	9	2, 5, 8, 13, 18, 19
Stratiomyidae	0.1 → 1.4	4	1–3, 5
Coleoptera – Haliplidae			
Haliplus fuscatus Clark	0.1	1	3
– Dytiscidae			
Allodessus bistrigatus (Clark)	0.1 → 24.6	15	2, 3, 5, 14, 16, 18–21, 24
Antiporus gilberti Clark	0.1 → 24.6	25	1–5, 18, 21, 24, 25
Cybister tripunctatus Olivier	0.2 → 1.1	2	2, 3
Necterosoma penicillatum (Clark)	0.1 → 37.5	9	4, 7, 14, 16–19, 21
Megaporus howitti Clark	0.1 → 24.6	28	1–5, 14, 16, 18, 21, 24, 25
Sternopriscus multimaculatus (Clark)	0.1 → 27.1	43	1–5, 7, 9, 12, 14, 16–18, 19, 21, 24, 25
Rhantus suturalis MacLeay	0.2 → 5.6	6	2, 3, 16, 21
Eretes australis (Erichson)	0.2	1	2
Hydaticus variegatus Watts	0.2 → 53.8	12	3, 4, 8, 14, 15, 17–19
Hydaticus consanguineus Aube	0.1	1	3
– Hydrophilidae			
Berosus approximanus Fairmaire	0.1 → 24.6	12	1, 3–5, 14, 16, 18
Berosus macumbensis Blackburn	0.3 → 9.1	9	1, 5, 18, 21
Berosus munitipennis Blackburn	0.1 → 149.4	52	1–10, 12, 14, 16–21, 24, 25
Berosus nutans MacLeay	0.2 → 0.3	2	3
Enochrus sp. (nr. *andersoni* Blackburn)	0.2 → 1.3	2	2, 24
Helochares australis (Blackburn)	1.1	1	2
Limnoxenus macer (Blackburn)	0.5 → 1.3	2	21, 25
Paroster sp.	1.9	1	5

[1] Key to lake numbers given in Table 1.

fresh and subsaline waters, and *B. australiensis* was only occasionally found in hyposaline pans. *B. nichollsi* was more characteristic of these pans; it was identified as the subspecies *buchananensis*, known previously only from Lake Buchanan, central north Queensland (Geddes, 1981). *Parartemia minuta* was the characteristic anostracan of the meso- and hypersaline lakes, as in other inland saline lakes (Timms, 1987; Williams & Kokkinn, 1988), though its salinity range is widest in the Paroo.

Notostraca

Triops australiensis was found regularly in sub- and hyposaline waters in autumn and early winter collections. At these higher salinities individuals were only about half the size as those in the nearby freshwater pans. Notostracans have not previously been recorded for specific Australian saline lakes, though Williams (1981b) mentions their occasional presence in mildly saline waters. The upper salinity limit is the highest for Australia and is similar to that for *Lepidurus lynchi* in western Canada (Hammer, 1986).

Table 8. Miscellaneous animals of the Paroo Lakes.

Number of species	Mean salinity range (g l^{-1})	Salinity records	Lakes from which recorded [1]
Hirudinea			
– unidentified leech	0.2	2	2
Rotifera [2]			
Brachionus plicatilis Muller	6.7 → 76.0	16	7, 8, 12–14, 16–20
Hexarthra cf. *fennica* (Levander)	0.2 → 76.0	30 +	4, 6–8, 11, 13–20, 22–24
Asplanchna sp.	0.1 → 5.2	4 +	2, 3, 18, 21
Keratella tropica (Apstein)	0.1 → 6.8	1 +	20
Arachnida – Hydracarina			
Arrenurus spp.			
(inc. *A. balladoniensis* Halik)	0.1 → 10.1	14	1–5, 24
Diplodontus spp.	10.1	1	5
Eylais spp.	0.1 → 13.2	28	1–5, 16–18
Hydrachna spp.	0.1 → 19.3	23	1–4, 7, 14, 18–21, 24
Limnesia spp.	0.2 → 0.3	2	2, 3
Piona cumberlandensis (Rainbow)	0.3	1	3
Mollusca – Gastropods			
Glyptophysa aliciae Reeve	0.1 → 1.1	9	2, 3
Isidorella newcombi	0.1 → 4.7	24	2–5, 7, 21, 24
Adams & Angas			
Physa sp.	0.1 → 4.9	7	2–4
Amphibia			
Tadpoles of *Limnodynastes* and	0.1 → 9.3	13	1–3, 5, 12, 18, 21, 24, 25
Notoden or *Neobatrachus*			
Pisces			
Leiopotherapon unicolor Gunther	0.1 → 0.3	6	3
Cyprinus carpio Linnaeus	5.5	1	16

[1] Key to lake numbers given in Table 1.
[2] Lakes Willeroo and Yandaroo not listed.

Conchostraca

Clam shrimps are not usually found in saline waters, but two species regularly enter hyposaline lakes to 11.2 g l^{-1} (Table 6). Only Geddes *et al.* (1981) mention conchostracans in Australian saline lakes, with an upper salinity limit of 4.8 g l^{-1}. Three of the four species listed were very common in temporary fresh waters in the area (unpublished data), though *Limnadia* sp.a seemed to be characteristic of the hyposaline pans. It occurred in autumn and winter, whereas *Cyzicus* was found in all seasons.

Cladocera

Of the 26 species encountered (Table 6), only a few were common and widespread. *Daphniopsis queenslandensis*, an inland form of a genus characteristic of mesosaline lakes, was generally found in winter-spring and is the most salt tolerant cladoceran in the Paroo (to 71 g l^{-1}). The other halobiont is *Moina baylyi*, now known to be widespread in inland mesosaline waters (Williams & Kokkinn, 1988). Usually it occurred in spring and summer and rarely was found with *Daphniopsis*. Compared to data for L. Eyre (Bayly, 1974; Williams, 1990), *D. queenlandensis* has a wider salinity range in the Paroo, and *M. baylyi* a narrower range.

A further 7 species entered mesosaline waters. *Daphnia carinata*, and to a lesser degree *Echinisca carinata*, were the most common, but the most interesting are new alonid chydorids being described by Frey (1991). All penetrate salinities as high as or higher than those from which they have been recorded elsewhere in Australia.

Copepoda

Eight taxa were found, with *Boeckella triarticulata* the most common in fresh to hyposaline waters, *Microcyclops varicans* in fresh waters and *Apocyclops dengizicus* and *Metacyclops platypus* prominent in hypo- to hypersaline lakes. An unknown species of *Microcyclops* was restricted to quite hypersaline lakes. The harpacticoid *Schizopera* sp. was the most euryhaline, with an upper salinity limit of 177.5 g l^{-1}, but it was not very widespread.

The penetration of the widespread *B. triarticulata* into hyposaline lakes is typical throughout Australia, though the field salinity range is less in the Paroo (Bayly, 1969; De Deckker & Geddes, 1980). The absence of the halobionts, *Calamoecia clitellata* and *C. salina*, is more significant, as these are characteristic species of southern and western saline lakes, but not apparently in the eastern inland (De Deckker & Geddes, 1980; Geddes *et al.*, 1981; Timms, 1987; Williams, 1981a, 1984, 1990). The cyclopoids are wide ranging in southern and inland Australia and the salinity ranges in the Paroo are about equivalent. *Schizopera* occurs in other saline waters in Australia (R. Hamond, pers. com.).

Ostracoda

Eleven species occurred in the collections, four of them new (Table 6). This fauna is about as rich as in L. Eyre (Williams, 1990), but not as diverse as in lakes nearer the coast (e.g. 16 species in southeast South Australia, De Deckker & Geddes, 1980; 20 species in southwest Western Australia, Geddes, *et al.*, 1981). *Diacypris* spp. were the most widespread, and between them had a remarkably wide salinity range, but were most

common at higher salinities (> 50 g l^{-1}). *Reticypris* spp. generally ocurred at intermediate salinities (20–80 g l^{-1}), *Heterocypris* sp. at lower salinities (3–20 g l^{-1}) and *Cyprinotus* at even lower salinities (3–10 g l^{-1}). Both *Trigonocypris globulosa* and a new mytilocyprinid occurred over a wide range of salinities and in fact had wider salinity ranges than reported elsewhere. These two species, together with *Reticypris walbu* and perhaps the new species, are components of an ostracod fauna characteristic of inland Australia (P. De Deckker, pers. com.), while the remainder occur in saline lakes over much of Australia.

Ephemeroptera

Mayflies are not regarded as saline lake inhabitants (Hammer, 1986), and this is their first mention for Australian saline lakes. *Tasmanocoenis tillyardi* and particularly *Cloeon* sp. were relatively common in the more persistent subsaline waters, and penetrated some hyposaline lakes, but only to 4.8 g l^{-1} (Table 7).

Odonata

Six species were recorded (Table 7), but more were probably present judging from the adults caught hawking near the lakes and from literature records. The damselfly *Austrolestes annulosus* was most frequently encountered; it had an upper salinity of 37.5 g l^{-1}, the maximum recorded for an odonate (Hammer, 1986). Three other species, all dragonflies, entered hyposaline waters, but to a lesser extent.

Other Australian studies report fewer odonates, generally of different taxa, and with lower salinity tolerances. The most similar are the Eyre Peninsula lakes, where both damselfly and dragonfly nymphs occur to 26.7 g l^{-1} (Williams, 1984).

Hemiptera

Waterbugs, particularly *Anisops* spp., *Agraptocorixa* spp., and *Micronecta* spp. were common components in most hyposaline and mesosaline lakes (Table 7). All three genera were usually represented by adults and juveniles, and so are halophilic. *Anisops thienemanni*, the dominant notonectid, was found all year, but its congeners were largely confined to the cooler months. Of the three species of *Agraptocorixa*, *A. hirtifrons* typically occurred in more saline waters, but was largely restricted to cooler months. *Agraptocorixa eurynome* was found all year round. *Micronecta* spp., though quite euryhaline, were generally found below 10 g l^{-1} (67 of 81 records), but with 11 occurrences at 10–30 g l^{-1} and 3 at 31–54 g l^{-1}.

Compared to other Australian salt lake districts, notonectids and corixids are more important and generally have higher salinity limits in the Paroo (De Deckker & Geddes, 1980; Geddes *et al.*, 1981; Timms, 1987; Williams, 1981a; Williams & Kokkinin, 1988). This confirms a singular observation by Ettershank *et al.* (1966) that some hemipterans can live in quite saline waters in inland NSW, more so than in southern Victoria. However, species richness and composition is about the same for at least the Victorian lakes, though is much more diverse than for other inland lakes studied.

Trichoptera
Although 3 of the 4 species found entered hyposaline waters (Table 7), only *Triplectides?australis* was common. It inhabited cases constructed from a variety of materials, including sheep dung, and was found in all seasons, whereas the others had characteristic cases and were encountered mainly in the cooler months.

Caddises are not common in saline lakes in Australia or elsewhere (Hammer, 1986). The Paroo fauna is the richest yet in Australia, though salinity ranges do not match the figures for *Oecetes?australis* in Victoria (Bayly & Williams, 1966; Timms, 1981).

Pyralidae
Moth larvae became more common in hyposaline lakes as the study progressed (1 record in 1988, 4 in 1989, 9 in 1990) and were most noticeable in spring when food plants (*Lepilaena* sp. and *Ruppia* sp.) grew. Their upper salinity record of 37.5 g l^{-1} is almost double that recorded for pyralids in Victoria (Williams, 1981a), the only other region where their presence has been recorded in salt lakes.

Chironomidae
Seven species were recorded (Table 7), though certainly more occur as the benthos was not sampled adequately. Two species predominate: *Chironomus tepperi* which occurred mainly in hyposaline pans in spring and *Tanytarsus barbitarsis* which lived in meso- and hypersaline waters at all seasons. Both are widespread, *C. tepperi* in temporary fresh waters (Edward, 1964), and *T. barbitarsis* in saline lakes in Victoria (Timms, 1983; Williams, 1981a), southern South Australia (Kokkinn, 1986), L. Eyre (Williams & Kokkinn, 1988) and L. Buchanan where it was wrongly recorded as *Rheotanytarsus* sp. (Timms, 1987). For both species, the upper salinity limits in the Paroo are the highest known.

Culicidae
Mosquito larvae occurred occasionally, but mainly in the hyposaline pans in spring. More than 4 species listed (Table 7) may be involved, as the dominant species, *Anopheles annulipes*, is a species complex and other species are known in the arid zone after flood rains (E. Marks, pers. com.).

Species richness is about the same in the Paroo saline lakes as in some other Australian areas, but upper salinity tolerances may be less, for Williams (1981a) reports *Aedes vigilax* from western Victorian saline lakes at 23 g l^{-1}.

Ceratopogonidae
These are not an important component of the plankton or littoral of the Paroo lakes (though they could be common in the benthos), whereas many Australian studies show them to be common in hypersaline lakes. Interestingly, they are not in L. Eyre (Williams & Kokkinn, 1988) and are unimportant in L. Buchanan (Timms, 1987), two other inland lakes. The salinity range in the Paroo is much narrower than in southern Australia.

Coleoptera
Nineteen species were recorded, 11 in saline waters (Table 7). *Berosus munitipennis* was the most common and euryhaline species, with many

records to 149 g l^{-1} and one supplementary collection at 255 g l^{-1} from Barton's Creek which flows into Horseshoe Lake. This species is widespread in Australia and common in inland saline lakes (Timms, 1987; Timms & Watts, 1987; Watts, 1978; Williams & Kokkinn, 1988). Seven species (Table 7) were common in mesosaline waters. Many breeding populations of a hydrophilid, probably *B. munitipennis*, were found in salinities from 3–255 g l^{-1}, and some breeding dytiscids, possibly *Necterosoma penicillatum*, occurred in waters to 19 g l^{-1}.

The Coleoptera fauna of the Paroo region is somewhat less diverse than in Victorian salt lakes around Colac (19 species, Timms & Watts, 1987) but is largely of similar composition. The main differences are the prominence of the inland species *Hydaticus variegatus*, the abundance and wider salinity range of *Antiporus gilberti*, and scarcity of *Enochrus* sp. (near *andersoni*) in the Paroo. On the other hand, the Paroo fauna is much more diverse than that of either L. Eyre (Williams & Kokkinn, 1988) or L. Buchanan (Timms, 1987), probably because these large lakes offered a much more homogeneous environment than the numerous and variable lakes of the Paroo.

Rotifera

Only two species, the ubiquitous *Brachionus plicatilis* and *Hexarthra* cf. *fennica*, were common in the Paroo lakes, though there were a few others in hyposaline lakes, and many in the two freshwater lakes (Table 8). Few studies of Australian saline lakes consider rotifers, but those that do (e.g. Williams, 1981a) stress the importance of *B. plicatilis* and *H. fennica*. Salinity ranges for these species in the Paroo are the highest for Australia, but are below maximum ranges (*ca* 100 g l^{-1}) recorded for the world (Hammer, 1986).

Hydracarina

Water-mites were common in the freshwater lakes and in subsaline and some hyposaline lakes. Six genera are involved, with *Eylais* and *Hydrachna* the most widespread and salt tolerant (Table 8). Water-mites are not mentioned in other Austra-

lian studies, but elsewhere they are known from waters of similar salinities (Hammer, 1986).

Gastropoda

Of the three species found (Table 8), *Glyptophysa aliciae* and *Isidorella newcombi* are widespread in inland waters and *Physa* sp. is introduced. The latter two taxa occur in mildly hyposaline waters. *G. aliciae* and *Physa* were restricted to the nearly permanent waterbodies, but *I. newcombi* lived as well in three other lakes which dried regularly for many months. It seems the adults aestivate among plant roots and debris. A significant omission from the gastropod fauna, as well as that of L. Eyre (Williams & Kokkinn, 1988), is *Coxiella salina*. It apparently cannot survive severe and prolonged drying (Williams & Mellor, 1991).

Amphibia

Two species of tadpoles were found in fresh to hyposaline lakes (Table 8). From adults found near these lakes, the species involved were probably *Limnodynastes tasmaniensis* and *Notoden bennettii*. Tadpoles are not mentioned as being in even low salinity waters in other Australian studies or overseas (Hammer, 1986).

Pisces

Despite a variety of fish living in inland Australia (Glover & Sim, 1978), only two occurred in the Paroo lakes, probably because almost all lakes are ephemeral and there is virtually no access to permanent water even in flood. However this did not prove a major barrier to the European carp, as juveniles appeared in the isolated Burkanoko system in 1990 (though not in the two previous years).

Community structure

Momentary species richness, *i.e.* the number species collected on given visit to a lake, decreases with increasing salinity, with the relationship being more or less linear when TDS is scaled logarithmically (Fig. 2). The use of a log TDS scale has the further advantage of clearly show-

284

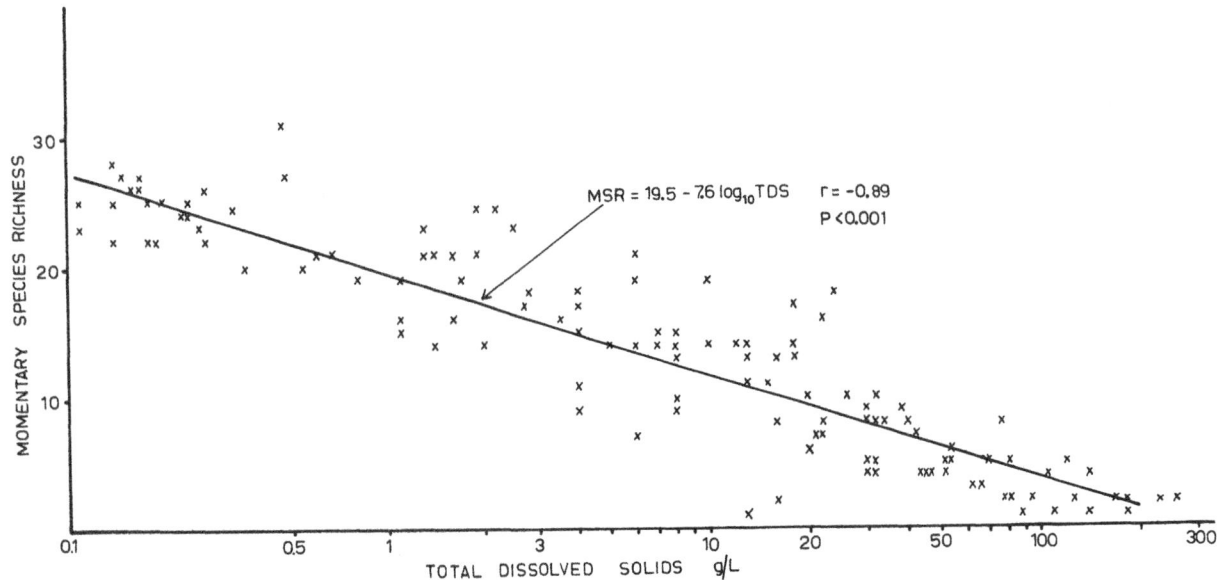

Fig. 2. Relationship between momentary species richness and salinity in the Paroo lakes.

ing patterns on species richness at low salinities. The slope is the same everywhere, *i.e.* between 0.1 and 1.0, 1 and 10, and between 10 and 100 g l^{-1}. There are no indication of boundaries, say at 3 g l^{-1} (the so-called fresh-salt water boundary, Williams, 1981a) or beyond 50 g l^{-1}.

It should be noted that although Fig. 2 is based on 133 data points, really there are 25 independent sets of data points (each set representing a lake), so that any limited segment of the salinity range could be strongly influenced by 'atypical' lakes.

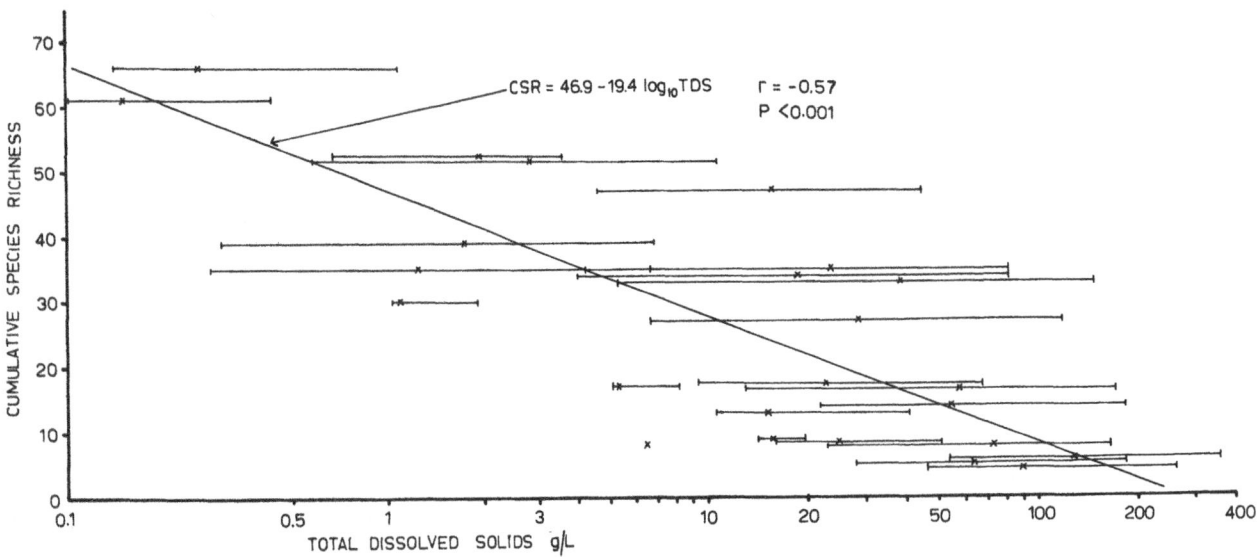

Fig. 3. Relationship between cumulative species richness and salinity in the Paroo lakes. Range and arithmetric mean are given for each lake.

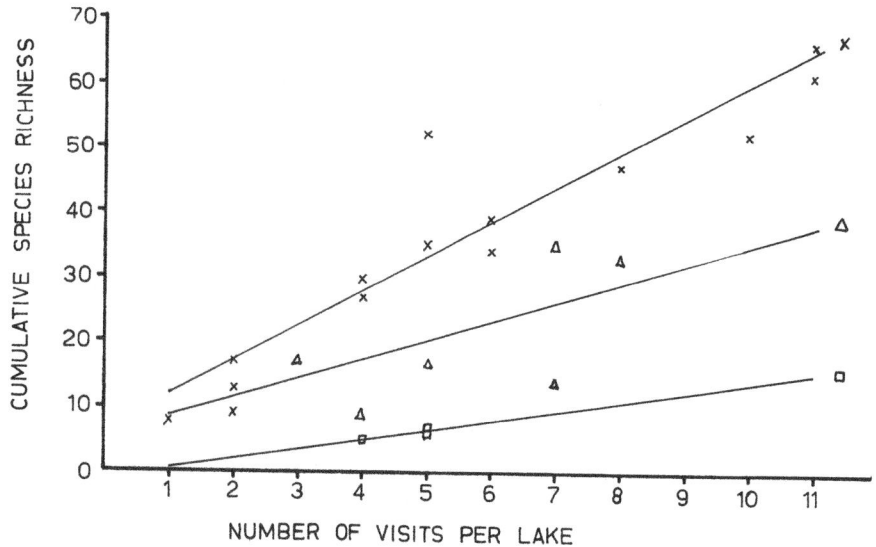

Fig. 4. Relationship between cumulative species richness and the number of visits to each lake. Salinity of lakes indicated as follows: x, $<20\,\mathrm{g\,l^{-1}}$; △, 21–$60\,\mathrm{g\,l^{-1}}$; □, $>61\,\mathrm{g\,l^{-1}}$. Regression equations differ according to salinity range: (i) 0–$20\,\mathrm{g\,l^{-1}}$ $y = 6.45 + 5.26\,x$; (ii) 21–$60\,\mathrm{g\,l^{-1}}$ $y = 0.41 + 3.73\,x$; (iii) $>61\,\mathrm{g\,l^{-1}}$ $y = -1.0 + 1.50\,x$.

Table 9. Taxonomic representation at various salinities

Lake	Mean salinity $(\mathrm{g\,l^{-1}})$	Cumulative number of species	Proportion of		
			Crustaceans	Insects	Others
Willeroo	0.15	61	25	59	16
Yandaroo	0.26	66	27	58	15
Pirillee	1.1	30	23	60	17
'Gypsum'	1.3	35	37	51	12
'Freshwater'	1.8	39	38	51	10
Taylors	2.0	52	35	54	11
'Ballymere'	2.9	52	38	52	10
'Strathern'	5.4	17	41	53	6
'Rainbar'	6.7	8	38	38	25
'near Pelora'	10.1	13	54	46	0
'Woolshed'	14.5	9	78	11	11
Burkanoko	14.6	27	44	41	15
'Gidgee'	16.2	47	38	49	13
'Sth Nichebulka'	19.3	34	41	50	9
'Avondale'	23.1	17	59	24	17
Mere	24.7	35	46	43	11
'Flowing Bore'	25.7	9	67	33	0
'Lower Bell'	29.4	27	56	30	14
'Middle Bell'	39.8	33	48	42	10
'Kings Bore'	55.8	14	57	29	14
Horseshoe	58.6	17	53	35	12
'Bells Bore'	64.7	5	60	20	20
'Barakee'	75.2	7	57	29	14
Utah	91.0	5	60	20	20
Nichebulka	129.8	6	50	33	17

286

Table 10. Prominent animals in the Paroo lakes.

	Fresh (<0.5 g l^{-1})	Subsaline (0.5–3.0 g l^{-1})	Hyposaline (3.0–20 g l^{-1})	Mesosaline (21–50 g l^{-1})	Hypersaline (>51 g l^{-1})
Dominant	*Micronecta* spp, *Anisops thienemanni* *Anisops gratus* *Austrolestes annulosus* *Boeckella triarticulata*	*B. triarticulata* *Micronecta* spp. *A. thienemanni*	*Daphniopsis queenslandensis* *Apocyclops dengizicus*	*D. queenslandensis* *A. dengizicus*	*P. minuta* *Diacypris* spp.
Subdominant	mites *Triplectides ?australicus*	*Agraptocorixa* spp. *A. gilberti* *M. howitti*	*S. multimaculatus* *A. thienemanni* *Trigonocypris globulosa* *Micronecta* spp. *Diacypris* spp.	*Parartemia minuta* *S. multimaculatus* *B. munitipennis* *Reticypris* spp. *A. thienemanni* *Micronecta* spp.	*A. dengizicus* *H. cf. fennica* *B. plicatilis*
Common	*Agraptocorixa* spp. *Megaporus howitti* *Sternopriscus multimaculatus* *Daphnia carinata* *Microcyclops varicans* *Chironomus tepperi* *Isidorella newcombi* *Allodessus bistrigatus* *Xanthagrion erythroneurum* *Antiporus gilberti* *Diplacodes bipunctata* *Cloeon* sp.	*Cyprinotus* sp. *D. carinata* *T. ?australicus* *C. tepperi* *S. multimaculatus* *Cyzicus* sp. *A. annulosus* *Calamoecia* spp. *Branchinella* spp.	*Brachionus plicatilis* *D. carinata* *Heterocypris* sp. *Agraptocorixa* spp. *Berosus munitipennis* *Limnadia* sp. a *M. howitti*	*Hexarthra* cf. *fennica* *Necterosoma penicillatum* *D. carinata*	*D. queenslandensis* *Tanytarsus barbitarsis*

Cumulative species richness, *i.e.* total number of species caught in each lake over the study period, also decreases significantly with increasing salinity (Fig. 3). In both cases there is considerable variability at any given salinity due to many factors including lake size and efficiency of sampling effort. Figure 4 partly illustrates the latter by indicating that lakes with low sampling effort tend to have lower species numbers. The slope is steepest for the less saline lakes though clearly the true relationship between species richness and number of visits is not linear. In an analysis of partial correlations, when this relationship is removed, the correlation coefficient between cumulative species richness and $\log_{10}$ salinity increases from -0.57 to -0.68. This still does account for the unmeasured influence of the relatively less efficient sampling effort in freshwater lakes as described earlier.

As salinity increases crustaceans become more important in lake communities and insects less important (Table 9), the change occurring in the 10-20 g l^{-1} range. Although there is no overall detectable change in the 'other taxa' category, its components change, as mites and gastropods are only present at lower salinities while rotifers dominate at higher salinities.

These changes in dominant classes with increased salinity is even more apparent when the prominent animals in each salinity class is examined (Table 10). A variety of insects dominate at lower salinities, while at higher salinities a less diverse array of crustaceans is dominant.

Discussion

The saline lakes of the Paroo are small, ephemeral and their basins arose by deflation or damming due to wind action. They contrast with most other Australian saline lakes, which are large, isolated, episodic lakes of tectonic origin (e.g. Lakes Eyre, Buchanan), episodic playas in relictual drainage systems (as in inland Western Australia), smaller but clustered permanent to semi-permanent lakes of volcanic origin (e.g. western Victoria) or ephemeral but reliably filled small lakes near the coast (e.g. southeast South Australia). Perhaps the closest geomorphological and hydrological analogues of the Paroo lakes are the smaller lakes in southwest Western Australia studied by Geddes *et al.* (1981), and those away from the coast on the Eyre Peninsula, South Australia (Williams, 1984). Both of these analogues however, lie in relatively more benign climatic zones than the Paroo (Gaffney, 1975), and so contain water more reliably. Furthermore the Paroo lakes tend fill first, if at all, in autumn, rather than in winter, so imposing a different seasonal regimen than in southern Australia.

Nevertheless, despite their different physical characteristics, the Paroo lakes are typical desert saline lakes in that they exhibit wide temperature and salinity fluctuations and have quite clear and alkaline waters (Hammer, 1986). Moreover, as in most Australian lakes, their waters are dominated by Na and Cl ions and these ions show increasing importance at higher salinities (Hart & McKelvie, 1986). The only unusual feature is that of sulphate dominance in some lakes of lower salinity.

Like most other Australian saline lakes, the fauna is dominated by crustaceans, though insects are relatively important in meso- and hyposaline waters. In the Paroo lakes the prominent crustaceans are *Parartemia minuta, Apocyclops dengizicus, Daphniopsis queenslandensis*, and of ostracods *Diacypris, Reticypris*, and *Heterocypris*, and to a lesser extent *Trigonocypris globulosa*. Important insects include *Tanytarsus barbitarsis* and *Berosus munitipennis* in hypersaline lakes; these and the beetles *Sternopriscus multimaculatus, Necterosoma penicillatum* and *Megaporus howitti*, and the hemipterans *Anisops thienemanni, Micronecta* spp. and *Agraptocorixa* spp. occur in meso- and hyposaline lakes. Other less important taxa in hypo- and subsaline lakes include damsel nymphs (especially *Austrolestes annulosus*), caddis larvae (mainly *Triplectides australis*), moth larvae, and a greater variety of beetles and hemipterans.

The salt lakes of the Paroo exhibit important biogeographical differences from other salt lake areas in Australia. In such comparisons, it should be realised that data on species occurrences and species richness are strongly influenced by sampling effort and sampling procedures, among other factors. These have been very different in the various Australian studies, so while comparisons of lakes within this study has some basis, comparisons between lake areas may be less reliable.

While most species in the Paroo lakes are ubiquitous or shared with saline lakes in southern Australia, there are many which are endemic to the inland or far more important there than elsewhere. These include most of the common crustaceans, viz. *Parartemia minuta, Daphniopsis queenslandensis, Moina baylyi, Reticypris walbu, Trigonocypris globulosa*, and a new mytilicyprinind ostracod, all of which also occur in Lake Buchanan or Lake Eyre or both as well as the Paroo and perhaps in other inland areas. A few insects, including *Hydaticus variegatus*, belong to this category as well. Just as significant is the absence of many characteristically southern species, including *Parartemia zietziana, Calamoecia salina, C. clitellata, Haloniscus searli, Austrochiltonia* spp., ostracods like *Australocypris* and *Mytilocypris*, and the snails *Coxiella* spp. *Parartemia* and the ostracods have close relatives in the Paroo, but the others do not have equivalents. The absent forms are apparently unable to survive desiccation in unreliable habitats (Williams, 1984), a proposal that seems reasonable under the present climate of the Paroo and certainly for the hypothesized past climatic regimes for inland Australia (e.g. De Deckker, 1986).

As indicated above many similarities are obvious with the fauna of the Lake Buchanan complex in tropical Queensland (Timms, 1987). However it now seems that Buchanan's fauna is best categorized as an inland one, rather than a tropical one. It does have a few (unimportant) insects limited to the tropics, but its common species are those crustaceans (e.g. *Parartemia minuta, Trigonocypris globulosa, Reticypris walbu, Daphniopsis* sp.) now recognised as characteristic of the inland. Thirty-six of Buchanan's 53 species also occur in the Paroo, while 15 of the 22 in Lake Eyre occur in the Paroo, suggesting a inland component to Australia's salt lake fauna.

Overall, species richness in the Paroo lakes is at least as high as it is in most other saline lake districts in southern and western Australia, so that the depauperate fauna of Lake Buchanan (Timms, 1987), and particularly Lake Eyre (Williams, 1990), gives a biased impression of the faunal diversity of inland salt lakes. While these particular episodic lakes may be of little importance as evolutionary loci (Williams, 1984), a special faunal component of salt lakes has evolved in inland waters. Lakes like those of the Paroo, with their wide range of salinities and habitats grading towards reliability, may be important centres.

Many taxa have their highest recorded field salinity tolerances for Australia or even the world in the Paroo. These include *Parartemia minuta*, *Triops australiensis*, conchostracans, *Daphniopsis queenslandensis*, *Trigonocypris globulosa*, *Austrolestes annulosus*, notonectids, corixids, pyralid larvae, and various beetles and water mites. Perhaps this reflects an appropriate response to the harsh environment in the Paroo. The relative importance of insects (for Australia) may be associated with the possible stressful influence on crustaceans of high summer temperatures and low oxygen tensions; the absence of higher forms (e.g. amphipods, isopods) may be due to their lack of aestivating mechanisms.

Dominant, sub-dominant and common taxa in each of the five salinity classes are listed in Table 10. Few species are unique to one salinity class, except for the freshwater end, and even most of these (e.g. *Anisops gratus*, mites, *Cloeon* sp. in the freshwater class, and *Cyzicus* sp. in the subsaline class) extend as less important species in adjacent classes. The greatest overlap in prominent species between classes is in the lists for hyposaline and mesosaline lakes. A contributing reason for this could be the wide salinity fluctuations experienced in these lakes, and the associated presence of euryhaline species. Another anomaly is also explained by wide salinity fluctuation, and this is the appearance in the me-

sosaline list of species (e.g. *Daphnia carinata*) whose upper tolerance is less than the designated salinity range for mesosaline lakes. The presence of such in the list is due to their dominance when such lakes are less saline.

Acknowledgements

Sincere thanks are due to the Avondale College Foundation for financial support, to numerous field assistants, especially Alec Gaszik, Andrew North, Grant Murray and John Vosper, who endured heat or floods yet helped more than once, to the landholders of the Paroo, bar one, who were very helpful and understanding, and to the following taxonomists for their identifications: M. Brock and M. Cassanova (plants), L. Forro (*Moina*), D. Frey (Chydoridae), M. Geddes (Anostraca, Conchostraca), R. Hamond (Harpactoida), J. Harris (fish), M. Harvey (Hydracarina), G. Ingram (Amphibia), I. Lansbury (Hemiptera), E. Marks (Culicidae), J. Martin (Chironomidae), D. Morton (Cyclopoida), A. Neboiss (Trichoptera), V. Sergeev (*Daphniopsis*), R. Shiel (Rotifera), P. Suter (Ephemeroptera), J. Walker (Gastropoda), J. A. L. Watson (Odonata), and C. H. S. Watts (Coleoptera). I am also grateful to Professor W. D. Williams for his valuable comments on the manuscript.

References

Bayly, I. A. E., 1969. The occurrence of calanoid copepods in athalassic saline waters in relation to salinity and ionic proportions. Verh. int. Ver. Limnol. 17: 449–455.

Bayly, I. A. E., 1974. The plankton of Lake Eyre. Aust. J. Mar. Freshwat. Res. 27: 661–665.

Bayly, I. A. E. & W. D. Williams, 1966. Chemical and biological studies on some saline lakes of south-east Australia. Aust. J. mar. Freshwat. Res. 17: 177–228.

De Deckker, P., 1983. Australian salt lakes: their history, chemistry, and biota – a review. Hydrobiologia 105: 231–244.

De Deckker, P., 1986. What happened to the Australian aquatic biota 18 000 years ago? In P. De Deckker & W. D. Williams (eds), Limnology in Australia. CSIRO & Dr W. Junk Publishers, Melbourne & Dordrecht: 487–496.

De Deckker, P. & M. C. Geddes, 1980. Seasonal fauna of ephemeral saline lakes near the Coorong Lagoon, South Australia. Aust. J. mar. Freshwat. Res. 31: 677–699.

Edward, D. H. D., 1964. The biology and taxonomy of the Chironomidae of south western Australia. PH.D. thesis, University of Western Australia.

Ettershank, G., M. Fuller & E. J. Brough, 1966. Hemiptera from saline waters in inland Australia. Aust. J. Sci. 29: 144–145.

Frey, D. G., 1991. A new genus of alonine chydorid cladocerans from athalassic saline waters of New South Wales, Australia. Hydrobiologia 224: 11–48.

Gaffney, D. O., 1975. Rainfall deficiency and evaporation in relation to drought in Australia. Presented to 46th Anzaas Congr., Canberra 1975, Bur. Met., Melbourne.

Geddes, M. C., 1981. Revision of Australian species of *Branchinella* (Crustacea: Anostraca). Aust. J. mar. Freshwat. Res. 32: 253–295.

Geddes, M. C., P. De Deckker, W. D. Williams, D. W. Morton & M. Topping, 1981. On the chemistry and biota of some saline lakes in Western Australia. Hydrobiologia 82: 201–222.

Glover, C. J. M. & T. C. Sim, 1978. A survey of Central Australian ichthyology. Aust. Zool. 19: 245–256.

Hart, B. T. & I. D. McKelvie, 1986. Chemical Limnology in Australia. In P. De Deckker & W. D. Williams (eds), Limnology in Australia, CSIRO, Melbourne/Dr W. Junk Publishers, Dordrecht: 3–31.

Hammer, U. T., 1986. Saline Lake Ecosystems of the World. Dr W. Junk Publishers, Dordrecht, 616 pp.

Johnson, M., 1980. The origin of Australia's salt lakes. Rec. geol. Surv. N.S.W. 19: 221–266.

Kokkinn, M. J., 1986. Osmoregulation, salinity tolerance and the site of ion excretion in the halobiont chironomid, *Tanytarsus barbitarsis* Freeman. Aust. J. mar. Freshwat. Res. 37: 243–250.

Timms, B. V., 1981. Animal communities in three Victorian lakes of differing salinity. Hydrobiologia 81: 181–193.

Timms, B. V., 1983. A study of benthic communities in some shallow saline lakes of western Victoria, Australia. Hydrobiologia 105: 165–177.

Timms, B. V., 1987. Limnology of Lake Buchanan, a tropical saline lake, and associated pools, of North Queensland. Aust. J. mar. Freshwat. Res. 38: 877–884.

Timms, B. V. & C. H. S. Watts, 1987. Water beetles of salt lakes near Colac, Victoria. Bull. Aust. Soc. Limnol. 11: 1–7.

Watts, C. H. S., 1978. A revision of the Australian Dytiscidae (Coleoptera). Aust. J. Zool. Suppl. Ser. No. 57: 1–166.

Williams, W. D., 1981a. The limnology of saline lakes in Western Victoria. Hydrobiologia 82: 233–259.

Williams, W. D., 1981b. The Crustacea of Australian inland waters. In A. Keast (ed.), Ecological Biogeography of Australia. Dr W. Junk Publishers, The Hague: 1103–1138.

Williams, W. D., 1984. Chemical and biological features of salt lakes on the Eyre Peninsula, South Australia, and an explanation of regional differences in the fauna of Australian salt lakes. Verh. int. Ver. Limnol. 22: 1208–1215.

Williams, W. D., 1990. Salt lakes: The limnology of Lake Eyre. In M. J. Tyler, C. R. Twidale, M. Davies & C. B. Wells (eds), Natural History of the North East Deserts. Royal Society of South Australia, Adelaide: 85–99.

Williams, W. D. & M. J. Kokkinn, 1988. The biogeographical affinities of the fauna in episodically filled salt lakes: a study of Lake Eyre South, Australia. Hydrobiologia 158: 227–236.

Williams, W. D. & M. Mellor, 1991. Ecology of *Coxiella*. a prosobranch snail of Australian inland waters. Palaeogeography, Palaeoclimatology, Palaeoecology 84: 339–355.

Williams, W. D., K. Walker & G. W. Brand, 1970. Chemical composition of some inland surface waters and lake deposits of New South Wales, Australia. Aust. J. mar. Freshwat. Res. 21: 103–116.

Conservation of salt lakes

W. D. Williams
Department of Zoology, University of Adelaide, G.P.O. Box 498, Adelaide, S.A. 5001, Australia

Key words: Conservation, salt lakes, Aral Sea, Mono Lake, Akrotiri Lake, RAMSAR

Abstract

Salt lakes have a variety of important uses and values, including especially both economic and scientific ones. These uses and values have been and are increasingly subject to degradation from a variety of impacts: diversion of inflows, pollution, agricultural practices, and introduction of exotic species are among the more important. Recognition of these impacts upon salt lakes has led to some international and national measures for their conservation, but considerably more effort in this direction is needed. Against this background, Mono Lake, California, USA, and the Aral Sea, central Asia, are discussed as two localities which bring into sharp focus the various matters discussed in the paper. Finally, attention is drawn to the need to conserve the Akrotiri Salt Lake, Cyprus.

Introduction

There are many definitions of the term conservation. For present purposes, that used by the International Union for the Conservation of Nature (IUCN) is suitable and, despite the objections that can and have been raised against it, it is adopted here. It states that: 'Conservation is the management of human use of the biosphere so that it may yield the greatest sustainable benefit to present generations while maintaining its potential to meet the needs and aspirations of future generations' (IUCN, 1980). In the present context, the parts of the biosphere in focus are inland saline waters.

To a much greater degree than their abundance, distribution and importance would suggest, salt lakes – with only a few notable exceptions – have not attracted much attention from conservationists, and relatively few are the subject of active conservation. There is little doubt that this reflects the general lack of attention accorded salt lakes by limnologists – itself a reflection of the fact that salt lakes are often located far from where most limnologists live (Williams, 1986). Perhaps this relative inaccessibility itself has suggested that salt lakes need little protection (and therefore that most salt lakes need little protection).

But, despite the lack of attention given them, salt lakes are significant components of the biosphere (on a global basis, inland salt water is not markedly less in volume than inland fresh waters: according to recent estimates (Shiklomanov, 1990), 0.006 percent of total global water as compared with 0.007 percent), and they have a number of important uses and values. To a considerable degree, however, these uses and values have been and are being degraded. The agents causing such degradation are many and varied, and their impact is accelerating. Even so, the conservation of salt lakes has generally not been given high priority by governments, although, as indicated, some notable exceptions occur.

The present paper aims to draw attention to the

need to conserve salt lakes more actively. It does this by outlining their values and uses, and the nature of destructive impacts upon them. The extent and type of protective measures which have been and can be put in place are briefly considered. Finally, to bring into sharper focus the general discussion, two salt lakes where significant changes have recently occurred following the impact of man, viz. Mono Lake, California, USA, and the Aral Sea, central Asia, are discussed in more detail. Brief attention is directed to Akrotiri Salt Lake, Cyprus, in an epilogue.

Uses and values of salt lakes

Seven reasons have been put forward in support of the conservation of particular aquatic systems: economic, cultural, aesthetic, recreational, scientific, educational and ecological. All of these, though at different degrees of applicability according to locality, can be advanced to support measures to conserve salt lakes. They are discussed *seriatim* below. More formal classifications and evaluations of uses and values exist to support the conservation of 'wetlands' in general, but the emphasis of these for the most part is upon fresh and coastal waters, and the more simple approach taken here will suffice for the present discussion. However, in any evaluation of a specific salt lake with conservation measures in mind, recourse to these more formal classifications and evaluations will be necessary. An introduction to them is provided by, for example, Adamus & Stockwell (1983), Claridge (1991), Dugan (1990), Foster (1978), Marble & Gross (1984) and Stone (1991).

Economic uses

An extensive account of the economic uses of salt lakes and their associated hydrological systems has been given by Hammer (1986) and Williams & Kokkinn (1988). Here, all that need be given is a summary listing of these uses.

1. As a source of minerals. Salt (NaCl) is particularly important, but others have become of increasing importance, such as uranium, lithium and zeolites.

2. As a source of fresh water by diversion of inflowing rivers.

3. As a source of power. Heliothermal ponds which involve an upper freshwater layer of water and a lower salt layer, and which trap incoming solar energy, have become increasingly important as sources of hot water from which power may then be generated.

4. In moderately saline lakes, as a source of fish of commercial importance.

5. As places where living organisms may be cultured for the production of fine chemicals, protein and other biochemicals. Of particular importance in this respect are *Dunaliella* (a source of β-carotene and glycerol), and *Spirulina* (protein).

6. As places where *Artemia* can be cultured. *Artemia* adults and cysts are important food items in the aquacultural industry.

Cultural values

Salt lakes have not had as large a cultural value for mankind as have fresh waters – reflecting their relative isolation –, but a number of important exceptions to that statement exist. Thus, the Aral and Caspian Seas figure prominently in the classical literature of central Asia, as does the Dead Sea in the history of the Middle East and eastern Europe (Nissenbaum, 1979). The Aral region in particular is one of the ancient centres where civilization and agriculture are presumed to have arisen. Primitive irrigation was practised in its basin as early as the sixth century B.C. No doubt many particular salt lakes of North and South America had special local cultural significance also. The cultural value of Mono Lake, California, to the indigenous Owens Valley Paiute Indians, for example, has been noted by Patten *et al.* (1987).

In this context, brief mention may be made of the cultural significance of flamingos, birds characteristically associated with salt lakes. Flamingo skins, flesh and eggs have long been trade items around the shores of the Mediterranean, and the ancient Phoenicians traded flamingo tongues which, according to the Roman Pliny, were a delicacy 'without which no Roman banquet was complete'.

Aesthetic values

Attitudes to the aesthetic values of salt lakes have been divided and have changed with time. Certainly, many early European explorers did not rate the aesthetic appeal of salt lakes highly when they first sighted them in remote areas. According to Serventy (1985), the Australian explorer Eyre, on first sighting Lake Eyre, stared in horror at 'one vast, low and dreary waste'. A later explorer, Warburton, noted that 'Lake Eyre was dry – terrible in its death-like stillness and the vast expanse of its unbroken sterility'. Modern perceptions are somewhat more progressive, as no doubt local perceptions have long been. Thus, Dulhunty (1975), writing of these early descriptions of Lake Eyre, said: 'But what damning descriptions they all are! No mention of the joy of experiencing its sea-salt freshness ... nothing of its scientific wonder ... no word of the unbelievable mirages ... nor of the exhilarating sight, and feel, of the beauty ...'

On another continent, Mono Lake, California, is perceived as a lake of outstanding beauty, a feature not lessened by the recently exposed tufa columns. Lake Nakuru, Kenya, and the Etosha Pan, Namibia, and their associated flamingo and wildlife populations, provide the focus of an important tourist industry drawing visitors worldwide who come to photograph and observe their beauty. Flamingos, both at Lake Nakuru and elsewhere, have long been regarded as birds of great beauty and grace.

Recreational values

The cultural and aesthetic values of many salt lakes mean that many are visited on a recreational basis to experience these values at first hand. But salt lakes have several recreational values in addition to those of a passive kind. Fishing, swimming and sailing are frequent recreational activities associated with lakes of moderate salinity. Perhaps, given the unproven therapeutic value of salt spas and lakes, the use of salt lakes in this way is better regarded as a form of recreation than as an economic use. The use of certain salt lakes when dry as the loci for attempts on land-speed records may also be regarded as a recreational value of salt lakes.

Scientific values

Salt lakes are of particular value to a number of disciplines. To ecologists, they are of value because of their habitat homogeneity, discreteness, low taxonomic diversity, and value as a source of material for microecosystem studies (Williams, 1972; Collins, 1977; Hammer, 1978; Vareschi, 1987). To physiologists, they are of interest because of the nature of biological adaptations to the environmental extremes operating within salt lakes (high salinity, low oxygen concentrations, high light exposure). To biochemists, they are of interest because of the enzyme mechanisms used by halophiles, and the mechanisms by which Halobacteria fix light energy. And to evolutionary biologists, they are of interest because, *inter alia*, stromatolites, a particular sort of microbial/sediment assemblage, appear to be amongst the oldest known form of life on earth (3000 million years BP). The interest and value of salt lakes to non-biological science, especially geochemistry, is equally as wide. And, of course, the sensitivity of salt lakes to relatively small climatic changes means that palaeolimnological studies of salt lakes have also attracted considerable scientific interest, an interest recently catalysed by impending global climatic change.

Educational values

Closely allied to many scientific values of salt lakes, of course, are educational values. At a time of increasing global climatic change, this value should not be undervalued in regions where salt lakes are close to educational institutions. At my own institution, class exercises based on field observations of salt lakes are an important part of courses. And it is my opinion that microecosystems derived from salt lakes have the potential to play a most important rôle as teaching tools: their simplicity, ease of manipulation, and the wide range of experiments that can be undertaken using them are outstanding in this respect (Williams, 1991).

Ecological value

Not least amongst the values of salt lakes, though the most difficult to measure, is their value as an integral part of the biosphere. Their biological diversity and ecological processes cannot be excluded from global diversity and biospheric processes with any certainty that exclusion will not have profound repercussions. Changes in the nature of the Aral Sea referred to in more detail below are the most indicative evidence of this sort. Already, widespread unfavourable climatic and regional environmental changes have followed man-made alteration to this lake. One important ecological value of salt lakes that should receive particular mention is their rôle as feeding, refuge and breeding sites for many migratory or nomadic bird species. The loss of certain salt lakes of value in this respect may pose very serious threats to the continued viability of the bird species in question.

The impact of humans upon salt lakes

Our use of salt lakes and of resources in their drainage basins has had impacts upon them that are significant, diverse, comprehensive and mostly irreversible. Almost without exception, these impacts have been deleterious. In short, we have already irreparably damaged an important part of our biosphere.

Impacts have been many and diverse, of short or long-term duration, affecting part of the biota or the ecosystem as a whole, of limited extent or totally destructive. Effects have reflected this diversity, but with comprehensive overlap: that is, different impacts often have similar effects, in particular those, the most ubiquitous, resulting in increased salinities. Particular impacts are often site specific, but global generalities are easily discerned. Because of this comprehensive overlap, and in an attempt to avoid unnecessary repetition, the following discussion considers both impacts and effects without attempting to separate them too widely.

For ease of discussion, the impacts are considered as those which primarily (a) act upon the catchment or drainage basin, (b) involve diversion of inflowing waters, (c) result in the addition of unnatural waste products or pollutants, (d) directly affect the biota, (e) cause physical change to the nature of the lake basin, and (f) will follow global climatic and associated changes. Not discussed separately as an impact, though its importance (and thus effect) can scarcely be overestimated, is human ignorance: the perception that salt lakes have limited uses and values, are expendable 'wastelands', and do not merit serious consideration as sites of conservation interest (see Williams (1986) for an extended discussion of this subject).

Catchment/drainage basin activities

All lakes reflect catchment events, and in this respect salt lakes are particularly sensitive because their catchments are frequently in semi-arid regions where habitats respond quickly to perturbations. Two events of significance in the present context are grazing by wild and domesticated mammals, and more direct changes to the natural vegetation imposed by man.

The effects of grazing, particularly overgrazing, become manifest in changes to run-off patterns and increases in sediment loads in run-off (e.g. Grainger, 1990). In Australia, for example, overgrazing by the rabbit, an introduced species, has caused profound changes to both the biological and physical nature of lake catchments, and these changes compound those caused by the grazing of domesticated mammals (chiefly sheep and cattle). Most important of the physical changes are the formation of animal tracks and breakage of protective surface crusts; both events lead to mobilization of surface particles and erosion.

Changes to the nature of vegetation on catchments by more direct human activity have been equally if not more significant. The clearing of deep-rooted natural vegetation (trees) and its replacement by shallow-rooted grasses and crop species frequently has led to changes in local hydrology. These changes lead to changes in the salinity, composition and seasonality of run-off (e.g. Pereira, 1973; Holmes & Talsma, 1981). Thus, underlying groundwaters (often saline) may approach the land surface more closely and ultimately to a position where capillary action alone causes it to reach the surface. In certain regions, such as central Asia, capillary action may begin when the water table is as low as 10 m below the ground level. Evaporation then acts to increase salinity and subsequent run-off adds the saline water (or precipitated salts) to the local drainage terminus, that is, the lake. The obvious result, of course, is that lake salinity increases. Western Australia provides many examples of this phenomenon. Salinity increases are by no means confined to natural salt lakes, so that many previously freshwater lakes and rivers in areas where induced salinization is occurring now have elevated salinities (e.g. Lake Toolibin, Western Australia; Froend et al., 1987). Clear evidence of this is to be seen in the dead stumps of trees both in and marginal to many Australian lakes now markedly saline (e.g. Lake Tallinga, South Australia) and in observed spatial and temporal patterns of salinity change in rivers draining cleared areas (e.g. Blackwood River, Western Australia).

Overgrazing, vegetation clearance and salinization frequently lead to severe erosion and overall 'desertification' (land degradation) in semi-arid catchments. Already, some $1-2.5 \times 10^6$ ha of the Aral Sea catchment is at hazard from this phenomenon, especially in the delta areas of the Amu- and Syr-Darya (Koust, 1991), and its catchment is far from unique in this respect.

In this context, it should also be recognised, as Stine (1991) has stressed recently, that a lake and its catchment are closely linked geomorphologically, so that a change in the lake can instigate secondary changes in the lake catchment. Thus, at Mono Lake California, the fall in water-level has forced the main inflowing streams to incise as much as 10 m. This incision, in turn, has resulted in a fall in the water-table over a considerable area with the consequential loss of many wetlands in the catchment. Stine also wrote that there are critical levels for geomorphological impact (akin to critical biological levels, for example, in salinity). These critical geomorphological levels may represent threshhold levels which, once passed, do not permit the restoration of original conditions should transgression (rather than regression) occur.

Diversion of inflows

Very large salt lakes frequently have more or less continuous inputs of fresh water. As indicated, this has long been recognized as a useful resource and considerable use of it has been made. As long as this use was of limited extent, no major impact upon the hydrological budget of the salt lake occurred. With increasing use, especially following the growth of human populations in semi-arid regions, major impacts followed. Some important lakes in this context are the Caspian, Aral and Balkash in central Asia, Lop Nor and Qinghai in China, and Pyramid and Mono Lakes in the USA. Diverted water is used for a variety of purposes: from Mono Lake, most is used for domestic purposes in Los Angeles; from other lakes, irrigation is often an important use.

A special case involving water diversion is provided by Kara-Bogaz-Gol Lake in Turkmenia. This was a large salt lake connected to the Caspian Sea by a narrow channel. Formerly, significant quantities of water flowed from the moderately saline Caspian Sea into the lake where it evaporated to create a large and highly saline water-body. Beginning in 1980, when the water level of the Caspian was at a very low level, an essentially 'blind' dam was built across the entrance to the Kara-Bogaz-Gol. Subsequently, the lake more or less disappeared. Following the expression of some concern on this matter, small amounts of water were and are now allowed to flow through to the lake, but the 'reconstituted' lake is smaller and different from the old one.

The diversion of water by direct drainage has also led to the disappearance of many interesting salt lakes. This is particularly so in agricultural areas, as for example, the Seewinkel pans in Austria (Metz & Forró, 1991) and pans in the Coto Doñana, Spain (Montez & Martino, 1987). Others have been destroyed by the excessive diversion (extraction) of underground water. The 'axalapazcos' of the Mexican plateau, whose existence largely depends or depended upon the existence of underground supplies, provide cases in point (Alcocer & Escobar, 1990).

Conversely, in some inland situations, water diversion may also give rise to salt lakes – though in the main such salt-water bodies are far from natural. (Left out of account here are solar salt ponds derived from diverted sea-water.) The most notable example is the Salton Sea in California. This was created when the Colorado River broke the banks of a man-made channel in 1905 and flooded a large depression in southern California (Stanley, 1966). The salinity of the lake has increased since 1905 and it now contains many introduced marine species, including barnacles. Less notable, but more numerous examples are provided by so-called 'evaporating basins' designed to reduce river salinities by diversion of saline agricultural drainage water. These basins may have been artificially created but are often flooded natural wetlands. There are many examples of such basins near the lower reaches of the River Murray, Australia, and also in California. They are far from natural, and may receive major inputs of water at times when natural salt lakes in the region are dry, and they may accumulate unnaturally high concentrations of toxic elements. Unnatural salt lakes of this sort are not further discussed here given the paucity of information on them (but see, for example, Chilcott et al., 1990).

The primary effects of significant water diversion from salt lakes upon the lakes themselves are obvious and two-fold: water volume decreases, and salinity increases. Each has many consequential effects.

The decrease in volume is accompanied by a decrease in lake area, especially in shallow lakes, and this in turn may expose large areas of the former lake bed. Sometimes a significant transfer of salt and sediment particles from the bed to surrounding parts of the drainage basin occurs. Lower lake levels may also cause the destruction of shallow and deltaic areas which may have provided important refuge areas and otherwise have been of conservation significance. The delta of the Volga in the Caspian Sea, for example, is regarded as the most important site for waterfowl conservation in the region, with up to 750 000 waterfowl present in mid-winter (Finlayson, 1991). Lowered lake levels may lead to the destruction of islands which formerly served to protect breeding populations from terrestrial predators, and the emergence of formerly submerged objects such as the tufa at Mono Lake, California. In the case of the Aral Sea, where very large areas of the former lake bed now lie exposed, regional climatic change has been attributed to the decreased area and volume of the lake, and the increased area of exposed lake-bed.

Increases in salinity lead to several chemical, physical and biological changes. Thus, increased salinity values may exceed the solubility products of certain dissolved salts leading to their precipitation and thus an alteration in the ionic composition of the remaining solution. Increased salinities also cause decreases in oxygen solubility (Sherwood et al., 1991). Increased densities may lead to changes in many physical phenomena, including seasonal patterns of thermal (and chemical) stratification (which, of course, are also influenced by decreased depths). Perhaps more obvious than physico-chemical changes are changes in the composition of the biota. Whilst salinity may not be as important as a direct determinant of the biota of highly saline lakes as once thought (Williams et al., 1990), it is certainly true that good correlations exist between salinity and species composition, richness and diversity in moderately saline lakes. Thus, as the salinity tolerances of indigenous species are exceeded, these species are replaced by more tolerant species until their tolerance is exceeded.

Pollution

The fact that salt lakes are the termini of closed hydrological systems has not prevented the discharge of a wide variety of pollutants to rivers flowing into them or to the lakes themselves when it has been economically convenient to do so. Mostly, it seems that loadings relate more to criteria erected to protect open, freshwater systems, and thus pollutant concentrations in salt lakes often reflect progressive accumulation (Williams, 1981).

Almost the whole range of pollutants discharged to fresh waters is also discharged to salt lakes or their influent rivers. Little point is served by the provision of details, but a few examples will illustrate the general statement. Lake Colongulac, Victoria, Australia, receives the effluent from a sewage plant located on its bank, and adjacent salt lakes, nutrients in agricultural runoff (Williams, 1981). High concentrations of certain metals occur in some Bolivian salt lakes with nearby mining activities (Beveridge et al., 1985). Lake Maryut, Egypt, has high concentrations of tin in its sediments (Aboul Dahab et al., 1990). High concentrations of organochloride residues are found in Kenya's rift valley lakes (Lincer et al. 1981). And many salt lakes are used as dumps for domestic and other garbage.

There can be little doubt that the effect of these pollutants on salt lakes is essentially the same as it is on fresh waters – though the actual evidence is thin: additional nutrients promote algal growth, high organic loadings decrease diversity but increase biomass, and poisons decrease both diversity and biomass. The modifying effects of salinity have yet to be determined fully.

Direct impacts on the biota

In several salt lakes, the fauna largely represents purposive or serendipitous introductions. The Caspian, Aral and Salton Seas provide the most notable examples. In many more examples, individual components of the fauna have been introduced. Thus, fish have been introduced in several moderately saline Canadian (Rawson, 1946) and Bolivian lakes (Hammer, 1986). Fish have also been introduced into some moderately saline Australian lakes where they cannot breed but where populations are maintained by stocking. Not all attempts to introduce fish into saline lakes, of course, have proved successful. Of introduced invertebrates, various species and subspecies of *Artemia* have been spread worldwide. In the main, initial introductions were confined to coastal solar salt fields. There have been no attempts to control these largely *ad hoc* introductions either individually or at the governmental level despite the danger they pose to the regional genetic diversity of *Artemia* and the value of this. Geddes & Williams (1987) drew attention to the danger of these introductions, and the following motion was passed at a recent meeting of *Artemia* specialists (Sorgeloos et al., 1987):

'the 2nd International Symposium on *Artemia*, meeting in Antwerp in September 1985, resolves that all possible measures be taken to ensure that the genetic resources of natural *Artemia* populations are conserved; such measures include the establishment of gene-banks (cysts), close monitoring of inoculation policies, and where possible the use of indigenous *Artemia* for inoculating *Artemia*-free waters'.

Persoone & Sorgeloos (1980), amongst others, had earlier pleaded for the conservation of all remaining natural habitats containing *Artemia*.

Leaving aside introductions, direct impacts on faunal species are few. However, exploitation of flamingo populations, either to provide meat or eggs, poses a direct threat to the survival of some South American species (Hurlbert & Flores, 1988).

Physical impacts on lake basins

Salt lake sediments, as noted, frequently contain minerals of commercial value (e.g. salt, soda, lithium, zeolites) and the mining of these frequently either physically damage the natural structure of the lake basin and/or indirectly lead to long-term changes in lake chemistry (and consequential biological effects). The derivation of useful salts from brines may also lead to physical change; often, this takes the form of dividing the basin into separate regions using low banks or levees separating waters of different salinity. Dredging activities may lead to the physical damage of some basins, as for example in the Caspian Sea.

Climatic and atmospheric changes

Finally in this consideration of hazards facing salt lakes, brief reference should be made to possible changes in global climatic patterns and to changes in the ozone concentration in the upper atmosphere. Because salt lakes represent a sensitive balance between many climatic parameters (e.g. evaporation, rainfall, temperature), relatively small changes in these will cause large changes to the natural character of salt lakes. This has already been recognised by workers in Canada (Hammer, 1990). The effects of climatic change could take various forms. Rising sea-levels would flood many coastally located athalassic salt lakes (e.g. those in southeastern South Australia). Particularly in danger in this respect are the many athalassic saline lakes occurring on small oceanic islands (e.g. Laysan Lagoon, Caspers, 1968). Increased aridity would lead to increased average salinities and ultimately to desiccation. Increased rainfall would lead to decreased average salinities and in extreme cases to the conversion of closed to open drainage systems. Changes to the seasonal patterning of climatic events would also lead to fundamental ecological changes. A major feature of this potential problem is that climatic changes are likely to occur, it is claimed, far too rapidly to permit the biota of salt lakes fully to adapt naturally to them (Parsons, 1990). On the other hand, it is possible that previous natural climatic changes have also been too rapid to permit tandem evolution. Certainly, there is evidence that some past climatic changes have been rapid, and a growing number of geologists are of the opinion that some significant global environmental changes occurred over relatively small periods of time (up to a few thousand years).

As for decreased ozone concentrations in the atmosphere, the problem here is that these allow more ultra-violet radiation to reach the surface of salt lakes and excessive exposure to such radiation is deleterious to living tissues. The plankton of lakes cannot stand increased exposure for any length of time (Traulich & Wagner, 1989). In deep lakes, however, avoidance is possible by sinking to lower depths, but this, of course, is not possible in shallow lakes. Since many salt lakes are shallow and already receive large amounts of ultra-violet radiation, the hazard is obvious.

Conservation measures

One of the essential first steps in the protection and conservation of any habitat is a recognition of its values, and the provision of a clear statement of these and the position of the habitat, its major features, hazards facing it, and knowledge available concerning it. A pioneer attempt to do this for inland waters (fresh and saline) worldwide formed the basis of 'Project Aqua' (Luther & Rzóska, 1971). 'Project Aqua' has been succeeded by more comprehensive wetland inventories amongst which those produced under the aegis of the International Waterfowl and Wetlands Research Bureau [IWRB] are notable. Comprehensive inventories of wetlands on all continents, however, do not yet exist. Even so, available inventories and similar publications have provided the basis for many important measures to conserve wetlands already. Several are important sources of information on salt lakes aside from any conservation value they may have (e.g. that by Scott, 1989, on wetlands of southeast Asia).

Measures to conserve salt lakes operate on

a variety of organizational levels: at the local, regional (State, Province), national and international level; and at the non-governmental (public-interest) and governmental level. No comprehensive documentation of local, regional and national measures available to conserve salt lakes is given here, of course. However, it should be pointed out that regional and national governmental bodies with an interest in conservation are in general increasingly open to specific proposals on sites of interest. Pressure from non-governmental bodies is often an important catalyst in this context. To a not inconsiderable degree it is the responsibility of the scientific community to draw the attention of both sorts of body to particular salt lakes of conservation value.

It need scarcely be added that the responsiveness, sophistication and legislative power of regional and national governmental bodies, and the enthusiasm, credibility and the extent to which non-governmental bodies are informed, differ from country to country. In general, the more affluent countries have well-developed bodies of both types, so that in Australia, for example, the formal responsibility for conservation is vested in a variety of governmental bodies at State and Federal level (Bridgewater, 1991), and there are also various regional and national public-interest groups. Formal responsibility at the federal level is largely the province of the Australian National Parks and Wildlife Service; national non-governmental conservation bodies include the Australian Conservation Foundation and the Worldwide Fund for Nature (Australia). The value of non-governmental bodies should not be underestimated, and the Worldwide Fund for Nature (Australia) played an important rôle in promoting the publication of two influential books concerning Australian fresh and saline wetlands and supporting their conservation (McComb & Lake, 1988, 1990). Unfortunately, many salt lakes occur in countries that are not as affluent as Australia and lack well-developed conservation bodies and formal mechanisms to provide for conservation. Non-governmental bodies can be especially important in such countries. The Kalahari Conservation Society of Botswana, for example, has been active in drawing the attention of the international limnological community to threats to the Makgadikgadi pans posed by mining.

Of perhaps greater importance here is to note that international measures are now available promoting the conservation of wetlands. They should be used more effectively for salt lakes than they apparently have been. Measures which involve both non-governmental bodies and international agreements are important. Amongst the more influential non-governmental bodies with particular interests in wetland conservation rather than conservation issues in general, mention is made of the International Waterfowl and Wetlands Research Bureau (IWRB) (of which the Wetland Management Group is an active component), the International Association for Limnology (SIL) (of which the Working Group on Conservation specifically targets lake conservation issues), and the International Lake Environment Committee (ILEC).

The most important international measure for the conservation of wetlands (fresh and saline) is 'The Convention on Wetlands of International Importance Especially as Waterfowl Habitat'. The convention is widely known as the RAMSAR Convention after the name of the city in Iran where the Convention was adopted in 1971. At present (March 1991), there are 60 Contracting Parties with 508 listed sites of total area $> 30 \times 10^6$ ha. Australia was the first nation to become a party to the convention. In order for a wetland to be identified as a site of international importance, it must meet one of the following three general criteria (ANPWS, 1991):

(1) Be a representative or unique wetland
   - be a particularly good representative example of a natural or near-natural wetland, characteristic of the appropriate biogeographical region;
   - be a particularly good representative example of a natural or near-natural wetland, common to more than one biogeographical region;
   - be a particularly good representative example of a wetland, which plays a substantial

hydrological, biological or ecological rôle in the natural functioning of a major river basin or coastal system, especially where it is located in a trans-border position;
 – be an example of a specific type of wetland, rare or unusual in the appropriate biogeographical region.

(2) Be significant on the basis of contained biota
 – support an appreciable assemblage of rare, vulnerable or endangered species or subspecies of plant or animal, or an appreciable number of individuals of any one or more of these species;
 – be of special value for maintaining the genetic and ecological diversity of a region because of the quality and peculiarities of its flora and fauna;
 – be of special value as the habitat of plants or animals at a critical stage of their biological cycle;
 – be of special value for one or more endemic plant or animal species or communities.

(3) Be significant in the support of waterfowl
 – regularly support 20 000 waterfowl;
 – regularly support substantial numbers of individuals from particular groups of waterfowl, indicative of wetland values, productivity or diversity;
 – where data on populations are available, regularly support 1% of the individuals in a population of one species or subspecies of waterfowl.

Application of these criteria is made through a set of guidelines which enable the Conference of the Contracting Parties to assess the suitability of wetlands for inclusion on the List of Wetlands of International Importance. Whilst the RAMSAR Convention was originally formulated with the conservation needs of waterfowl in mind, its bailiwick has now been greatly extended to cover many wetlands of ecological but not necessarily ornithological significance. Nonetheless, the number of saline lakes listed is few, though many non-listed salt lakes clearly meet one or more of the criteria documented above. Thus, of the 39 Australian sites listed, fewer than five could be regarded as saline although saline lakes are widespread and important throughout the country.

Also of some value at the international level as a means of conserving salt lakes is the 'World Heritage Convention'. Parties to this Convention list national features that have outstanding and universal natural and cultural values. Whilst the Convention does not specifically target the conservation of wetlands, many listed features do include wetlands considered to be globally significant in a natural and cultural sense (i.e. of conservation significance). Once listed, parties to the Convention have a responsibility to protect the listed feature. Features proposed for listing must be comprehensively documented, have been the subject of national consultation, be nominated by the country (party) concerned, and be evaluated by IUCN. It may be added that Parties to the Convention have an obligation to protect, so far as possible, all features of World Heritage significance irrespective of whether they have been listed or not.

The sorts of national and international conservation measures indicated above, it may be added, have not yet adequately resolved the many special difficulties and problems associated with the conservation of inherently variable ecosystems (such as salt lakes). Maintenance of current conditions, the status quo, in a given locality may be a quite unnatural phenomenon.

Mono Lake, California

Many of the issues considered above are brought into sharp focus by a detailed consideration of Mono Lake, California. This lake has a variety of values and uses, a threat to its current status is present, some effects have already resulted from changes induced by man, and measures to protect and stabilise the lake are underway in response to the threat. The lake is relatively well-known scientifically, but a general discussion of it can be based upon Patten et al. (1987) and Stine (1991) who provide an authoritative summary of previous work.

Mono Lake is a large, deep lake in central California to the east of the Sierra Nevada (38° 00′ N, 119° 00′ W). In 1986, its altitude above sea-level was ~ 1,945 m, its area, ~ 150 km², and its mean depth, ~ 18 m. Surface salinity was ~ 84 g l⁻¹.

The values of the lake are many and diverse. The lake was of cultural significance to the Paiute Indians (for whom it also served as a food resource) and the past and present aesthetic appeal of the lake and its environs has attracted and continues to attract many sightseers. In addition to passive enjoyment of the lake, visitors also boat, swim, birdwatch and walk along its shores. Because of its high salinity and low taxonomic diversity, amongst other reasons, the lake continues to attract scientific and educational attention. A major invertebrate species inhabiting it, *Artemia monica* Verill, is endemic to the lake. The lake is an important feeding and breeding site for several bird species. It is particularly important as a feeding site on the migration pathway of Wilson's phalarope, the eared grebe and several other shorebirds. Economically, the lake is of value because of its production of *Artemia* and because its inflows are fresh.

It is this last value of the lake, the value of its inflowing fresh waters, which provides the only significant threat to the lake. Beginning in 1941, the water authorities for Los Angeles began diverting water from the lake for domestic supplies. As a result, the lake level began to drop steadily from 1956 m asl to its present level of ~ 1945 m asl, *i.e.* a drop of some 11 m. The effects have been an almost two-fold increase in salinity, from about 48 to ~ 95 g l⁻¹, the exposure of considerable areas of the former lake bed, and the connection of islands (e.g. Negit Island) to the mainland. The effects have had consequential impacts. The connection of the islands allowed predators (coyotes) access to the colony of California gulls. And the exposure of the former lake bed now causes alkali dust storms during windy conditions. The ecological impacts of the salinity increase have not been fully documented, but physiological studies have shown that over the full range of salinity increase already undergone in the lake growth and reproductive rates for much of

the biota have been reduced (Dana & Lenz, 1986; Herbst, 1988; Herbst *et al.*, 1988). Further increases (e.g. to 120 g l⁻¹) are likely to lead to decreased phytoplankton and phytobenthic productivity, and reduced population densities of the major invertebrates in the lake (*A. monica* and *Ephydra hians* Say). Depending on the degree of decrease or reduction, this could reduce the value of the lake as a feeding station for migrating birds.

The past and predictable effects of water diversion have sparked considerable debate. In response to public concern on the fate of the lake, the California Department of Water Resources convened an 'Interagency Mono Lake Task Force' in 1978 charged with developing a plan of action to preserve the lake considering social and economic factors. The Task Force recommended that diversions be curtailed and the lake level raised. Further concern led to the formation, following a congressional directive, of the Mono Basin Ecosystem Study Committee (Patten *et al.*, 1987). After the passage of the California Wilderness Act in 1984, the Mono Basin National Forest Scenic area was established by Congress and the management of the basin placed with the U.S. Forest Service. The wider attention of the international limnological community was noted in 1982, when the following resolution was forwarded to President Reagan, governors and senators of respective states, and to the Mayor of Los Angeles.

'Because inland salt lakes are of interest to a variety of scientific disciplines;
And because Mono Lake, California, [and Pyramid Lake, Nevada,] are of scientific interest, play critical rôles in the support of several bird, fish and other animal populations, but are seriously threatened as viable environments by continued water diversions in the States of California [and Nevada];
We, participants of the Second International Symposium on Athalassic (Inland) Saline Lakes, meeting under the aegis of the Societas Internationalis Limnologiae urge responsible government agencies and municipalities in the

States of California [and Nevada], and the U.S. Federal Government to take into account scientific consideration about the value of these lakes when arriving at water resource decisions affecting them.'

The resolution was supported by summaries of the scientific value of Mono Lake.

The present position (1990) is that a local judge has ordered Los Angeles to limit its diversion of water from Mono Lake until the level of the lake has risen to an acceptable level, and a state agency (the State Water Resources Control Board) is studying whether changes need to be made in Los Angeles' licence to divert water from Mono Lake. Its ruling on Mono basin water rights is expected in 1993.

The Aral Sea

A consideration of the present situation in the Aral Sea, likewise, will serve to focus the earlier discussion. A considerable amount of material has recently been published on this lake, but much of it is of a general nature and rather little published material exists which documents the precise nature of limnological and wider ecological changes that have taken place since the lake began to change significantly in response to human activity. The primary scientific literature on the lake, moreover, is somewhat inaccessible to the international limnological community and most, of course, is in Russian. The necessarily brief account given here is drawn largely from the following recent references: Micklin (1988), Aladin & Khlebovich (1989), Glazovsky (1990), Anon. (1990, 1991), Williams & Aladin (1991), Aladin & Williams (in press), and various unpublished documents.

The lake lies in south central Asia (between 43°24' and 46°53' north and 58°12' and 61°59' east), and is fed by two major rivers, the Amu- and Syr-Darya. Prior to 1960, its major physico-chemical features had been more or less stable for a considerable time, with a water-level at ~ 53 m asl. At that time, the surface area of the lake was $\sim 68\,000$ km^2, its volume was 1090 km^2, its mean depth was 16 m, and its mean salinity was ~ 10 g l^{-1}. Lake level was maintained by an annual inflow of ~ 50 km^3 from rivers and 9 km^3 from rain.

In its natural state, the lake had a number of important uses and values (some evident only with hindsight). Although fish production was not high ($\sim 44\,000$ ton per annum), commercial fishing was a significant activity for several lakeside communities. The moderate amounts of water diverted from inflowing rivers for irrigation and other uses was of great economic significance in the semi-arid regions through which the rivers flowed. The deltas of the Amu- and Syr-Darya were important habitats for resident and migrating birds, and these and other shallow parts of the lake were important foci of regional biodiversity (including terrestrial species able to use the only moderately saline water in marginal regions of the lake). As the fourth largest lake in the world, the Aral Sea was of considerable scientific value. It was also an integral part of central Asian culture and history. And, not least, the lake served to meliorate the severely continental climate of the region.

The major threats to the lake have been twofold: the introduction of exotic species, and the excessive diversion of inflowing waters.

Since 1927, numerous introductions of animal species have taken place. Most of these have been on an *ad hoc* basis. Up to 1957, the introductions were entirely fish, but since then other groups have been introduced. Details need not be given, but it should be noted that the effects of these introductions have been greatly to change the nature of the Aral zooplankton, benthos and fish community. Biodiversity in the lake was never high; introductions simply decreased it.

The diversion of inflowing waters has been equally profound and certainly much more obvious. From its semi-stable state prior to 1960, the water-level has now dropped some 14 km (to ~ 39 m asl), the water surface has decreased to 37 000 km^2, and lake volume has dropped to 340 km^3. Lake salinity, on the other hand, has increased to 30 g l^{-1}. The mean annual input to the lake for the period 1980–1989 was ~7 km³ (range: <1 to 22).

The effects of the changes brought about by water diversion have been marked. Falling water-levels have led to the loss of important shoal areas and islands in the south of the lake, destroyed the shallow deltas of the Amu- and Syr-Darya, converted islands elsewhere in the lake to peninsulas, and exposed large expanses of the former lake-bed. Each of these effects has had profound biological impacts (mostly involving loss in biodiversity), and in addition the newly exposed areas of former lake-bed are now the sources of salt and sand blown from the area in dust storms during windy conditions. The decreased area of free water, and increased area of lake bed, have caused local micro-climatic changes such that the climate is now a more 'continental' one.

Falling water levels have been accompanied by increasing salinities – from ~10 g l⁻¹ in 1980, through 11.1 g l⁻¹ in 1970, 16.5 g l⁻¹ in 1980, and now ~30 g l⁻¹ (1990). Many of the species present have been unable to tolerate such increases in salinity and have become extinct or are in the process of becoming so. This perhaps is not as devastating as it might first appear: recall that most of the fauna at least results from introductions.

Associated with these falls in water-level and increases in salinity have been several secondary impacts: pollution of groundwater from excessive use of pesticides and defoliants on irrigated cotton crops, medical problems (a general fall in public health, and an increase in child mortality and birth malformations), lowered crop productivity, and salinization of soils and rivers. It is these secondary impacts, taken with the primary ones, which are generally referred to as constituting the 'Aral Sea crisis'.

Predictions of what the end result of the water diversions will be depend upon the quantities of diverted water to be permitted in the future. Two scenarios advanced by P. P. Micklin (personal communication, 24 February 1991) for the year 2010 (~20 years hence) and based upon various Soviet sources are that: if only 16 km³ of water is left to flow into the lake annually, then lake area will reduce to ~22 000 km², water-level will be 31.5 m asl, lake volume will be 140 km³, and salinity will be ~100 g l⁻¹; if 30 km³ of water is allowed to flow into the lake annually, then lake area will reduce to 38 000 km², water-level will be 37.5 m asl, lake volume will be 310 km³, and salinity will be ~40 g l⁻¹. Both scenarios, it may be noted, involve values of inflow water substantially greater than presently reach the lake (0–5 km³ per annum), but still below that volume (~50 km³ per annum) which reached the lake before its water-level began to drop. These scenarios, for simplicity of calculation, assumed a single water-body. However, the lake divided into two in 1988.

Widespread regional, national and international concern over the deteriorating ecological situation in the Aral Sea basin and in the lake itself has provoked a number of reactions at various levels. Of particular importance so far as international knowledge is concerned have been two recent symposia at which the 'Aral Sea crisis' was the main issue for discussion. The first was a joint USA/USSR meeting held in Indiana, USA, 14–19 July 1990; the second was a more open meeting held at Nukus, then in the USSR, 2–5 October 1990. A number of constructive resolutions and suggestions, as measures to conserve the lake, arose at these symposia. Additionally, the United Nations Environmental Programme, in cooperation with former Soviet scientists, has set up a working group of 'experts' to consider proposals which address various aspects of the crisis and which, it is hoped, will meliorate the more significant negative features of it. A plan incorporating an overall approach was recently put before the working party in Moscow (Anon, 1991). It will not be easy to implement this plan (and strong political decision-making will be necessary).

Epilogue

There are many salt lakes worldwide which have conservation significance, which are threatened by one or more hazards, but which are apparently not the object of any serious conservation effort.

Any attempt to document them is premature, but it would be remiss not to use this opportunity to draw attention to at least one of them. Accordingly, attention is drawn to the Akrotiri Salt Lake in Cyprus. Very little has been published on this lake despite its importance.

The Akrotiri Salt Lake occurs in southwestern Cyprus near Limassol. It has a maximum surface area (in winter) of 9.4 km², a maximum depth of 1 m, and the surface of the lake when full is 1.7 m below sea-level. Salinity varies from <50 to >200 g l⁻¹ each year, and the lake dries in summer. It is separated from the sea in the east and west by a broad ridge of sand and shingle which in former times was breached by a channel. The last time there was a direct connection between the lake and the sea was in the early nineteenth century. Akrotiri Salt Lake is important as a refuge and resting place for migratory birds; as many as 150×10^6 birds pass through Cyprus in autumn, and a lesser number in spring, *en route* from the Palaeartic to Africa and *vice versa*. The lake, of course, is especially important for aquatic species, notably flamingos, the more so since it is one of a small and decreasing number of natural wetlands still left in the eastern Mediterranean. Note in this connection the extensive loss and degradation of Egyptian wetlands, particularly in the Nile delta. A major factor in the loss has been continuous land reclamation (Hollis & Jones, 1991); the addition of pesticides, heavy metals and nutrients (from sewage) is implicated in their degradation (Burgis & Symoens, 1987).

An important potential threat to the continued existence of the lake in its present condition is posed by the construction of a dam on the Kouris River, 5 km northwest of the lake, as part of a so-called 'Southern Conveyor Project'. The dam will store water to be used for the most part elsewhere on the island, and little water will be allowed to 'spill'. Below the dam, therefore, the Kouris River will dry up. The lake receives water from rainfall (annual mean ~0.5 m), groundwater, surface drainage (including a direct connection from the Kouris River), and the sea when small amounts are added by waves during storms. Inputs to local groundwater from the Kouris River

will clearly change after the dam is complete, and this change will undoubtedly be reflected in the groundwater input to Akrotiri Salt Lake. Changes in this and the input from surface drainage may be expected to impact upon the lake. It is said that the influence of the dam on the lake 'is expected to be minor', but this seems far from certain and substantive hydrological data are not available to prove it.

The World Bank, the organization financing the 'Southern Conveyor Project', has called for a statement on the ecological implications of the scheme (in line with its normal procedures), and the preparation of this by a locally-based group is presently in hand. The importance of the lake is also recognized by the Cyprus Ornithological Society and the Association for the Protection of the Cyprus Environment. A recommendation has been made that the lake and its environs be declared a Nature Reserve. Thus, measures are in place which could provide for the conservation of the lake. These, however, may be less robust than they seem or than is desirable, and it is hoped that this epilogue, which will draw the attention of the international limnological community to the lake and the present situation there, will serve to strengthen any case to conserve the lake.

Acknowledgements

This paper was largely written whilst I took part in a Workshop on Wetland Conservation and Management, 11–15 February 1991, Newcastle, Australia, and the Second United Nations Environmental Programme Expert Group Meeting on the Project on Assistance for Preparation of an Action Plan for Conservation of the Aral Sea, 18–24 February 1991, Moscow, Russia. These meetings provided an appropriate diurnal backdrop for my nocturnal writing, and I thank in general many Australian, Russian, French, American and other colleagues who contributed indirectly (and largely unknowingly) to my thoughts. I thank in particular for invitations and support Dr Peter Bridgewater, Director, Australian National Parks and Wildlife Service, Canberra, and Dr S. Morozov, Director, Centre for International Projects, Moscow. For specific comments on Mono Lake I thank Dr David Herbst, Sierra Nevada Aquatic Research Laboratory, University of California, and Dr J. Jehl, Hubbs Sea World Research Institute, San Diego. Not all of these comments have been accepted. Specific comments on some Victorian (Australian) lakes have been provided by Dr I. A. E-. Bayly, Monash University, and he is thanked too. Dr Scott Stine, Lamont-Doherty Geological Observatory of Columbia University, made some general remarks of considerable insight and I have taken the liberty of incorporating them in the text. They are acknowledged and I thank him for them and for a reprint of his 1991 paper. Miss Sandra Lawson, Secretary, Department of Zoology, University of Adelaide, is thanked for her usual care in the preparation of the final manuscript at short notice.

References

Aboul Dahab, O., M. A. El-Sabrouti & Y. Halim, 1990. Tin compounds in sediments of Lake Maryut, Egypt. Envir. Pollut. 63: 329–344.

Adamus, P. R. & L. T. Stockwell, 1983. A Method for Wetland Functional Assessment: Vol. I. Critical Review of Evaluation Concepts. U.S. Department of Transportation, Federal Highway Administration, Office of Research and Management, Washington, D.C.

Aladin, N. V. & V. V. Khlebovich, 1989. [Hydrobiological Problems of the Aral Sea]. Proc. Zool. Inst. USSR Academy of Sciences, Leningrad, vol. 189. [In Russian]

Aladin, N. V. & W. D. Williams. The Aral Sea. Gleneagles Publishing, Adelaide (in press).

Alcocer, J. & E. Escobar, 1990. The drying up of the Mexican Plateau Axalapazcos. Salinet 4: 34–36.

Anon., 1990. The Aral Sea crisis and ways to meet this change. Supreme Soviet of the Union of Soviet Socialist Republics Committee on Ecology and Rational Use of Natural Resources. Moscow.

Anon., 1991. Outlines of the conception of conservation and restoration of the Aral Sea and normalization of the ecological, sanitary, medical, biological and socioeconomic situation in the Aral region. USSR Academy of Sciences/ USSR Committee for Conservation of Nature, Moscow. [Unpublished Report]

Australian National Parks and Wildlife Service [ANPWS], 1991. The Convention on Wetlands of International Importance Especially as Waterfowl Habitat. ANPWS, Canberra.

Beveridge, M. C. M., E. Stafford & R. Coutts, 1985. Metal concentrations in the commercially exploited fishes of an endorheic saline lake in the tinsilver province of Bolivia. Aquacult. Fish. Mgmt, 1: 41–53.

Bridgewater, P., 1991. Wetland conservation challenges in Oceania. In Proceedings of a Workshop on Wetlands Conservation and Management, Newcastle, Australia, 11–15 February 1991. Australian National Parks and Wildlife Service, Canberra.

Burgis, M. J. & J. J. Symoens, eds, 1987. African Wetlands and Shallow Water Bodies. ORSTOM, Paris.

Caspers, H., 1968. Biology of a hypersaline lagoon on a tropical atoll island (Laysan) Proc. Symp. Recent Adv. Trop. Ecol., 326–333.

Chilcott, J. E., D. W. Westcot, A. L. Toto & C. A. Enos, 1990. Water quality in evaporation basins used for the disposal of agricultural subsurface drainage water in the San Joaquin Valley, California, 1988 and 1989. California Regional Water Quality Control Board, Central Valley Region Report, December 1990, 1–48.

Claridge, G., 1991. An overview of wetland 'values': A necessary preliminary to wise use. In Proceedings of Workshop on Wetland Conservation and Management, Newcastle, Australia, 11–15 February 1991. Australian National Parks and Wildlife Service, Canberra.

Collins, N. C., 1977. Ecological studies of terminal lakes – their relevance to problems in limnology and population biology. In D. C. Greer (ed.), Desertic Terminal Lakes. Utah Water Res. Lab., Logan, Utah.

Dana, G. L. & P. H. Lenz, 1986. Effects of increasing salinity on an *Artemia* population from Mono Lake. Oecologia 68: 428–436.

Dugan, P. J., 1990. Wetland Conservation: A Review of Current Issues and Required Action. IUCN, Gland, Switzerland.

Dulhunty, R., 1975. The Spell of Lake Eyre. Lowden, Kilmore.

Finlayson, C. M., 1991. Conservation of the Volga delta. IWRB News, 5: 3.

Foster, J. H., 1978. Measuring the social value of wetland benefits. In P. E. Greeson, J. R. Clark & J. E. Clark (eds), Wetland Functions and Values: The State of Our Understanding. Proceedings of the National Symposium on Wetlands. Lake Buena Vista, Florida. Am. Wat. Resour. Ass. Minneapolis: 7–10 November 1984: 84–92.

Froend, R. H., E. M. Heddle, T. D. Bell & A. J. McComb, 1987. Effects of salinity and waterlogging on the vegetation of Lake Toolibin, Western Australia. Aust. J. Ecol. 12: 281–289.

Geddes, M. C. & W. D. Williams, 1987. Comments on *Artemia* introductions and the need for conservation. In P. Sorgeloos, D. A. Bengtson, W. Decleir & E. Jaspers (eds), *Artemia* Research and its Applications. Universa Press, Wetteren.

Glazovsky, N. F., 1990. [The Aral Crisis]. Academy of Science, Moscow. [In Russian]

Grainger, A., 1990. The Threatening Desert. Controlling Desertification. Earthscan, London.

Hammer, U. T., 1978. The saline lakes of Saskatchewan. I. Background and rationale for saline lakes research. Int. Revue ges. Hydrobiol. 63: 173–177.

Hammer, U. T., 1986. Saline Lake Ecosystems of the World. Dr W. Junk Publishers, Dordrecht.

306

Hammer, U. T., 1990. The effects of climatic change on the salinity, water levels and biota of Canadian prairie saline lakes. Verh. int. Ver. Limnol. 24: 321–326.

Herbst, D., 1988. Scenarios for the impact of changing lake levels and salinity at Mono Lake: Benthic ecology and the alkali fly, *Ephydra (Hydropyrus) hians* Say (Diptera: Ephydridae). Section 2, Appendix D-2. In Botkin, D. *et al.* (eds), The Future of Mono Lake. University of California Water Resources Centre Report No. 68.

Herbst, D., F. P. Conte &V. J. Brookes, 1988. Osmoregulation in an alkaline salt lake insect, *Ephydra (Hydropyrus) hians* Say (Diptera: Ephydridae) in relation to water chemistry. J. Insect Physiol. 34: 903–909.

Hollis, G. E. & T. A. Jones, 1991. Europe and the Mediterranean basin. In M. Finlayson & M. Moser (eds), Wetlands. Facts on File, Oxford: 27–56.

Holmes, J. W. & T. Talsma, 1981. Land and Stream Salinity. Elsevier, Amsterdam.

Hurlbert, S. H. & E. Flores, 1988. Nesting and conservation of flamingoes in the central Andes. Abstract of paper at IV Int. Symp. Salt Lakes, Banyoles, Spain, 28 May 1988.

International Union for the Conservation of Nature (IUCN), 1980. World Conservation Strategy. IUCN, Gland, Switzerland.

Koust, G. S., 1991. Evaluation of desertification in the south and east Aral region. Proceedings of Second Meeting of UNEP/USSR Working Group for the project 'Assistance for the Preparation of an Action Plan for the Conservation of the Aral Sea', Moscow.

Lincer, J. L., D. Zalkind, L. H. Brown & J. Hopcraft, 1981. Organochlorine residues in Kenya's rift valley lakes. J. appl. Ecol. 18: 157–172.

Luther, H. & J. Rzóska, 1971. Project Aqua: A Source Book of Inland Waters Proposed for Conservation. IBP Handbook, 21. Blackwell, Oxford.

Marble, A. D. & M. Gross, 1984. A method for assessing wetland characteristics and values. Landscape Planning 11: 1–17.

McComb, A. J. & P. S. Lake, 1988. The Conservation of Australian Wetlands. Surrey Beatty & Sons, Sydney.

McComb, A. J. & P. S. Lake, 1990. Australian Wetlands. Angus & Robertson, Sydney.

Metz, H. & L. Forró, 1991. The chemistry and crustacean zooplankton of the Seewinkel pans: A review of recent conditions. Hydrobiologia 210: 25–38.

Micklin, P. P., 1988. Desiccation of the Aral Sea: A water management disaster in the Soviet Union. Science 241: 1170–1176.

Montes, C. & P. Martino, 1987. La lagunas salinas espanolas. In Bases Cientificas para la Proteccion de los humendales en Espana. Dept. de Ecologia, Universidad Autonoma de Madrid.

Nissenbaum, A., 1979. Life in the Dead Sea – Fables, allegories, and scientific research. BioScience 29: 153–157.

Parsons, P. A., 1990. Biodiversity and climatic change. Proceedings of an International Conference on Conservation of Genetic Resources for Sustainable Development, Rxros, Norway, September 1990.

Patten, D. T. *et al.* [The Mono Basin Ecosystem Study Committee], 1987. The Mono Basin Ecosystem. Effects of Changing Lake Level. National Academy Press, Washington, D.C.

Pereira, H. G., 1973. Land Use and Water Resources. Cambridge University Press, Cambridge.

Persoone, G. & P. Sorgeloos, 1980. General aspects of the ecology and biogeography of *Artemia*. In G. Persoone, P. Sorgeloos, O. Roels & E. Jaspers (eds), The Brine Shrimp *Artemia*. Vol. 3: 3–24. Universa Press, Wetteren.

Rawson, D. S., 1946. Successful introduction of fish in a large saline lake. Can. fish. Culturalist, Nov. 1946.

Scott, D. A., 1989. A Directory of Asian Wetlands. IUCN, Gland, Switzerland & Cambridge, UK.

Serventy, V., 1985. The Desert Sea. The Miracle of Lake Eyre in Flood. Macmillan, Melbourne.

Sherwood, J. E., F. Stagnitti, M. J. Kokkinn & W. D. Williams, 1991. Dissolved oxygen concentrations in hypersaline waters. Limnol. Oceanogr. 36: 235–250.

Shiklomanov, I. A., 1990. Global water resources. Nat. Resour. 26: 34–43.

Sorgeloos, P., D. A. Bengtson, W. Decleir & E. Jaspers, 1987. *Artemia* Research and its Applications. Universa Press, Wetteren.

Stanley, M. de, 1966. The Salton Sea Yesterday and Today. Triumph Press, Los Angeles.

Stine, S., 1991. Geomorphic, geographic, and hydrographic basis for resolving the Mono Lake controversy. Envir. Geol. Wat. Sci. 17: 67–83.

Stone, A., 1991. Economic evaluation of wetlands. In Proceedings of a Workshop on Wetlands Conservation and Management, Newcastle, Australia, 11–15 February 1991. Australian National Parks and Wildlife Service, Canberra.

Traulich, B. & G. Wagner, 1989. Detektion von UV-B-Strahlung durch *Halobacterium halobium*. Akad. Natursch. Landschaftsplf. (ANL) Laufener Seminarbeiträge 3/88: 62–66.

Vareschi, E., 1987. Saline lake ecosystems. In E.-D. Schultz & H. Zwölfer (eds), Potentials and Limitations of Ecosystem Analysis. Springer Verlag, Berlin: 347–363.

Williams, W. D., 1972. The uniqueness of salt lake ecosystems. In S. Kajak & A. Hillbricht-Illkowska (eds), Productivity Problems of Freshwaters. Polish Academy of Science, Warsaw: 349–361.

Williams, W. D., 1981. Problems in the management of inland saline lakes. Verh. int. Ver. Limnol. 21: 688–692.

Williams, W. D., 1986. Limnology, the study of inland waters: A comment on perceptions of studies of salt lakes, past and present. In P. De Deckker & W. D. Williams (eds), Limnology in Australia. CSIRO & Dr W. Junk Publishers, Melbourne and Dordrecht.

Williams, W. D., 1991. Saline lake microcosms (microecosystems) as a method of investigating ecosystem attributes. Verh. int. Ver. Limnol. 24: 1134–1138.

Williams, W. D. & N. V. Aladin, 1991. The Aral Sea: Recent limnological changes and their conservation significance. Aquat. Conserv. 1: 3–23.

Williams, W. D., A. J. Boulton & R. G. Taaffe, 1990. Salinity as a determinant of salt lake fauna: a question of scale. Hydrobiologia 197: 257–266.

Williams, W. D. & M. J. Kokkinn, 1988. Wetlands and aquatic saline environments. AWRC Research Project 84/160, Completion Report. Dept. Resources and Energy, Canberra.

Microcosm analysis of salinity effects on coastal lagoon plankton assemblages

Glenn M. Greenwald & Stuart H. Hurlbert
Department of Biology, San Diego State University, San Diego, California 92182-0057, USA

Key words: microcosm, phytoplankton, zooplankton, estuarine plankton, coastal lagoon, salinity

Abstract

A microcosm experiment was conducted to assess the effects of salinity on coastal lagoon plankton assemblages. Five salinity levels were replicated four-fold in 380 l fiberglass tanks. Salinity levels used were 0, 8.5, 17, 34 and 51 ppt, or 0, 25, 50, 100 and 150 percent seawater. These were achieved by mixing concentrated lagoon water and tapwater in different proportions. Tanks were inoculated with plankton collected from San Dieguito Lagoon (Del Mar, San Diego County, California) and other fresh and saline waterbodies in the area. Selected physical-chemical variables, phytoplankton, zooplankton, and other invertebrate populations were monitored on five sampling dates over a 114 day period (13 August– 5 December 1986).

Total phytoplankton abundance increased with salinity, for salinities > 17 ppt. Most taxa showed marked effects of salinity, though the pattern of the effects often varied greatly from date to date. Chlorophytes tended to be most abundant at 51 ppt. Pyrrhophytes were most abundant at 0 or 51 ppt, and least abundant at 8.5 or 17 ppt. Cryptophytes increased with increasing salinity. Euglenophytes exhibited no salinity effect on any date. Bacillariophytes were most abundant at 8.5–34 ppt and least abundant at 51 ppt, with individual taxa showing maxima at 0–17 ppt (*Navicula, Synedra*), 8.5–34 ppt (*Surirella, Amphora*), and 34 ppt (*Cylindrotheca*).

Total zooplankton abundance decreased with salinity, for salinities > 17 ppt. The dominant taxa were protozoans, rotifers, cladocerans, and copepods, and all but the first group showed strong salinity effects. Protozoan abundance was unaffected by salinity. Rotifers were most abundant at 0 ppt (*Keratella, Filinia*) or 8.5 ppt (*Brachionus*). With few exceptions, cladocerans (*Alona, Ceriodaphnia, Scapholeberis*) were found only at 0 ppt. Abundance of calanoid copepods decreased with increasing salinity, with individual taxa showing maxima at 0 ppt (*Diaptomus*), 8.5–17 ppt (*Pseudodiaptomus, Eurytemora*), and 34 ppt (*Acartia*). Cyclopoid copepods were most abundant at 17 ppt, with individual taxa showing maxima at 0 ppt (*Eucyclops*), 8.5 ppt (*Halicyclops*), and 17 ppt (*Oithona*). Harpacticoid copepods (*Cletocamptus, Tachidius*) were most abundant at 17–34 ppt. Ostracods and mosquito (*Culex*) larvae were most abundant at 8.5 ppt and absent at 34 and 51 ppt. Polychaetes generally were most abundant at 17–34 ppt, and water boatmen (*Trichocorixa*) at 8.5–34 ppt. Various physical and chemical variables also showed significant variations with salinity. Tending to increase with salinity were temperature, ammonia and orthophosphate concentrations. Decreasing with salinity were pH, dissolved oxygen and silica concentrations. The causes and interrelationships of these salinity effects are discussed.

Introduction

Despite longstanding recognition of the role of salinity as a primary influence on plankton in various types of aquatic ecosystems (e.g. Beadle, 1943, Braarud, 1951, 1962; Carpelan, 1957, 1964; Caspers, 1952; Cronin *et al.*, 1962; Day *et al.*, 1989; Gauthier, 1928; Hammer, 1986; Hedgpeth, 1959; Javor, 1989; Ketchum, 1983; Löffler, 1961; McLachlan, 1961; Provasoli, 1958; Smayda, 1958; Rawson & Moore, 1945; Remane & Schlieper, 1971), there has been almost no experimental study of the effects of salinity on plankton communities. There have been numerous descriptive studies of how plankton biotas and communities vary among lakes of different salinities or along salinity gradients in coastal lagoons and estuaries. There also have been numerous experimental studies on the short-term physiological responses of individual species, maintained in isolation from other species, to variation in salinity. But in the entire aquatic biology literature, we find only three studies, all of limited scope, two relating to lakes and one to estuaries, that have attempted to determine the responses of natural or semi-natural plankton assemblages to different experimentally imposed salinities. Melack (1985) conducted pilot experiments with laboratory microcosms containing Mono Lake (California) plankton. He found that phytoplankton abundance and productivity decreased and *Artemia* mortality increased as salinity was increased from 90 to 135 g l^{-1}. Galat and Robinson (1983) used large (47 m^3), greenhouse-maintained microcosms to assess effects on Pyramid Lake (Nevada) zooplankton of salinities ranging from 5.6 to 11 ppt. They found negative effects of increased salinity on *Ceriodaphnia* and *Acanthocyclops*. Klos (1988) described the setting up of four 13 m^3 microcosms, one each at 0, 5, 10 and 30 ppt, to study salinity effects on plankton of Rhode Island estuaries, but data presented were few and inconclusive.

Since microcosms have been used extensively to assess the effects of many other variables (e.g. predators, grazers, macronutrients, trace metals, pollutants, light, vertical mixing) on estuarine or other coastal planktonic assemblages, their non-use for salinity studies is quite surprising.

This lack of experimentation at the community level has created an anomalous situation. Reviews of our knowledge of these variably saline ecosystems (e.g. Day *et al.*, 1989; Hammer, 1986; Ketchum, 1983) always emphasize, in general terms, the importance of salinity as an influence on them. Yet they are unable to tell us what the salinity effects are, except, at best, on a taxon-by-taxon basis or for extreme salinities (e.g. ⩾ 50 ppt). The primary literature simply contains no clear information on how salinity affects, for example, primary production, nitrogen cycling, the relative abundances of diatoms, dinoflagellates and cyanophytes, or of protozoans, rotifers and crustaceans, and so on.

The present experimental study was undertaken in order to begin filling this yawning lacuna. It was carried out in conjunction with a descriptive study (Greenwald & Hurlbert, in prep.) of the plankton of San Dieguito Lagoon in San Diego County, California. Its specific objective was to determine how salinities of 0–51 ppt influenced the structure of semi-natural phyto- and zooplankton assemblages in outdoor microcosms.

San Dieguito Lagoon, 28 ha in area and one of several sources for plankton inocula used in this study, is located in Del Mar, California at the end of the 70 km long watershed of the San Dieguito River. As with most coastal lagoons of California, man's activities have greatly altered its form, hydrological regime, trophic state and biota (Marcus, 1989). Since completion of dams creating two reservoirs (Lake Hodges in 1947; Lake Sutherland in 1954), the watershed functionally has been about 18 km long and 114 km^2 in area. Only during winters of unusually heavy rain is there any overflow from the reservoirs mentioned above. Salinity of the lagoon in recent years has ranged from <1 to 60 ppt. High salinities are generally the result of closure of the lagoon mouth during the long rainless summers typical of this region. During 1985–86 the observed salinity range for the lagoon was 14–32 ppt, the phytoplankton was dominated by euglenophytes, dinophytes and bacillariophytes, and the zooplankton was dominated by ciliates and copepods (Greenwald & Hurlbert, in prep.). Further information on the lagoon is given by Carpelan (1969), Greenwald (1989), Greenwald & Hurlbert (in prep.), Mudie *et al.* (1976), and various unpublished reports cited by these authors.

Methods

Experimental design and tanks

The experimental study was carried out using five different salinity levels, each replicated four times. Salinity levels used were 0, 25, 50, 100, and 150 percent of local seawater salinity, or, nominally, 0, 8.5, 17, 34, and 51 ppt ($= g kg^{-1}$). These salinity levels were selected on the basis of a pilot project and the salinity range in San Dieguito Lagoon during recent years. Actual salinity for the freshwater treatment was ca. 0.5 ppt. A hand refractometer (American Optical, Model No. 10419) was used in establishing and monitoring salinity levels of the other treatments. Actual salinities of these treatments probably were ca. 5–7 percent higher than nominal ones, given the bias of this NaCl-calibrated instrument (Hurlbert, Gonzalez & Hart, unpub. data).

The experimental units used were 380 l capacity fiberglass tanks with conical (120°) bottoms and cylindrical upper portions (see Fig. 2 in Greenwald, 1989). These tanks were set on the ground in an open area near the San Dieguito River in Del Mar. The tanks were covered with 2.5 cm mesh fruit tree netting to keep out squirrels, birds, and other large animals.

Establishment of salinity levels

Hypersaline water for creating different salinity treatments was obtained by pumping lagoon water (ca. 34 ppt; Table 1) into 24 of the experimental tanks and then allowing them to evaporate from June to August 1986. Commercial sodium hypochlorite (NaOCl) solution was introduced into the tanks to sterilize the lagoon water

Table 1. Comparison of some properties of the two types of waters used to create the different salinity levels or treatments.

Variable	San Dieguito Lagoon (10 July 1986)[a]	Lake Miramar, effluent from filtration plant (August 1986)[b]
Plankton density ($\mu g C l^{-1}$)	121	0.0[c]
Tripton density ($\mu g C l^{-1}$)	??	0.0[c]
pH	8.1	8.25
Salinity (ppt)	33.8	0.47
NH_4-N ($mg l^{-1}$)	0.19	0.07
NO_3-N ($mg l^{-1}$)	<0.01	2.11
ortho PO_4-P ($mg l^{-1}$)	0.05	<0.01
poly PO_4-P ($mg l^{-1}$)[d]	no data	0.08
SiO_2 ($mg l^{-1}$)	0.38	8.63
Ca^{+2} ($mg l^{-1}$)	400	49.6
Mg^{+2} ($mg l^{-1}$)	1350	21.6
Na^+ ($mg l^{-1}$)	10,500	59
K^+ ($mg l^{-1}$)	380	4.6
HCO_3^- ($mg l^{-1}$)	142	126
SO_4^{-2} ($mg l^{-1}$)	2700	153
Cl^- ($mg l^{-1}$)	19,000	61.9

[a] Data are from Greenwald (1989) and represent means based on values for 5 sampling stations in the lagoon. Lagoon water was concentrated 3-fold before being used to establish treatments (see text). Values for the 7 major ions are based on assumption that their concentrations were the same as in standard seawater.

[b] Data from City of San Diego Water Quality Lab. Analyses are for a composite of daily samples taken throughout the month. Value for NH_4, however, is for a single sample collected on August 8.

[c] Values presume that esentialy all particulate matter was removed at water filtration plant.

[d] Total acid hydrolyzable phosphate minus orthophosphate.

so that the different salinity treatments would not differ initially with respect to the densities of microorganisms present. After the eight week evaporation period, the water in the tanks had a salinity of about 110 ppt. The water from 20 of them was then transferred to the remaining four and the 20 tanks were then scrubbed clean and rinsed out.

Tapwater (Table 1) was used to fill the 20 empty tanks and allowed to age in the sun for 14 days to remove chlorine and chloramine compounds. The desired experimental salinity levels were then obtained by mixing the aged tap water and concentrated lagoon water in appropriate ratios. In all tanks the water level was set at about 10 cm below the tank rim, corresponding to a water volume of about 320 l and a water depth of 64 cm. Salinity and water levels were maintained by periodic addition of distilled or deionized water. Assignment of treatments to tanks was conducted using a randomized block design and subjective intervention to rearrange 'undesirable randomizations'.

Inoculation of tanks

Seven days after establishment of salinity levels all units were equally inoculated with plankton on 13 August. Inoculations were repeated on 16 August, 27 August, 15 October, and 7 November. Sources of inocula were several locations in San Dieguito Lagoon (0–48 ppt) and several locations in each of several other waterbodies within 15 km of it: San Dieguito River (0–17 ppt), Lake Hodges (0 ppt), Los Peñasquitos Lagoon (6–44 ppt), and San Elijo Lagoon (0–65 ppt) (see Table 6 in Greenwald, 1989, for more detail)

The use of 'seed' inocula from multiple sources of varying salinity and on multiple dates was intended to help attain fuller representation of the plankton species present in San Dieguito Lagoon and other water bodies of the region and to simulate an acceleration of those natural dispersal processes that result in colonization of natural waterbodies. At each source location, larger plankters were obtained by hand-towing a 35 μm mesh plankton net and smaller plankters were obtained in unfiltered water. The multiple samples taken from a given waterbody were composited in a bucket for immediate transport to the experimental site. A separate bucket was used for each waterbody.

Aliquots were dispensed from the bucket with an 80 ml vial into all experimental tanks. Multiple aliquots from each bucket were placed in each experimental tank, by making repeated circuits of the 20-tank array with our dispensing vial and bucket. On average, a total of 2–3 l of plankton concentrate was added to each tank per inoculation date.

No attempt was made to document what spe-

cies or densities were present in these inocula. The rarer species would not have been detected, and the objectives of the experiment required only that the inocula were the same for all salinity treatments.

Sampling and monitoring regime

Sampling and monitoring were carried out at midday (1100–1300 h local time) on six dates: 13 and 23 August, 13 September, 3 October, 24 October, and 5 December 1986. These dates corresponded to days 0, 10, 31, 51, 72, and 114 after the first inoculation on 13 August. Not all variables were monitored on every date.

Physical and chemical variables were measured at, or with samples taken from, a tank depth of 30 cm. The purpose of these measurements was to document the general conditions under which the experiment was conducted as well as to test whether salinity itself might alter other physical-chemical variables. Temperature and dissolved oxygen were measured with a YSI temperature/dissolved oxygen probe (Model No. 51-B) and hydrogen ion activity with a Beckman Select-Mate pH meter. Nitrate, ammonium, orthophosphate, and silica concentrations were measured only on the first inoculation date (13 August) and on 24 October, using the methods of Technicon Industrial Systems (1973a, b, 1977a, b, c).

Phytoplankton methods

The phytoplankton and planktonic protozoans were sampled using a 1 cm diameter $\times$ 80 cm long PVC tube. A sample was taken by slowly lowering the tube in vertical position to a point about 10 cm above the tank bottom. Seven such samples, dispersed uniformly over the area of the water surface, were taken from each tank and combined to form a single composite sample. After this was thoroughly mixed, an 80 ml subsample was taken, preserved with 1% Lugol's solution and refrigerated.

Phytoplankton samples were analyzed using a

modified Utermohl method (Likens & Wetzel, 1979). Identifications were made using a variety of references, in most cases could not be made to species, and in some cases were only made to class or division. Generally, 20 ml of each 80 ml sample was placed in the settling chamber and allowed to settle overnight, but multiple settlings of smaller volumes were used in samples with much detritus. Counting was done using a Leitz inverted binocular compound microscope with phase contrast.

Smaller organisms were counted at 400 × magnification and larger ones were counted at 100 × magnification. In most cases, crossed diameters were counted on each settled chamber; these equalled 5 percent of the total chamber area at 400 × magnification, and 21 percent at 100 ×. Each taxon was assigned a regular geometric shape (Beers et al., 1977), and, with an ocular micrometer, estimates were made of its length, width, and depth to determine average volume. As the depth of a phytoplankter was not always evident under the microscope, depth values were sometimes approximated by reference to Beers et al. (1977). From this determination of approximate volume, an average individual carbon weight was estimated for each taxon (given in Table 2 in Greenwald 1989) using equations from Beers et al. (1977).

The Utermohl method is unsatisfactory for the picoplankton (0.2–2 μm) and no attempt was made to obtain information on such organisms. Phototrophic picoplankton rarely constitutes a large fraction of total phytoplankton biomass except in oligotrophic waters (Stockner, 1988).

Zooplankton methods

The zooplankton sampling device was an 11 cm diameter × 74 cm long clear plexiglass tube sampler, with a bottom flap valve, which at full capacity held 7 l. A 4-l sample was taken by lowering it rapidly in vertical position into the tank, quickly raising it out of the tank and decanting its contents into a bucket. This was repeated at 5 locations in each tank to produce a 20 l composite sample, representing about 6 percent of the tank volume.

The 20 l sample was filtered through a 35 μm mesh plankton net and the zooplankton preserved in 80 ml plastic vials with 1 percent Lugol's solution and refrigerated. The 20 l of net-filtered water was poured back into the tank, and the net, sampler and bucket carefully rinsed before proceeding to the next tank.

Zooplankton samples were analyzed using a Sedgwick-Rafter cell and compound microscope. Larger zooplankters were counted in 55 percent of the sample at 40 × magnification and small zooplankters in 20 percent of the sample at 100 × magnification. Identifications were made using a variety of references and with the assistance of colleagues. Except in the case of protozoans and larval forms, most zooplankters could be identified at least to genus. In each sample, lengths of up to 10 to 20 individuals were measured for each taxon. These lengths were used to estimate the dry mass of each taxon for each sample by using length-mass regression formulae or values obtained or extrapolated from Miller (1966), Doohan & Rainbow (1971), Krylov (1973), Kudrinskaya & Yushko (1973), Rosen (1981) and Culver et al. (1985) (see Table 3 in Greenwald 1989 for details). Carbon masses were then approximated by multiplying the calculated dry masses by a factor of 0.40 (Beers et al., 1977). For the smaller taxa with less variable lengths, such as rotifers and copepod nauplii, estimates of mean individual dry mass were obtained from Dumont et al. (1975) and Bottrell et al. (1976), and then converted to carbon mass using the factor of 0.40.

For the copepods, counts were made separately for the nauplii and the copepodids, but these two groups were defined by an operational criterion. 'Nauplii' were designated to include from the first naupliar to approximately the third copepodid stage. 'Copepodids' were designated to include from approximately the fourth through sixth copepodid stages. These definitions were used because the copepods usually could not be identified to species until the fourth copepodid stage. Throughout the rest of this paper, 'nauplii' and 'copepodids' are used in this sense.

Other invertebrates

Observations were also made on each sampling date for such macroinvertebrates as mosquito larvae, corixids, and polychaetes (specifically, the tubes of the latter). These organisms were not actually collected, but were counted, to the nearest ten individuals when numbering more than that, as they were observed in each tank's water column, or on its side or bottom. This technique was possible because the turbidity of the tanks was normally low.

Data analysis

For physical and chemical variables, arithmetic means were used as measures of central tendency. ANOVA was used to test for differences among treatment means and a separate ANOVA was carried out for each sampling date. Numerator and denominator degrees of freedom, in all F tests were 4 and 15, respectively.

Analyses of the phytoplankton and zooplankton were carried out for both individual taxa and for grouped taxa. For individual species, individual genera, and total nauplii, analyses were carried out on abundance expressed as number per ml or per l, while for suprageneric taxa or other composite categories analyses were carried out on abundance expressed as carbon mass per l. The latter approach accounted for disparities in size among the taxa being grouped together.

For all plankton abundance data, geometric means were used as measures of central tendency. Data sets were predominantly positively skewed and we wished to moderate the influence on the means of the occasional extreme high values, to increase the normality of distributions and to reduce the heterogeneity of variances.

In order to avoid problems posed by values of zero, a value of 1.0 was added to each raw count datum prior to calculating geometric mean numerical densities on a per ml or per l basis. For grouped taxa, where density was being expressed as carbon mass per l, the value added to each raw datum was not 1.0 but rather the carbon mass of the smallest individual observed among the taxa being grouped.

Converted to a per ml or per l basis, these 1.0 values and carbon mass values become the constants given on our individual graphs (Figs 2, 3, 6, 7, 8, 10), e.g. the '0.005' in the graph for total phytoplankton (Fig. 2). When a taxon is completely absent from all samples for a given treatment on a given date, the geometric mean graphed will equal this constant, as we have not fully back-transformed the geometric means by subtracting these constants from them. That is, zero abundance in all four samples for all five treatments on a given date is represented by a horizontal line at $y = $ constant, e.g. at $y = 0.005$ for total phytoplankton.

For all plankton data, Kruskal-Wallis tests were used to test for differences among salinity treatments. For each taxon or composite group a separate Kruskal-Wallis test was carried out for each sampling date. As in the case of the date-by-date ANOVAs applied to the data for physical and chemical variables, these successive date-by-date Kruskal-Wallis tests are not independent of each other. Provided one is aware of that, this statistical approach poses no special problems of interpretation. The validity and desirability of the approach is briefly argued further by Mead (1988) and Soto and Hurlbert (1991).

To portray directly the effect of salinity on the gross taxonomic composition of the plankton, two graphical approaches have been utilized. Phytoplankters were grouped into six major taxa (divisions) and the zooplankters into seven major taxa (phyla, classes, orders). The relative abundances of the taxa in each tank on each date were then represented using Whittaker (rank-abundance or dominance diversity) curves, and their mean relative abundances in each treatment were represented using percent composition graphs.

Results and discussion

Presentation of data

All results, including those of our statistical tests, are presented graphically (Figs 1–10). Virtually

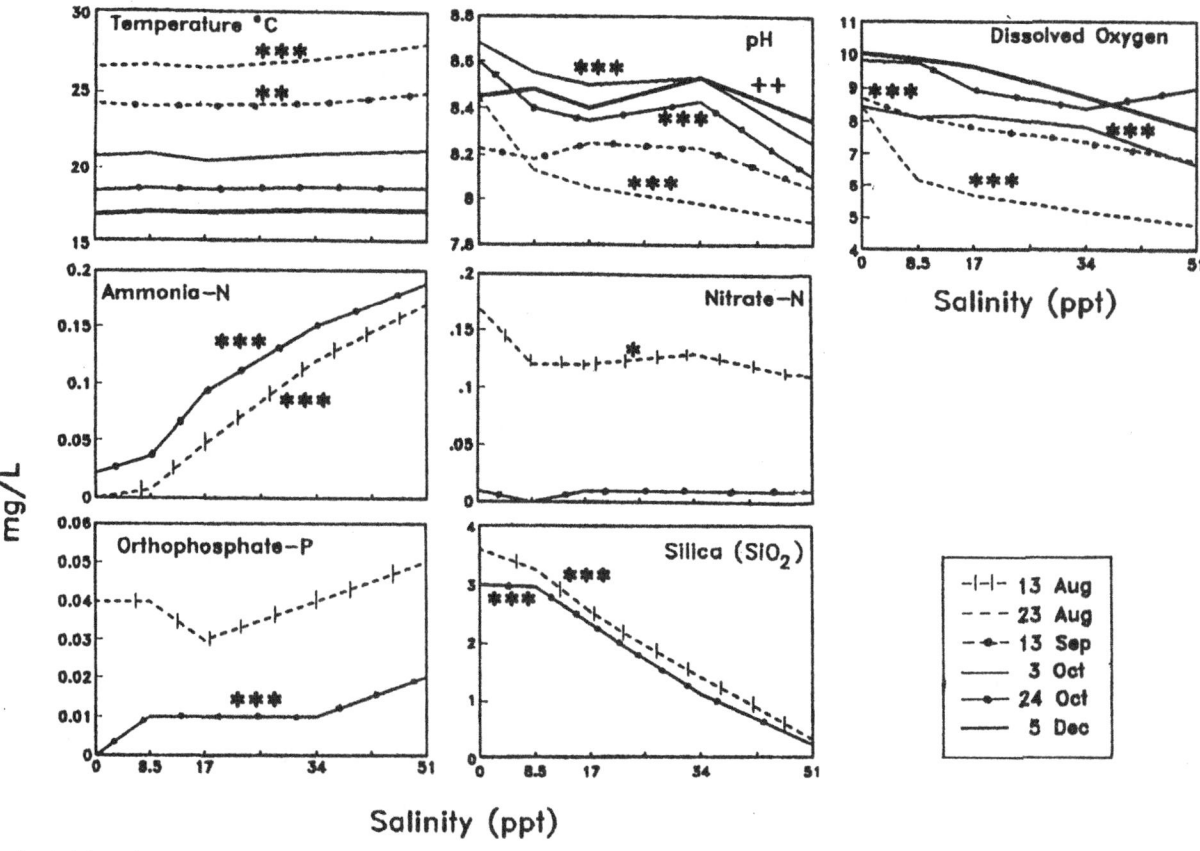

Fig. 1. Arithmetic means of some selected physical and chemical variables in the different salinity treatments. *P* values for date-by-date ANOVAs are indicated by symbols as follows: no symbol, $P>0.2$; +, $0.1<P\leq0.2$; ++, $0.05<P\leq0.1$; *, $0.01<P\leq0.05$; **, $0.001<P\leq0.01$; ***, $P\leq0.001$.

all physical and chemical variables (Fig. 1), most phytoplankton taxa (Figs 2, 3), most zooplankton taxa (Figs 6, 7, 8), and three other invertebrate taxa (Fig. 11) exhibited statistically significant effects of salinity. For the organisms, these effects usually represented more than one order of magnitude variation in abundance among treatments. The variable nature of the effects on individual taxa produced correspondingly large effects on the ecological and taxonomic structure of the plankton assemblages (Figs 4, 5, 9).

The principal graphical approach used (Figs 1, 2, 3, 6, 7, 8, 10) was selected because it portrays most clearly the nature of the salinity effect and how that varies from date to date. It does not show quite so clearly how individual taxa changed in abundance over time. It also does not show the variability among tanks within treatments, appre-

ciation of which can valuably supplement the more condensed and formal statistical approaches. Whittaker curves showing the absolute and relative abundances of the major taxa in each tank on each date (Figs 5, 9) partially compensate for the above limitations; and graphs (Fig. 4) showing directly how taxonomic composition changed over time provide yet another perspective.

The statistical significance of the ANOVA or Kruskal-Wallis test on a given date is indicated by a symbol (see caption for Fig. 1) placed in close proximity to the line representing that date. In a few cases (e.g. Figs 6, 7) where the lines representing different dates are closely coincident, these symbols are presented in a column, in chronological order, so as to avoid ambiguity.

314

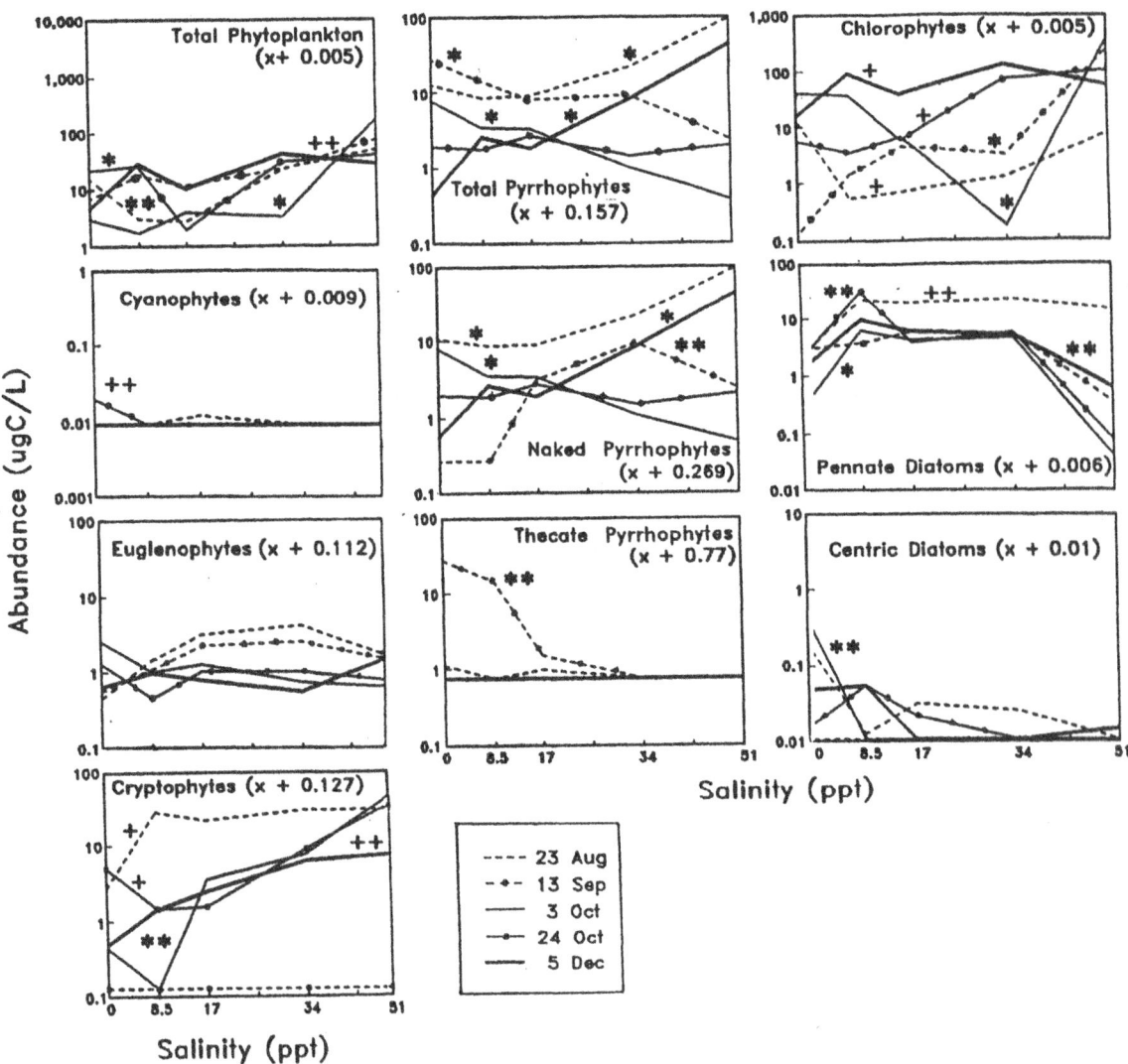

Fig. 2. Effects of salinity on geometric mean abundances of major phytoplankton taxa. Symbols represent results of date-by-date Kruskal-Wallis tests; P-values as in Fig. 1. Explanation of constants added to data (x + 0.005, etc.) is given in the text.

Some generalizations and caveats

As will become more apparent later, 'salinity effect' is a term of some ambiguity. This is true both in general and in the context of this experiment. Here we note only that the date-to-date variation in the nature of the effect was itself different from one response variable to another. For example, the effect on *Navicula* C (Fig. 3) or on *Ceriodaphnia* (Fig. 7) was much the same on all dates, whereas the effect on total pyrrhophytes (Fig. 2) or total cyclopoids (Fig. 8) varied markedly from date to date. This generally greater variation of the multispecific or higher taxa is attributable, in part, to changes in their species composition over time, with more salinity tolerant species dominating on some dates and less tolerant species dominating on others. For such multispecific taxa, generalization about their responses to salinity will clearly be difficult.

This date-to-date variation in the nature of effects also may contain information about the directness of the mechanism whereby salinity exerted its effect. Where the same response is seen

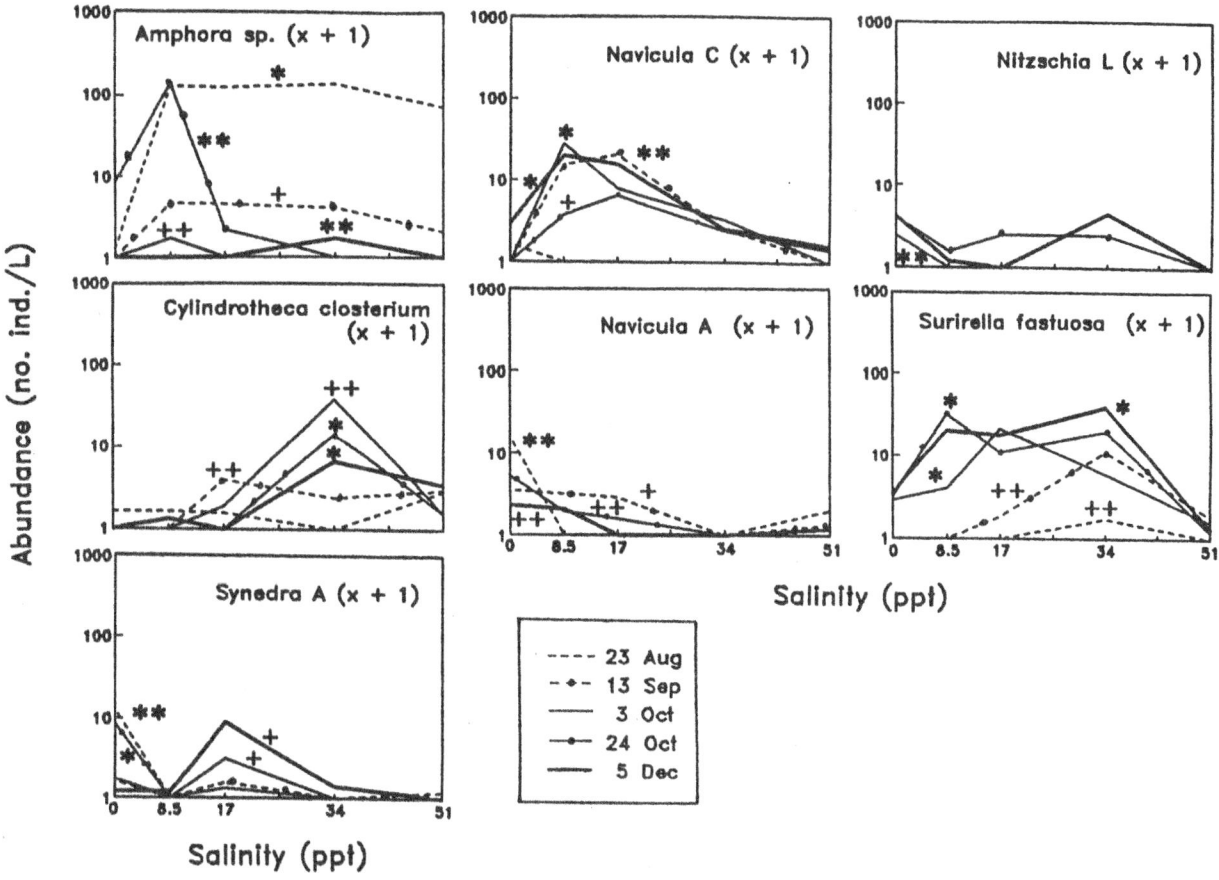

Fig. 3. Effects of salinity on geometric mean abundances of major diatom taxa. P-values and constants as explained in Figs 1 and 2 and the text.

date after date, as for the cladoceran species (Fig. 6), one suspects the primary mechanism is very direct, such as osmoregulatory stress. Where the response is more variable, as with *Surirella* (Fig. 3), one suspects a correspondingly greater role of indirect mechanisms. One category of such would be those involving salinity-induced changes in the populations of predators and competitors that interact with a given species.

Discussion and interpretation of the observed 'salinity effects' must be prefaced by two caveats. First, to document the effect of an experimental variable is to imply nothing about the mechanism whereby it was produced. This may have been direct or indirect. While we have great interest in knowing what the mechanism was in each case, we can in fact reach few concrete conclusions on

the matter. The experiment itself provides no direct evidence on mechanisms. These can in some cases tentatively be inferred from existing information. More generally, conclusive determination of mechanisms requires, for each response variable, much additional experimentation.

A second caveat concerns the difference between 'salinity' as it is usually construed and as it was defined by our operations. Salinity usually refers to the sum of the concentrations of either the seven commonest inorganic ions plus silica or of all inorganic ions, the difference between these two measures often being less than the error in analytical procedures.

The most obvious way to assess salinity effects would be to create salinity levels or treatments that differed in the summed abundances of the

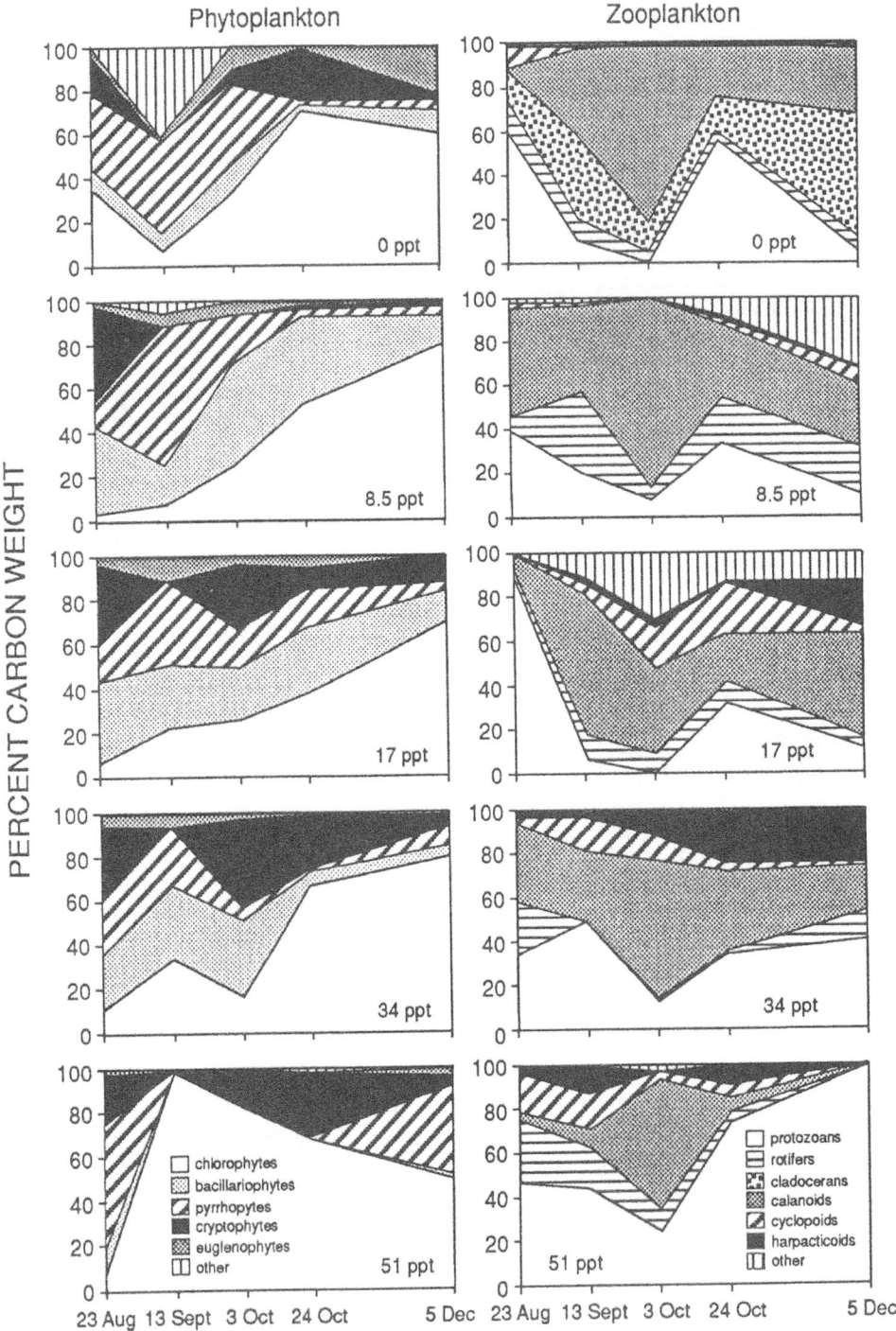

Fig. 4. Effects of salinity on taxonomic composition of phytoplankton and zooplankton. The vertical width of the polygon for each taxon represents the percentage contributed to either the total phytoplankton carbon (left column) or to the total zooplankton carbon (right column).

seven major ions but not in their relative proportions, in nutrient concentrations, or in the concentrations of any other substances. But the

procedures we used to create different salinity levels did not do this.

The tapwater and concentrated lagoon water

used to establish our treatments differed in many properties other than the summed concentrations of the major ions. We did not analyze samples for these on the specific dates we collected the lagoon water (25 June) or added the tapwater (6 August). A surrogate comparison is nevertheless possible using analyses for lagoon water collected on 10 July and analyses for an August multi-day composite sample of processed water at the reservoir from which the tapwater used was predominantly derived (Table 1).

One can see how the water chemistries of the treatments initially must have differed, given that they were established by mixing concentrated lagoon water and tapwater in different proportions. Clearly the ionic proportions were quite different in the different 'salinity' treatments, just as they always differ among natural waters of widely varying salinities. The amounts of nutrients (N, P, Si) added to tanks likewise must have varied among 'salinity' treatments in a regular manner, as surely did those of other unmeasured elements.

Thus in this report, the term, 'salinity effects' is generally used to designate the combined effects of salinity and a suite of covariates. The separate effects of salinity and these covariates cannot be formally distinguished. From a statistical point of view, this represents confounding caused by inadequate control of extraneous variables. From the point of view of natural correlates of salinity variation, however, this confounding to some extent represents a realistic aspect of the experiment, as is discussed later.

Physical and chemical variables

The observed salinity effects on temperature, pH, and dissolved oxygen concentrations probably represented the direct effects of salinity on the physical and physico-chemical properties of water. The salinity effects on nutrient concentrations, however, were primarily a consequence of our procedures for creating the different salinity levels, and not an effect of salinity *per se* on the cycling of nutrients in the tank ecosystems.

Temperature

Daytime water temperatures of all treatments declined with time, from an August mean of 26.7 °C to a December mean of 16.9 °C (Fig. 1). More surprisingly, on the two warmest sampling dates, there were slightly but significantly higher temperatures at the high salinities.

This latter effect is perhaps best explained as the result of the dependence of specific heat on salinity. As the salinity of a solution increases the amount of heat required to raise its temperature by 1 °C decreases. Seawater (35 ppt) for example, has a specific heat about 7 percent lower than that of freshwater (Cole, 1983). Whether heat input was principally by conduction through the tank walls or by direct solar insolation of the water, equal inputs of heat would have caused greater temperature increases in the high salinity treatments.

Lack of a salinity effect on temperature during the colder months could have been due to lower day-night temperature fluctuations at that time and reduced heating rates during the morning and midday hours. The effect of differences in specific heat on water temperature might be expected to be greatest when the heating (or cooling) rates were greatest. Our temperature measurements were all made about midday. We suspect temperature differences among treatments would have been less likely to be detected at other times of the day.

Greater evaporative cooling of low salinity tanks and higher phytoplankton densities in high salinity tanks (Fig. 2) are other factors that may have contributed to temperature differences among treatments.

Oxygen

On all dates dissolved oxygen concentrations tended to decrease with increasing salinity, though significant differences among treatments were found only on the first three dates (Fig. 1). This decrease reflected the fact that solubility of gases in water decreases as salinity increases. Saturation concentrations of oxygen in seawater (35 ppt) are about 20 percent lower than those in fresh water (Cole, 1983).

Oxygen levels were also strongly affected by temperature, the warmer sampling dates yielding lower values than the cooler ones. This was primarily a consequence of the inverse relationship of oxygen and water temperature. Exposed to the wind and undoubtedly subject to strong convectional circulation as a result of heat exchange through the tank walls, the tank waters probably were close to 100 percent saturation most of the time.

Oxygen concentrations can be strongly influenced by biological processes, especially in nutrient rich microcosms such as ours. These processes undoubtedly account for some departures of the oxygen curves from those expected on the basis of the known effects of temperature and salinity. In general these biological processes may have moderated the negative relationship between salinity and oxygen concentrations. At higher salinities, the oxygen-generating phytoplankters were generally more abundant (Fig. 2) and the oxygen-consuming zooplankters generally less abundant (Fig. 6).

The oxygen-salinity relationship thus favored by those organisms was the reverse of that fostered by the direct influence of salinity on saturation concentrations. The direct influence predominated in the microcosms and could be expected to do so in natural situations where the water column is well mixed and nutrient levels not too high. Where nutrient levels are high and turbulence and currents are weak, however, biological processes will become increasingly more important relative to salinity in determining actual oxygen concentrations.

The lowered oxygen concentrations at high salinities could have directly influenced phytoplankters and zooplankters, especially those species in which oxygen requirements would have been increased at high salinities by increased osmoregulatory activities. Low oxygen concentrations at high salinities have the potential to be more stressful for some species than the direct toxic or osmoregulatory stresses posed by the ions themselves.

pH

The overall tendency of pH was to decrease as salinity increased, and on four dates the differences among treatments were significant (Fig. 1). Such a trend has been observed in many saline waterbodies, both coastal and inland (Hammer, 1986). We have also observed a temporal decline in pH that accompanied a small temporal increase in salinity (from 0.5 to 2.5 ppt) resulting from evaporative concentration in freshwater microcosms (Soto & Hurlbert, 1991).

The cause of the pH-salinity relationship appears to be strictly physical-chemical and involves the dissociation of the bicarbonate ion (Amit & Bentor, 1971; Krumgalz, 1980). High salinities inhibit this dissociation, thereby lowering the concentration of the hydroxide ion and, consequently, the pH.

As in the case of oxygen concentrations, biological processes can strongly influence pH. And also as in the case of the oxygen data, these processes would have been expected to favor a salinity-pH relationship just the reverse of that found. At the highest salinities, the high densities of CO_2-consuming phytoplankters (Fig. 2) and the low densities of CO_2-generating zooplankters (Fig. 6) would have been expected to have produced the highest pH values, not the lowest.

Two other conspicuous patterns in the pH data were (1) the general increase in pH values from summer to winter, and (2) the negligible or inconsistent response of pH to salinity over the intermediate portion (8.5 to 34 ppt) of the salinity range. We see no clear physical-chemical nor biological explanations for those patterns but suspect that biological processes were responsible.

pH influences chemical and biological processes, just as it is influenced by them. It is likely that the salinity-induced pH changes in turn affected various plankton populations by, for example, altering the availability of inorganic carbon and possibly other nutrients.

Nutrients

The significant variation of N, P, and Si concentrations among treatments (Fig. 1) almost certainly was due both to the manner in which the experimental salinity levels were established and to real effects of salinity on nutrient cycling. The primary interpretational problem derives from the fact that nutrient concentrations were not the same in the tapwater and concentrated lagoon water that were used in different proportions to create the salinity levels. A secondary problem is that we only have surrogate data (Table 1) on what the actual nutrient concentrations were.

The surrogate data suggest that our procedures caused initial N and Si concentrations to decrease with increasing salinity and initial P concentration to vary relatively little among treatments. Initial N and Si levels should have been approximately an order of magnitude higher at the lowest salinity (0 ppt) than at the highest (51 ppt). This calculation takes into account the relatively minor amounts of N, P and Si likely to have been contributed by the particulate matter (plankton plus tripton) in the lagoon water. It also takes into account the fact that lagoon water was concentrated (by evaporating 66 percent of its volume) before it was used to create the experimental salinities.

Si decreased markedly with increasing salinity in both the early and later parts of the experiment (Fig. 1). This most likely simply reflected the initial Si gradient produced by our procedures. It, in turn, may have been responsible for the negative 'salinity effect' on diatoms which on most dates were least abundant at the highest salinity (Fig. 2).

Patterns for N were more complex (Fig. 1). Nitrate levels decreased markedly from August to October, but on neither date did they vary much among treatments. Ammonium, in contrast, increased over time and on both dates showed a strong positive correlation with salinity. This correlation seems to represent a true salinity effect, as the reverse correlation would have been expected on the basis of initial total N concentrations (Table 1). Salinity somehow may have altered the balance between ammonium production through bacterial degradation on the one hand and ammonium depletion by algal uptake and bacterial nitrification on the other. Zooplankton

excretion was not likely a direct factor as zooplankton biomass decreased with increasing salinity (Fig. 6). The correspondingly smaller amount of N bound up in zooplankton biomass at high salinities meant, however, that a larger percentage of the total N present was available for other phases, such as the dissolved one.

Phosphate levels declined over time, showed little variation among treatments in August, and a significant increase with salinity in October (Fig. 1). Uncertainty as to the relative P levels in the lagoon water and tapwater inhibit interpretation of the October salinity effect. It might represent an imbalance between bacterial release and algal uptake similar to that postulated in explanation of the data for ammonium.

Phytoplankton

Total phytoplankton
Total phytoplankton abundance was significantly affected by salinity on most sampling dates (Figs 2, 5). The most consistent pattern was an increase in abundance as salinity increased from 17 to 51 ppt. At lower salinities (0–17 ppt), the effects were more variable. Minimum phytoplankton abundance usually was found at 8.5 or 17 ppt. If data are averaged over all dates, however, total phytoplankton abundance is found to be invariate over the salinity range of 0–17 ppt.

The above pattern is the opposite of that shown by total crustacean zooplankton (Fig. 6). Low grazing pressure by zooplankters thus may have been a factor favoring the high phytoplankton standing crops observed at the higher salinities. Greater availability of ammonium and orthophosphate ions (Fig. 1) might also have favored high phytoplankton abundance at high salinities, though, as discussed earlier, it was unlikely that the *total* amounts of N and P in a tank increased with salinity.

A third factor is the direct effect of salinity itself. It is possible that for highly diverse assemblages, maximal phytoplankton standing crop or productivity will be favored by salinities in the range of 34 to 51 ppt because in this range the

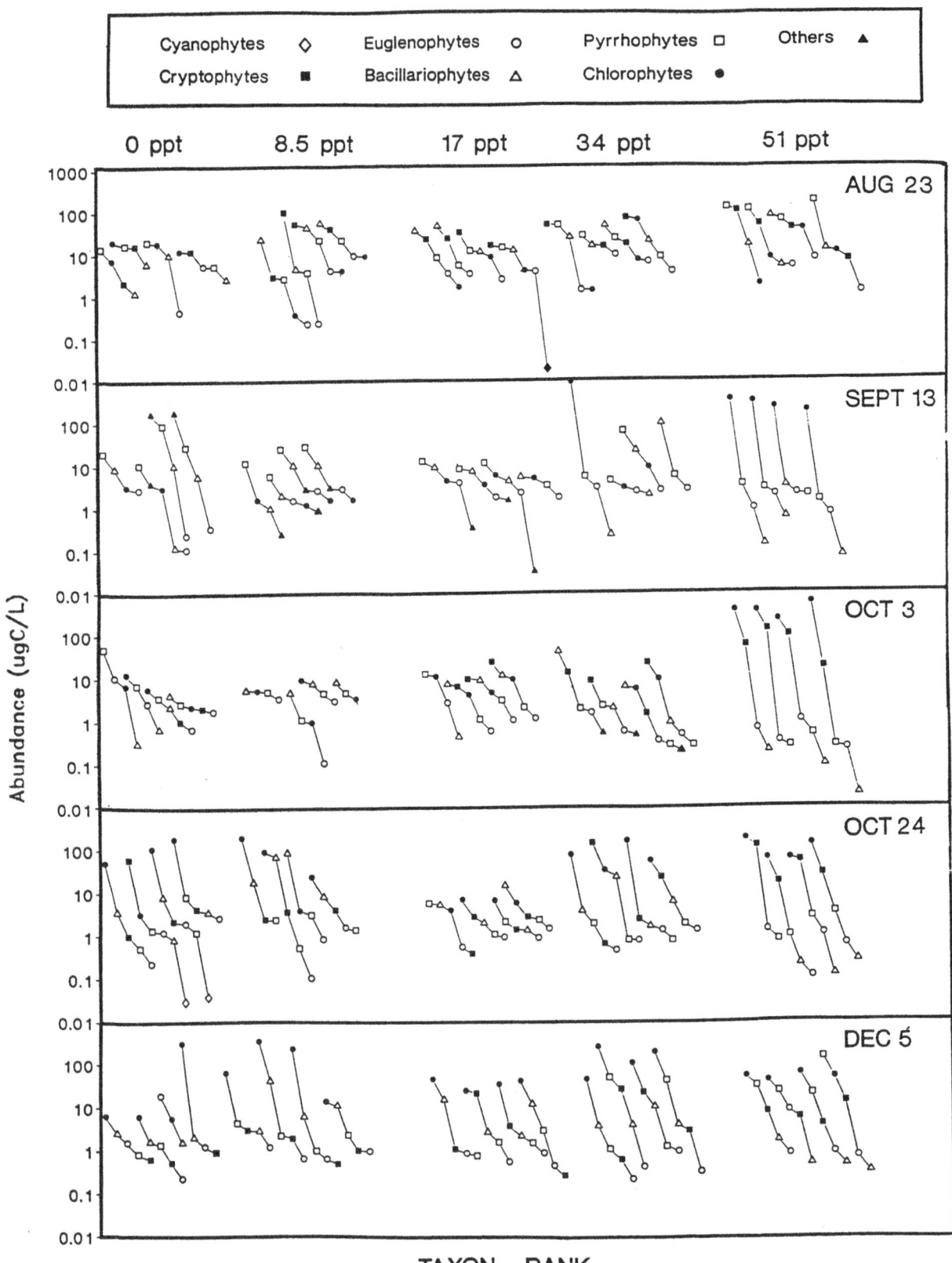

Legend:
Cyanophytes ◇ Euglenophytes ○ Pyrrhophytes □ Others ▲
Cryptophytes ■ Bacillariophytes △ Chlorophytes ●

0 ppt 8.5 ppt 17 ppt 34 ppt 51 ppt

Abundance (ugC/L)

AUG 23
SEPT 13
OCT 3
OCT 24
DEC 5

TAXON RANK

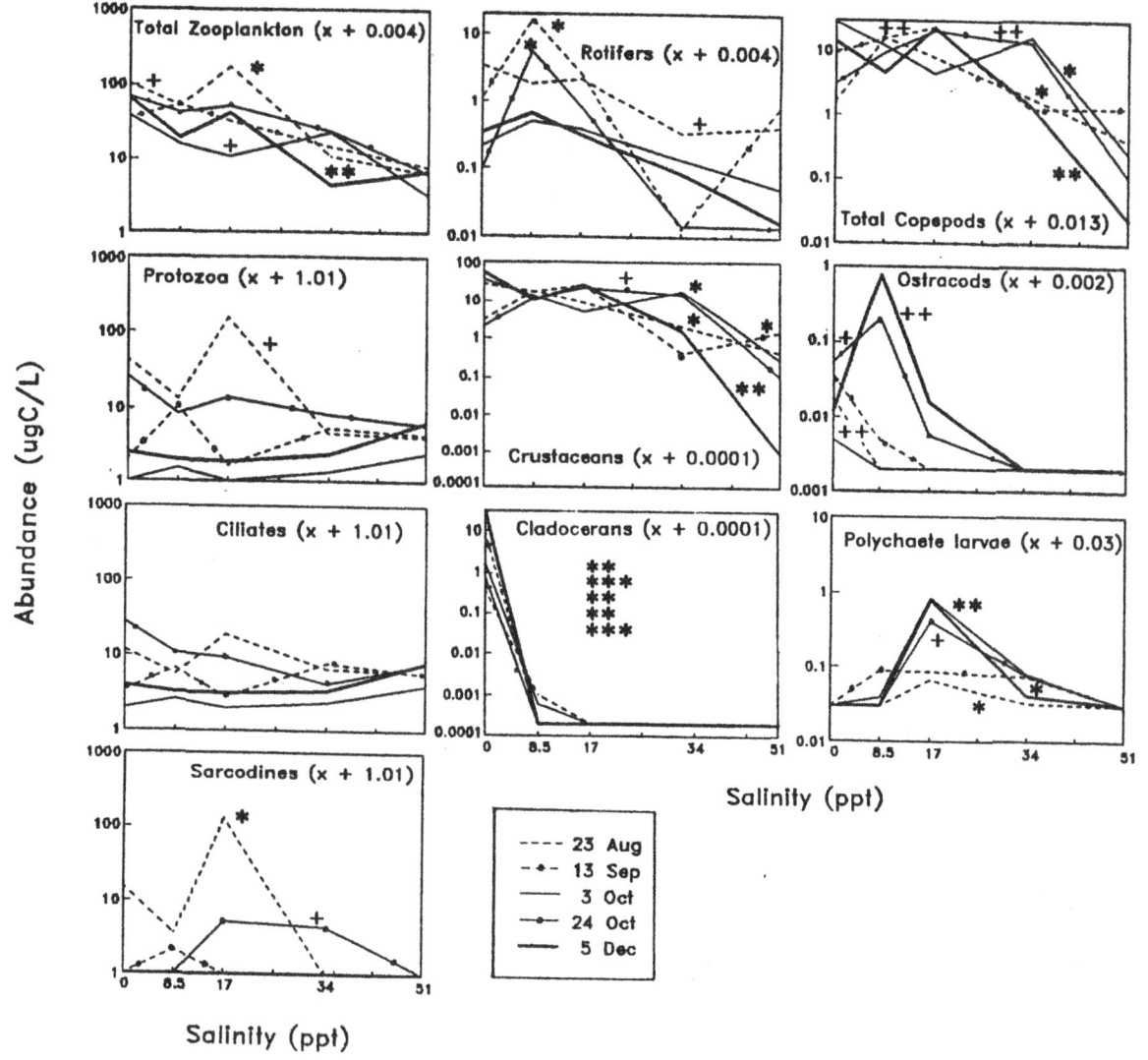

Fig. 6. Effects of salinity on geometric mean abundances of major zooplankton taxa. P values and constants as explained in Figs 1 and 2 and the text.

energetic costs of osmoregulation are minimized. A test of this hypothesis would require that grazing intensity and nutrient content of microcosms be held constant over the range of experimental salinities. A test is *not* possible by reference to the abundant literature on the salinity optima of individual species: these optima span a much wider salinity range than the 0–51 ppt used here. No prior experiments with diverse algal assemblages have been conducted, except for Melack's (1985) pilot experiment at salinities > 90 ppt. A review of the literature for saline lakes suggests maximal phytoplankton primary productivities are found at salinities < 50 ppt (Hammer, 1986: 259). Such

---

Fig. 5. Whittaker curves showing phytoplankton community structure, by major groups, for each tank (from left to right: A, B, C, or D) in the salinity treatment indicated at the top of the figure.

a relationship may not hold for phytoplankton standing crop, however.

We now discuss the responses to salinity of the individual phytoplankton taxa, discussing them in approximate order of abundance.

Chlorophytes

The majority of chlorophytes were smaller forms (3–12 μm in diameter) which we were mostly unable to identify. They were initially scarce relative to other algae but dominated the 51 ppt treatment by September 13 and all other treatments by October 24 (Figs 4, 5). On two sampling dates, chlorophyte abundance showed a strong and fairly regular increase with salinity, while at other times other statistically significant patterns were present (Fig. 2).

Pyrrhophytes

Pyrrhopytes, or dinoflagellates, were, with diatoms and cryptophytes, a dominant group at the beginning of the experiment at all salinity levels (Figs 4, 5). Our counts included heterotrophic forms, if they were present, as well as the autotrophic ones. The naked and thecate forms differed in their responses to salinity. The thecate forms were scarce or absent on all dates except September 13 when they were abundant at the lowest salinities but undetected at the highest (Fig. 2). The naked forms were present in most treatments on all dates and exhibited a significant salinity effect on every date except September 13. On three dates the trend was for naked dinoflagellates to increase with salinity, but on a fourth date the opposite but statistically significant trend was observed. Some of this inconsistency may have reflected differences among the responses of different dinoflagellate species in the constantly changing composition of the assemblage. The effects of salinity on the group as a whole, however, are probably produced indirectly via salinity effects on nutrient levels and grazers, rather than via direct physiological effects. As a group, dinoflagellates are common in estuarine waters and are known for their euryhalinity (Smayda, 1983). On the other hand, they are very readily grazed by copepods such as the *Acartia* and

Oithona present in our tanks (Berggreen *et al.*, 1988; Hayward & McGowan, 1979; Stoecker & Sanders, 1985; Uchima & Hirano, 1986).

Bacillariophytes

Diatoms were represented by at least 34 taxa, 5 of which were centric and 29 of which were pennate. Centric diatoms were represented mostly by *Cyclotella* spp. and *Odontella* spp., were relatively scarce at low salinities and were absent at higher salinities (Fig. 2). Only on October 3, when *Cyclotella* dominated, was the negative relationship between total centric diatoms and salinity significant.

Pennate diatoms were much more abundant and showed salinity effects on most dates (Fig. 2). For the group as a whole, the typical pattern was for abundance to be markedly lower at 0 and 51 ppt than at intermediate salinities. The same pattern was exhibited, albeit more irregularly, by *Amphora* spp. and *Surirella fastuosa* (Fig. 3), but most individual taxa achieved high abundances over a narrower range of salinities, e.g. *Synedra* 'A' at 0 ppt, *Navicula* 'A' at 0–17 ppt, *Navicula* 'C' at 8.5–17 ppt, and *Cylindrotheca closterium* (Ehrenberg) at 34 ppt (Fig. 2). These were the dominant diatom taxa and their variation among salinity treatments was generally significant statistically.

While the patterns shown by individual taxa perhaps do not merit speculative interpretation, it is appropriate to suggest some factors that may help explain the salinity response of pennate diatoms collectively (Fig. 2). The depressed abundance of diatoms at 0 ppt is correlated with the presence of certain crustacean zooplankters (*Scapholeberis kingii* Sars, *Ceriodaphnia reticulata* (Jurine), and *Diaptomus oregonensis* Lillj.) that were completely absent at higher salinities (Figs 7, 8). Grazing by these zooplankters thus may have been a factor. These dominant pennates, however, were mostly 30–50 μm in length which is larger than these zooplankters might be expected to handle easily.

Another factor may have been a faster sinking rate for diatoms at the lowest salinities, where the lower specific gravity provided less buoyancy.

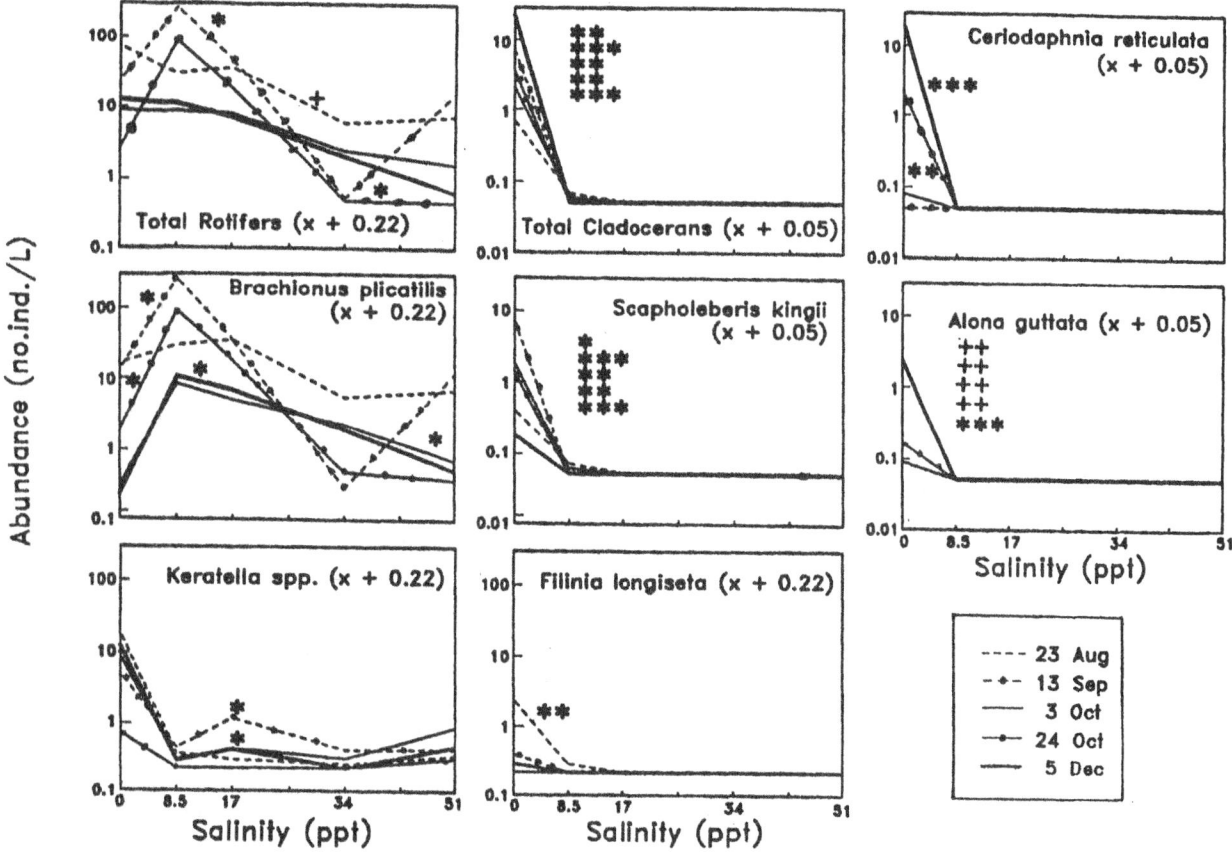

Fig. 7. Effects of salinity on geometric mean abundances of major rotifer and cladoceran taxa. P values and constants as explained in Figs 1 and 2 and the text.

Depression of diatom abundance at 51 ppt has two reasonable explanations. First, Si availability was very low at that salinity (Fig. 1), and there is much evidence that low Si concentrations can limit diatom abundance (Doering *et al.,* 1989, give a good recent summary). Second, the direct physiological effects of salinities on the order of 51 ppt may be strong and negative. Maximal rates of photosynthesis and cell division for many estuarine, neritic and oceanic diatom taxa are obtained at salinities below 35 ppt, with diminished rates occurring at salinities of 35 ppt or only slightly higher (Williams, 1964; Nakanishi & Monsi, 1965; Quasim *et al.,* 1972; Terada & Ichimura, 1979; Tanaka *et al.,* 1983; Brand, 1984). Field studies also suggest high salinities to be unfavorable for diatoms. In salt-producing ponds in San Francisco Bay, California, with salinities ranging from 27 to 111 ppt, most diatom

species were limited to salinities < 40 ppt (Carpelan, 1957). When salinities were > 30 ppt in Ucatena Island Pond, Massachusetts, pelagic diatoms were not found (Hurlbut, 1963). A review of the literature on saline lakes shows that the number of diatom species decreases greatly at salinities > 50 ppt (Hammer, 1986), though this tells us little about how total diatom abundance may vary with salinity.

Cryptophytes
These algae were initially quite abundant in all treatments except 0 ppt (Fig. 2). On the second sampling date they were not found at any salinity. On the three subsequent dates their abundance increased with increasing salinity over the range 17–51 ppt, while exhibiting no consistent correlation with salinity at salinities < 17 ppt.

As much as any algal group, the cryptophytes

324

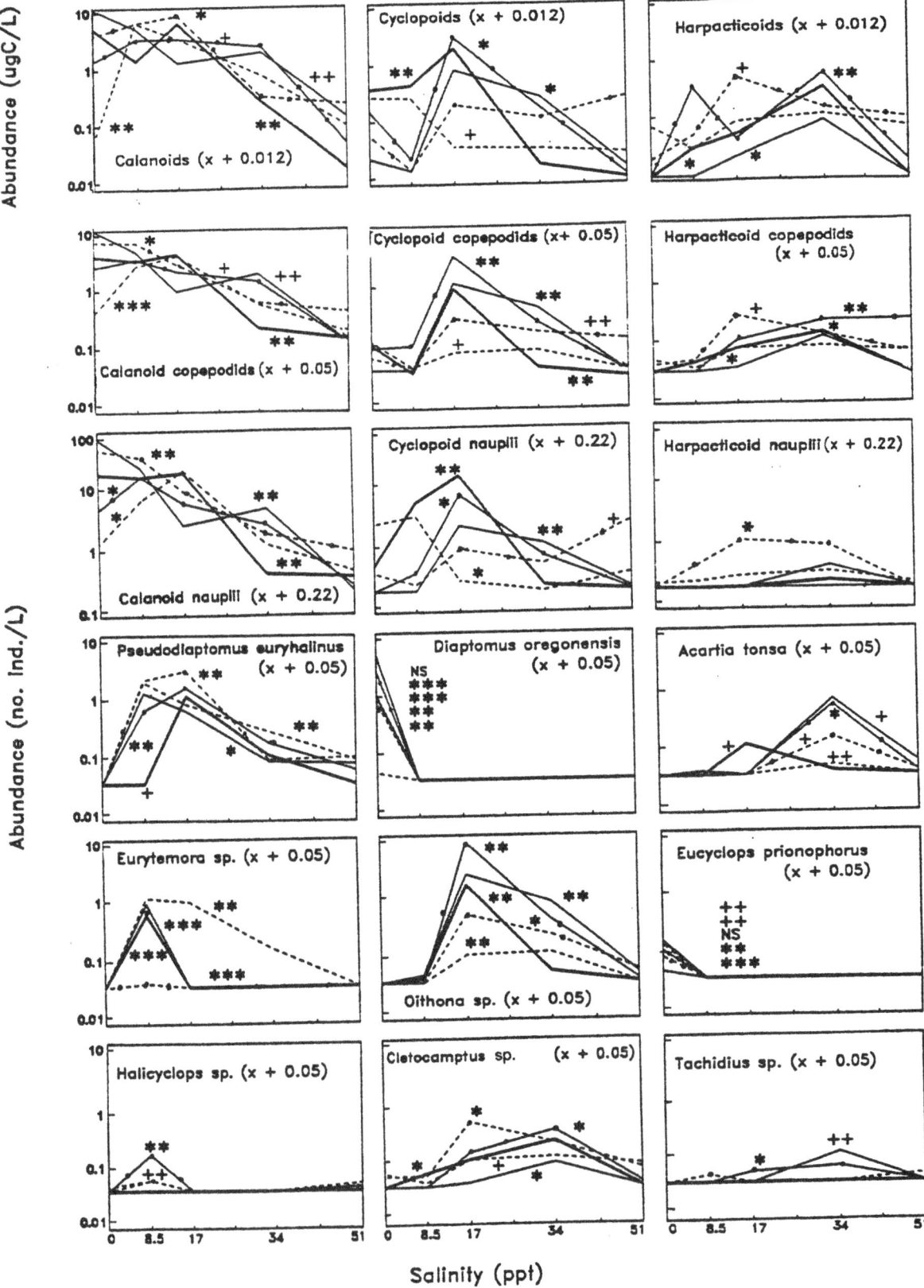

Abundance (ugC/L)

Abundance (no. Ind./L)

Calanoids (x + 0.012)

Cyclopoids (x + 0.012)

Harpacticoids (x + 0.012)

Calanoid copepodids (x + 0.05)

Cyclopoid copepodids (x+ 0.05)

Harpacticoid copepodids (x + 0.05)

Calanoid nauplii (x + 0.22)

Cyclopoid nauplii (x + 0.22)

Harpacticoid nauplii (x + 0.22)

Pseudodiaptomus euryhalinus (x + 0.05)

Diaptomus oregonensis (x + 0.05)

Acartia tonsa (x + 0.05)

Eurytemora sp. (x + 0.05)

Oithona sp. (x + 0.05)

Eucyclops prionophorus (x + 0.05)

Halicyclops sp. (x + 0.05)

Cletocamptus sp. (x + 0.05)

Tachidius sp. (x + 0.05)

Salinity (ppt)

were responsible for the tendency of total phytoplankton abundance to increase with increasing salinity over the range 17–51 ppt.

Many types of zooplankters readily feed on these naked flagellates. We suspect reduction of zooplankton abundance at the higher salinities was a factor permitting the high cryptophyte densities at those salinities.

Other taxa

Euglenophytes were uncommon and cyanophytes very rare and for neither group were any certain salinity effects detected (Fig. 2). The euglenophytes were represented by at least three unidentified taxa and the cyanophytes by *Anabaena* spp. and *Spirulina* sp.

Taxonomic composition

As a consequence of the responses of many individual taxa to salinity, marked changes in the structure of the phytoplankton assemblage were apparent (Figs 4, 5). The main trends may be summarized as follows.

Initially the treatments fell into three categories, according to the percent composition by major taxa (Fig. 4). At 0 ppt, dinoflagellates (pyrrhophytes) and chlorophytes dominated, at 51 ppt dinoflagellates and cryptophytes dominated, and at the three intermediate salinities (8.5, 17 and 34 ppt), dinoflagellates, diatoms and cryptophytes were co-dominants. Very quickly at 51 ppt and more gradually at other salinities, chlorophytes came to dominate in all treatments. This rise to dominance was slowest at 17 ppt. On the penultimate sampling date, 5 major taxa each constituted > 6 percent of total phytoplankton biomass at 17 ppt, while only 2–3 major taxa constituted > 6 percent of total phytoplankton biomass at the other four salinities. On the final sampling date, chlorophytes dominated all treatments, with strong co-dominance achieved only by dinoflagellates at 51 ppt.

As in the interpretation of changes in abundance of individual taxa, we note that 'salinity effects' on taxonomic composition and diversity are likely to reflect the imposed variations among treatments in nutrient concentrations. It is possible, for example, that at high salinities diatoms would have constituted a greater percentage of total phytoplankton biomass than they did, had these higher salinities not been established with lower Si concentrations.

Zooplankton

Total zooplankton

Total zooplankton abundance was generally greatest at 0–17 ppt and decreased at higher salinities (Fig. 6). The maxima at 0–17 ppt were primarily a reflection of the maxima of the dominant metazoan groups such as cladocerans, rotifers, and copepods. The 0 ppt treatment was presumably physiologically optimal for the freshwater forms, while 51 ppt was well above the optimal salinity for estuarine and marine forms.

The decline of total zooplankton abundance with salinity does not seem related to food supplies. Phytoplankton standing crop tended to increase with salinity, and dominant phytoplankters at the high salinities seemed to be of palatable sorts.

Predation was another possible factor. The largest predators, mosquito larvae (*Culex* sp.) and water boatmen (*Trichocorixa* sp.), were most common at lower salinities. The copepod *Acartia tonsa* Dana occurred almost exclusively at 34 ppt (Fig. 8). *Acartia* copepodids are omnivores and can feed on crustacean and other zooplankters (Miller, 1983). It may therefore have been responsible for the decrease, observed on 4 out of 5 sampling dates, in total zooplankton abundance as salinity increased from 17 to 34 ppt. At least this seems a more plausible explanation than one postulating a marked increase in physiological stress with the salinity increase from 17 to 34 ppt.

Fig. 8. Effects of salinity on geometric mean abundances of dominant copepod taxa. P values and constants as explained in Figs 1 and 2 and the text. Dates indicated as in earlier figures.

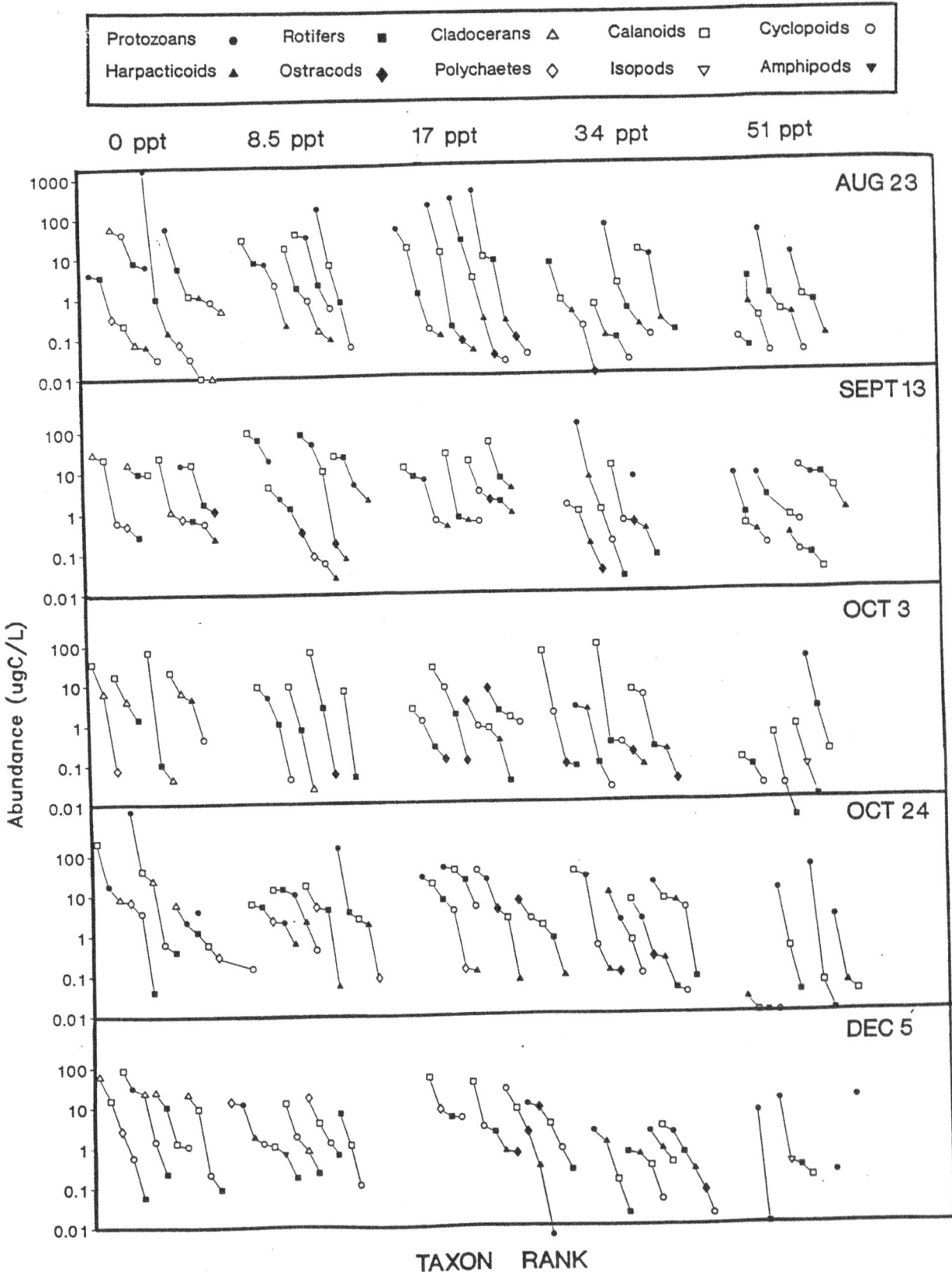

Legend:
Protozoans ● Rotifers ■ Cladocerans △ Calanoids ☐ Cyclopoids ○
Harpacticoids ▲ Ostracods ◆ Polychaetes ◇ Isopods ▽ Amphipods ▼

0 ppt   8.5 ppt   17 ppt   34 ppt   51 ppt

AUG 23

SEPT 13

OCT 3

OCT 24

DEC 5

Abundance (ugC/L)

TAXON   RANK

Protozoans

Protozoans were a major component of the zooplankton (Fig. 4), but there was little evidence of the effects of salinity on total protozoa abundance (Fig. 6). The dominant groups were the ciliates and sarcodines, and neither group showed a consistent, clear effect of salinity, though sarcodines were never observed at all at 51 ppt and achieved very high densities at 17 ppt on the first sampling ate. Ciliates were represented by four unidentified taxa, and sarcodines by three.

Rotifers

Total rotifer abundance showed evidence of a salinity effect only on three dates, usually showing maximal abundance at 8.5 ppt and a general decline in abundance at higher salinities (Fig. 6). Five taxa of rotifers were distinguished, with the most common being *Brachionus plicatilis* Müller, *Keratella* spp., and *Filinia longiseta* (Ehrenberg) (Fig. 9). These taxa exhibited variable effects of salinity.

Brachionus plicatilis generally was the dominant rotifer, and, although found in all salinity treatments on all dates, usually had its highest densities at 8.5 ppt (Fig. 7).

Keratella, represented by *K. valga* (Ehrenberg) and *K. cochlearis* (Gosse), was most abundant at 0 ppt and scarce at higher salinity levels (Fig. 7). *Filinia longiseta* was collected only at 0 and (rarely) 8.5 ppt, and usually in small numbers (Fig. 7).

Some rotifers, especially *B. plicatilis* and *Keratella* spp. are known to be very euryhaline and capable of tolerating the full salinity range of this experiment (Carpelan, 1957; Hammer, 1986; Remane & Schlieper, 1971). Their relative scarcity at 51 ppt was perhaps a function of the increased metabolic costs of adapting to this salinity. The restriction of *Filinia* to 0 ppt likewise probably was a direct effect of salinity. Nevertheless, rotifers in general are preyed on by various cladocerans and copepods (Williamson, 1983) and compete with these same groups, so it seems likely that such interactions influenced the patterns of rotifer abundance. We note, for example, that the highest rotifer mean densities were observed for *B. plicatilis* at 8.5 ppt, a salinity at which cladocerans, diaptomid copepodids and cyclopoid copepodids were extremely rare (Fig. 8). *Diaptomus oregonensis* in particular is suspected to compete with and, perhaps, prey upon rotifers (Bergquist *et al.*, 1985).

Cladocerans

These were almost always present in the 0 ppt tanks and very rarely found at higher salinities. The most common species were *Ceriodaphnia reticulata*, *Alona guttata* Sars, and *Scapholeberis kingii* (Fig. 7). *Daphnia pulex* (deGeer) was found occasionally in some 0 ppt tanks. All these are basically freshwater forms and their absence from salinities >0 ppt probably represents a direct physiological effect of salinity. There are however reports of *C. reticulata* occurring at salinities up to 6 ppt, of *D. pulex* at up to 30 ppt, and of *S. kingii* at up to 40 ppt (Hammer, 1986).

Marine cladocerans, such as species of *Podon* or *Evadne*, were never observed in any of the tanks. Probably this reflected their absence from our inocula. Neither was observed in our yearlong survey of the plankton of San Dieguito Lagoon (Greenwald and Hurlbert, in prepn).

Copepods

Collectively these tended to be uniformly abundant over the salinity range of 0–17 ppt and much reduced in abundance at 51 ppt (Fig. 6).

Calanoid, cyclopoid, and harpacticoid copepods were present, and each of these groups and each species in them showed a different response to salinity (Fig. 8). (Recall that the first naupliar to third copepodid instars were designated 'nauplii', while fourth copepodid to sixth copepodid instars were designated as 'copepodids'. Data in Fig. 8 for individual genera and species are for 'copepodids' only.)

Fig. 9. Whittaker curves showing zooplankton community structure, by major groups, for each tank on each date. Each line represents a tank (from left to right: A, B, C, or D) in the salinity treatment indicated at the top of the figure.

Calanoids collectively tended to have their greatest densities at 0–17 ppt, cyclopoids at 17 ppt, and harpacticoids at 17–34 ppt. The scarcity of copepods at 51 ppt is most reasonably interpreted as a direct effect of salinity on their physiology, possibly exacerbated by the reduced supply of diatoms available for food at 51 ppt (Fig. 2).

For each calanoid species the salinity of maximum abundance tended to be the same over all sampling dates (Fig. 8). *Diaptomus oregonensis* was found only at 0 ppt. *Eurytemora* sp. (probably *E. affinis* Poppe) was most abundant at 8.5 ppt. *Pseudodiaptomus euryhalinus* Johnson was most abundant at 8.5–17 ppt, though, befitting its name, it was also found at 51 ppt on all dates and has been observed in California coastal lagoons at salinities as high as 66 ppt (Hedgpeth, 1959). *Arcatia tonsa* was mostly found only at 34 ppt.

There exist few comparative data that might be used in analyzing these results. *D. oregonensis* seems to be known only from freshwaters (Balcer *et al.*, 1984). In short-term (6 h) tolerance trials with species congeneric with those in this study, Tundisi and Tundisi (1968) observed good survival (90–100 percent) of *Pseudodiaptomus acutus* over the salinity range 0–20 ppt and of *Acartia lilljeborgei* over the range 13–20 ppt. In other short term (8 d) assays of *Acartia tonsa* survival at salinities of 0–36 ppt, Lance (1963) observed maximal survival over the range 11–36 ppt.

The apparent 'partitioning' of the salinity spectrum by these calanoids in this experiment is not likely the reflection of some past evolutionary specialization for survival in specific narrow segments of this spectrum. Rather it is more likely a result of intense prey-predator and competitive interactions both among the calanoid species as well as with other invertebrates in the tanks. With the exception of *D. oregonensis*, all the calanoids present in our tanks are at least moderately euryhaline and capable of reproducing and surviving well at salinities where, in this experiment, they were rare or absent. For example, *A. tonsa* has been reported in estuaries and coastal lagoons ranging from 0.3 to 80 ppt in salinity and is one of the commonest zooplankters in such coastal waters over much of the globe (summaries in Lance 1963 and Day *et al.*, 1989). Yet in our study it was always scarce or absent except at 34 ppt.

The cyclopoids 'divided up' the salinity spectrum in much the same way as the calanoids (Fig. 8). *Eucyclops prionophorus* Kiefer was found only at 0 ppt, *Halicyclops* sp. almost exclusively at 8.5 ppt, and *Oithona* sp., the most abundant, almost exclusively at 17–34 ppt.

Harpacticoids were represented by two taxa, *Cletocamptus* sp. and *Tachidius* sp., which, as far as the data show, both tended to be most abundant at 17–34 ppt. As harpacticoids are mostly benthic in habit, it is likely that our sampling procedure markedly underestimated their true densities.

Other taxa

Two other groups, ostracods and polychaete larvae, were often present in our plankton samples.

Ostracods, representing only or predominantly a single species of *Cypridopsis*, were collected in low numbers at 0–17 ppt but were not found at higher salinities (Fig. 6). Both their low numbers and their high within-treatment variability perhaps reflect the inefficiency of our plankton sampler for those essentially benthic organisms.

Polychaete larvae of undetermined species showed, on 4 of 5 dates, a sharp peak in abundance at 17 ppt and were completely undetected at 0 and 51 ppt (Fig. 2). In the 17 ppt tanks their abundance increased over time. Presumably only larvae were introduced in our inocula and the larvae observed in our samples were the offspring or later descendants of these inoculated individuals.

Taxonomic composition

As with the phytoplankton, the diverse responses of individual zooplankton taxa to salinity resulted in changes in gross taxonomic composition that are best illustrated directly. The general trends are shown in Fig. 4, but the Whittaker curves (Fig. 9) are important in showing the often considerable variability among tanks at a given salinity level.

On the first sampling date, protozoans tended to dominate at all salinity levels, except at 8.5 ppt where calanoid copepods were slightly more abundant (Fig. 4). As other zooplankton taxa increased in abundance, protozoans tended to become relatively less important. By the final sampling date, however, protozoans constituted a percentage of the zooplankton that increased – 5, 10, 11, 40 and 99 percent – with salinity level.

On most dates after the first, and at all but the highest salinity, calanoids usually were the single most abundant zooplankton group. Rotifers were most important at 8.5 ppt and 51 ppt where they represented, on average, 14 and 16 percent of the zooplankton biomass, respectively, but they also averaged 5–10 percent of the zooplankton at all other salinities. Other major taxa constituted important fractions (e.g. >5 percent, averaged over all dates) over a more restricted range. By this criterion, cladocerans were important only at 0 ppt, and cyclopoid and harpacticoid copepods only at 17, 34 and 51 ppt. It would be interesting to compare these compositional trends with those for a spectrum of lakes of differing salinities or for a salinity gradient along an estuary or coastal lagoon. So far as we are aware, no such systematically gathered data sets exist, however.

The general diversity structure of the zooplankton assemblages was highly variable among tanks and among dates and few relationships with salinity were apparent, except for the stronger tendency for dominance by a single taxon at 51 ppt (Fig. 9).

Macroinvertebrates

Macroinvertebrate taxa found included *Culex* sp. (mosquito larvae), *Trichocorixa* sp. (water boatman), *Gammarus* sp. and unidentified polychaetes. Salinity was found to exert strong effects on most of these groups.

As indicated by the presence of their tubes formed from detritus and sediment, polychaete adults (Fig. 10), like their larvae (Fig. 6), tended to be most abundant at 17 ppt. In contrast to the larvae, however, the adults (tubes) were also abundant at 34 ppt and, on two dates, at 8.5 ppt. Polychaete reproduction may have been reduced at 8.5 and 34 ppt, relative to that at 17 ppt, or survival of polychaete larvae may have been reduced at 8.5 and 34 ppt by interactions with other planktonic organisms representing their food supply, competitors and predators.

Though we did not identify the polychaetes in our tanks, polychaetes in the genera *Polydora* and *Capitella* in neighboring Los Peñasquitos Lagoon were found to disappear when salinities were

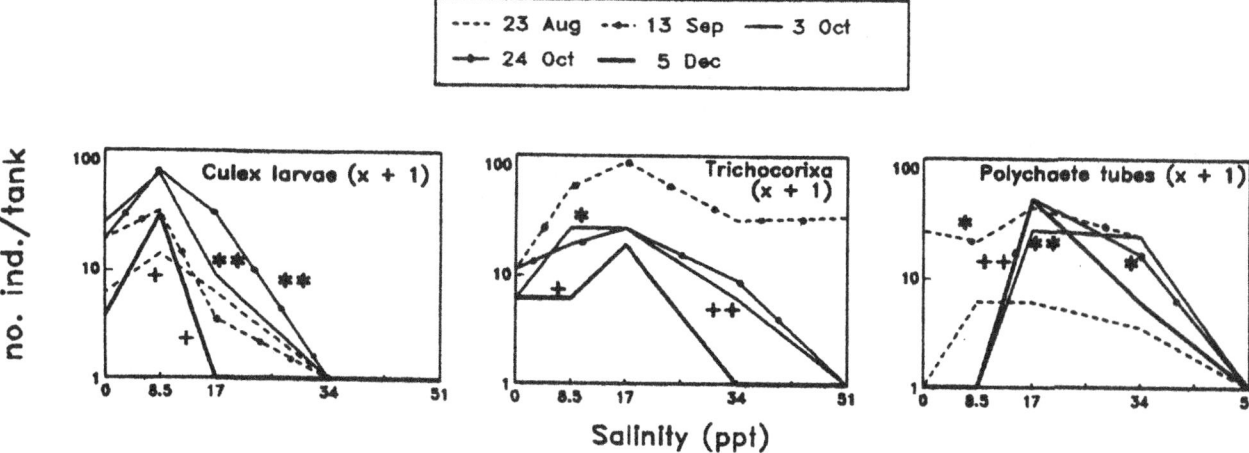

Fig. 10. Effects of salinity on geometric mean abundances of macroinvertebrates. P values and constants as explained in Figs 1 and 2 and the text.

under 10 ppt or over 40 ppt (Greenwald & Britton, 1987).

Amphipods (*Gammarus* sp.) appeared in the zooplankton sample for tank A at 8.5 ppt on the final sampling date (Fig. 9) and were observed to be very abundant in the benthos of that tank when it was later drained. They were not found in other tanks. This probably reflected only their great scarcity in the original inocula. The population in the above tank conceivably represented the progeny of a single female.

Culex sp. larvae were most abundant at 0–8.5 ppt, and were absent at salinities greater than 17 ppt. These larvae graze on algae, protozoa, and other organisms, and some effects of such grazing activities probably occurred in this experiment though they cannot be ascertained from our data. Reduced numbers of *Culex* larvae at 17 and 34 ppt may have reflected negative interactions with *Trichocorixa* populations.

Trichocorixa sp., ranging from 1–4 mm in length, generally was most abundant at 8.5–17 ppt and was never observed at 51 ppt (Fig. 10). These insects are mostly browsers of the microflora and microfauna associated with surfaces. They also can prey on some zooplankters such as *Artemia*, however (Reynolds, 1975; Wurtsbaugh & Berry, 1990). It is possible they affected some zooplankton populations in our tanks though we cannot identify such effects. As the adults can fly and move from one tank to another, for example in search of better food supplies, it is possible for populations to persist in tanks where salinity levels did not allow successful reproduction or larval survival. Nearby tanks with favorable salinities for reproduction can function as a continuous source of colonists for the other tanks, which consequently may develop rather different communities than they would in the absence of the source tanks.

Comparisons with and generalization to natural lagoons

The extrapolation of the findings of this experiment to natural events in San Dieguito Lagoon

and other coastal lagoons is not easy. While a microcosm experiment may allow unequivocal demonstration of the effects of a specific factor such as salinity, the greater complexity and larger scale of most natural systems precludes any assurance that the findings of the experiment are directly applicable to these natural systems.

Some differences existed between San Dieguito Lagoon and the tanks in physical-chemical characteristics of the water (Greenwald & Hurlbert, in preparation). Lagoon values generally were higher for ammonia and phosphate, but lower for pH. Temperature, dissolved oxygen, nitrate, and silica ranges for the tanks were, however, similar to those for the lagoon. The silica trend found in the tanks (Fig. 1) paralleled that in the lagoon, where there was an inverse relationship between salinity and silica concentration over time. Salinity in the lagoon was 14 ppt on one sampling date and 32–34 ppt on the other three.

Total phytoplankton densities in the lagoon at 32–34 ppt ($11–64 \ \mu g \ C \ l^{-1}$) were similar to those in the tanks at 34 ppt. The taxonomic composition was very different, however. In the lagoon, euglenophytes dominated (77–92 percent) on two dates, dinoflagellates dominated (50–96 percent) on two dates, diatoms were always a minor component (0.1–9 percent), and chlorophytes and cryptophytes were essentially absent. This contrasts, in particular, with the scarcity of euglenophytes and abundance of chlorophytes in the 17–34 ppt tanks (Figs 3, 4).

Total zooplankton densities in the lagoon were $11–57 \ \mu g \ C \ l^{-1}$ when salinity was 32–34 ppt (Greenwald and Hurlbert, in preparation), values which closely embrace the range of mean values observed for the 34 ppt tanks (Fig. 6). On the date the lagoon was at 14 ppt, however, total zooplankton measured $1371 \ \mu g \ C \ l^{-1}$, perhaps reflecting the recent influx of nutrient-rich freshwaters. On all dates the lagoon zooplankton biomass was strongly dominated by ciliates (82–99 percent) with calanoid copepods, mostly *Acartia tonsa*, in second place (0.1–12 percent), cyclopoid copepods (*Oithona* spp., *Oncea* sp.) in third place (0–2 percent), and polychaete larvae in fourth (<<1 percent) (Greenwald and Hurlbert, in preparation). The primary contrast between the lagoon zooplankton and the zooplankton assemblages in the 17–34 ppt tanks (Fig. 4) was the greater dominance in the lagoon of protozoans over copepods. This may have been due to the abundance of planktivorous fishes in the lagoon (Greenwald, 1985). Fish predation is a well known regulator of plankton assemblages in many aquatic ecosystems, and one of its most universal effects is a shift toward dominance by smaller zooplankters.

We considered adding fish to the tanks, but decided against it because of the difficulty of realistically imitating the highly variable San Dieguito Lagoon fish population densities, and also because larger tanks would have been required. It would be desirable in some future such experiments to incorporate the factor of fish predation into the experimental design.

In addition to the absence of fish, the tank ecosystems differed in major ways from San Dieguito Lagoon and coastal lagoons in general. These included a higher ratio of solid surfaces to water volume, reduced spatial heterogeneity and turbulence, and the absence of tidal circulation, stream inflow, birds and most benthic invertebrates. These represent the compromises inherent in the experimental microcosm approach. We cannot say in any specific way how they may have influenced our results. On the other hand, the mere *possibility* that the patterns in our results are unreflective of the effects of salinity in nature should not, by itself, cause them to be dismissed.

It is likely our results may apply more closely to coastal lagoons closed off from the ocean than to open ones or to estuaries. The planktonic assemblages of such closed coastal lagoons are tremendously variable, of course, and the compositions of the planktonic assemblages in the tanks seem in no way atypical of what is found in nature.

One aspect of our methodology that should particularly favor generalizability of our results was our inoculation procedure. By obtaining our inocula from a variety of waterbodies representing a wide salinity spectrum and by inoculating the experimental tanks at several points in time, we allowed a very large number of species the

opportunity to colonize and develop populations in the tanks. These species probably represented the great majority of planktonic genera found in our local coastal lagoons and estuaries. With a less intensive inoculation protocol, the probability of the results being strongly influenced by the chance absence of a functionally influential, say, zooplankton species would be higher. If we had obtained inocula on a single occasion and from only a few sites, for example, conceivably half as many crustacean species would have been represented in the tanks as were found. For future studies of this sort we recommend intensive inoculation protocols similar to those we used.

Covariation of salinity, nutrients and ionic composition

The nutrient content and ionic composition of the microcosms varied with salinity as a consequence of procedures used to create the salinity levels. This created severe interpretational ambiguities, some already noted in our discussion of the phytoplankton data. This covariation also raises important issues of general methodology that have not been addressed previously in the literature.

The confounding of salinity effects and nutrient effects is perhaps the weakest aspect of this study. It is true that to some extent this reflects correlations that often exist in nature, such as the negative one between salinity and silica along many estuarine gradients (Smayda, 1983) or in the vicinity of freshwater inflows into saline lakes. Nevertheless, our understanding both of effects and of the mechanisms producing them should advance most rapidly by carefully controlling for factors other than salinity.

Our procedures did not do this completely nor have those of previous workers (Galat & Robinson, 1983; Klos, 1988; Melack, 1985). If there is interest in nutrient effects as well as those of salinity, then ideally some sort of factorial design is called for. Salinity can be manipulated by addition of various pure salts in appropriate proportions. Major nutrients can likewise be added directly as wished. Additional treatments could be established by way of dilution or evaporative concentration, the procedures used in the few previous studies of salinity effects in microcosms (see Introduction). Responses to such dilution/evaporation treatments will reflect effects not only of changes in salinity and major nutrients but also of changes in concentrations of all other substances present (e.g. trace metals, organic compounds, etc.) in the original water.

Our procedures also did not control for variation in ionic composition. Major ion concentrations were not measured in the tanks. Nevertheless, it was evident from their concentrations in the lagoon water and tapwater (Table 1) used to create the salinity levels that ionic proportions were different at different salinities. For example, the ratio of calcium to other cations, individually and collectively, must have decreased greatly with increasing salinity. The ratio of bicarbonate to other anions, individually and collectively, would have behaved in the same way. The possibility therefore exists that some of the observed 'salinity effects' were due not to change in the sum of the major ions but to change in their relative proportions.

We would argue that this uncertainty should not be dealt with in salinity experiments by attempting to keep ionic proportions constant across salinity levels. We adduce three reasons. First, variation in ionic proportions invariably accompanies variation in salinity in nature. This is an inevitable consequence of the differing solubilities of salts and the differential loss of ions via precipitation reactions as waters are concentrated. Thus in our experiment the variation in ionic proportions across salinity levels was of the same sort as that typically found along the salinity gradient in estuaries and lagoons or among lakes ranging from freshwater to highly saline ones (of the sodium chloride type). In principle one could create freshwater or low salinity microcosms with the same ionic proportions (at least initially) present in highly saline (e.g. 35–100 ppt) microcosms. But these would necessarily be ionic proportions, e.g. very low calcium/sodium ratios, extremely atypical of freshwaters. Their use thus

would only result in the replacement of one set of interpretational difficulties by another.

A second factor diminishing the value of any attempt to keep ionic proportions constant is the difficulty of doing this completely. Even if we accomplished it for the major ions, we would not be sure that salinity correlated variations in, say, the lithium/sodium ratio or the borate/bicarbonate ratio were not responsible for observed 'salinity effects'.

Finally, any attempt to keep ionic proportions invariate over a wide range of salinities would be foiled by the strong influence that biological processes can exert on the concentrations of some major ions, especially calcium, sulfate and bicarbonate. These biological processes are likely to be different at different salinities, regardless of what other variables are or are not controlled for.

The above considerations indicate that experiments to assess the effects of 'salinity' cannot realistically be designed in a way to exclude confounding by covariation in ionic proportions. The importance of variations in ionic proportions will be most clearly assessed by experiments that vary those proportions while maintaining salinity constant. Results from such experiments can then be used to help distinguish the separate effects of salinity and ionic proportions in salinity experiments such as ours.

Conclusions

This experiment was informative in demonstrating clear trends in the abundance of many planktonic taxa along a salinity gradient. The extent to which the observed trends are generalizable and reflect how salinity affects plankton communities in nature is unknown. It can be determined only by further experimental studies of this sort carried out under different conditions and protocols and by more systematic and intensive comparative analyses of planktonic communities along natural salinity gradients than have been carried out in the past.

The experimental analysis of the effects of

salinity and ionic composition on aquatic ecosystems is in its infancy. There is a rich lode to
be mined by those willing to engage in such
studies. Multi-investigator teams are desirable so
that the considerable labor involved can be shared
and so that benefits can be maximized by the
gathering of information on a large number of
processes and groups of organisms.

Acknowledgements

Many persons contributed to this project. We
especially acknowledge the support provided by
Sandra Britton to almost every aspect of the
project, including especially preparation of the
figures. Thanks are also extended to: J. Verfaillie,
S. Grothey and B. Ellis for nutrient analyses; to
F. Reid, K. Lange, J. Beers, A. Fleminger,
J. Knapp, E. Brooks, D. Por, M. White, D. Soto,
and F. Jara for taxonomic assistance; to D.
Dexter, W. Hazen, B. Hemmingsen, and E. Keen
for technical and editorial advice; to I. A. E.
Bayly, D. B. Herbst and B. Sullivan for critical
comments on the manuscript; to D. Kent and the
Hubbs-Sea World Research Center for loan of
the tanks; to S. Nicosea, D. Hammons and the
City of Del Mar Public Works Department for
providing a secure site for the experiment and
assistance with water pumping; to L. Keil for
major assistance with field sampling; to B.
Blakistone, D. Bryant, P. Dickerson, E. Howe,
H. Howe, R. Howe, M. Irvine, D. A. Kreager,
D. Roberts, and J. Varnell for assistance in the
field or in data management and analysis; and to
J. Zimmer for general logistical support and
advice.

Partial funding of this project was provided by
Grant No. 86-22 from the Fish and Wildlife
Advisory Commission of the County of San Diego
Department of Planning and Land Use. This ar-
ticle is based on a thesis submitted by the senior
author in partial fulfillment of the requirements
for a Master of Science degree at San Diego State
University.

References

Amit, O. & Y. K. Bentor, 1980. pH-dilution curves of saline
waters. Chemical Geol. 7: 307–313.

Balcer, M. D., N. L. Korda & S. I. Dodson, 1984. Zooplankton of the Great Lakes. Univ. Wisconsin Press, Madison,
Wisconsin, 175 pp.

Beadle, L. C., 1943. An ecological survey of some inland
saline waters of Algeria. J. linn. Soc., Zool. 41: 218–242.

Beers, J. R., F. M. H. Reid & G. L. Stewart, 1977. Microplankton in the central gyre of the north Pacific Ocean,
Part II: Population structure and abundance, IMR Reference 77-1. Institute of Marine Resources, University of
California, San Diego, La Jolla, California, USA, 481 pp.

Berggreen, U., B. Hansen & T. Kiorboe, 1988. Food size
spectra, ingestion and growth of the copepod Acartia tonsa
during development: implications for determination of
copepod production. Mar. Biol. 99: 341–352.

Bergquist, A. M., S. R. Carpenter & J. C. Latino, 1985. Shifts
in phytoplankton size structure and community composition during grazing by contrasting zooplankton assemblages. Limnol. Oceanogr. 30: 1037–1045.

Bottrell, H. H., A. Duncan, Z. M. Gliwicz, E. Grygierek,
A. Herzig, A. Hillbricht-Ilkowska, H. Kurasawa, P.
Larsson & T. Weglenska, 1976. A review of some problems
in zooplankton production studies. Norw. J. Zool. 24:
419–456.

Braarud, T., 1951. Salinity as an ecological factor in marine
phytoplankton. Physiol. Plant. 4: 28–34.

Braarud, T., 1962. Species distribution in marine phytoplankton. J. oceanogr. Soc. Japan, 20th Anniversary Volume: 628–649.

Brand, L. W., 1984. The salinity tolerance of forty-six marine
phytoplankton isolates. Estuar. coast. shelf Sci. 18: 543–
556.

Carpelan, L. H., 1957. Hydrobiology of the Alviso Salt Ponds.
Ecology 38: 375–390.

Carpelan, L. H., 1964. Effects of salinity on algal distribution.
Ecology 45: 70–77.

Carpelan, L. H., 1969. Physical characteristics of Southern
California coastal lagoons. In: A. A. Castanares and F. B.
Phleger (eds), Lagunas Costeras, un Simposio. Universidad
Nacional Autonoma de Mexico, Mexico: 319–334.

Caspers, H., 1952. Untersuchungen über die Tierwelt von
Meeresalinen an der bulgarischen Küste des Schwarzen
Meeres. Zool. Anz. 148: 243–259.

Cole, G. A., 1983. Textbook of limnology, 3d edn. C. V.
Mosby, St. Louis, Missouri.

Cronin, L. E., J. C. Daiber & E. M. Hulburt, 1962. Quantitative seasonal aspects of zooplankton in the Delaware
River estuary. Chesapeake Sci. 3: 63–90.

Culver, D. A., M. M. Boucherle, D. J. Bean & J. W. Fletcher,
1985. Biomass of freshwater crustacean zooplankton from
length-weight regressions. Can. J. Fish. aquat. Sci. 42:
1380–1390.

Day, J. W. Jr., C. A. S. Hall, W. M. Kemp & A. Yáñez

334

Arancibia, 1989. Estuarine ecology. Wiley-Interscience, New York, New York, USA, 558 pp.

Doohan, M. & V. Rainbow, 1971. Determination of dry weights of small aschelminthes (< 0.1 μg). Oecologia 6: 380–383.

Doering, P. H., C. A. Oviatt, L. L. Beatty, V. F. Banzon, R. Rice, S. P. Kelly, B. K. Sullivan & J. B. Frithsen, 1989. Structure and function in a model coastal ecosystem: silicon, the benthos and eutrophication. Mar. Ecol. Progr. Ser. 52: 287–299.

Dumont, H. J., I. Van de Velde & S. Dumont, 1975. The dry weight estimate of biomass in a selection of Cladocera, Copepoda, and Rotifera from the plankton, periphyton, and benthos of continental waters. Oecologia 19: 75–97.

Galat, D. L. & R. Robinson, 1983. Predicted effects of increasing salinity on the crustacean zooplankton community of Pyramid Lake, Nevada. Hydrobiologia 105: 115–131.

Gauthier, H., 1928. Recherches sur les faune des eaux de l'Algerie et de la Tunisie. Minerva, Alger, 419 pp.

Greenwald, G. M., 1985. Final report: San Dieguito Lagoon fish/plankton project. Department of Biology, San Diego State University, San Diego, California, USA. Prepared for the County of San Diego Department of Planning and Land Use, Fish and Wildlife Advisory Commission, project no. 84-25, San Diego, California, USA, 33 pp.

Greenwald, G. M., 1989. Effects of salinity on coastal lagoon plankton assemblages. M.S. Thesis, San Diego State University, San Diego, California, 114 pp.

Greenwald, G. M. & S. L. M. Britton, 1987. Los Peñasquitos Lagoon biological monitoring program. Ecological Research Associates, Del Mar, California, USA. Prepared for the Los Peñasquitos Lagoon Foundation, San Diego, California, USA, 74 pp.

Hammer, U. T., 1986. Saline lake ecosystems of the world. Dr W. Junk Publishers, Dordrecht. 616 pp.

Hayward, T. L. & J. A. McGowan, 1979. Pattern and structure in an oceanic zooplankton community. Am. Zool. 19: 1045–1055.

Hedgpeth, J., 1959. Some preliminary considerations of the biology of inland mineral waters. Arch. Oceanogr. Limnol. (Rome) 11 (Suppl.): 111–141.

Hurlbut, E. M., 1963. The diversity of phytoplankton populations in oceanic, coastal and estuarine regions. J. mar. Res. 21: 81–93.

Javor, B., 1989. Hypersaline environments: microbiology and geochemistry. Springer-Verlag, New York, 328 pp.

Ketchum, B. H. (ed.), 1983. Estuaries and enclosed seas. Elsevier, New York, 500 pp.

Klos, E., 1988. An experimental estuarine salinity gradient. In Proceedings of the Oceans '88 Conference, Baltimore, Maryland: 1529–1535.

Krumgalz, B., 1980. Salt effect on the pH of hypersaline solutions. In A. Nissenbaum (ed.), Hypersaline brines and evaporitic environments. Elsevier, New York, New York, USA: 73–83.

Krylov, V. V., 1973. Relation between wet formalin weight of copepods and copepod body length. Oceanology 8: 723–727.

Kudrinskaya, O. I. & L. N. Yushko, 1973. Determination of the weight/length ratio in massively growing forms of copepods in Kremunchug Reservoir. Gidrobiol. Zh. 6: 100–104.

Lance, J., 1963. The salinity tolerance of some estuarine planktonic copepods. Limnol. Oceanogr. 8: 440–449.

Likens, R. G. & G. E. Wetzel, 1979. Limnological Analyses. W. B. Saunders Company, Philadelphia, Pennsylvania, USA, 357 pp.

Löffler, H., 1961. Beitrage zur Kenntnis der Iranischen Binnengewasser, II. Regionale-limnologische Studie mit besonderer Berucksichtigung Crustaceen fauna. Int. Revue ges. Hydrobiol. 46: 309–406.

Marcus, L., 1989. The coastal wetlands of San Diego County. California State Coastal Conservancy, 65 pp.

Mead, R., 1988. The design of experiments. Cambridge University Press, New York, 620 pp.

Melack, J. M., 1985. The ecology of Mono Lake, California. National Geographic Society Research Reports 20: 461–470.

McLachlan, J., 1961. The effect of salinity on growth and chlorophyll content in representative classes of unicellular marine algae. Can. J. Microbiol. 7: 399–406.

Miller, J. K., 1966. Biomass determinations of selected zooplankters found in the California Cooperative Oceanic Fisheries Investigations. SIO Reference 66-15. Marine Life Research Group, Scripps Institution of Oceanography, University of California, San Diego, California, USA, 16 pp.

Mudie, P. J., B. M. Browning & J. M. Speth, 1976. The natural resources of San Dieguito and Batisquitos Lagoons, Coastal Wetland Series No. 12. State of California Department of Fish and Game, Long Beach, California, USA, 128 pp.

Nakanishi, M. & M. Monsi, 1965. Effect of variation in salinity on photosynthesis of phytoplankton growing in estuaries. J. Fac. Sci., Univ. Tokyo 9: 209–215.

Provasoli, L., 1958. Nutrition and ecology of protozoa and algae. Ann. Rev. Microbiol. 12: 279–308.

Quasim, S. Z., P. M. A. Bhahattathiri & V. P. Devasy, 1972. The influence of salinity on the rate of photosynthesis and abundance of some tropical phytoplankton. Mar. Biol. 12: 200–206.

Rawson, D. S. & J. E. Moore, 1945. The saline lakes of Saskatchewan. Can. J. Res. 22: 141–201.

Remane, A. & C. Schlieper, 1971. Biology of brackish waters, 2nd edition. John Wiley & Sons, New York, 322 pp.

Reynolds, J. D., 1975. Feeding of corixids (Hemiptera) in small alkaline lakes of central B.C. Verh. int. Ver. Limnol. 19: 3073–3078.

Rosen, R. A., 1981. An energy budget for adult Brachionus plicatilis (Muller) (Rotatoria). Oecologia 13: 351–362.

Smayda, T., 1958. Biogeographical studies of marine phytoplankton. Oikos 9: 158–191.

Smayda, T. J., 1983. The phytoplankton of estuaries. In B. H. Ketchum (ed.), Ecosystems of the world 26: Estuaries and enclosed seas. Elsevier, New York.: 65–102.

Soto, D. & S. H. Hurlbert, 1991. Long-term experiments on calanoid-cyclopoid interactions. Ecol. Monogr. 61: 245–265.

Stockner, J. G., 1988. Phototrophic picoplankton: An overview from marine and freshwater ecosystems. Limnol. Oceanogr. 33: 765–775.

Stoecker, D. K. & N. K. Sanders, 1985. Differential grazing by Acartia tonsa on a dinoflagellate and a tintinnid. J. Plankton Res. 7: 85–100.

Tanaka, N., M. Sugiyama & K. Ohwada, 1983. Ecological studies of phytoplankton In Ajo Bay with special reference to the relation between growth and salinity. Bull. Plankton Soc. Japan 30: 1–10.

Technicon Industrial Systems, 1973a. Ammonia in water and seawater. Industrial method no. 154-71W, Technicon Industrial Systems, Tarrytown, New York, USA, 22 pp.

Technicon Industrial Systems, 1973b. Orthophosphate in water and seawater. Industrial method no. 155-71W, Technicon Industrial Systems, Tarrytown, New York, USA, 3 pp.

Technicon Industrial Systems, 1977a. Individual/simultaneous determination of nitrogen and/or phosphorus in BD acid digests. Industrial method no. 329-74W/B, Technicon Industrial Systems, Tarrytown, New York, USA, 9 pp.

Technicon Industrial Systems, 1977b. Nitrate and nitrite in water and seawater. Industrial method no. 158-71W, Technicon Industrial Systems, Tarrytown, New York, USA, 4 pp.

Technicon Industrial Systems, 1977c. Silicates in water and seawater. Industrial method no. 186-72W/B, Technicon Industrial Systems, Tarrytown, New York, USA, 2 pp.

Terada, T. & S. Ichimura, 1979. Phytoplankton photosynthesis in an eutrophic estuary with special reference to salinity gradient. La Mer 17: 171–177.

Tundisi, J. & T. M. Tundisi, 1968. Plankton studies in a mangrove environment V. Salinity tolerances of some planktonic crustaceans. Bolm Inst. Oceanogr. S. Paulo 17: 57–65.

Uchima, M. & R. Hirano, 1986. Food of Oithona davisae (Copepoda: Cyclopoida) and the effect of food concentration at first feeding on the larval growth. Bull. Plankton Soc. Japan 33: 21–28.

Williams, R. B., 1964. Division rates of salt marsh diatoms in relation to salinity and cell size. Ecology 45: 877–880.

Williamson, C. E., 1983. Invertebrate predation on planktonic rotifers. Hydrobiologia 104: 383–396.

Wurtsbaugh, W. A. & T. S. Berry, 1990. Cascading effects of decreased salinity on the plankton, chemistry, and physics of the Great Salt Lake (Utah). Can. J. Fish. aquat. Sci. 47: 100–109.

The manufacturer's authorised representative in the EU is Springer
Nature Customer Service Centre GmbH, Europaplatz 3, 69115 Heidelberg,
Germany. If you have any concerns regarding our products, please
contact ProductSafety@springernature.com

Printed and bound by CPI Group (UK) Ltd, Croydon, CR0 4YY

23/04/2026

02095657-0002